Further Titles of Interest

K. H. Nierhaus, D. N. Wilson (eds.)

Protein Biosynthesis and Ribosome Structure

ISBN 3-527-30638-2

R. J. Mayer, A. J. Ciechanover, M. Rechsteiner (eds.)

Protein Degradation

ISBN 3-527-30837-7 (Vol. 1)
ISBN 3-527-31130-0 (Vol. 2)

G. Cesareni, M. Gimona, M. Sudol, M. Yaffe (eds.)

Modular Protein Domains

ISBN 3-527-30813-X

S. Brakmann, A. Schwienhorst (eds.)

Evolutionary Methods in Biotechnology

ISBN 3-527-30799-0

Protein Fo
Handbook
Edited by J.
T. Kiefhabe

Contents

Part I, Volume 1

Protein Folding Handbook. Part II. Edited by J. Buchner and T. Kiefhaber

ISBN: 3-527-30784-2

13
The Role of Trigger Factor in Folding of Newly Synthesized Proteins

Elke Deuerling, Thomas Rauch, Holger Patzelt, and Bernd Bukau

13.1
Introduction

The efficient folding of newly synthesized proteins in vivo requires the assistance of molecular chaperones [1, 2]. Well established is the role of cytosolic chaperones in protein folding, which mainly act posttranslationally, e.g., the *E. coli* chaperones DnaK and GroEL with their respective DnaJ/GrpE and GroES co-chaperones (see also Chapters 14 and 20). During the past few years, however, the discovery of ribosome-associated chaperones that associate with nascent polypeptides as soon as they emerge from the ribosomal exit tunnel has shed new light on the process of de novo protein folding and has expanded the repertoire of chaperones forming a functional network to support productive protein folding in vivo. The coupling of protein biosynthesis with protein folding via ribosome-associated chaperones seems to be an evolutionarily conserved principle that exists in all organisms even though many different types of chaperones are involved (see also Chapter 12). So far *E. coli* trigger factor is the best studied of these chaperones. In this chapter we will summarize the current knowledge about the in vivo role of trigger factor in protein folding, we will describe in detail the structure-function relationship of trigger factor, and we will put emphasis on biochemical methods used to analyze trigger factor in vitro.

13.2
In Vivo Function of Trigger Factor

13.2.1
Discovery

Trigger factor was first identified in 1987 by W. Wickner and coworkers as a factor triggering the translocation-competent conformation of pro-outer-membrane protein A (proOmpA) for its import into membrane vesicles [3]. Trigger factor was shown to be an abundant cellular protein that associates in a 1:1 stoichiometry

Protein Folding Handbook. Part II. Edited by J. Buchner and T. Kiefhaber

ISBN: 3-527-30784-2

with 70S ribosomes via the large ribosomal subunit [4]. These results were initially interpreted to indicate a secretion-specific chaperone function for trigger factor during the shuttling of secretory precursor proteins. However, it was shown thereafter that *E. coli* cells depleted of trigger factor did not show any secretion defects [5].

Years later, trigger factor was simultaneously rediscovered by three groups who investigated different processes related to protein folding in vivo [6–8]. A search by G. Fischer and coworkers for a ribosome-associated peptidyl-prolyl *cis/trans* isomerase (PPIase) led to the identification of trigger factor [6]. Purified trigger factor catalyzed the prolyl-peptide bond isomerization in tetrapeptides and in refolding RNase T1. J. Luirink and coworkers used an in vitro translation system to investigate the interactions of the signal recognition particle (SRP) with a variety of nascent polypeptides by chemical cross-linking [8]. Besides cross-links of nascent chains to SRP, they found that trigger factor efficiently cross-links to nascent chains of preprolactin and other secretory substrates. When searching for *E. coli* chaperones that associate with nascent β-galactosidase produced in an in vitro transcription/translation system, B. Bukau and colleagues discovered that ribosome-associated trigger factor interacts with both cytosolic and secretory nascent polypeptides during protein biosynthesis [7]. The association of trigger factor with the ribosome proved to be sensitive to the translational state of the ribosome, as the release of the nascent polypeptides by puromycin caused dissociation of trigger factor from the ribosome. It was therefore proposed that trigger factor is a folding catalyst acting co-translationally during the biogenesis of proteins on ribosomes [6, 7, 9].

13.2.2
Trigger Factor Cooperates With the DnaK Chaperone in the Folding of Newly Synthesized Cytosolic Proteins

The deletion of the trigger factor gene *tig* in *E. coli* results neither in defects in growth at temperatures between 15 °C and 42 °C nor in protein folding [10, 11]. A similar finding was made when the *dnaK* gene, encoding the major Hsp70 protein in *E. coli*, was deleted. Δ*dnaK* cells are viable between 15 °C and 37 °C and have only mild protein-folding defects of newly synthesized proteins [10–13]. Δ*dnaK* cells are, however, not viable at temperatures above 37 °C, probably because DnaK is essential to prevent misfolding at elevated temperatures and to repair heat-damaged proteins. Genetic evidence clarified why the loss of trigger factor or DnaK has no phenotype at permissive temperature and no consequence on the de novo folding of proteins. The simultaneous deletion of the *tig* and *dnaK* gene leads to synthetic lethality at 30 °C and above [10, 11]. Thus, either trigger factor or DnaK is sufficient for cell viability at permissive temperatures, but the loss of both leads to cell death, indicating that these two chaperones cooperate in the folding of newly synthesized polypeptides. A physical interaction between DnaK and newly synthesized proteins was demonstrated by co-immunoprecipitation experiments [10, 11]. In wild-type *E. coli* cells about 5–18% of newly synthesized proteins interact transiently with DnaK. This level increased two- to threefold (26–36%) in the absence of trigger factor, indicating that upon the loss of trigger factor, the cells make more intensive use of the DnaK chaperone [10, 11]. A fraction of these proteins is in *statu nascendi*, indicating that DnaK interacts co- and posttranslationally. Moreover, it was shown that deletion of the *tig* gene induces the heat shock response in *E. coli*, thereby leading to a compensatory increase in the steady-state levels of chaperones including DnaK [14]. Thus, DnaK may serve as a potent backup system in the absence of trigger factor.

13.2.3
In Vivo Substrates of Trigger Factor and DnaK

The depletion of DnaK in the absence of trigger factor leads to the aggregation of cytosolic proteins in *E. coli*. The isolation of aggregated cytosolic proteins from *E. coli* cells is achieved by a centrifugation step after cell lysis and repetitive washing steps of the pellet containing aggregates with NP40 to remove membrane contaminations [15]. Through two-dimensional SDS-PAGE more than 340 aggregated protein species in DnaK-deficient Δ*tig* cells could be detected (Figure 13.1A) [14]. Quantitative comparison of total cellular proteins with aggregated proteins in these cells revealed that approximately 10% of the total cellular cytosolic proteins are aggregation-prone at 37 °C, a condition at which growth of these chaperone-deficient cells was not yet impaired. In contrast, only about 1% of the cytosolic proteins aggregated in DnaK-depleted tig^+ cells [14]. Although the total amount of aggregated proteins differed significantly in DnaK-depleted tig^+ and tig^- cells, the aggregation-prone protein species in these cells were similar. This finding suggests that trigger factor and DnaK share overlapping substrate pools. In cells lacking only trigger factor, no protein aggregates could be detected [10]. Proteins that are misfolded in Δ*tig* cells are likely to be rescued by the DnaK system, since the protein level of DnaK is elevated about two- to three-fold in these cells.

The aggregated protein species found in cells lacking trigger factor and DnaK range in their molecular weights from 16 kDa up to 167 kDa, whereby large-sized multi-domain proteins (> 60 kDa) are enriched. It is unclear which features of large-sized proteins render them vulnerable to misfolding and aggregation during de novo folding. Unfolded conformers of large proteins expose statistically more hydrophobic surface patches than do small proteins and as a consequence may have a greater chance of undergoing intra- and intermolecular aggregation. Furthermore, hydrophobic interdomain surfaces may become exposed and thus may nucleate aggregation during co- and posttranslational folding. Finally, large-sized proteins may fold more slowly to the native state, which increases the timeframe during which aggregation of folding intermediates may occur.

Ninety-four different aggregated proteins isolated from DnaK-depleted Δ*tig* cells have been identified by mass spectrometry [14], including many essential proteins such as the elongation factor EF-Tu and the RNA-polymerase subunit RpoB, which

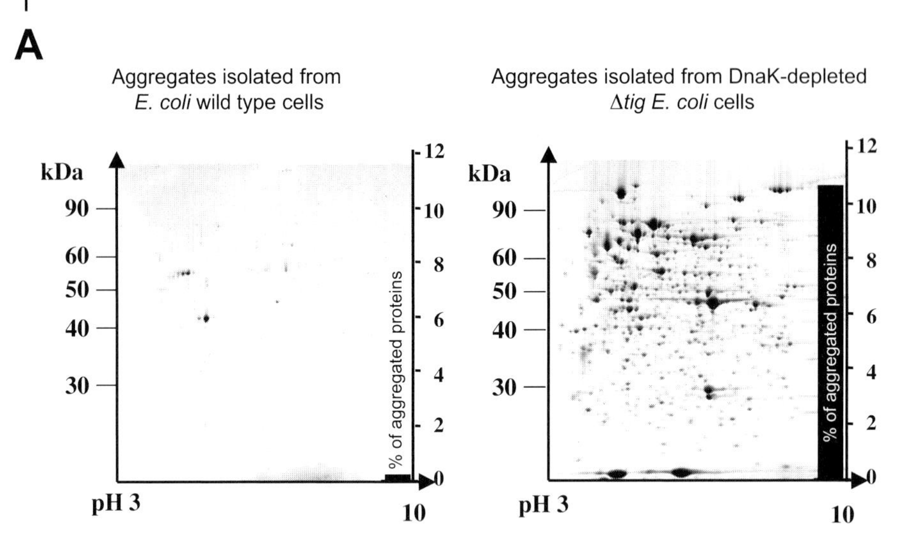

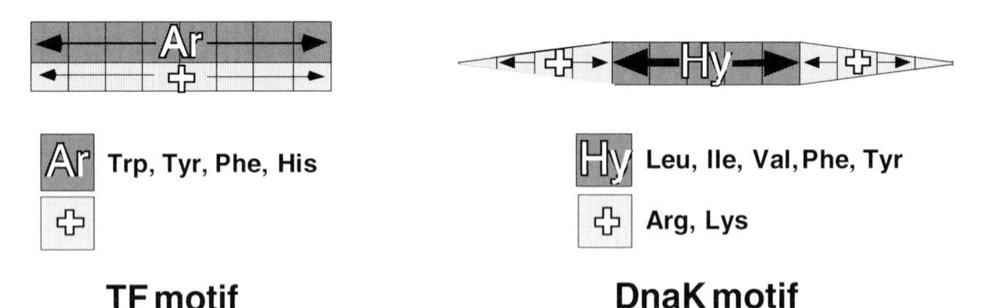

Fig. 13.1. Trigger factor and DnaK substrates. (A) Aggregated cytosolic proteins (isolated from cells grown at 37 °C in LB) were separated by 2D gel electrophoresis and Coomassie stained. Data were taken with permission from Ref. [14]. (B) Trigger factor and DnaK overlap in their binding specificities. The recognition motifs of both chaperones are shown schematically. TF and DnaK data were taken with permission from Ref. [17].

could explain the synthetic lethality of $\Delta tig\Delta dnaK$ mutant cells. The identified proteins are involved in a large variety of cellular processes, including transcription, translation, and metabolism. They do not show any common features regarding pI, specific secondary structural elements, or content of prolines, which could depend on the PPIase activity of trigger factor. Interestingly, 70% of the identified proteins were also identified as thermolabile proteins that aggregate in $\Delta dnaK$ cells under heat shock conditions at 45 °C [16].

13.2.4 Substrate Specificity of Trigger Factor

Since trigger factor and DnaK work by completely different mechanisms, it was unclear how they were able to support folding of the same protein pool in the cytosol. The mapping of potential binding sites of trigger factor and DnaK in natural substrate proteins using cellulose-bound peptide libraries (for protocol, see Chapter 15) revealed that DnaK and trigger factor share a high overlap (77%) in binding peptides [14]. This finding is consistent with the overlapping features of the substrate-binding motifs of both chaperones (Figure 13.1B) [17].

Trigger factor binds to peptides enriched in aromatic and basic amino acids, whereas peptides with acidic amino acids are disfavored [17]. The trigger factor binding motif consists of eight amino acids whereby the positioning of basic and aromatic residues within this motif seems not to be crucial for binding (Figure 13.1B). Interestingly, prolyl residues do not contribute to the binding of trigger factor to peptides [17]. Moreover, trigger factor binds to the protein substrate RNase T1 independently of prolyl residues [18]. These results were unexpected since trigger factor is a PPIase and catalyzes the *cis/trans* isomerization of peptidyl-prolyl peptide bonds in peptides and protein substrates [6, 7]. The prolyl-independent substrate recognition and the PPIase activity involve the same substrate-binding pocket in the PPIase domain of trigger factor (see Section 13.3.1), suggesting that trigger factor scans nascent polypeptide chains for stretches of aromatic and basic amino acids and may isomerize peptidyl-prolyl peptide bonds within these stretches when present. However, since the majority of binding peptides (55%) do not even contain a proline [17], the isomerization reaction may not be the major activity of trigger factor when interacting with nascent polypeptide chains. Mapping of the potential trigger factor binding sites (identified by peptide library screening) within the three-dimensional structures of the substrate proteins does not reveal any enrichment within specific secondary structure elements or clustering of binding sites near the N- or C-terminus of proteins or within or outside of signal sequences. On average, trigger factor binding sites appear every 32 amino acids in a protein and are mostly localized in the interior of the native protein structures.

A related but not identical binding motif is recognized by DnaK. This chaperone recognizes a short stretch of five amino acid residues with hydrophobic character, among which leucine is especially favored (Figure 13.1B; see also Chapter 15). This hydrophobic core is flanked by positively charged amino acids, whereas negative charges are excluded [19, 20]. Trigger factor and DnaK thus both prefer to interact with hydrophobic and positively charged stretches in proteins [17, 21]. This overlap in binding specificity might be required so that both chaperones, despite their different mechanisms of action, can protect similar hydrophobic stretches in unfolded protein species and therefore promote folding of the same substrate pool.

While investigating the interaction of trigger factor and DnaK with nascent polypeptides in an in vitro transcription/translation system (see protocols, Section 13.5), it was shown that trigger factor and DnaK compete for cross-linking to a shared binding site in a short nascent polypeptide chain of proOmpA [14]. DnaK, however, is not efficiently cross-linked to this chain in the presence of trigger factor. This is in agreement with data showing that two to three times more newly synthesized proteins associate with DnaK in the absence of trigger factor [10, 11]. Chaperone interaction with nascent polypeptides is therefore likely to follow a hierarchical order imposed by the positioning of trigger factor adjacent to the polypeptide exit tunnel on the large ribosomal subunit (see Section 13.3.4) [22, 32].

13.3
Structure–Function Analysis of Trigger Factor

13.3.1
Domain Structure and Conservation

E. coli trigger factor consists of 432 amino acids with a molecular weight of 48 kDa. To date, no high-resolution structure of the full-length trigger factor protein is available. Limited proteolysis of full-length trigger factor with proteases revealed a compactly folded, protease-resistant domain spanning residues 145–247 (Figure 13.2A). This domain has homology to the FK506-binding protein (FKBP) type of PPIases (see also Chapter 10) [6, 7, 23]. The main structural element of trigger factor's PPIase domain is a five-stranded, antiparallel β-sheet. Recently, the NMR structure of the PPIase domain of the *Mycoplasma genitalium* trigger factor was solved and it confirmed the FKBP-like fold of this domain [24]. An alignment of the PPIase domains of trigger factor homologues from different bacteria with human FKPB12 shows that the six aromatic key residues forming the FKBP12 substrate-binding pocket are conserved within the trigger factor family (Figure 13.2C). In contrast to other FKBP-like proteins, trigger factor cannot bind the PPIase inhibitor drug FK506. This is due to structural differences in the binding pocket that narrow the active site thereby causing steric clashes with FK506 [24]. Besides its catalytic activity, the PPIase domain mediates the binding specificity of trigger factor [17]. It was shown that the isolated PPIase domain displays a similar binding pattern towards membrane-coupled peptides and hence a binding specificity similar to full-length trigger factor. The structural features of the PPIase domain are compatible with the determined binding specificity of trigger factor for peptides that are enriched in aromatic and basic residues. The ring of aromatic side chains is located within a groove (Figure 13.2C) that is surrounded by negatively charged amino acids. While the aromatic amino acid residues could mediate hydrophobic interactions with aromatic amino acids enriched in peptides that are bound by trigger factor, the negative charges could explain the preference of trigger factor for basic amino acids within the substrate. However, a 10-fold higher concentration of the PPIase domain had to be used to obtain binding signals on peptide libraries with comparable intensity as compared to full-length trigger factor [17]. This finding indicates a role of the flanking domains of trigger factor in substrate binding, e.g., by forming additional contacts to the backbone of the substrate or by modulating the conformation of the PPIase domain. This hypothesis is sup-

B

		TF-signature
E. coli	27-67	AVKSELVNVAKKVRIDGFRKGKVPMNIVAQRYGASVR-QDVL
H. influenza	27-67	AVREEFKRAAKNVRVDGFRKGHVPAHIIEQRFGASIR-QDVL
B. subtilis	26-67	ALDDAFKKVVKQVSIPGFRKGKIPRGLFEQRFGVEALYQDAL
S. pyrogenes	27-68	ALDKAFNKIKKDLNAPGFRKGHMPRPVFNQKFGEEVLYEDAL
H. pylori	27-67	RYDKIAQKIAQKVKIDGFRRGKVPLSLVKTRYQAQIE-QDAQ
R. prowazekii	27-67	DIQKELLDLTKKVKVAGFRAGKVPVSIVKKKYGTSVR-HDII
Synechocystis sp.	27-68	VYENVVKKLTRTVNIPGFRRGKVPRAIVIQRLGQSYIKATAI
M. genitalium	30-69	TQKKLVGEMAKSIKIKGFRPGKIPPNLASQSINKAELMQKSA
		GFRxGxxP

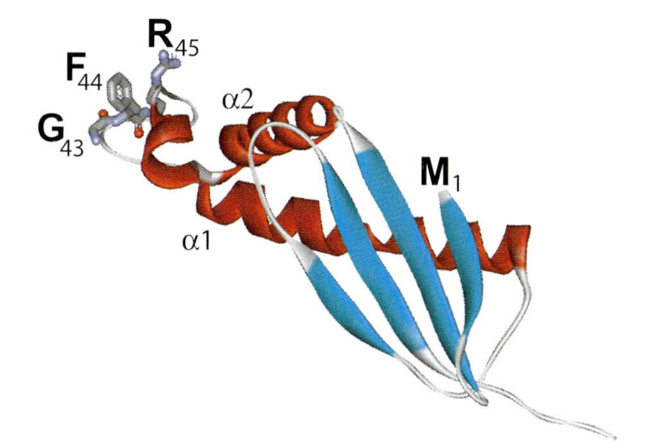

Fig. 13.2. Structure-function relationship of *E. coli* trigger factor. (A) Domain organization of *E. coli* trigger factor. (B) Alignment of the conserved region within the N-terminal trigger factor domain from different bacteria. The conserved TF signature is highlighted (top). Crystal structure of N-terminal domain in ribbon representation based on Kristensen et al. [28] (bottom).

C

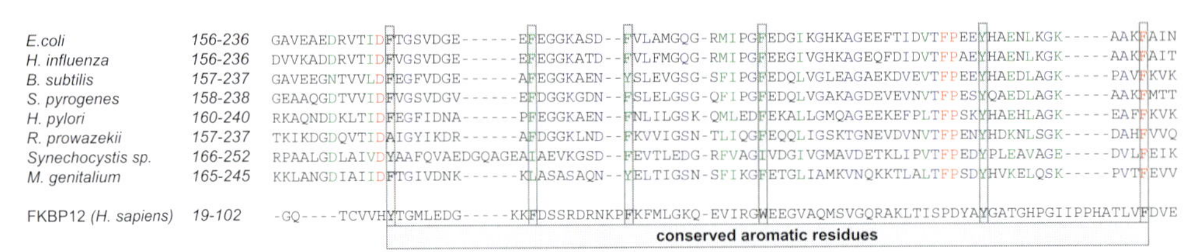
E.coli 156-236 GAVEAEDRVTIDFTGSVDGE------EFEGGKASD--FVLAMGQG-RMIPGFEDGIKGHKAGEEFTIDVTFPEEYHAENLKGK-----AAKFAIN
H. influenza 156-236 DVVKADDRVTIDFVGSVDGE------EFEGGKATD--FVLFMGQG-RMIPGFEEGIVGHKAGEQFDIDVTFPAEYHAENLKGK-----AAKFAIT
B. subtilis 157-237 GAVEEGNTVVLDFEGFVDGE------AFEGGKAEN--YSLEVGSG-SFIPGFEDQLVGLEAGAEKDVEVTFPEEYHAEDLAGK-----PAVFKVK
S. pyrogenes 158-238 GEAAQGDTVVIDFVGSVDGV------EFDGGKGDN--FSLELGSG-QFIPGFEDQLVGAKAGDEVEVNVTFPESYQAEDLAGK-----AAKFMTT
H. pylori 160-240 RKAQNDDKLTIDFEGFIDNA------PFEGGKAEN--FNLILGSK-QMLEDFEKALLGMQAGEEKEFPLTFPSKYHAEHLAGK-----EAFFKVK
R. prowazekii 157-237 TKIKDGDQVTIDAIGYIKDR------AFDGGKLND--FKVVIGSN-TLIQGFEQQLIGSKTGNEVDVNVTFPENYHDKNLSGK-----DAHFVVQ
Synechocystis sp. 166-252 RPAALGDLAIVDYAAFQVAEDGQAGEAIAEVKGSD--FEVTLEDG-RFVAGIVDGIVGMAVDETKLIPVTFPEDYPLEAVAGE-----DVLFEIK
M. genitalium 165-245 KKLANGDIAIIDFTGIVDNK------KLASASAQN--YELTIGSN-SFIKGFETGLIAMKVNQKKTLALTFPSDYHVKELQSK-----PVTFEVV
FKBP12 (H. sapiens) 19-102 -GQ----TCVVHYTGMLEDG-----KKFDSSRDRNKPFKFMLGKQ-EVIRGWEEGVAQMSVGQRAKLTISPDYAYGATGHPGIIPPHATLVFDVE
conserved aromatic residues

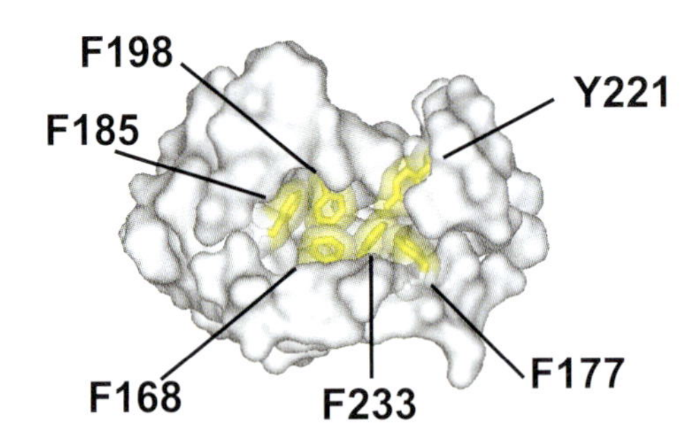
F198
F185
Y221
F168
F233
F177

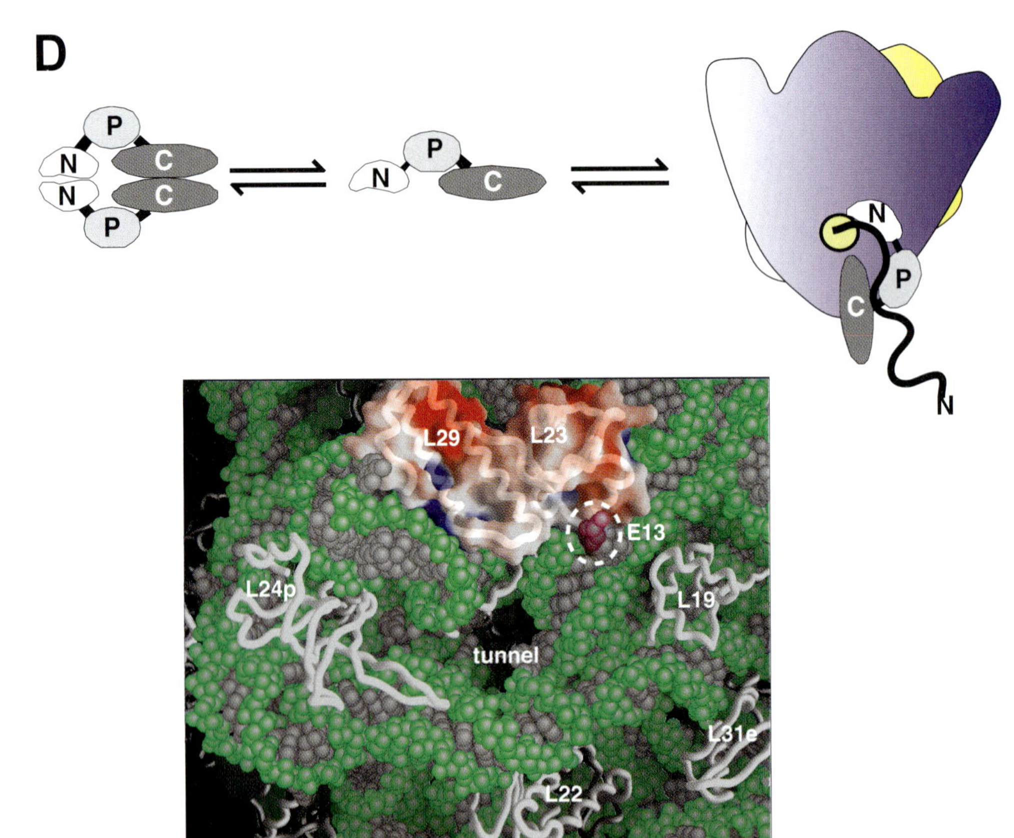
D
P
N
C
N
C
P
P
N
C
N
P
C
N
L29
L23
E13
L24p
L19
tunnel
L31e
L22

ported by the finding that the PPIase domain alone cannot efficiently promote peptidyl-prolyl isomerization-limited folding of the model protein RNase T1 [25, 26].

The N-terminal part of trigger factor (amino acids 1–144) forms another structural and functional module (Figure 13.2A,B) that is necessary and sufficient for the specific binding of trigger factor to ribosomes [27]. The N-terminus contains a compactly folded domain comprising the first 118 amino acids that mediates binding of trigger factor to the ribosome. Sequence alignments of several trigger factor homologues revealed a highly conserved GFRXGXXP motif termed TF-signature within this domain (Figure 13.2B) [22]. Evidence that the TF signature is involved in the binding to ribosomes came from mutational analysis. Introduction of three alanine residues at positions 44–46 (FRK, see Figure 13.2B) within the *E. coli* TF signature results in a strong ribosome-binding deficiency. The crystal structure of the N-terminal domain was recently solved and supports the biochemical data [28]. This domain consists of a four-stranded, antiparallel β-sheet flanked by two long α-helices (Figure 13.2B). Importantly, both helices are linked via a connecting loop that forms a protruding tip. This twisted loop contains the TF signature and shows high structural flexibility (Figure 13.2B) [28]. The molecular details of the interaction of this TF-signature loop with the ribosome remain to be determined.

The function of the C-terminal domain, which constitutes nearly half of the trigger factor protein, is unknown. This domain is poorly conserved among trigger factor proteins and no homology to any other protein is known [29]. For *E. coli* trigger factor it has been shown that the high affinity of trigger factor for unfolded RNaseT1 is severely decreased in the absence of the C-terminal domain [26], indicating that this domain itself participates in substrate binding or modulates the substrate interactions of the other domains.

So far trigger factor has been found in prokaryotes but not in the cytoplasm of eukaryotic cells [1] or archaea [30]. A potential eukaryotic trigger factor homologue is found in *Arabidopsis thaliana*, where it is most likely targeted to the chloroplast. The *A. thaliana* trigger factor has a predicted mass of 65 kDa and displays a total identity of 21% and similarity of 42% to *E. coli* trigger factor. Fragments of at least

Fig. 13.2. (C) Alignment of the trigger factor PPIase domains from different bacteria with the human FKBP12 homologue; conserved aromatic residues are highlighted (top); homology modeling of trigger factor's PPIase domain based on the yeast FKBP12 structure using INSIGHT II. The stick representation shows the conserved aromatic residues of the FKBP-like binding pocket (yellow), numbered by their position in full-length *E. coli* trigger factor. Data were taken with permission from Ref. [17]. (D) Model of trigger factor's three-state equilibrium. Trigger factor is monomeric when associated with the ribosome. The conformation might differ from the conformation of monomeric non-ribosome-associated trigger factor. In solution, TF follows a monomer-dimer equilibrium. The magnification of the ribosome (bottom) shows a space-filling model of the ribosomal polypeptide exit region of the 50S subunit from *H. marismortui*. L23/L29 are shown in a surface charge distribution illustration (GRASP) [57]. The surface-exposed E13 (corresponding to E18 in *E. coli*) involved in the interaction with trigger factor is shown in purple and circled. Other proteins at the exit are visible by their Cα traces. Data were taken with permission from Refs. [31] and [22].

four additional putative trigger factor homologues exist in barley (*Hordeum vulgare*), sugar beet (*Beta vulgaris*), wheat (*Triticum aestivum*), and an insect herbivory (*Medicago truncatula*). These potential trigger factor homologues are 50–65% identical to the *A. thaliana* protein and include the signature motif GFRPxxxxP that is very similar to their bacterial homologues (Figure 13.2B). A potential targeting sequence and the fact that, when compared to eubacterial trigger factor homologues, the *A. thaliana* protein shows the highest similarity to the trigger factor of the cyanobacterium *Synechocystis* spp. argue for a localization in chloroplasts.

It remains elusive why trigger factor is conserved in plant chloroplasts but not in mitochondria. Mitochondria have a much lower protein synthesis activity than chloroplasts, which may render the presence of a ribosome-associated chaperone unnecessary. Alternatively, mitochondria may have evolved a different chaperone system that acts as a functional equivalent of trigger factor.

13.3.2 Quaternary Structure

Experiments using size-exclusion chromatography, glutaraldehyde cross-linking (see protocols, Section 13.5), and ultracentrifugation show that non-ribosome-associated trigger factor is in a monomer-dimer equilbrium with a dissociation constant of dimeric trigger factor of approximately 18 μM [31]. Cross-linked dimeric trigger factor species were observed only in the absence of ribosomes, whereas only monomeric trigger factor was discovered in association with the ribosome, indicating that trigger factor is a monomer when attached to ribosomes. Consistent with this finding is that under saturating conditions trigger factor binds ribosomes in an apparent 1:1 stoichiometry [4, 7, 31]. Opposing results were obtained from a study based on calculations from neutron-scattering experiments with trigger factor–ribosome complexes, suggesting that trigger factor may bind to the ribosome as a homodimer [32]. Association of trigger factor with ribosomes can be monitored by the incubation of purified components and subsequent isolation of trigger factor–ribosome complexes by sucrose cushion centrifugation and analysis of the pellet by SDS-PAGE (see Section 13.5). Quantitative analysis of the binding data revealed a dissociation constant for the trigger factor–ribosome complex of approximately 1 μM [31, 33].

Glutaraldehyde cross-linking using recombinant trigger factor fragments (see Section 13.5) demonstrates that the N-terminal and C-terminal domains individually form cross-linking products corresponding to dimeric species, whereas no oligomeric species was detected for the isolated PPIase domain. In contrast to the full-length protein, however, much higher concentrations of the fragments were necessary to observe efficient cross-linking, which indicates that in the full-length protein, contacts between the N- and C-terminal domains of the monomers contribute to dimerization. Taken together, trigger factor exists in three states: the non-ribosome dimer and monomer states, which are in fast equilibrium, and the ribosome-associated monomer state. Based on the determined dissociation constants for the trigger factor dimer and the trigger factor–ribosome complex together with the already known in vivo concentrations of trigger factor (50 μM) [11] and ribosomes (20 μM) [4], 39% of the cellular trigger factor is a free dimer, 26% a free monomer, and 35% is associated with ribosomes as monomer. Furthermore, about 90% of all ribosomes in the *E. coli* cytosol are found in association with trigger factor [31]. The in vivo function of the trigger factor dimer formation remains elusive. The dimeric trigger factor species might be a storage form, in which the protease-sensitive ribosome attachment site of trigger factor is protected. It was observed earlier that trigger factor's N-terminal domain is cleaved by endogenous proteases within the TF signature [22]. The fast monomer/dimer equilibrium would then ensure a permanent presence of functional monomeric trigger factor and saturation of ribosomes with trigger factor in the cell. Alternatively, non-ribosome-associated trigger factor might fulfill a yet unknown cellular function as a dimeric species in vivo.

13.3.3
PPIase and Chaperone Activity of Trigger Factor

Trigger factor carries out at least two activities. In vitro, both activities have been characterized. Trigger factor catalyzes the peptidyl-prolyl isomerization of chromogenic tetrapeptides [6, 7] and the peptidyl-prolyl isomerization-limited refolding of the model substrate RNase T1 [6]. The high catalytic efficiency of trigger factor towards the protein substrate results from its tight binding to protein substrates with a K_D of approximately 0.7 μM [25]. In contrast, the affinity of trigger factor for peptides is orders of magnitude lower (K_D of approximately 120 μM) [17, 24]. Moreover, trigger factor acts as a typical chaperone in vitro. It binds to unfolded proteins including denatured RNaseT1, D-glyceraldehyde-3-phosphate dehydrogenase (GAPDH), lysozyme, and bovine carbonic anhydrase [25, 34, 35] and prevents aggregation of these unfolded proteins upon dilution from the denaturant. Furthermore, it was shown that trigger factor supports the refolding of chemically denatured GAPDH in vitro [34] (see also Section 13.5). A recent study addressed the question of whether the activity of trigger factor as a ribosome-associated folding factor relies on both its chaperone and its PPIase activity [36]. The trigger factor mutant protein TF F198A, carrying a point mutation in its PPIase domain, fully retained its chaperone activity but lacked the catalytic PPIase function in vitro. Expression of the TF F198A-encoding gene complemented the synthetic lethality of Δ*tig*Δ*dnaK* cells and prevented global protein misfolding at temperatures between 20 °C and 34 °C in these cells. This indicates that the PPIase activity is not essential for the function of trigger factor in folding of newly synthesized proteins at permissive temperature.

It is possible that the PPIase function of trigger factor might be required under specific conditions or for specific substrates. In *Bacillus subtilis* the simultaneous deletion of the *tig* gene and the *ppiB* gene encoding for the second cytosolic PPIase in *Bacillus* inhibits growth under starvation conditions in minimal media lacking certain amino acids [37]. It is, however, unclear whether this phenotype relates to trigger factor's PPIase or chaperone function. Fischer and colleagues reported on the trigger factor mutant protein F233Y, which has reduced PPIase activity towards peptides [38]. When the F233Y mutant was used to complement a Δ*tig* strain, a reduced survival rate on plates after storage for 25 days at 6 °C was observed compared to wild-type trigger factor [39]. This might suggest an involvement of trigger factor in cell viability at very low temperatures. However, the interpretation of this finding is difficult since 6 °C is below the physiological growth temperature of *E. coli*. Very recently, it was reported that the maturation of the extracellular protease SpeB of *Streptococcus pyogenes* is influenced by the PPIase activity of its trigger factor [40]. However, a database search revealed that no SpeB homologue exists in *E. coli* [36].

13.3.4 **Importance of Ribosome Association**

The atomic structures of the large (50S) and small (30S) ribosomal subunits of bacterial and archaeal ribosomes were recently solved [41–45]. The structure of the *Haloarcula marismortui* large ribosomal subunit revealed a tunnel approx. 100 Å in length with an average diameter of 15 Å, through which the nascent polypeptide extends while still connected to the tRNA in the peptidyl-transferase center [42, 46]. The tunnel is long enough to accommodate a 30- to 35-amino-acid segment of a growing polypeptide chain [47] and it is, in principle, wide enough to allow a peptide to acquire a helical structure [46]. The tunnel exit, where the nascent polypeptide emerges, is surrounded by a protein ring of at least five ribosomal proteins including L32, L22, L24, L29, and L23 (Figure 13.2D). Through use of the engineered trigger factor mutant TF-D42C with a UV-activatable cross-linker located directly adjacent to the TF signature, trigger factor was cross-linked to the neighboring ribosomal proteins L23 and L29 of the large ribosomal subunit (Figure 13.2D; see also protocols, Section 13.5) By using neutron-scattering experiments, Nierhaus and coworkers confirmed that trigger factor is located near the ribosomal proteins L23 and L29 [32]. Analyses of *E. coli* strains carrying mutations in ribosomal protein genes provided evidence that trigger factor binds to L23 but does not associate with L29. Binding of trigger factor occurs via the conserved residue Glu18 of L23, and mutational alteration of Glu18 resulted in a strong binding deficiency for trigger factor [22] (Figure 13.2D). The association of trigger factor with L23 is crucial for its association with nascent polypeptides and its in vivo function as a chaperone for newly synthesized proteins. This was demonstrated by using an in vitro transcription/translation system (see protocols), where employing L23 mutant ribosomes leads to a drastic reduction of the cross-linking efficiency of trigger factor towards nascent polypeptides compared to wild-type ribosomes. *E. coli* strains carrying L23 mutant ribosomes showed conditional lethality at 34 °C in the absence of the cooperating DnaK chaperone [22], providing evidence that the interaction with L23 is essential for the chaperone function of trigger factor in vivo. Ribosomal protein L23 thus plays a key role in linking two fundamental cellular events: translation and chaperone-assisted protein folding. Interestingly, the signal recognition particle (SRP), which is important for the export of proteins, was also found to cross-link to L23 in both prokaryotes and eukaryotes [48–50]. The ribosomal exit site thus provides a platform that recruits proteins needed to control the early interactions of nascent polypeptide chains. It is not clear how the different partners in such a system could bind without steric conflict and in a coordinated fashion. Recent data suggest that trigger factor and SRP compete for the binding to L23 [49], which raises the question of how the activities of these proteins at the ribosome can be coordinated.

The association of trigger factor with the ribosomal exit site and the nascent polypeptide chains raises interesting questions regarding the dynamic nature of these interactions. So far the kinetics of the interactions of trigger factor with ribosomes and substrates has been investigated exclusively by using purified components and vacant ribosomes. Fluorescently labeled trigger factor was used to investigate its interaction with vacant ribosomes. The affinity of trigger factor to these ribosomes is rather low and is similar to its affinity for unfolded protein substrates (K_D of both is approximately 1 μM) [25, 33, 51]. The association and dissociation of trigger factor with vacant ribosomes are rather slow processes, with an average lifetime of the complex of about 20 s at 20 °C [33]. In contrast, the substrate binding is highly dynamic, with an estimated lifetime of a trigger factor–substrate complex of about 100 ms [51]. The difference in the dynamics might be relevant for trigger factor's function. In principle, the long lifetime of the trigger factor–ribosome complex would allow the chaperone to remain bound while a complete protein or at least a protein domain is synthesized. Since the interaction with the polypeptide substrate is much faster, several cycles of substrate binding and release may occur during this timeframe. However, the most relevant parameters, the association and dissociation rates of the trigger factor complex with translating ribosomes and nascent chains, have not been determined so far. These rates may differ significantly from those mentioned above when, e.g., ongoing translation causes structural changes at the ribosomal docking site for trigger factor, thereby affecting the affinity for trigger factor. In addition the binding of trigger factor to ribosomes and nascent chains could be allosterically coupled.

13.4
Models of the Trigger Factor Mechanism

How does trigger factor promote the folding of newly synthesized proteins? To account for the folding activity of trigger factor, two alternative working models are conceivable (Figure 13.3): the "looping model," where trigger factor remains ribosome-associated while scanning the nascent polypeptides, or the "cycling model,"

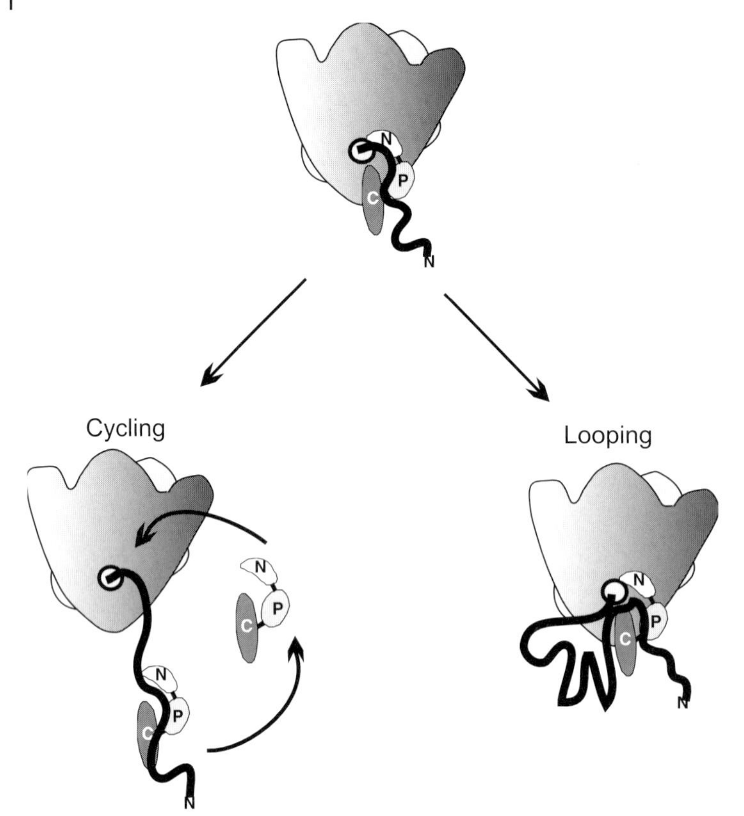

Fig. 13.3. Model of the folding of newly synthesized proteins in *E. coli*. Ribosome-associated trigger factor associates with all nascent polypeptides emerging from the ribosomal exit tunnel. The mechanism of trigger factor action is unclear. Two models are conceivable: the cycling model and the looping model (see text for detail). Beyond the interaction of newly synthesized proteins with trigger factor, at least a subset of proteins needs further chaperoning by DnaK and GroEL to reach the native conformation (not depicted here).

where trigger factor escorts the nascent chain as it moves away from the exit site. The cycling model predicts that binding of trigger factor to a nascent polypeptide substrate causes the release of trigger factor from the ribosome and its transient co-migration with the nascent chain (Figure 13.3). After dissociation from its substrate, trigger factor may rebind to the same or another ribosome and start a new round of substrate interaction. Such a mechanism has two possible consequences. First, trigger factor may protect the growing polypeptide from premature and possibly incorrect folding by allowing a longer stretch of amino acids, i.e., a folding domain, to emerge from the ribosomal tunnel before folding starts upon trigger factor dissociation. Second, trigger factor may spatially separate the ribosome from the folding process and thereby avoid potential clashes of folding intermediates with the ribosomal surface. Interestingly, it was shown recently that nascent polypeptides cross-link to the ribosomal protein L23 in the absence of trigger factor [52, 53], a finding that indicates a shielding function of the chaperone. The two- to threefold molar excess of trigger factor over ribosomes in *E. coli* and the high frequency of trigger factor binding sites in protein sequences (every 30–40 residues) [17] are in agreement with the demands of the cycling model. As soon as trigger factor dissociates from the ribosome binding site together with its bound nascent chain, a second trigger factor molecule can immediately occupy the ribosome for the next nascent chain interaction.

The looping model predicts that trigger factor stays bound to the ribosome during its interaction with the nascent peptide. This would create a ternary complex in which the nascent chain is tethered to the translating ribosome via trigger factor. Due to the ongoing synthesis, this would result in a "looping out" of the nascent polypeptide and, as proposed in this model, prevent folding of the looped-out polypeptide stretch. The dissociation of trigger factor from the nascent chain, preferably after a folding domain has emerged, will then allow the folding of the domain as a discrete entity. Dissociation of trigger factor may be triggered by several possible mechanisms. First, trigger factor may dissociate with a rate governed solely by its substrate- and ribosome-binding properties. Second, the folding process may generate a mechanical force that "strips" trigger factor from the polypeptide chain. Third, trigger factor dissociation may be controlled by a mechanism that couples the folding process to the translation process. Accordingly, the translation status of the ribosome determines the dissociation rates of trigger factor–substrate and trigger factor–ribosome complexes.

In summary, the mechanism of trigger factor in assisting the folding of nascent polypeptide chains is elusive and can be understood only by detailed kinetic analysis of the involved interactions. A molecular understanding of the trigger factor function would be propelled by the availability of an atomic structure of trigger factor in association with ribosomes.

13.5 Experimental Protocols

13.5.1 Trigger Factor Purification

Trigger factor carrying a C-terminal hexa-histidine tag shows properties identical to wild-type trigger factor. Therefore, all trigger factor variants and fragments are purified as hexa-histidine-tagged versions, which can be overexpressed and purified in large quantities by the following two-step purification protocol. We recommend conducting the entire purification (starting with cell lysis) within one day since the trigger factor protein is prone to proteolysis during the purification procedure, especially within the surface-exposed TF signature motif.

Cell Growth Conditions, Overexpression, and Cell Lysis *E. coli* cells carrying the *tig* gene on a pDS56 vector [27] and either a chromosomal or plasmid-encoded *lacIq* gene are grown at 30 °C in Luria broth (LB), supplemented with 100 μg mL^{-1} ampicillin for plasmid selection, to an OD_{600} of 0.6. Induction of trigger factor expression is then started by addition of 500 μM IPTG. Two hours later, cells are harvested, shock-frozen in liquid nitrogen, and stored at −20 °C. One and a half liters of culture can yield up to 100 mg of trigger factor protein.

The cell pellet is thawed and resuspended in 25 mL ice-cold French press buffer (50 mM Tris/HCl pH 7.5, 20 mM imidazole pH 7.5, 200 mM NaCl, 1 mM EDTA) supplemented with either 1 mM PMSF or Complete protease inhibitor according to the manufacturer (EDTA-free, Roche). Cells are lysed with a French press two times at 8000 psi pressure. Cell debris is separated from the soluble fraction by centrifugation at 20 000 g for 30 min.

Step 1: Ni-NTA Purification The Ni/NTA purification is done as a batch purification. Six to eight milliliters of Ni-NTA agarose (QIAGEN) is equilibrated with five volumes of cold French press buffer in a suction filter. The supernatant from the centrifugation step is supplemented with 10 mM $MgCl_2$ (to complex free EDTA), and this solution is incubated with the equilibrated Ni-NTA agarose on ice for 15 min with gentle stirring. The solution is then passed through the suction filter. The agarose is washed with at least half a liter of ice-cold washing buffer (50 mM Tris/HCl pH 7.5, 20 mM imidazole pH 7.5, 500 mM NaCl). Next, the salt concentration is reduced by the application of five volumes of ice-cold low-salt buffer (50 mM Tris/HCl pH 7.5, 20 mM imidazole pH 7.5, 25 mM NaCl). The protein is then eluted from the Ni-NTA agarose with three volumes of cold elution buffer (50 mM Tris/HCl pH 7.5, 250 mM imidazole pH 7.5, 25 mM NaCl).

Step 2: Anion-exchange Chromatography Since the Ni-NTA elution buffer contains low salt, the Ni-NTA eluate can be directly applied to an anion-exchange column (e.g., ResourceQ6, Pharmacia) at 4 °C and eluted with a salt gradient (low-salt buffer [50 mM Tris/HCl pH 7.5, 25 mM NaCl, 1 mM EDTA] to high-salt buffer [50 mM Tris/HCl pH 7.5, 1 M NaCl, 1 mM EDTA]). Under these conditions, trigger factor elutes at 150 mM NaCl (15% high-salt buffer). Time of elution of trigger factor fragments or mutants may vary. Trigger factor-containing fractions from the anion-exchange chromatography step are analyzed by SDS-PAGE, pooled, and dialyzed against 2 L of storage buffer (20 mM Tris/HCl pH 7.5, 100 mM NaCl, 1 mM EDTA) overnight at 4 °C. Typically, one purification yields about 6–12 mL of a 4–8 mg mL^{-1} trigger factor solution.

13.5.2
GAPDH Trigger Factor Activity Assay

The chaperone activity of trigger factor can be monitored by following the aggregation or refolding of chemically denatured glyceraldehyde-3-phosphate dehydrogenase (GAPDH) in the presence of trigger factor [34].

Prevention of Unfolded GAPDH Aggregation Equal amounts of freshly prepared GAPDH solution in GAPDH buffer (250 μM GAPDH, Sigma G-2267, in 0.1 M Ca phosphate pH 7.5, 1 mM EDTA, 5 mM DTT) and denaturation buffer (6 M GdnHCl, 5 mM DTT) are mixed and kept for 1 min at 4 °C. Denatured GAPDH is diluted to a final concentration of 2.5 μM into the GAPDH buffer containing trigger factor. Immediately after dilution, aggregation of GAPDH is measured in a spectrofluorometer as an increase in light scattering at 620 nm. In the absence of any additional factor, GAPDH aggregation is complete within 10 min. The addition of stoichiometric amounts of trigger factor suppresses GAPDH aggregation completely and no increase in light scattering is observed during this time period [34].

GAPDH Refolding Assay GAPDH is denatured by mixing equal amounts of GAPDH solution and denaturation buffer (see above) for 1 min at 4 °C. The denatured GAPDH protein is diluted to a final concentration of 2.5 μM into cold GAPDH buffer containing trigger factor at a concentration of 2.5 μM. The solution is incubated for 30 min at 4 °C and is then shifted to room temperature, where GAPDH refolding starts. Within 3–4 h, aliquots are taken from the solution and GAPDH activity is measured as follows. Final concentrations of 0.67 mM glyceraldehyde-3-phosphate (GAP, Sigma G-5251) and 0.67 mM β-nicotinamide adenine dinucleotide (NAD, Sigma N-0632) are premixed in GAPDH buffer. An aliquot from the GAPDH/trigger factor solution is diluted 1:5 into the mixture and absorption at 340 nm is measured immediately in a photometer. After 4–5 min, the reaction is complete. The rate constants of the reactions are determined by a single-exponential fit of the absorption data. GAPDH activity is best represented as the percentage of non-denatured GAPDH activity, i.e., by dividing the measured rate constant by the rate constant determined for the same amount of non-denatured GAPDH enzyme.

13.5.3
Modular Cell-free *E. coli* Transcription/Translation System

The in vitro transcription/translation system allows the synthesis of any target protein in vitro based on a plasmid-encoded target gene transcribed from a T7 promoter. The basic principle of this system is outlined in Figure 13.4. For the generation of a homogenous population of arrested nascent polypeptides, an antisense oligonucleotide and RNaseH are added to produce an mRNA lacking the stop

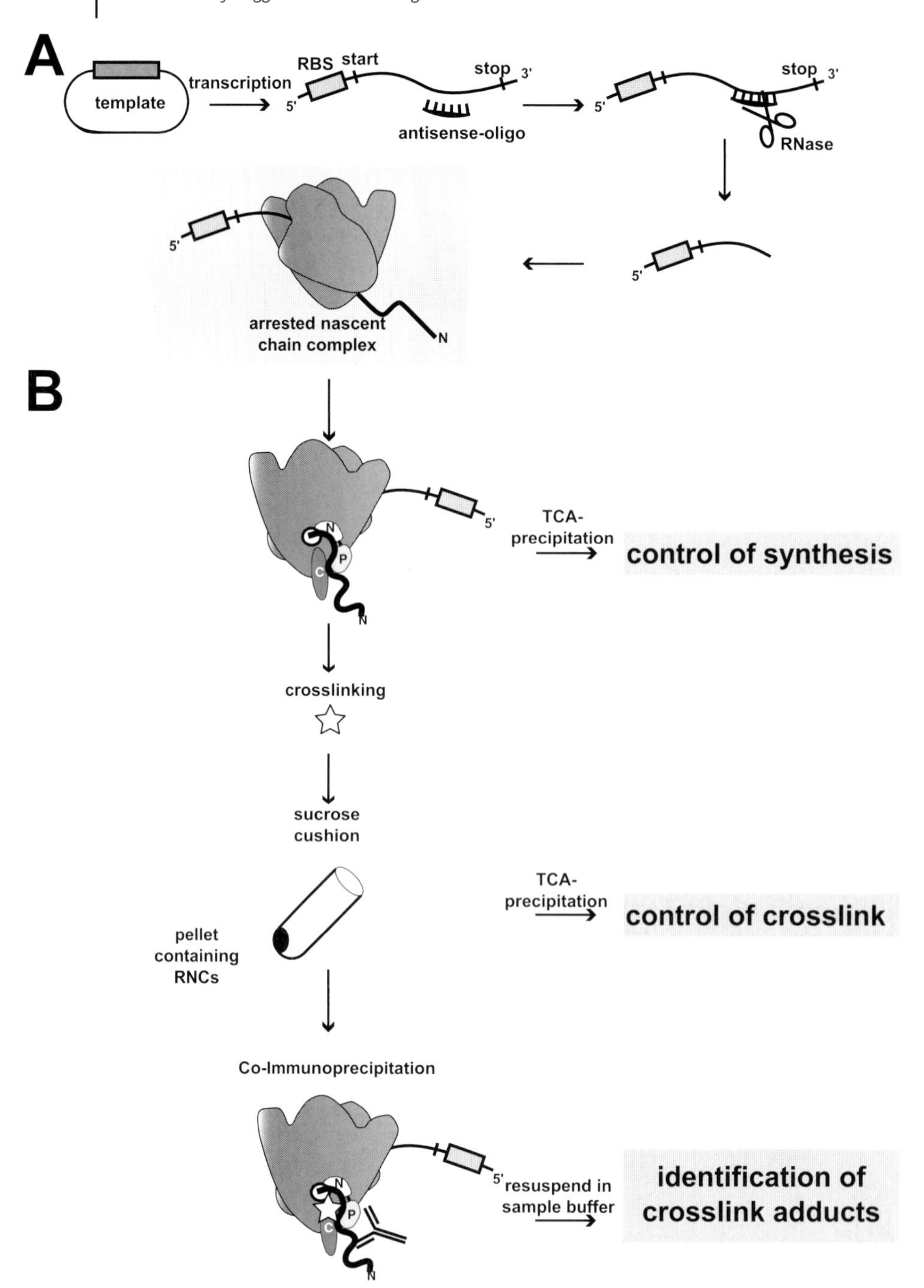

Fig. 13.4. Schematic outline of the in vitro transcription/translation system. (A) Generation of stalled ribosomes by truncation of the mRNA transcript. (B) Chemical cross-linking between radioactively labeled arrested nascent chains and trigger factor and subsequently co-immunoprecipitation.

codon. The addition of a chemical cross-linker allows subsequent monitoring of the interaction of arrested ^{35}S-radio-labeled nascent polypeptides with trigger factor.

The transcription/translation system consists of a subfraction of *E. coli* lysate, termed "translation-active fraction," and high-salt-washed ribosomes. The system is derived from Δ*tig* cells and thus allows for the addition of trigger factor or, if appropriate, of other chaperones. The modular setup makes it easy to test mutant ribosomes or chaperones using the same translation-active fraction. All components can be mixed on ice, and the in vitro transcription/translation reaction is initiated by the addition of ^{35}S-methionine and a shift to 37 °C.

Preparation of the components for in vitro translation essentially follows the protocol described by Zubay [54] with slight modifications. The preparation of components is carried out in three steps:

1. cell culturing and preparation of the S30 extract,
2. isolation of the S150 extract and ribosomes, and
3. separation of the extract by sucrose density centrifugation.

The *E. coli* wild-type strain MC4100, or the derived strain MC4100Δ*tig*, is used as a source for the extracts. To prevent RNase contamination, glassware and centrifuge tubes should be washed with DEPC water. In addition, all buffers should be freshly prepared with the same water and supplemented with proteinase inhibitors (PMSF or Complete EDTA-free, Roche).

Cell Culturing and Preparation of the S30 Extract A 100-mL preculture is grown overnight in LB at 37 °C. Both preculture and main culture are grown in the absence of antibiotics to acquire fully active ribosomes. In the MC4100Δ*tig* strain an integrated kanamycin-resistance cassette replaces the *tig* gene and is stable even in the absence of selective pressure [10].

The preculture is diluted into S30 medium to an OD_{600} of 0.08 and the culture is grown at 37 °C to an OD_{600} of 1.2. Two liters of culture should be grown in a 5-L flask for optimal aeration growth. At the desired OD_{600} cultures are cooled on ice and centrifuged (JA-10 rotor, 6000 g, 15 min, 4 °C) and the cell pellets are resuspended in S30 buffer (1 mL of buffer per 1 g of cell pellet). Cells are lysed using a French press equipped with a 40-mL pressure chamber at 8000 psi. Cells are lysed twice and 1 μL of DTT (1 M) is added per 10 mL of lysate. The cell lysates are then centrifuged (30 000 g, 4 °C, 30 min) and the resulting supernatant is the S30 extract. The extract can be used immediately or frozen in liquid nitrogen and stored at −80 °C. (S30 medium (per liter): 5.6 g KH_2PO_4, 28.9 g K_2HPO_4, 10 g yeast extract, 1% glucose, pH 7.4; S30 buffer: 50 mM triethanolamine-OAc (pH 8.0), 50 mM KOAc, 15 mM $Mg(OAc)_2$, 1 mM DTT, 0.5 mM PMSF.)

Isolation of the S150 Extract and Ribosomes The S30 extract is centrifuged in a Ti 50 rotor at 150 000 g at 4 °C for 3 h to separate ribosomes and membrane fragments from soluble cytosolic proteins. Two-milliliter aliquots of the supernatant (the S150 extract) can be frozen in liquid nitrogen and stored at −80 °C. The yield of S150 extract from 4 L of culture was about 16 mL in the case of MC4100 (48 g L^{-1}).

The remaining pellet is used for ribosome isolation. These pellets are usually overlayed with a yellow viscous layer containing membrane and cellular debris. This layer can be removed by washing the pellet briefly with RL buffer. The clear pellet is suspended overnight in 5 mL RL buffer. The suspended ribosomes are again centrifuged (20 min, 30 000 g, 4 °C) and the remainder of the cell debris is found in the pellet. The supernatant of this centrifugation step containing the ribosomes is diluted in RL buffer to a final volume of 20 mL and the final KOAc concentration is adjusted to 1 M. This ribosome-containing fraction is loaded onto a 40-mL 20% sucrose cushion in high-salt buffer and is centrifuged at 150 000 g at 4 °C for 4 h (in a Ti 45 rotor). The resulting ribosomal pellet is suspended in ribosome-buffer and the concentration is determined by measuring the absorption at 260 nm. Absorption of 1.0 equals the ribosome concentration of 23 pmol mL^{-1} or 23 nM. The A_{260}/A_{280} ratio should be ~2. Ribosomes were aliquoted, frozen in liquid nitrogen, and stored at −80 °C.

RL buffer:	50 mM triethanolamine-OAc (pH 8.0)
	100 mM KOAc
	6 mM $Mg(OAc)_2$
	4 mM β-mercaptoethanol
Sucrose cushions:	20% (w/v) sucrose in high-salt buffer
High-salt buffer:	50 mM triethanolamine-OAc (pH 8.0)
	1 M KOAc
	6 mM $Mg(OAc)_2$
	4 mM β-mercaptoethanol
Ribosome buffer:	50 mM triethanolamine-OAc (pH 8.0)
	50 mM KOAc
	6 mM $Mg(OAc)_2$
	4 mM β-mercaptoethanol

Isolation of the Translation-active Fraction from the S150 Extract To obtain a concentrated and efficient translation-active fraction, the S150 extract is separated on a preformed sucrose gradient [55, 56]. To prepare the gradients, 13-mL centrifuge tubes are filled with 6 mL 10% sucrose solution and afterwards 6 mL of the 30% sucrose solution is slowly injected with a syringe below the 10% layer. Linear gradients are subsequently generated by the "Gradient Master" (Life Technologies), which rotates the tilted tubes until the gradients are formed. The tubes containing the sucrose gradient are then cooled at 4 °C for 1 h. Six-hundred-fifty microliters of the S150 extract is loaded onto each gradient and the centrifugation is carried out (in a SW41 rotor) at 39 000 rpm at 4 °C for 19 h. After the sedimentation equilibrium is reached, the centrifuge is stopped (without braking to prevent mixing of the gradients). The bottoms of the tubes are punctured with a needle and the gradients are fractionated while recording the UV absorption at 280 nm. Twenty-one fractions of 600 μL can be collected per gradient. The recorded absorption data allows comparison of the gradients and comparable fractions can be pooled. The fractions can be frozen in liquid nitrogen and stored at −80 °C. Because not every fraction is sufficiently active for the in vitro transcription/translation system, each fraction is tested for its ability to transcribe and translate a target gene, and active fractions are pooled. The end concentration of the MC4100 extract should be roughly 5.4 g L^{-1}. Five hundred microliter aliquots of the concentrated and active fractions are frozen in liquid nitrogen and stored at −80 °C.

Sucrose solutions:	10% or 30% (w/v) sucrose
	50 mM HEPES/KOH (pH 7.5)
	50 mM KOAc (pH 7.5)
	5 mM $Mg(OAc)_2$
	1 mM DTT
	0.5 mM PMSF

The in Vitro Transcription/Translation Reaction Several additional components (cofactor mix) must be added to the purified ribosomes and translation-active fraction to generate the active transcription/translation system. The final volume of one reaction is 25 μL. The coding DNA template (under control of a T7 promotor) and T7 RNA polymerase were added on ice. Addition of [^{35}S]-methionine and a shift of the samples to 37 °C start the transcription/translation reaction. The produced mRNA is translated in the presence of [^{35}S]-methionine to synthesize a radioactively labeled protein or nascent chain.

Cofactor mix:

Co-factor	Final concentration
PEG	3.2% (w/v)
19-amino-acid mix (no methionine)	0.1 mM each
DTT	2 mM
Nucleotide mix, pH 7.5	2.5 mM ATP
	0.5 mM (CTP, UTP, GTP)
Phosphoenol-pyruvate	5 mM
Putrescine	8 mM
Creatine phosphate	8 mM
Creatine phosphokinase	40 μg mL^{-1}

10× compensation buffer:

	Final concentration
HEPES (pH 7.5)	310 mM
KOAc (pH 7.5)	540 mM
$Mg(OAc)_2$	89 mM
Spermidine	8 mM
H_2O (DEPC)	Adjust volume

The final transcription/translation reaction (25 μL) is composed of the following components (use DEPC water to adjust the volume to 25 μL):

	Concentration	Amount (μL)	Final concentration
Translation-active fraction		4.0–6.0	
Ribosomes	5 μM	1.0	100 nM
Template DNA		1.0	
T7 RNA polymerase	20 U	0.5	0.4 U μL^{-1}
Cofactor mix		6.25	
[^{35}S]-methionine	15 μCi μL^{-1}	0.5	0.3 μCi (0.3 μM)
Compensation buffer (10×)		2.5	

The optimal concentration of the template DNA was individually determined for each product and it varied between 50 and 200 ng μL^{-1}. The ideal ion and buffer concentration was determined to be 40 mM HEPES (pH 7.5), 70 mM KOAc (pH 7.5), 10 mM $Mg(OAc)_2$, and 0.8 mM spermidine. The ion and buffer concentration is adjusted using the 10× compensation buffer, taking into account the buffer content of ribosomes (50 mM triethanolamine-OAc, 50 mM KOAc, 6 mM $Mg(OAc)_2$), the translation-active fraction (50 mM HEPES, 50 mM KOAc, 5 mM $Mg(OAc)_2$), and proteins added that contain additional buffers and salts. Proteins added to the system should be dialyzed against translation buffer (50 mM HEPES, 50 mM KOAc, 5 mM $Mg(OAc)_2$) prior to use.

For a given experiment, components that are identical to each reaction (except the [^{35}S]-methionine) are mixed into a master mix, and this mix is kept on ice. Aliquots of the master mix can be added to the different samples with varying DNA templates or protein components. The [^{35}S]-methionine is added last and the samples are mixed well using a pipette. Shifting the samples from ice to 37 °C starts the reaction, and the samples are incubated for 45 min. After this time, cooling the samples on ice and addition of 3.2 μL chloramphenicol solution (25 mg mL^{-1} in ethanol) stop the transcription/translation reaction.

The transcription/translation efficiency is determined by TCA precipitation of all proteins and subsequent separation by SDS-PAGE and detection of radioactively labeled chains by autoradiography. One volume of 10% TCA is added to the desired sample, and the proteins are precipitated for at least 30 min on ice. Precipitated proteins are collected by centrifugation (16 000 g, 4 °C, 10 min) and are resuspended in 25 μL alkaline sample buffer (pH 9), denatured for 5 min at 99 °C, and used for SDS-PAGE. In addition, the translation efficiency can be determined using a scintillation counter to measure the amount of [^{35}S]-methionine incorporated into the newly synthesized protein.

Synthesis of Arrested Ribosome–Nascent Chain Complexes (RNCs) Translation can be arrested in the *statu nascendi* if translation is carried out from an mRNA lacking a stop codon. In vitro such an mRNA can be generated by addition of an oligonucleotide that anneals with the mRNA and subsequent digestion with RNaseH, which cleaves the DNA-mRNA hybrid upstream of the stop codon (Figure 13.4). The translating ribosomes stop at the end of the mRNA and cannot proceed. The *E. coli* ssrA system has the potential to release the arrested ribosomes; therefore, an anti-*ssrA* oligonucleotide can be added to the reaction that targets and inactivates the 10Sa RNA of the *E. coli ssrA* machinery.

The optimal concentration of template DNA, oligonucleotides, RnaseH, and anti-*ssrA* oligonucleotides must be determined by titration for each nascent chain. In addition to the components described above, one 25-μL reaction contains the following additives (example):

	Final concentration
Oligonucleotide	0.5–2 ng[1]
RNaseH	0.04 U
Anti-*ssrA* oligonucleotide (5′-ttaagctgctaaagcgtagttttcgtcgtttgcgacta-3′)	0.12 μg

[1]Optimal concentration of the oligonucleotides must be determined for each nascent chain.

TCA		10% in ddH_2O
Alkaline SDS sample buffer:	Solution 1:	0.2 M Tris-Base
		0.02 M EDTA pH 8.0
	Solution 2:	8.3% SDS
		83.3 mM Tris-Base
		29.2% glycerol
		0.03% bromphenol blue

The sample buffer is made by mixing solution 1:solution 2:water in the ratio 5:4:1.

Cross-linking of Trigger Factor to Nascent Polypeptides The interaction of proteins (like chaperones) with nascent chains can be studied using cross-linking experiments, which employ different chemical nonspecific cross-linking agents (DSS, EDC) to create a covalent bond between the nascent chain and the protein of interest. For a cross-linking experiment, the sample reaction can be scaled up 10-fold (i.e., 250 μL transcription/translation reaction). One volume of the reaction (i.e., 25 μL) is used to assess the efficiency of the transcription/translation reaction, two volumes are used to assess efficiency of cross-linking, and the remaining seven volumes are used for co-immunoprecipitation analysis (see below). Samples are loaded immediately on a gel for comparison.

Cross-linking With DSS Disuccinimidyl suberate (DSS, Pierce #21555) is a highly mobile NHS ester that preferentially reacts with amines. Therefore, the protein buffers should not contain any primary amines (e.g., Tris). HEPES is used as buffer since it is also compatible with the transcription/translation system. After stopping the transcription/translation reaction, 2.5 μL of DSS (25 mM in DMSO) are added per 25-μL reaction, and cross-linking is performed for 30 min at RT. To quench the free DSS, Tris/HCl (pH 7.5) is added to a final concentration of 50 mM and quenching is carried out for 15 min at RT. The quenched cross-linking reaction can then be precipitated with TCA and analyzed by SDS-PAGE and autoradiography, or it can be centrifuged through a 20% sucrose cushion to collect ribosomal pellets plus cross-linking partners for further analysis.

Cross-linking With EDC EDC (1-ethyl-3-(3-dimethylaminopropyl); Pierce #22980) is a carbodiimide that couples carboxyl and amino groups via an amide bond. HEPES should be used once again for EDC cross-linking steps. A transcription/translation reaction (25 μL) is supplemented with 5 μL EDC (480 mM freshly prepared with ddH_2O). After cross-linking at 30 °C for 30 min, free EDC is quenched by adding 1/10 volume of quench buffer and incubating samples on ice for 15 min.

EDC:	480 mM in ddH_2O
Quench buffer:	1 M glycine
	100 mM $NaHCO_3$
	100 mM β-mercaptoethanol, pH 8.5

Isolation of RNCs After the transcription/translation reaction and cross-linking, the samples are centrifuged (16 000 g, 5 min, 4 °C) to remove potential protein aggregates generated during the procedure from the solution. To isolate RNCs the supernatants are subsequently loaded onto chilled sucrose cushions (cushions should be 2–3× volume of the loaded sample), and the samples are centrifuged in either a TLA-100 or a TLA-100.2 rotor, depending on the volume, at 75 000 rpm at 4 °C for 90 min. The supernatant is rapidly removed after the centrifugation and the clear ribosomal pellet can be resuspended in either 100 μL of ribosome buffer (plus 0.3 mg mL^{-1} chloramphenicol and Complete EDTA-free, Roche) for co-immunoprecipitation or in sample buffer for immediate analysis by SDS-PAGE.

Ribosome buffer:	50 mM triethanolamine-OAc (pH 8.0)
	100 mM KOAc
	15 mM $Mg(OAc)_2$
	1 mM DTT
Sucrose cushion:	20% sucrose
	50 mM HEPES/KOH (pH 7.5)
	100 mM KOAc (pH 7.5)
	10 mM $Mg(OAc)_2$
	4 mM β-mercaptoethanol
	0.3 mg mL^{-1} chloramphenicol

Co-immunoprecipitation of the Cross-linking Products Co-immunoprecipitation of the cross-linking products under denaturating conditions is carried out at 4 °C with antibodies directed against the target protein, e.g., trigger factor. The ribosomal pellet from at least seven reactions (175 μL of the original transcription/translation reaction) is used for immunoprecipitation. The suspended pellet is mixed with 50 μL protein A Sepharose suspension (0.125 g mL^{-1} equilibrated with RIPA buffer) and 5 μL (depending on the quality of the antibody) of antibody; the mixture is diluted with RIPA buffer to a final volume of 1 mL. The binding is allowed to proceed for at least 1 h at 4 °C with constant mixing. The protein A Sepharose with the bound antibody-protein complexes is pelleted (1600 *g*, 4 °C, 1 min) and washed twice with 1 mL RIPA buffer and once with 1 mL PBS. Finally, the pellet is resuspended in 30 μL of sample buffer and denatured at 100 °C prior to SDS-PAGE analysis.

Protein A Sepharose CL-4 B: 0.125 mg mL^{-1} equilibrated with RIPA buffer

RIPA buffer:	150 mM NaCl
	50 mM Tris/HCl (pH 7.5)
	0.5 mM EDTA
	1% NP40
	0.5% Doc
	0.1% SDS
PBS:	137 mM NaCl
	2.7 mM KCl
	10 mM Na_2HPO_4
	2 mM KH_2PO_4
	pH 7.4

13.5.4 Isolation of Ribosomes and Add-back Experiments

High-salt-washed ribosomes can be purified from Δ*tig* *E. coli* strains and used for in vitro rebinding assays with purified trigger factor. The purification protocol is identical to the one described above for the transcription/translation system, except that LB medium can be used for the main culture and that the sucrose cushion centrifugation can be repeated, to yield very pure ribosomes. For cross-linking experiments with trigger factor D42C-BPIA (see below), a HEPES-ribosome buffer can be used to suspend the ribosomes. This prevents quenching of the cross-linker by Tris.

HEPES-ribosome buffer:	25 mM HEPES pH 7.5
	6 mM $MgCl_2$
	30 mM NH_4Cl

Binding Trigger Factor to Ribosomes in Vitro Purified trigger factor can be rebound to high-salt-washed ribosomes in vitro and the bound protein can be separated from soluble trigger factor by sucrose cushion centrifugation. After the centrifugation, the ribosomes and bound proteins are found in the pellet, while soluble proteins remain in the supernatant.

In a typical binding experiment, 6 μM trigger factor is added to 2 μM ribosomes and the final volume is adjusted to 60 μL using one of the ribosome buffers described above. To allow trigger factor to bind to the ribosomes, the sample is incubated at 30 °C for 30 min, and the reaction mixture is then layered on top of a 120-μL 20% sucrose cushion prepared with ribosome-buffer. The volume of the cushion should be 3× volume of the sample. Samples are centrifuged for 70 min at 75 000 rpm and 4 °C in a TLA-100 rotor.

Eighty microliters of the resulting supernatant is mixed with 20 μL of 5× SDS sample buffer, and the rest of the supernatant is removed. The clear ribosomal pellet is suspended in 50 μL 1× SDS sample-buffer. After denaturing the samples at 100 °C for 5 min, 30 μL of the supernatant and 15 μL of the pellet are used for SDS-PAGE. A typical gel depicting the rebinding of trigger factor to *E. coli* ribosomes is shown in Figure 13.5.

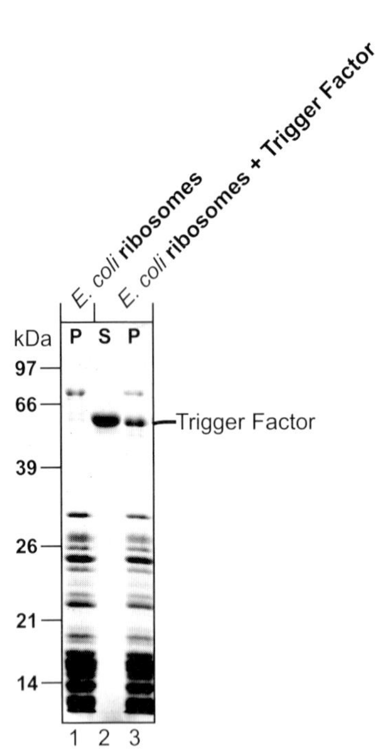

Fig. 13.5. Rebinding of purified trigger factor to ribosomes. Two micromoles of purified ribosomes were incubated with 6 μM trigger factor under physiological salt conditions. Ribosome–trigger factor complexes were purified by sucrose cushion centrifugation. Ribosomal pellets (P) and supernatants (S) were analyzed by SDS-PAGE and Coomassie staining.

13.5.5 Cross-linking Techniques

Labeling TF-D42C Protein With BPIA The TF-D42C mutant of trigger factor is purified from MC4100Δ*tig* cells as described above. To reduce the protein, 500 μL (3–4 mg mL^{-1}) is mixed with 50 μL of immobilized TCEP disulfide reducing gel (Pierce #77712) and the sample is incubated at RT for 10 min with gentle shaking. The immobilized TCEP can be removed by centrifugation and transfer of the protein-containing supernatant into a new tube (twice).

One micromole of the chemical cross-linker BPIA (4-(2-iodoacetamido)benzophenone; Molecular Probes) is added to the protein and the sample is incubated at 30 °C for 30 min. The BPIA should be protected against extensive light exposure during the entire process. Addition of 2 μL β-mercaptoethanol (14.4 M) and cooling on ice stop the labeling process, and the labeling efficiency can be checked by mass spectrometry. The labeled protein is dialyzed against the cross-linking buffer and is frozen in liquid nitrogen and stored at −80 °C.

Cross-linking buffer: 25 mM HEPES pH 7.5
50 mM NaCl

Cross-linking TF-D42C BPIA to Ribosomes in Vitro The TF-D42C BPIA protein (6 μM) is incubated with ribosomes (2 μM) as described above. After incubating the sample for 30 min at 30 °C (in the dark), the tubes are placed on ice, are opened, and are UV irradiated using a UV lamp at 365 nm for 10 min. The cross-linked samples can then be analyzed as described above, and the bands corresponding to cross-linking adducts can be cut out of the gel for analysis by mass spectrometry.

Cross-linking Trigger Factor With Glutaraldehyde Trigger factor dimers can be cross-linked using the nonspecific cross-linker glutaraldehyde. Different concentrations of trigger factor protein are incubated for 15 min at 30 °C in a glutaraldehyde buffer with varying NaCl concentrations. Cross-linking is initiated by addition of 0.1% glutaraldehyde (Sigma) and is stopped after 10 min by adding 100 mM Tris pH 7.5 for 10 min at RT. Soluble proteins are precipitated by mixing the samples with one volume 10% TCA and incubating for 60 min on ice. After centrifugation (16 000 *g*, 4 °C, 10 min), the pellet is analyzed by SDS-PAGE.

Glutaraldehyde buffer: 20 mM HEPES pH 7.5
100 mM–2 M NaCl
1 mM EDTA

References

1 B. Bukau, E. Deuerling, C. Pfund, and E. A. Craig. Getting newly synthesized proteins into shape. *Cell* **2000**, *101*, 119–122.

2 F. U. Hartl, and M. Hayer-Hartl. Molecular chaperones in the cytosol: from nascent chain to folded protein. *Science* **2002**, *295*, 1852–1858.
3 E. Crooke, and W. Wickner. Trigger factor: a soluble protein that folds pro-OmpA into a membrane-assembly competent form. *Proc. Natl. Acad. Sci. USA* **1987**, *84*, 5216–5220.
4 R. Lill, E. Crooke, B. Guthrie, and W. Wickner. The "Trigger factor cycle" includes ribosomes, presecretory proteins and the plasma membrane. *Cell* **1988**, *54*, 1013–1018.
5 B. Guthrie, and W. Wickner. Trigger factor depletion or overproduction causes defective cell division but does not block protein export. *J. Bacteriol.* **1990**, *172*, 5555–5562.
6 G. Stoller, K. P. Ruecknagel, K. H. Nierhaus, F. X. Schmid, G. Fischer, and J.-U. Rahfeld. A ribosome-associated peptidyl-prolyl *cis/trans* isomerase identified as the trigger factor. *EMBO J.* **1995**, *14*, 4939–4948.
7 T. Hesterkamp, S. Hauser, H. Lütcke, and B. Bukau. *Escherichia coli* trigger factor is a prolyl isomerase that associates with nascent polypeptide chains. *Proc. Natl. Acad. Sci. USA* **1996**, *93*, 4437–4441.
8 Q. A. Valent, D. A. Kendall, S. High, R. Kusters, B. Oudega, and J. Luirink. Early events in preprotein recognition in *E. coli*: interaction of SRP and trigger factor with nascent polypeptides. *EMBO J.* **1995**, *14*, 5494–5505.
9 B. Bukau, T. Hesterkamp, and J. Luirink. Growing up in a dangerous enviroment: a network of multiple targeting and folding pathways for nascent polypetides in the cytosol. *Trends Cell Biol.* **1996**, *6*, 480–486.
10 E. Deuerling, A. Schulze-Specking, T. Tomoyasu, A. Mogk, and B. Bukau. Trigger factor and DnaK cooperate in folding of newly synthesized proteins. *Nature* **1999**, *400*, 693–696.
11 S. A. Teter, W. A. Houry, D. Ang, T. Tradler, D. Rockabrand, G. Fischer, P. Blum, C. Georgopoulos, and F. U. Hartl. Polypeptide flux through bacterial Hsp70: DnaK cooperates with Trigger Factor in chaperoning nascent chains. *Cell* **1999**, *97*, 755–765.
12 B. Bukau, and G. C. Walker. Cellular Defects caused by deletion of the *Escherichia coli dnaK* gene indicates roles for heat shock protein in normal metabolism. *J. Bact.* **1989**, *171*, 2337–2346.
13 T. Hesterkamp, and B. Bukau. Role of the DnaK and HscA homologs of Hsp70 chaperones in protein folding in *E. coli*. *EMBO J.* **1998**, *17*, 4818–4828.
14 E. Deuerling, H. Patzelt, S. Vorderwülbecke, et al. Trigger Factor and DnaK possess overlapping substrate pools and binding specificities. *Mol. Microbiol.* **2003**, *47*, 1317–1328.
15 T. Tomoyasu, A. Mogk, H. Langen, P. Goloubinoff, and B. Bukau. Genetic dissection of the roles of chaperones and proteases in protein folding and degradation in the Escherichia coli cytosol. *Mol Microbiol* **2001**, *40*, 397–413.
16 A. Mogk, T. Tomoyasu, P. Goloubinoff, S. Rüdiger, D. Röder, H. Langen, and B. Bukau. Identification of thermolabile *E. coli* proteins: prevention and reversion of aggregation by DnaK and ClpB. *EMBO J.* **1999**, *18*, 6934–6949.
17 H. Patzelt, S. Rudiger, D. Brehmer, et al. Binding specificity of Escherichia coli trigger factor. *Proc Natl Acad Sci USA* **2001**, 98, 14244–14249.
18 C. Scholz, M. Mücke, M. Rape, A. Pecht, A. Pahl, H. Bang, and F. X. Schmid. Recognition of protein substrates by the prolyl isomerase trigger factor is independent of proline residues. *J. Mol. Biol.* **1998**, *277*, 723–732.
19 S. Rüdiger, L. Germeroth, J. Schneider-Mergener, and B. Bukau. Substrate specificity of the

DnaK chaperone determined by screening cellulose-bound peptide libraries. *EMBO J.* **1997**, *16*, 1501–1507.

20 X. Zhu, X. Zhao, W. F. Burkholder, A. Gragerov, C. M. Ogata, M. Gottesman, and W. A. Hendrickson. Structural analysis of substrate binding by the molecular chaperone DnaK. *Science* **1996**, *272*, 1606–1614.

21 S. Rüdiger, A. Buchberger, and B. Bukau. Interaction of Hsp70 chaperones with substrates. *Nat Struct Biol.* **1997**, *4*, 342–349.

22 G. Kramer, T. Rauch, W. Rist, S. Vorderwülbecke, H. Patzelt, A. Schulze-Specking, N. Ban, E. Deuerling, and B. Bukau. L23 protein functions as a chaperone docking site on the ribosome. *Nature* **2002**, *419*, 171–174.

23 I. Callebaut, and J. P. Mornon. Trigger factor, one of the Escherichia coli chaperone proteins, is an original member of the FKBP family. *FEBS Lett* **1995**, *374*, 211–215.

24 M. Vogtherr, D. M. Jacobs, T. N. Parac, M. Maurer, A. Pahl, K. Saxena, H. Ruterjans, C. Griesinger, and K. M. Fiebig. NMR solution structure and dynamics of the peptidyl-prolyl cis-trans isomerase domain of the trigger factor from Mycoplasma genitalium compared to FK506-binding protein. *J Mol Biol* **2002**, *318*, 1097–1115.

25 C. Scholz, G. Stoller, T. Zarnt, G. Fischer, and F. X. Schmid. Cooperation of enzymatic and chaperone functions of trigger factor in the catalysis of protein folding. *EMBO J.* **1997**, *16*, 54–58.

26 T. Zarnt, T. Tradler, G. Stoller, C. Scholz, F. X. Schmid, and G. Fischer. Modular Structure of the Trigger Factor Required for High Activity in Protein Folding. *J. Mol. Biol.* **1997**, *271*, 827–837.

27 T. Hesterkamp, E. Deuerling, and B. Bukau. The amino-terminal 118 amino acids of *Escherichia coli* trigger factor constitute a domain that is necessary and sufficient for binding to ribosomes. *J. Biol. Chem.* **1997**, *272*, 21865–21871.

28 O. Kristensen, and M. Gajhede. Chaperone binding at the ribosomal exit tunnel. *Structure (Camb)* **2003**, *11*, 1547–1556.

29 T. Hesterkamp, and B. Bukau. Identification of the prolyl isomerase domain of *Escherichia coli* trigger factor. *FEBS Letters* **1996**, *385*, 67–71.

30 A. J. L. Macario, M. Lange, B. K. Ahring, and E. Conway de Macario. Stress genes and proteins in the Archaea. *Microbiol. Mol. Biol. Rev.* **1999**, *63*, 923–967.

31 H. Patzelt, G. Kramer, T. Rauch, H. J. Schönfeld, B. Bukau, and E. Deuerling. Three-Sate Equilibrium of *Escherichi coli* Trigger Factor. *Biol. Chem.* **2002**, *383*, 1611–1619.

32 G. Blaha, D. N. Wilson, G. Stoller, G. Fischer, R. Willumeit, and K. H. Nierhaus. Localization of the trigger factor binding site on the ribosomal 50S subunit. *J Mol Biol* **2003**, *326*, 887–897.

33 R. Maier, B. Eckert, C. Scholz, H. Lilie, and F. X. Schmid. Interaction of trigger factor with the ribosome. *J Mol Biol* **2003**, *326*, 585–592.

34 G. C. Huang, Z. Y. Li, J. M. Zhou, and G. Fischer. Assisted folding of D-glyceraldehyde-3-phosphate dehydrogenase by trigger factor. *Protein Sci* **2000**, *9*, 1254–1261.

35 C. P. Liu, and J. M. Zhou. Trigger factor-assisted folding of bovine carbonic anhydrase II. *Biochem Biophys Res Commun* **2004**, *313*, 509–515.

36 G. Kramer, H. Patzelt, T. Rauch, T. A. Kurz, S. Vorderwülbecke, B. Bukau, and E. Deuerling. Trigger Factor's peptidyl-prolyl cis/trans isomerase activity is not essential for the folding of cytosolic proteins in Escherichia coli. *J. Biol. Chem.* **2004**, *in press*.

37 S. F. Göthel, C. Scholz, F. X. Schmid, and M. A. Marahiel. Cyclophilin and Trigger Factor from *Bacillus subtilis* Catalyze in Vitro

Protein Folding and are Necessary for Viability under Starvation Conditions. *Biochemistry* **1998**, *37*, 13392–13399.

38 T. Tradler, G. Stoller, K. P. Rücknagel, A. Schierhorn, J.-U. Rahfeld, and G. Fischer. Comparative mutational analysis of peptidyl prolyl *cis/trans* isomerases: active sites of *Escherichia coli* trigger factor and human FKBP12. *FEBS Letters* **1997**, *407*, 184–190.

39 C. Schiene-Fischer, J. Habazettl, T. Tradler, and G. Fischer. Evaluation of similarities in the cis/trans isomerase function of trigger factor and DnaK. *Biol Chem* **2002**, *383*, 1865–1873.

40 W. R. Lyon, and M. G. Caparon. Trigger factor-mediated prolyl isomerization influences maturation of the Streptococcus pyogenes cysteine protease. *J Bacteriol* **2003**, *185*, 3661–3667.

41 N. Ban, P. Nissen, J. Hansen, M. Capel, P. B. Moore, and T. A. Steitz. Placement of protein and RNA structures into a 5 Å-resolution map of the 50S ribosomal subunit. *Nature* **1999**, *400*, 841–847.

42 N. Ban, P. Nissen, J. Hansen, P. B. Moore, and T. A. Steitz. The complete atomic structure of the large ribosomal subunit at 2.4 Å resolution. *Science* **2000**, *289*, 905–920.

43 J. Harms, F. Schluenzen, R. Zarivach, A. Bashan, S. Gat, I. Agmon, H. Bartels, F. Franceschi, and A. Yonath. High resolution structure of the large ribosomal subunit from a mesophilic eubacterium. *Cell* **2001**, *107*, 679–688.

44 F. Schluenzen, A. Tocilj, R. Zarivach, et al. Structure of functionally activated small ribosomal subunit at 3.3 angstroms resolution. *Cell* **2000**, *102*, 615–623.

45 B. T. Wimberly, D. E. Brodersen, W. M. Clemons, Jr., R. J. Morgan-Warren, A. P. Carter, C. Vonrhein, T. Hartsch, and V. Ramakrishnan. Structure of the 30S ribosomal subunit. *Nature* **2000**, *407*, 327–339.

46 P. Nissen, J. Hansen, N. Ban, P. B. Moore, and T. A. Steitz. The structural basis of ribosome activity in peptide bond synthesis. *Science* **2000**, *289*, 920–930.

47 B. Hardesty, and G. Kramer. Folding of a nascent peptide on the ribosome. *Prog Nucleic Acid Res Mol Biol* **2001**, *66*, 41–66.

48 M. R. Pool, J. Stumm, T. A. Fulga, I. Sinning, and B. Dobberstein. Distinct modes of signal recognition particle interaction with the ribosome. *Science* **2002**, *297*, 1345–1348.

49 R. S. Ullers, E. N. Houben, A. Raine, C. M. ten Hagen-Jongman, M. Ehrenberg, J. Brunner, B. Oudega, N. Harms, and J. Luirink. Interplay of signal recognition particle and trigger factor at L23 near the nascent chain exit site on the Escherichia coli ribosome. *J Cell Biol* **2003**, *161*, 679–684.

50 S. Q. Gu, F. Peske, H. J. Wieden, M. V. Rodnina, and W. Wintermeyer. The signal recognition particle binds to protein L23 at the peptide exit of the Escherichia coli ribosome. *Rna* **2003**, *9*, 566–573.

51 R. Maier, C. Scholz, and F. X. Schmid. Dynamic association of trigger factor with protein substrates. *J Mol Biol* **2001**, *314*, 1181–1190.

52 G. Eisner, H. G. Koch, K. Beck, J. Brunner, and M. Muller. Ligand crowding at a nascent signal sequence. *J Cell Biol* **2003**, *163*, 35–44.

53 R. S. Ullers, E. N. G. Houben, A. Raine, C. M. Ten Hagen-Jongman, M. Ehrenberg, J. Brunner, B. Oudega, N. Harms, and J. Luirink. Interplay of SRP and Trigger Factor at L23 near the nascent chain exit site on the *E. coli* ribosome. *J. Mol. Biol.* **2003**, *in press.*

54 G. Zubay. In vitro synthesis of protein in microbial systems. *Annu. Rev. Genet.* **1973**, *7*, 267–287.

55 E. Schaffitzel, S. Rüdiger, B. Bukau, and E. Deuerling. Functional dissection of Trigger Factor and DnaK: Interactions with nascent polypeptides and thermally denatured proteins.

Biological Chemistry **2001**, *382*, 1235–1243.

56 M. Behrmann, H.-G. Koch, T. Hengelage, B. Wieseler, H. K. Hoffschulte, and M. Müller. Requirements for the translocation of elongation-arrested, ribosome-associated OmpA across the plasma membrane of *Escherichia coli*. *J. Biol. Chem.* **1998**, *273*, 13898–13904.

57 A. Nicholls, R. Bharadwaj, and B. Honig. Grasp – Graphical representation and analysis of surface properties. *Biophys. J.* **1993**, *64*, A166–A166.

14
Cellular Functions of Hsp70 Chaperones

Elizabeth A. Craig and Peggy Huang

14.1
Introduction

Hsp70s are ubiquitous molecular chaperones that function in a wide variety of physiological processses, including protein folding, protein translocation across membranes, and assembly/disassembly of multimeric protein complexes. Hsp70s rarely, if ever, function alone, but rather with J-protein (also known as Hsp40 or DnaJ-like protein) partners. All eukaryotic, and the vast majority of prokaryotic, genomes encode Hsp70 and J-proteins. Most, in fact, encode multiple Hsp70 and J-proteins, indicative of the evolution of Hsp70s/J-proteins such that they function in a variety of cellular processes. For example, the yeast genome encodes 14 Hsp70s (Figure 14.1), while the mouse genome encodes at least 12; the number of J-proteins is even more numerous, with the yeast and mouse genomes encoding approximately 20 and 23, respectively [1]. The larger number of J-proteins likely reflects the fact that a single Hsp70 can function with more than one J-protein, driving the function of an Hsp70 in a particular cellular role. In fact, in some cases, J-proteins act as specificity factors, modulating the activity of Hsp70s or delivering specific substrates to them.

This proliferation of Hsp70 and J-protein genes reflects the divergence in their biological functions, the subject of this chapter. In most cases, the basic biochemical properties of Hsp70s and J-proteins utilized when carrying out their functions are the same. Diverse cellular roles result from harnessing these basic biochemical properties in different ways. For example, by tethering a chaperone to a particular cellular location or substrate protein, or by evolving the substrate specificity of these chaperones, they are optimized to function in a particular cellular function.

The general structure of Hsp70s has been conserved [2, 3]: a highly conserved N-terminal ATPase domain, followed by a less conserved domain having a cleft in which hydrophobic peptides can bind, and a variable C-terminal 10-kDa region. ATP binding and hydrolysis in the N-terminus regulate the interaction of the C-terminus with unfolded polypeptides, with ATP hydrolysis stabilizing the interaction. J-proteins transiently interact with the ATPase domain of ATP-bound Hsp70 via their highly conserved J-domains and stimulate its ATPase activity, thus facili-

Protein Folding Handbook. Part II. Edited by J. Buchner and T. Kiefhaber

ISBN: 3-527-30784-2

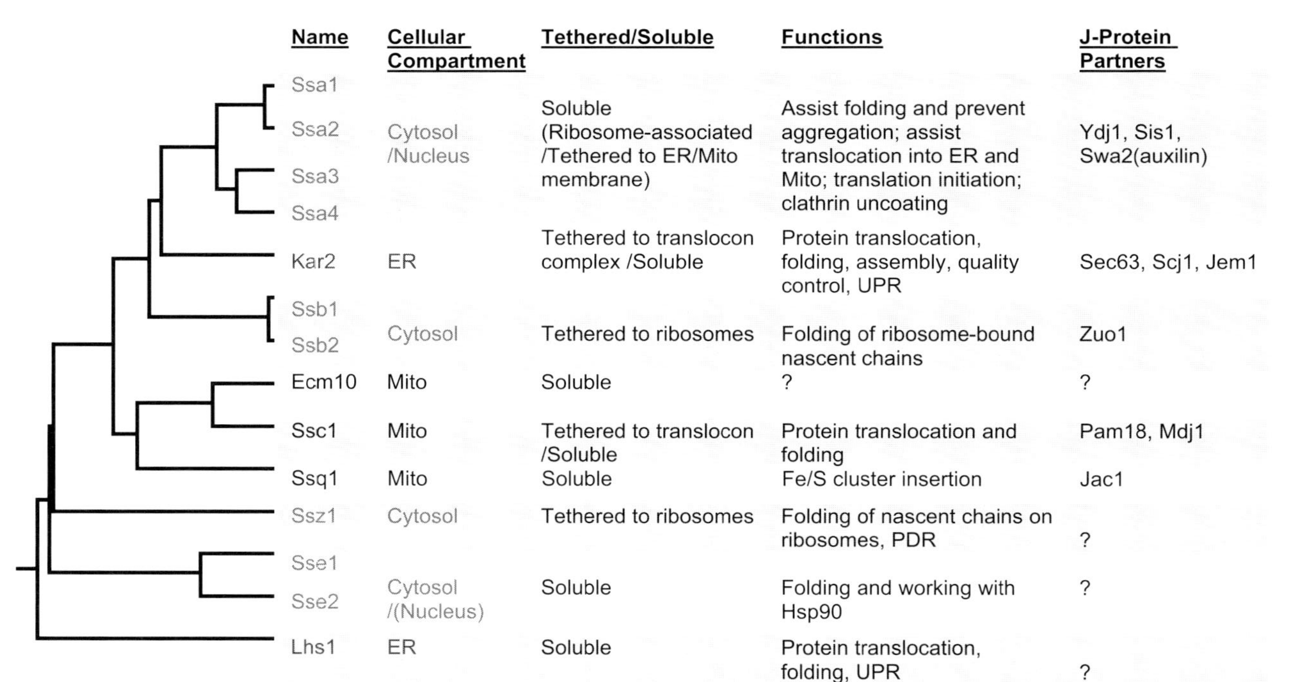

Fig. 14.1. Phylogenetic tree of Hsp70 proteins from the yeast *Saccharomyces cerevisiae*. Subcellular locations, functions, and J-protein partners are indicated for each Hsp70 family. The tree was created using MegAlign software from DNASTAR (Madison, WI) with the clustal method. See text for references regarding evidence for information on particular genes and their functions. Locations or functions that should be considered tentative assignments are in parentheses.

tating the chaperone cycle. Some J-proteins themselves are also able to bind polypeptide substrates, delivering them to Hsp70s. In addition to J-proteins, some Hsp70s also require nucleotide releasing factors to facilitate the exchange of ADP from ATP [2, 3]. For a more thorough discussion of Hsp70 and/or J-protein function, see Chapters 15 and 16.

In this chapter, we first focus on Hsp70/J-proteins' role in the folding of newly synthesized proteins and the recovery of proteins partially denatured by stress, a function carried out by the abundant soluble Hsp70/J-proteins found in all major cellular compartments. Secondly, we discuss Hsp70s/J-proteins, tethered to specific sites within the cell, that are involved in the early stages of protein folding and translocation across membranes. Thirdly, we discuss examples in which Hsp70/J-proteins function in specific cellular processes where folded proteins, rather than unfolded polypeptides, serve as substrates for Hsp70s/J-proteins. In such cases, specific effects on protein conformation, disassembly of multimeric protein complexes, and/or regulation of activity are the typical effects of chaperone action. At the end of the chapter, we discuss two examples in which a member of the yeast Hsp70 family has evolved a function distinct from that of a chaperone, raising the possibility that during evolution some such proteins have been usurped to play other, perhaps regulatory, roles.

14.2
"Soluble" Hsp70s/J-proteins Function in General Protein Folding

The "classic" notion of an Hsp70 is an abundant, likely stress-inducible, soluble protein that primarily facilitates the refolding of proteins partially denatured by stress or the folding of newly synthesized proteins. Such Hsp70s are ubiquitous in the cytosol of eubacteria, eukarya, and some archaea, as well as in the matrix of mitochondria and the lumen of the endoplasmic reticulum (ER).

14.2.1
The Soluble Hsp70 of *E. coli*, DnaK

The *E. coli* Hsp70 DnaK is an abundant protein even under optimal growth conditions, making up about 1% of cellular protein. It is also inducible when cells encounter a stress, such as a rapid temperature increase, at which point it can become 3% of cellular protein, making it the most abundant cellular chaperone [4]. DnaK works with its J-protein partner DnaJ and its nucleotide releasing factor GrpE to perform chaperone functions in the cell (see Chapter 15 for more discussion). The evidence that the DnaK system is critical for preventing the aggregation of cellular proteins upon heat stress and for assisting in the refolding of denatured proteins is incontrovertible [5–10]. At high temperatures, 15–25% of cytosolic proteins, comprising 150–200 different species of polypeptides, aggregate after a temperature upshift of cells lacking DnaK [4, 11].

DnaK also plays a role in the folding of newly synthesized proteins, but that function is not unique to DnaK and can be performed by other chaperones, particularly the ribosome-associated chaperone trigger factor (for more discussion about trigger factor, see Chapter 13). Cells lacking either DnaK or trigger factor are viable and little or no protein aggregation is seen [12, 13]. But a double mutant is inviable; upon depletion of trigger factor in cells lacking DnaK, a dramatic increase in aggregation of more than 40 proteins was observed prior to cell death [12, 13]. Experiments using pulse-labeling of newly synthesized proteins to track the flux of newly synthesized proteins through DnaK revealed that between 5% and 18% of newly made proteins, mainly polypeptides larger than 30 kDa, could be immunoprecipitated with DnaK-specific antibodies in wild-type cells [12]. But, in the absence of trigger factor, this number increased two- to threefold, up to as much as 36% [13]. Together these results suggest that DnaK acts as a chaperone for newly synthesized chains but that function can be taken over by other chaperones, particularly trigger factor, in DnaK's absence. In turn, DnaK is able to take over some of the chaperone activities of trigger factor in its absence. Such overlaps in chaperone function likely are common and reflect a general plasticity of the chaperone networks.

14.2.2 Soluble Hsp70s of Major Eukaryotic Cellular Compartments

14.2.2.1 Eukaryotic Cytosol

While the data for eukaryotic cells is less extensive than that available for *E. coli*, Hsp70s, working with J-protein co-chaperones, are thought to be important for protein folding in the cytosol of eukaryotes as well. In yeast the major cytosolic family of Hsp70s, the Ssa's, is most closely related to the abundant cytosolic Hsp70s of mammalian cells, Hsc70 and Hsp70. Ssa's are thought to participate in "bulk" protein folding in the cytosol, mainly in cooperation with the J-protein Ydj1. Perhaps the most compelling evidence that Ssa proteins are involved in protein folding in the cytosol is the fact that yeast cells carrying a temperature-sensitive *SSA* mutant accumulate newly synthesized ornithine transcarbamylase in a misfolded, but soluble, form after shift to the non-permissive temperature [14].

In mammalian tissue culture cells, Hsc70 has also been implicated in bulk folding of newly synthesized proteins. It has been found to interact with ~20% of newly synthesized polypeptides [15, 16]. In addition, overexpression of human Hsp70 and the J-protein Hdj2 enhances refolding of heat-inactivated luciferase in vivo [17]. Interestingly, in several eukaryotic model systems for protein aggregation, including overexpression of disease-related proteins such as huntingtin and ataxin, Hsp70s and J-proteins have been found to co-localize with such proteins and/or overexpression of the chaperones have been found to decreases their aggregations (reviewed in Ref. [3], also see Chapters 36 and 37 for more discussion).

While in many cases Hsp70s, along with their co-chaperone J-protein, are thought to function alone in protein folding, there are many indications that they also function cooperatively with other chaperone systems to assist posttranslational protein folding. In particular, Hsp70s are thought to function with the Hsp90 system, interacting with Hsp90 client proteins prior to interacting with Hsp90 itself [18]. In yeast both the Ssa and the Sse classes of Hsp70s have been implicated in such cooperation [19, 20]. These issues are described in detail in Chapter 23.

14.2.2.2 **Matrix of Mitochondria**

As perhaps expected of an organelle evolved from eubacteria, the major Hsp70 and its J-protein partner in the mitochondrial matrix (called Ssc1 and Mdj1, respectively, in yeast) are more closely related to the bacterial Hsp70 DnaK and its J-protein partner DnaJ than to other Hsp70s of the same species. A nucleotide release factor, related to *E. coli* GrpE, called Mge1 completes the prokaryotic-type chaperone triad of mitochondria. This chaperone triad is involved in the folding of proteins synthesized on cytosolic ribosomes and imported into the matrix [21–24] as well as of proteins encoded by mitochondrial DNA and translated on mitochondrial ribosomes [23, 25]. Ssc1 also has a second role: a portion of Ssc1 is tethered to the mitochondrial inner membrane and is critical for the process of protein translocation across membranes.

14.2.2.3 **Lumen of the Endoplasmic Reticulum**

Approximately 30% of the proteins synthesized on cytosolic ribosomes are translocated into ER, where posttranslational modification, folding, and assembly take place. These include secreted and cell-surface proteins as well as resident proteins of the endocytic and exocytic organelles. Hsp70s in the ER interact with these newly translocated proteins, assisting in their folding and assembly and preventing their aggregation. BiP (also known as GRP78) in mammalian cells and its yeast homologue, Kar2, are the best-studied ER lumenal Hsp70s that have been shown to interact with a number of secretory pathway proteins to promote their folding and assembly [26–28] (see also Chapter 16). BiP/Kar2 have also been found to play a role in retaining proteins in the ER that do not mature properly and in directing them to back across the ER membrane to be degraded by the proteasome, a process known as ER quality control (reviewed in Refs. [29–31]). In addition, the interaction between unfolded proteins and BiP/Kar2 has been shown to serve as a sensor for ER stress [32]. Working with ER-membrane-associated J-proteins (Sec63 and Mtj1), BiP/Kar2 are also involved in the early step of protein translocation into ER (see Sections 14.3.1 and 14.3.2 for more discussion).

In addition to Kar2, a second Hsp70, Lhs1 (also known as Ssi1 or Cer1), has also been found to function in the lumen of the ER in yeast. Unlike *KAR2*, *LHS1* is not an essential gene and is not induced by heat shock. However, Δ*lhs1* strains are cold-sensitive for growth and accumulate precursor forms of several secretory proteins [33, 34]. Lhs1 is thus believed to play important roles in protein translocation into the ER. Increased ER protein aggregation and degradation after heat stress have also been found in strains lacking Lhs1, suggesting a role in the refolding of denatured proteins in the ER as well [35]. Genetic analysis also strongly imply that it plays a role in unfolded protein response (UPR) in the ER [34, 36]. A number of genetic interactions have been reported between Lhs1 and Kar2 [33, 34, 37], indicating a complex relationship between these two Hsp70s. Recently, Steel et al.

reported that the ATPase activities of Kar2 and Lhs1 are coupled and coordinately regulated. It was shown that Lhs1 stimulates Kar2's ATPase activity by serving as a nucleotide exchange factor, which could be a novel function of Hsp70s [38]. Because the mammalian homologues of Kar2 and Lhs1, BiP and GRP170, both bind to unfolded immunoglobulin chains [39, 40], their chaperone functions might be coupled in vivo by this coordinated regulation.

14.3 "Tethered" Hsp70s/J-proteins: Roles in Protein Folding on the Ribosome and in Protein Translocation

Examples of Hsp70s that are targeted to particular positions within the cell have been known for some time. Of these, the best-studied is the tethering of organellar Hsp70s to the import channel, where they play critical roles in the import of proteins from the cytosol into both the ER and mitochondria. In these cases, Hsp70 binding to unfolded proteins provides a driving force for the import process by favoring vectorial movement of the translocating polypeptide into the organelle, preventing movement back towards the cytosol. In other cases, chaperones tethered to their site of action appear to play a role similar to that of soluble chaperones, but their targeting places them in very close proximity to their protein substrate. One such example is the ribosome-associated Hsp70/J-protein pair Ssb/Zuo1 of yeast, discussed in Section 14.3.2.

14.3.1 Membrane-tethered Hsp70/J-protein

Translocation across both the endoplasmic reticulum membrane and the mitochondrial inner membrane requires passage through channels whose narrow width demands that proteins be substantially unfolded. In both cases, organellar Hsp70s tethered to the translocation channel play a significant role in protein translocation into the ER lumen and the mitochondrial matrix. In the ER, Hsp70 is important for both SRP-dependent co-translational and posttranslational translocation [41, 42]. In mitochondria, Hsp70 is particularly important for translocation of proteins that are tightly folded [43] and thus need to be unfolded before translocation.

In the ER of yeast, Sec63, a polytopic membrane J-protein with a substantial cytosolic domain and a lumenal J-domain, is thought to tether the lumenal Hsp70 Kar2 to the channel, as well as to stimulate its ATPase activity [44]. In mitochondria, the roles of tether and J-protein are separated into (or at least shared by) two proteins, the tether being Tim44 [45–47] and the J-protein being Pam18 [48–50]. Both are associated with the import channel. Tim44 is a peripheral membrane protein, while Pam18 is an integral membrane protein with an N-terminal domain extending into the intermembrane space and a J-domain extending into the matrix.

Pam18 stimulates the ATPase activity of Ssc1; Tim44 does not, but also does not interfere with stimulation by Pam18.

While the exact mechanism of Hsp70 function at the channel remains controversial, there is no debate over the idea that Hsp70 makes use of the same basic biochemical properties utilized in other chaperone activities such as protein folding and binding of short hydrophobic peptide segments, which are regulated by ATP binding and hydrolysis, which in turn is modulated by J-proteins [51–53]. This binding drives vectorial movement of the polypeptide chain by sterically blocking movement back through the narrow channel. Minimally, the act of tethering to the channel allows an extremely high local concentration at the channel. It has also been argued that Hsp70 exerts a force upon the incoming chain, making use of an ATP-dependent conformational change to "pull" it into the organelle. Proof of such an unprecedented activity for an Hsp70 awaits more sophisticated biophysical experiments than have been carried out to date.

Interestingly, however, there is evidence for an additional and unusual role for Hsp70 at the translocon in the ER membrane: gating the channel [54]. The lumenal Hsp70 is responsible for sealing the translocon pore, thereby maintaining the permeability barrier of a non-translocating pore. Mechanistically, how this occurs and which domains of Hsp70 are involved have not been established.

Hsp70s and J-proteins also play a role in protein import into organelles from the cytosolic side. There are multiple reports, both in vivo and in vitro and in both yeast and mammalian cells, that indicate that soluble cytosolic chaperones facilitate translocation, presumably by preventing the premature folding of precursor proteins [51]. While soluble cytosolic chaperones play significant roles in this process, it is intriguing that the J-protein Ydj1 is farnesylated [55] and a portion is found to be associated with the membrane. Whether this association is important for aiding translocation is not clear. However, an intriguing means of targeting chaperone-associated proteins to mitochondria has been reported by Young et al. [56]. Tom70 is an outer mitochondrial membrane receptor for proteins with internal mitochondrial targeting signals that are present in certain mitochondrial proteins such as membrane carrier proteins of the inner membrane. Tom70 is also a tetratrico peptide repeat (TPR) domain-containing protein. The TPR domain of Tom70 interacts with the C-terminus of Hsp70, targeting it to the mitochondria and thus facilitating the interaction of Tom70 with the targeting signal of the protein. Thus, Tom70 serves to target cytosolic Hsp70 (and Hsp90 in mammalian cells), carrying its cargo to the mitochondrial outer membrane.

14.3.2
Ribosome-associated Hsp70/J-proteins

One can imagine more than one possible function for molecular chaperones associated with ribosomes. Three examples are discussed below. While none of them has been thoroughly mechanistically defined, they represent three ways that chaperones may function in protein translation and folding of newly synthesized polypeptides: directly interacting with the nascent chain, modulating protein translocation across membranes, and in the process of translation itself.

The advantage of tethering molecular chaperones near the exit site of the ribosome in terms of protein folding is perhaps the easiest to envision. As protein translation occurs in a vectorial fashion, a newly synthesized polypeptide emerging from the ribosome cannot fold until the entire domain is synthesized. Being exposed to the crowded cellular environment, ribosome-bound nascent chains are thus prone to misfolding and/or aggregation. In the yeast *S. cerevisiae*, two types of Hsp70s (Ssb and Ssz1) and one type of J-protein (Zuo1) have been found to nearly quantitatively associate with ribosomes, with Ssz1 and Zuo1 forming a surprisingly stable heterodimer known as RAC (ribosome-associated complex) [57–59]. The ribosome association of Ssb and RAC is believed to situate them in close proximity to nascent chains, thus facilitating their chaperone functions. Actually, Ssb is believed to bind to the ribosome near the polypeptide exit tunnel since it can be cross-linked to short ribosome-bound nascent chains with a length only slightly longer than that required to span the ribosome exit tunnel [60]. It has been shown that Ssb and Zuo1 bind ribosomes independently, whereas Ssz1 associates with ribosome via its interaction with Zuo1 [59].

Genetic analysis has implied that Ssb, Zuo1, and Ssz1 function together in the same biological pathway, as any combinatory deletions of the three cause the same phenotypes [60, 61]. A chaperone system containing two Hsp70s and one J-protein such as this had not been observed before. How do they function together to chaperone nascent chains on ribosomes? Current evidence suggests that this chaperone triad is actually a variation of the classic Hsp70/J-protein pair, in which the J-protein's role is played by RAC, the Zuo1-Ssz1 complex. Ssb is believed to be the main Hsp70 chaperone interacting with ribosome-bound nascent chains, since it can be cross-linked to nascent chains of between 54 and 152 amino acids [60, 62] and its peptide-binding domain has been shown to be important for its in vivo function [63]. In contrast, Ssz1, the other Hsp70 in this system, has not been observed to interact with nascent chains, and its entire putative peptide-binding domain is dispensable [60].

However, compelling evidence has shown the importance of Ssz1 in working with Zuo1 to assist Ssb's function, as Ssb's cross-link to nascent chain depends on the presence of a functional RAC [61]. In addition, Zuo1 carrying an alteration in its J-domain, even though bound to Ssz1, could not restore the cross-link by Ssb, indicating that Zuo1 is the J-protein partner for Ssb [61]. Collectively, it has been proposed that Ssz1 functions as a cofactor/regulator of Zuo1, modifying the ability of Zuo1 to stimulate Ssb's chaperone activity [60]. This model is supported by the recent finding that Ssb's ATPase activity can be stimulated by Zuo1 only when it is in complex with Ssz1 (P. Huang, personal communication). Genetic analysis also suggests more central roles for Ssb and Zuo1 than for Ssz1, since overexpression of Zuo1 or Ssb can partially compensate for the absence of Ssz1, but Ssz1 overexpression cannot compensate for the lack of either Zuo1 or Ssb [61]. It seems that Ssz1 has evolved to function differently from classic Hsp70s (see Section 14.6.1 for more discussion).

It should be noted that only around half of the total Ssb proteins are found to be ribosome-associated, whereas all of Zuo1 and Ssz1 are found on the ribosomes [57–59]. This suggests that Ssb might play additional roles in the cytosol or that it could move along with the elongating nascent chains to facilitate their folding. It was recently proposed that Ssb might cooperate with the yeast chaperonin system (TriC; for further detail, see Chapters 21 and 22) in the cytosol to fold a specific class of substrates containing WD domains [64].

A ribosome-associated J-protein, Mtj1p, from dog pancreas microsomes has recently been identified and characterized [65]. Mtj1p is thought to function very differently from Zuo1. It is an ER membrane protein having a single transmembrane domain; its J-domain, which extends into the ER lumen, interacts with the ER lumenal Hsp70 BiP. The large cytosolic domain of Mtj1p interacts with translating and non-translating ribosomes at low ionic strength, up to 200 mM KCl. Using truncation constructs, the ribosome-interacting region of Mtj1p was mapped to a highly charged N-terminal region (amino acids 176–194) in its cytosolic domain. Intriguingly, the interaction of Mtj1p with the ribosome appears to affect translation. Translation of a number of proteins in reticulocyte lysates is dramatically decreased when either Mtj1p's cytosolic domain or a peptide derived from its ribosome-interacting region is present. It has been proposed that Mtj1p acts during co-translational protein transport into ER to recruit both the ribosome and BiP to the translocon complex. In the process, it could help to modulate translation and facilitate any number of aspects of the translocation process, including: facilitating the handover of nascent polypeptides from the signal recognition particle (SRP) to the translocon complex, transmitting signals from the ribosome to BiP, or regulating lumenal gating of the translocon [65].

Ribosome association has also been observed for a portion of yeast Hsp70 Ssa and one of its J-protein partners, Sis1 [66, 67]. However, the exact nature of their ribosome association has not been established. The data indicate that Ssa interacts with Sis1 and poly(A)-binding protein (Pab1) via its C-terminal 10-kDa domain [66, 68, 69], and its interaction with Sis1 and Pab1 occurs preferentially on translating ribosomes [66]. Supporting a meaningful role for Sis1 on ribosomes, genetic interactions have been reported between mutations in genes encoding ribosomal subunit proteins and mutations in *SIS1* [66, 67]. A dramatic decrease in interaction between Pab1 and the translation initiation factor, eIF4G, has also been observed in strains carrying a temperature-sensitive allele of *SSA* growing at non-permissive temperature [66]. Collectively, the data suggest that Ssa and Sis1 may function in translation initiation on yeast ribosomes, possibly via interaction with Pab1. For further discussion of ribosome-associated chaperones see Chapter 12.

14.4
Modulating of Protein Conformation by Hsp70s/J-proteins

Although the examples are limited, Hsp70 and J-proteins have evolved in some instances to interact with proteins that are fully, or nearly fully, folded. In this manner, chaperones function in specific, and very diverse, cellular functions, including the biogenesis of Fe/S proteins, regulation of the heat shock response, and regulation of initiation of DNA replication. In some cases an abundant chaperone that functions in generic protein folding, interacting with many unfolded protein substrates, will also interact with one or a few specific folded proteins. The interaction of DnaK with the heat shock transcription factor σ^{32} and the function of mammalian Hsc70 in clathrin uncoating are such examples. In other cases, such as the biogenesis of Fe/S centers, Hsp70 and J-proteins have evolved to work in a single, specific function.

14.4.1 Assembly of Fe/S Centers

Although in vitro Fe/S clusters can be formed in proteins directly from elemental iron and sulfur, in vivo this task falls to highly conserved and complex assembly systems. Central to this system is a scaffold protein onto which a cluster is assembled and then transferred to a recipient protein [70, 71]. The scaffold, called Isu in eukaryotes and IscU in prokaryotes, is a substrate protein for a specialized Hsp70:J-protein pair (Ssq1:Jac1 in eukaryotes and Hsc66:Hsc20 in prokaryotes) [72, 73]. IscU/Isu proteins are small, highly conserved proteins, with the mammalian and bacterial proteins having about 70% amino acid identity. One of the stretches (LPPVK) of complete identity among the proteins from diverse organisms contains the binding site for the specialized Hsp70. The binding site for the J-protein has been elusive; the site may not be a continuous stretch of amino acids and thus is technically more difficult to define. In both eukaryotes and prokaryotes, the J-protein and the Hsp70 of this pair are very specialized. In fact, there is no evidence of the existence of substrate proteins in addition to the Fe/S cluster scaffold protein.

This J-protein:Hsp70 pair plays a very important role in assembly of Fe/S clusters, an essential cellular process [71, 74]. Bacteria have redundant systems for Fe/S cluster assembly, so the lack of Hsc66:Hsc20 has a fairly modest effect on the activity of Fe/S-containing enzymes in the cell [75]. However, *JAC1* is essential in most yeast strain backgrounds, and cells lacking Ssq1 have greatly reduced Fe/S-containing enzyme activity. Based on the extensive data available concerning Hsp70:J-protein function, it would be logical to propose that the chaperones alter the conformation of the scaffold proteins to facilitate either the assembly of the cluster or its transfer to a recipient protein. Recent results from the Lill laboratory [76], utilizing the labeling of Isu with radioactive iron, suggest that the chaperones may function in aiding cluster transfer. Strains depleted of Ssq1 or Jac1 accumulated higher than normal amounts of iron associated with Isu, while those having a decreased amount of proteins known to be needed for assembly of the cluster, such as the sulfur-transfer protein Nfs1, had lower levels. But, the mechanistic role of chaperones in the process remains to be clarified.

While it is clear that these chaperones involved in Fe/S cluster assembly are quite specialized, it is not so clear whether the abundant "generic" chaperones of the same cellular compartment can carry out this role as well. The data from bacteria suggest that the generic DnaK/DnaJ/GrpE system functions completely separately from the Hsc66/Hsc20 system [77]. But in yeast it appears that there is at least some functional overlap between the systems. Overexpression of the abundant Hsp70 of the mitochondrial matrix Ssc1, which also functions in protein translocation and folding of many matrix proteins, can substantially suppress the effects of the absence of Ssq1 [78]. Also, Ssq1 and Ssc1 appear to share the nucleotide release factor Mge1 [73]. Interestingly, it is not clear whether mammalian mitochondria have a specialized Hsp70 for Fe/S assembly, as no obvious homologue of Ssq1 has been found in these genomes. However, a homologue of Jac1 does exist. Perhaps in mammalian cells the major Hsp70, which is involved in protein translocation and general protein folding, also fulfils the role of facilitating Fe/S cluster assembly, using the Jac1 homologue as its alternative J-protein partner.

14.4.2
Uncoating of Clathrin-coated Vesicles

Regulated dynamic interaction among plasma membranes and cellular membrane compartments is accomplished by transport vesicles. Budding and fusion of vesicles require a series of assembly and disassembly cycles of the coat protein complex that provide the mechanical support for membrane bending and cargo selection. Hsc70, the constitutive form of Hsp70, has been known to play important roles in uncoating of clathrin-coated vesicles (CCVs), the major class of transport vesicles involved in endocytosis and synaptic transmission (for review, see Ref. [79]).

Hsc70, the major cytosolic Hsp70 of mammalian cells, promotes the disassembly of clathrin coats by binding to clathrin in an ATP-dependent manner, thereby releasing monomeric clathrin triskelia and other coat proteins from CCVs [80–82]. Like other members of the Hsp70 family, Hsc70 works with a J-protein partner in this process. The specialized type of J-protein involved is called auxilin, a name that encompasses brain-specific auxilin 1 and the ubiquitously expressed auxilin 2 (also called G-cyclin-associated kinase, or GAK) [83–88]. It is believed that auxilin binds to clathrin first and then recruits Hsc70 to CCVs via its J-domain, which then stimulates ATP hydrolysis and clathrin binding to Hsc70, resulting in release of clathrin monomers from CCVs [89–91].

Auxilin homologues have been identified in several species, including yeast, *C. elegans*, and mammals; they all contain a clathrin-binding domain and a C-terminal J-domain [87, 89, 90, 92–94]. Recent studies indicate that, as expected, the J-domain is important for interaction with Hsc70 [95] but that it also has features different from previously studied J-domains [96]. The structure of the bovine auxilin J-domain has been solved and was found to contain an extra N-terminal helix and a long loop inserted between helices I and II in addition to conserved J-domain features. Surface plasmon resonance analysis of auxilin mutants reveals that several positively charged residues in the long loop are also important for Hsc70 binding.

Recently, the results of a flurry of in vivo studies, including RNA interference in nematode cells [90] and dominant-negative interference experiments in mammalian cells [97], suggest that Hsc70s and auxilin play a more general role in clathrin dynamics than just uncoating of CCVs. Auxilin is released from CCVs during uncoating, but Hsc70 remains associated with free clathrin, possibly performing additional functions such as preventing clathrin-spontaneous polymerization or priming it for membrane recruitment by adaptor complex [98]. In addition, inactivation of auxilin in yeast also results in a phenotype similar to cells lacking clathrin [93].

Besides their functions with clathrin, Hsc70 and auxilin have recently been found to have unanticipated roles in other stages of vesicle transport. For example, Newmyer et al. [99] found that Hsc70 and auxilin can interact with GTP-bound dynamin, a protein well-known for its importance in vesicle formation. Two domains in auxilin were identified as being involved in dynamin binding; overexpression of these domains in vivo inhibited endocytosis without changing the dynamics or distribution of clathrin. It is thus suggested that Hsc70 and auxilin also participate in the early step of CCV formation [99]. In addition, Hsc70 has been suggested to play a role in neurotransmitter exocytosis. Mutations in Hsc70-4, a major cytosolic Hsp70 of *Drosophila*, were found to impair nerve-evoked neurotransmitter release in vivo. Genetic analysis suggested that Hsc70-4 might cooperate with cysteine string protein, a specialized J-protein on the membrane of secretary vesicles, to increase the Ca^{2+} sensitivity of neurotransmitter release upon vesicle fusion [100]. Additionally, a block in exocytosis has also been observed in squid synapses expressing dominant-interfering Hsc70 mutant proteins [95]. However, the critical protein substrates of Hsc70 during exocytosis are not yet known.

14.4.3
Regulation of the Heat Shock Response

While there are indications that Hsp70s play a role in regulation of the heat shock response in a variety of organisms, this role is by far best understood in bacteria. The heat shock response in *E. coli* is transcriptionally controlled by the heat shock promoter-specific transcription factor σ^{32}, a subunit of RNA polymerase (reviewed in Refs. [101, 102]). Upon heat shock, the level and activity of σ^{32} are rapidly increased due to the altered translatability of the σ^{32} gene (*rpoH*) and elevated stability of σ^{32} protein. The shutoff of the response results from the reversal of these effects, a decrease in the level and activity of σ^{32}. The *E. coli* Hsp70 system, DnaK/DnaJ/GrpE, has been found to regulate the heat shock response by directly associating with σ^{32} and modulating its degradation by proteases.

In non-heat-shocked cells, σ^{32} has an extremely short half-life of less than 1 minute. After heat shock, the half-life rapidly increases to, and remains at, eightfold, until the shutoff phase of the heat shock response begins [103, 104]. Several proteases are involved in σ^{32} degradation, including Lon (la), ClpAP and HslVU, and, most significantly, the ATP-dependent metalloprotease FtsH (previously known as HflB) [105–108]. The DnaK chaperone system was first implicated in the regulation of the heat shock response because strains carrying mutations in any one of the genes encoding components of the DnaK/DnaJ/GrpE system showed elevated expression of heat shock-inducible genes [109]. Later it was proposed that the DnaK and co-chaperones present σ^{32} to the proteases, thus negatively regulating the stability of σ^{32}. This model provides an attractive mechanism that links the level of σ^{32} to the level of unfolded proteins, the stress sensed by the DnaK chaperone system. Upon heat shock, less DnaK and DnaJ would be available for σ^{32} association and degradation, since more substrates (e.g., unfolded proteins) are competing for binding to the chaperones, resulting in the increased stability of σ^{32} [110]. This model predicts that the actual levels of the DnaK system proteins directly affect the heat shock response, which has been proven to be the case [111].

A significant amount of work has been done to understand the interaction between the DnaK system and σ^{32}, but exactly how the DnaK system modulates the degradation of σ^{32} by FtsH remains unclear. It has been reported that when binding to RNA polymerase, σ^{32} is resistant to FtsH-mediated degradation [111]. It is thus possible that the DnaK system affects the stability of σ^{32} by preventing its binding to the RNA polymerase, hence indirectly increasing the chance of σ^{32} degradation by FtsH. However, it has been shown that a mutant version of σ^{32} that cannot bind to RNA polymerase, but interacts normally with DnaK/DnaJ/GrpE, has a normal stability in the cells lacking DnaK or cells lacking both DnaJ and the DnaJ homologue ClbA [112], suggesting a more active role of the DnaK system in promoting the degradation of σ^{32}. It was recently reported that the in vitro interaction between DnaK and σ^{32} is highly temperature-sensitive, possibly due to a destabilized structural element in σ^{32} at elevated temperature [113], thus providing another way of regulating the stability of σ^{32} by the DnaK system.

The actual regions of σ^{32} that interact with DnaK and DnaJ, and thus are involved in regulation, remain elusive. In vitro, DnaK and the co-chaperone DnaJ bind independently to free σ^{32}, and it has been proposed that σ^{32} interacts similarly to any other substrate, with σ^{32} binding/release controlled by the ATP hydrolysis cycle [114]. Screening of a σ^{32}-derived peptide library for chaperone binding sites revealed two binding sites in the highly conserved RpoH box in σ^{32}, between residues 133 and 140 [115, 116]. Peptides corresponding to this region bind to DnaK and can be degraded by FtsH in vitro. However, introduction of alterations within these regions in the full-length σ^{32} protein did not affect DnaK binding in vitro or its degradation in vivo and in vitro [117]. Instead, these mutants of σ^{32} show decreased affinity for RNA polymerase and reduced transcriptional activity. It is thus still unclear whether the RpoH box in σ^{32} is directly involved in DnaK binding in vivo.

In addition to the turnover of σ^{32}, the DnaK system has also been suggested to modulate the activity of σ^{32} directly, since overexpression of DnaK and DnaJ in a Δ*ftsH* strain leads to reduced activity of σ^{32}, while the level of σ^{32} remains unchanged [108]. This type of modulation has also been reported in *Agrobacterium tumefaciens*, in which RpoH, the σ^{32} homologue, is normally quite stable and a change of the level of DnaK and DnaJ specifically changes the activity of RpoH [118]. However, little is known about how the activity of σ^{32} or RpoH is altered by the DnaK system.

14.4.4
Regulation of Activity of DNA Replication-initiator Proteins

Many plasmid-encoded DNA initiator proteins exist as monomers and dimers in equilibrium. Monomers activate initiation of DNA replication by binding to specific DNA-binding sites around the origin of replication, whereas dimers are inactive due to altered DNA-binding specificities. The role of the Hsp70 chaperone machinery in regulating the activity of DNA replication initiators was first demonstrated by the work of Wickner et al. [119–122] in the *E. coli* system of RepA, the replication-initiator protein of plasmid P1, motivated by earlier in vivo observations that mutations in *dnaK*, *dnaJ*, or *grpE* resulted in plasmid P1 instability [123]. The DnaK/DnaJ/GrpE chaperone machinery altered the equilibrium between dimer and monomer, destabilizing dimeric RepA in vitro and rendering it more active in origin binding [119, 120–122]. How DnaK/DnaJ/GrpE interacts with the folded RepA dimer and promotes its monomerization is still not fully understood. Both DnaJ and DnaK bind to a RepA dimer, but at different sites [119, 124]. It was proposed that DnaJ tags the RepA dimer for recognition by DnaK, which then stimulates the monomerization of RepA in an ATP-dependent reaction [121].

The DnaK chaperone system has also been shown to assist the monomerization of other replication initiation proteins, including RepA of plasmid P7 [122] and TrfA of RK2 plasmid [125]. Konieczny et al. [125] reported that ClpB, the Clp/Hsp100 family, cooperates with the DnaK chaperone system to convert TrfA dimers into active monomers. The activation of TrfA seems to be specific for bacterial chaperones (ClpB, DnaK, DnaJ, and GrpE) since the corresponding yeast mitochondrial homologues (Hsp78, Ssc1, Mdj1, and Mge1) did not activate TrfA in an in vitro system reconstituted with purified components.

Recent identification of similarities between RepA and the origin-recognition complex (ORC) subunit from eukaryotes and archaea raises the question as to whether Hsp70s play a more general role in DNA replication. Giraldo et al. [126] found that the C-terminal domain of ScOrc4p, a subunit of *S. cerevisiae* ORC, shares sequence and potential structural similarity with the N-terminal domain of RepA, including a common winged-helix domain and a leucine-zipper dimerization motif. Furthermore, they found that ScOrc4p can interact with yeast Hsp70 both in vivo and in cell lysates, although it is not clear with which Hsp70 it interacts [126]. Collectively, these results suggest Hsp70's role in regulating the assembly of active ORC in eukaryotes, possibly through a mechanism involving disassembly of ORC subunits.

In addition to monomerization of initiators for DNA replication, Hsp70/J-protein systems have long been known to play important roles in the assembly and activation of viral DNA replicative machinery in both prokaryotes and eukaryotes (see review in Ref. [127]). In the case of bacteriophage λ replication in *E. coli*, it was proposed that DnaJ interacts with DnaB-λP-λO-DNA complex and facilitates recognition by DnaK, which then stimulates DnaB helicase activity by catalyzing the release of λP from the initiation complex [120, 128–130]. In mammalian cells, Hsp70 and J-protein have also been found to enhance the origin binding of UL9, the origin-binding protein of herpes simplex virus type 1 [131]. In addition, hTid-1, a human homologue of DnaJ, was found to associate with UL9 and promote the multimer formation from dimeric UL9 [132]. The DNA binding and activity of helicase E1 from human papilloma virus is also promoted by human Hsp70 and J-protein [133, 134]. In summary, mounting evidence suggests that

Hsp70/J-protein systems interact with folded initiator proteins and play regulatory roles in DNA replication.

14.5
Cases of a Single Hsp70 Functioning With Multiple J-Proteins

The previous sections discuss many physiological roles of Hsp70s and their partner J-proteins (Figure 14.1). In some of the cases discussed, specific functions depend upon a functional specificity inherent in the Hsp70 itself. The Hsp70 evolved in Fe/S cluster biogenesis and the one evolved to bind to ribosomes, and to be regulated like ribosomal proteins, are examples of such specialized Hsp70s. Although dispersed in different sections of this chapter, there are also multiple examples of a single Hsp70 functioning with different J-proteins in very different physiological roles. Examples exist in mitochondria, the ER, and the cytosol of eukaryotes.

In mitochondria, the abundant Hsp70 Ssc1 is one such example. A portion of Ssc1 is tethered to the membrane and functions in translocation of proteins into the matrix with the membrane J-protein Pam18 [48–50]. The remainder of Ssc1 is soluble and functions with the soluble J-protein Mdj1 in the folding of imported and mitochondrially encoded proteins [21, 24]. In the ER, the predominant Hsp70, Kar2/BiP, functions with several J-proteins. Analogous to the case in the mitochondria, a portion of Kar2 is tethered to the translocon and acts with the integral membrane J-protein Sec63 in protein translocation. As a soluble protein in the lumen, it has the J-protein partners Scj1 and Jem1.

Multiple examples exist in the cytosol as well. The best-studied example in mammalian systems is the involvement of Hsc70 not only in protein folding and translocation into mitochondria but also in vesicle transport. In the former process, Hsc70 works with J-proteins Hdj1 and Hdj2, among others; in vesicle transport it works with auxilin and cysteine-string proteins (described in Section 14.4.2).

Similar examples exist in the yeast cytosol, where the Ssa class of Hsp70s has multiple J-protein partners, including Ydj1, Sis1, and Swa2 (yeast auxilin). Ydj1 and Sis1 are particularly interesting examples, in that the J-proteins show extensive similarity beyond their J-domains but are functionally distinct, in the sense that the functions of the essential Sis1 cannot be carried out by Ydj1. Interestingly, this functional specificity has been conserved, as the mammalian homologue of Sis1, Hdj1, can rescue a Δ*sis1* strain, but the homologue of Ydj1, Hdj2, cannot [135, 136]. The biochemical basis of this functional difference has yet to be defined.

14.6
Hsp70s/J-proteins – When an Hsp70 Maybe Isn't Really a Chaperone

As we learn more about the complex workings of cells, the idea has emerged that genes have been duplicated and evolved to serve additional functions. Such evolution includes the examples described above, as Hsp70 and J-proteins have evolved to serve specialized functions utilizing their inherent ability to interact with hydrophobic sequences in proteins in their peptide-binding clefts. However, examples have also been found in which Hsp70s have evolved to carry out functions that are no longer dependent on their ability to interact with substrate proteins. Below we discuss two examples in which an Hsp70 forms a heterodimer with another protein, perhaps serving a regulatory role.

14.6.1 The Ribosome-associated "Hsp70" Ssz1

Ssz1 (previously known as Pdr13) from *S. cerevisiae* is an example of a protein with sequence similarity to Hsp70s that clearly functions in ways not expected of an Hsp70 chaperone. Ssz1 is typically classified as an Hsp70 because it contains a conserved N-terminal ATPase domain that shares around 35% sequence identity with ATPase domains of Hsp70s across species and because its size of 60 kDa is close to that expected of an Hsp70. However, Ssz1's C-terminal region has little sequence similarity with known Hsp70s, rendering its overall sequence similarity to Hsp70s only 20–25%.

Ssz1's cellular properties also differ significantly from those of well-studied Hsp70s. First, although Ssz1 interacts with a J-protein, the ribosome-associated Zuo1, the complex, called the ribosome-associated complex (RAC), is surprisingly stable [59]. Typically, Hsp70–J-protein interactions are quite transient, while the RAC subunits cannot be separated without denaturation. But most compelling is the fact that Ssz1's C-terminal putative peptide-binding domain is not required for its in vivo function. The C-terminal truncated Ssz1, having only the ATPase domain, fully rescues the phenotypes of a Δ*ssz1* strain [60]. Up to now, there is no evidence to indicate that Ssz1 interacts with nascent chains or has any chaperone activity.

However, there is every indication that Ssz1 has an important in vivo role working with Zuo1, but as a component required for Zuo1's ability to act as the J-protein partner of the ribosome-associated Hsp70 Ssb. Both Ssz1 and Zuo1 are required for Ssb cross-link to nascent chains on the ribosomes [61]. Genetic analysis also suggests that Ssz1, Zuo1, and Ssb work in the same biological pathway, as the phenotypes of any single mutant are the same as cells lacking all three [60, 61]. Therefore, it has been proposed that Ssz1 functions as a modulator of Zuo1, priming it to interact with Ssb in chaperoning newly synthesized polypeptides on the ribosomes [60, 61] (see Section 14.3.2 on ribosome-associated Hsp70s). Consistent with this idea, Zuo1 stimulates Ssb's ATPase activity only when it is in complex with Ssz1, supporting the idea that Ssz1 functions as a modulator of a J-protein (P. Huang, personal communication).

The only other known in vivo activity of Ssz1 is its role in pleiotropic drug resistance (PDR), the resistance to multiple drugs with unrelated structures or modes of action [137]. PDR occurs because cells increase the efflux of drugs, primarily because of the increase in activity of the transcription factor Pdr1, which acts on a number of genes, including those encoding ABC transporters, which are responsible for drug efflux. Actually, Ssz1 (previously known as Pdr13) was first isolated from a genetic screen for genes that, when overexpressed from a multicopy plasmid, increased PDR in a Pdr1-dependent fashion [137]. Ssz1's activation of PDR is independent of Zuo1 or Ssb and does not require its C-terminal putative peptide-binding domain (H. Eisenman, personal communication). Although it remains an enigma as to how Ssz1 mediates Pdr1-dependent transcription, its ability to do so without its putative peptide-binding domain indicates that it does not act as a chaperone in this role.

14.6.2
Mitochondrial Hsp70 as the Regulatory Subunit of an Endonuclease

The major mitochondrial Hsp70, Ssc1, from *S. cerevisiae* serves as another example of an Hsp70 that carries out a function other than that of a conventional chaperone. Ssc1 forms a stable heterodimer with the endonuclease *Sce*I [138]. *Sce*I functions as a multisite-specific endonuclease in yeast mitochondria, where it cleaves more than 30 sites on mitochondrial DNA to induce homologous recombination upon zygote formation [139]. Ssc1 not only increased the thermostability and endonuclease activity of *Sce*I but also broadened its sequence specificity, as in the absence of Ssc1, *Sce*I is a unisite-specific endonuclease [140].

Although the mechanism by which Ssc1 alters the substrate specificity of *Sce*I is not thoroughly understood, progress has been made in understanding the interaction between Ssc1 and *Sce*I. The interaction between Ssc1 and *Sce*I is more stable in the presence of ADP than in the presence of ATP or in the absence of nucleotides [140]. *Sce*I was found to interact with Ssc1's ATPase domain; in addition, interaction of full-length Ssc1 could not be competed away by an excess of an unfolded protein substrate. Thus, *Sce*I is not a substrate of Ssc1 but rather a protein partner. However, only when full-length Ssc1 is complexed with *Sce*I does it confer multisite-specific endonuclease activity. It remains unclear whether Ssc1's ATPase activity plays any role in regulating *Sce*I's activity and specificity. However, no J-proteins seem to be required for Ssc1's effect on endonuclease activity, as in vitro reconstituted *Sce*I/Ssc1 heterodimers can recapitulate the multisite-specific endonuclease activity observed in mitochondrial extracts.

A similar function of mitochondrial Hsp70 in regulating endonuclease specificity has been found in another yeast strain, *Saccharomyces uvarum*. *Suv*I, the ortholog of the *Sce*I in *S. uvarum*, is a mitochondrial endonuclease that forms a stable heterodimer with the homologue of Ssc1, called mtHsp70, resulting in a change from a unisite-specific to a multisite-specific endonuclease [141]. In the absence of mtHsp70, *Suv*I and *Sce*I have the same unisite sequence specificity but show different multisite specificities upon binding to mtHsp70. These findings suggest that mtHsp70 can regulate the variety of sites cleaved by regulating the sequence specificity of endonucleases.

14.7 Emerging Concepts and Unanswered Questions

This chapter attempts to summarize the explosion of information and to outline emerging themes regarding the variety of functions of Hsp70s and J-proteins in the cell. Elegant biochemical experiments carried out over the past 15 years have established the basic biochemical parameters of the Hsp70:J-protein partnership and serve as the foundation for understanding the in vivo function of the myriad of individual J-protein:Hsp70 pairs that function in the cell. Clearly, much work remains, and certainly many surprises await us.

References

1 Ohtsuka, K. & Hata, M. (2000). Mammalian HSP40/DNAJ homologues: cloning of novel cDNAs and a proposal for their classification and nomenclature. *Cell Stress Chaperones* 5, 98–112.

2 Bukau, B. & Horwich, A. L. (1998). The Hsp70 and Hsp60 chaperone machines. *Cell* 92, 351–366.

3 Hartl, F. & Hayer-Hartl, M. (2002). Molecular chaperones in the cytosol: from nascent chain to folded protein. *Science* 295, 1852–8.

4 Hesterkamp, T. & Bukau, B. (1998). Role of the DnaK and HscA homologues of Hsp70 chaperones in protein folding in *E. coli*. *EMBO J* 17, 4818–4828.

5 Gaitanaris, G. A., Papavassiliou, A. G., Rubock, P., Silverstein, S. J. & Gottesman, M. E. (1990). Renaturation of denatured lambda repressor requires heat shock proteins. *Cell* 61, 1013–20.

6 Skowyra, D., Georgopoulos, C. & Zylicz, M. (1990). The *E. coli dnak* gene product, the hsp70 homologue, can reactivate heat-inactivated RNA polymerase in an ATP hydrolysis-dependent manner. *Cell* 62, 939–944.

7 Langer, T., Lu, C., Echols, H., Flanagan, J., Hayer, M. K. & Hartl, F. U. (1992). Successive action of DnaK, DnaJ, and GroEL along the pathway of chaperone-mediated protein folding. *Nature* 356, 683–689.

8 Schroder, H., Langer, T., Hartl, F.-U. & Bukau, B. (1993). DnaK, DnaJ and GrpE form a cellular chaperone machinery capable of repairing heat-induced protein damage. *EMBO Journal* 12, 4137–4144.

9 Ziemienowicz, A., Skowyra, D., Zeilstra-Ryalls, J., Fayet, O., Georgopoulos, C. & Zylicz, M. (1993). Both the Escherichia coli chaperone systems, GroEL/GroES and DnaK/DnaJ/GrpE, can reactivate heat-treated RNA polymerase. Different mechanisms for the same activity. *Journal of Biological Chemistry* 268, 25425–25431.

10 Szabo, A., Langer, T., Schroder, H., Flanagan, J., Bukau, B. & Hartl, F. U. (1994). The ATP hydrolysis-dependent reaction cycle of the Escherichia coli Hsp70 system DnaK, DnaJ, and GrpE. *Proc Natl Acad Sci USA* 91, 10345–9.

11 Mogk, A., Tomoyasu, T., Goloubinoff, P., Fudiger, S., Roder, D., Langen, H. & Bukau, B. (1999). Identification of thermolabile *Escherichia coli* proteins: prevention and reversion of aggregation by DnaK and ClpB. *EMBO J* 18, 6934–6949.

12 Teter, S. A., Houry, W. A., Ang, D., Tradler, T., Rockabrand, D., Fischer, G., Blum, P., Georgopoulus, C. & Hartl, F. U. (1999). Polypeptide Flux through Bacterial Hsp70: DnaK Cooperates with Trigger

Factor in Chaperoning Nascent Chains. *Cell* 97, 755–765.
13 Deuerling, E., Schulze-Specking, A., Tomoyasu, T., Mogk, A. & Bukau, B. (1999). Trigger factor and DnaK cooperate in folding of newly synthesized proteins. *Nature* 400, 693–696.
14 Kim, S., Schilke, B., Craig, E. & Horwich, A. (1998). Folding in vivo of a newly translated yeast cytosolic enzyme is mediated by the SSA class of cytosolic yeast Hsp70 proteins. *Proc Natl Acad Sci USA* 95, 12860–12865.
15 Thulasiraman, V., Yang, C.-F. & Frydman, J. (1999). In vivo newly translated polypeptides are sequestered in a protected folding environment. *EMBO J* 18, 85–95.
16 Eggers, D. K., Welch, W. J. & Hansen, W. J. (1997). Complexes between Nascent Polypeptides and Their Molecular Chaperones in the Cytosol of Mammalian Cells. *Molecular Biology of the Cell* 8, 1559–1573.
17 Michels, A. A., Kanon, B., Konings, A. W., Ohtsuka, K., Bensaude, O. & Kampinga, H. H. (1997). Hsp70 and Hsp40 chaperone activities in the cytoplasm and the nucleus of mammalian cells. *J Biol Chem* 272, 33283–9.
18 Pratt, W. B. & Toft, D. O. (2003). Regulation of signaling protein function and trafficking by the hsp90/hsp70-based chaperone machinery. *Exp Biol Med* 228, 111–133.
19 Chang, H. C. & Lindquist, S. (1994). Conservation of Hsp90 macromolecular complexes in Saccharomyces cerevisiae. *J Biol Chem* 269, 24983–8.
20 Liu, X. D., Morano, K. A. & Thiele, D. J. (1999). The yeast Hsp110 family member, Sse1, is an Hsp90 cochaperone. *J Biol Chem* 274, 26654–60.
21 Kang, P. J., Ostermann, J., Shilling, J., Neupert, W., Craig, E. A. & Pfanner, N. (1990). Requirement for hsp70 in the mitochondrial matrix for translocation and folding of precursor proteins. *Nature* 348, 137–143.
22 Westermann, B., Prip-Buus, C., Neupert, W. & Schwarz, E. (1995). The role of the GrpE homologue, Mge1p, in mediating protein import and protein folding in mitochondria. *EMBO J* 13, 1998–2006.
23 Westermann, B., Gaume, B., Herrmann, J. M., Neupert, W. & Schwarz, E. (1996). Role of the mitochondrial DnaJ homologue Mdj1p as a chaperone for mitochondrially synthesized and imported proteins. *Mol Cell Biol* 16, 7063–7071.
24 Liu, Q., Krzewska, J., Liberek, K. & Craig, E. A. (2001). Mitochondrial Hsp70 Ssc1: role in protein folding. *J. Biol. Chem.* 276, 6112–8.
25 Hermann, J., Stuart, R., Craig, E. & Neupert, W. (1994). Mitochondrial heat shock protein 70, a molcular chaperone for proteins encoded by mitochondrial DNA. *J. Cell Biol.* 127, 893–902.
26 Haas, I. G. & Wabl, M. (1983). Immunoglobulin heavy chain binding protein. *Nature* 306, 387–9.
27 Simons, J. F., Ferro-Novick, S., Rose, M. D. & Helenius, A. (1995). BiP/Kar2p serves as a molecular chaperone during carboxypeptidase Y folding in yeast. *J Cell Biol* 130, 41–49.
28 Hendershot, L., Wei, J., Gaut, J., Melnick, J., Aviel, S. & Argon, Y. (1996). Inhibition of immunoglobulin folding and secretion by dominant negative BiP ATPase mutants. *Proc Natl Acad Sci USA* 93, 5269–74.
29 Ellgaard, L. & Helenius, A. (2001). ER quality control: towards an understanding at the molecular level. *Curr Opin Cell Biol* 13, 431–7.
30 Brodsky, J. L., Werner, E. D., Dubas, M. E., Goeckeler, J. L., Kruse, K. B. & McCracken, A. A. (1999). The requirement for molecular chaperones during endoplasmic reticulum-associated protein degradation demonstrates that protein export and import are mechanistically distinct. *J Biol Chem* 274, 3453–60.
31 Plemper, R. K., Bohmler, S., Bordallo, J., Sommer, T. & Wolf, D. H. (1997). Mutant analysis links

the translocon and BiP to retrograde protein transport for ER degradation. *Nature* 388, 891–5.

32 Bertolotti, A., Zhang, Y., Hendershot, L. M., Harding, H. P. & Ron, D. (2000). Dynamic interaction of BiP and ER stress transducers in the unfolded-protein response. *Nat Cell Biol* 2, 326–32.

33 Baxter, B. K., James, P., Evans, T. & Craig, E. A. (1996). *SSI1* encodes a novel Hsp70 of the *Saccharomyces cerevisiae* endoplasmic reticulum. *Mol. Cell Biol.* 16, 6444–6456.

34 Craven, R. A., Egerton, M. & Stirling, C. J. (1996). A novel Hsp70 of the yeast ER lumen is required for the efficient translocation of a number of protein precursors. *EMBO J.* 15, 2640–2650.

35 Saris, N., Holkeri, H., Craven, R. A., Stirling, C. J. & Makarow, M. (1997). The Hsp70 homologue Lhs1p is involved in a novel function of the yeast endoplasmic reticulum, refolding and stabilization of heat-denatured protein aggregates. *J Cell Biol* 137, 813–824.

36 Tyson, J. R. & Stirling, C. J. (2000). LHS1 and SIL1 provide a lumenal function that is essential for protein translocation into the endoplasmic reticulum. *EMBO J* 19, 6440–52.

37 Hamilton, T. G. & Flynn, G. C. (1996). Cer1p, a novel Hsp70-related protein required for posttranslational endoplasmic reticulum translocation in yeast. *J Biol Chem* 271, 30610–3.

38 Steel, G. J., Fullerton, D. M., Tyson, J. R. & Stirling, C. J. (2004). Coordinated Activation of Hsp70 Chaperones. *Science* 303, 98–101.

39 Lin, H. Y., Masso-Welch, P., Di, Y. P., Cai, J. W., Shen, J. W. & Subjeck, J. R. (1993). The 170-kDa glucose-regulated stress protein is an endoplasmic reticulum protein that binds immunoglobulin. *Mol Biol Cell* 4, 1109–19.

40 Meunier, L., Usherwood, Y. K., Chung, K. T. & Hendershot, L. M. (2002). A subset of chaperones and folding enzymes form multiprotein complexes in endoplasmic reticulum to bind nascent proteins. *Mol Biol Cell* 13, 4456–69.

41 Panzner, S., Dreier, L., Hartmann, E., Kostka, S. & Rapoport, T. A. (1995). Posttranslational protein transport in yeast reconstituted with a purified complex of Sec proteins and Kar2p. *Cell* 81, 561–570.

42 Young, B. P., Craven, R. A., Reid, P. J., Willer, M. & Stirling, C. J. (2001). Sec63p and Kar2p are required for the translocation of SRP-dependent precursors into the yeast endoplasmic reticulum in vivo. *EMBO J* 20, 262–271.

43 Voisine, C., Craig, E. A., Zufall, N., von Ahsen, O., Pfanner, N. & Voos, W. (1999). The protein import motor of mitochondria: unfolding and trapping of preproteins are distinct and separable functions of matrix Hsp70. *Cell* 97, 565–574.

44 Corsi, A. K. & Schekman, R. (1997). The lumenal domain of Sec63p stimulates the ATPase activity of BiP and mediates BiP recruitment to the translocon in *Saccharomyces cerevisiae. J Cell Biol* 137, 1483–1493.

45 Schneider, H.-C., Berthold, J., Bauer, M. F., Dietmeier, K., Guiard, B., Brunner, M. & Neupert, W. (1994). Mitochondrial Hsp70/MIM44 complex facilitates protein import. *Nature* 371, 768–774.

46 Rassow, J., Maarse, A., Krainer, E., Kubrich, M., Muller, H., Meijer, M., Craig, E. & Pfanner, N. (1994). Mitochondrial protein import: biochemical and genetic evidence for interaction of matrix Hsp70 and the inner membrane protein Mim44. *J. Cell Biol.* 127, 1547–1556.

47 Liu, Q., D'Silva, P., Walter, W., Marszalek, J. & Craig, E. A. (2003). Regulated cycling of mitochondrial Hsp70 at the protein import channel. *Science* 300, 139–41.

48 D'Silva, P. D., Schilke, B., Walter, W., Andrew, A. & Craig, E. A. (2003). J protein cochaperone of the mitochondrial inner membrane required for protein import into the mitochondrial matrix. *Proc Natl Acad Sci USA* 100, 13839–44.

49 Mokranjac, D., Sichting, M., Neupert, W. & Hell, K. (2003). Tim14, a novel key component of the import motor of the Tim23 protein translocase of mitochondria. *EMBO J* 22, 4945–4956.

50 Truscott, K. N., Voos, W., Frazier, A. E., Lind, M., Li, Y., Geissler, A., Dudek, J., Muller, H., Sickmann, A., Meyer, H. E., Meisinger, C., Guiard, B., Rehling, P. & Pfanner, N. (2003). A J-protein is an essential subunit of the presequence translocase-associated protein import motor of mitochondria. *J Cell Biol* 163, 707–13.

51 Fewell, S. W., Travers, K. J., Weissman, J. S. & Brodsky, J. L. (2001). The action of molecular chaperones in the early secretory pathway. *Ann. Rev. Gen.* 35, 149–191.

52 Neupert, W. & Brunner, M. (2002). The protein import motor of mitochondria. *Nat Rev Mol Cell Biol* 8, 555–565.

53 Matouschek, A., Pfanner, N. & Voos, W. (2000). Protein unfolding by mitochondria. The Hsp70 import motor. *EMBO Rep* 5, 404–410.

54 Hamman, B. D., Henderschot, L. M. & Johnson, A. E. (1998). BiP maintains the permeability barrier of the ER membrane by sealing the lumenal end of the translocon pore before and early in translocation. *Cell* 92, 747–758.

55 Caplan, A. J., Tsai, J., Casey, P. J. & Douglas, M. G. (1992). Farnesylation of YDJ1p is required for function at elevated growth temperatures in *S. cerevisiae*. *J Biol Chem* 267, 18890–18895.

56 Young, J. C., Hoogenraad, N. J. & Hartl, F.-U. (2003). Molecular Chaperones Hsp90 and Hsp70 Deliver Proteins to the Mitochondrial Import Receptor Tom70. *Cell* 112, 41–50.

57 Nelson, R. J., Ziegelhoffer, T., Nicolet, C., Werner-Washburne, M. & Craig, E. A. (1992). The translation machinery and seventy kilodalton heat shock protein cooperate in protein synthesis. *Cell* 71, 97–105.

58 Yan, W., Schilke, B., Pfund, C., Walter, W., Kim, S. & Craig, E. A. (1998). Zuotin, a ribosome-associated DnaJ molecular chaperone. *EMBO J* 17, 4809–4817.

59 Gautschi, M., Lilie, H., Funfschilling, U., Mun, A., Ross, S., Lithgow, T., Rucknagel, P. & Rospert, S. (2001). RAC, a stable ribosome-associated complex in yeast formed by the DnaK-DnaJ homologues Ssz1p and zuotin. *Proc Natl Acad Sci USA* 98, 3762–7.

60 Hundley, H., Eisenman, H., Walter, W., Evans, T., Hotokezaka, Y., Wiedmann, M. & Craig, E. (2002). The in vivo function of the ribosome-associated Hsp70, Ssz1, does not require its putative peptide-binding domain. *Proc Natl Acad Sci USA* 99, 4203–8.

61 Gautschi, M., Mun, A., Ross, S. & Rospert, S. (2002). A functional chaperone triad on the yeast ribosome. *Proc Natl Acad Sci USA* 99, 4209–14.

62 Pfund, C., Lopez-Hoyo, N., Ziegelhoffer, T., Schilke, B. A., Lopez-Buesa, P., Walter, W. A., Wiedmann, M. & Craig, E. A. (1998). The Molecular Chaperone *SSB* from *S. cerevisiae* is a Component of the Ribosome-Nascent Chain Complex. *EMBO Journal* 17, 3981–3989.

63 Pfund, C., Huang, P., Lopez-Hoyo, N. & Craig, E. (2001). Divergent functional properties of the ribosome-associated molecular chaperone Ssb compared to other Hsp70s. *Mol. Biol. Cell* 12, 3773–3782.

64 Siegers, K., Bölter, B., Schwarz, J. P., Böttcher, U., Guha, S. & Hartl, F.-U. (2003). TRiC/CCT cooperates with different upstream chaperones in the folding of distinct protein classes. *EMBO J* 22, 5230–5240.

65 Dudek, J., Volkmer, J., Bies, C., Guth, S., Muller, A., Lerner, M., Feick, P., Schafer, K. H., Morgenstern, E., Hennessy, F., Blatch, G. L., Janoscheck, K., Heim, N., Scholtes, P., Frien, M., Nastainczyk, W. & Zimmermann, R. (2002). A novel type of co-chaperone mediates transmembrane recruitment of DnaK-

like chaperones to ribosomes. *EMBO J* 21, 2958–67.
66 Horton, L. E., James, P., Craig, E. A. & Hensold, J. O. (2001). The yeast hsp70 homologue Ssa is required for translation and interacts with Sis1 and Pab1 on translating ribosomes. *J Biol Chem* 276, 14426–33.
67 Zhong, T. & Arndt, K. T. (1993). The yeast *SIS1* protein, a DnaJ homologue, is required for initiation of translation. *Cell* 73, 1175–1186.
68 Qian, X., Li, Z. & Sha, B. (2001). Cloning, expression, purification and preliminary X-ray crystallographic studies of yeast Hsp40 Sis1 complexed with Hsp70 Ssa1 C-terminal lid domain. *Acta Crystallogr D Biol Crystallogr* 57, 748–50.
69 Qian, X., Hou, W., Zhengang, L. & Sha, B. (2002). Direct interactions between molecular chaperones heat-shock protein (Hsp) 70 and Hsp40: yeast Hsp70 Ssa1 binds the extreme C-terminal region of yeast Hsp40 Sis1. *Biochem J* 361, 27–34.
70 Frazzon, J. & Dean, D. R. (2003). Formation of iron-sulfur clusters in bacteria: an emerging field in bioinorganic chemistry. *Current Opinion in Chemical Biology* 7, 166–173.
71 Lill, R. & Kispal, G. (2000). Maturation of cellular Fe–S proteins: an essential function of mitochondria. *Trends in Biochemical Sciences* 25, 352–356.
72 Hoff, K. G., Ta, D. T., L., T. T., Silberg, J. J. & Vickery, L. E. (2002). Hsc66 substrate specificity is directed toward a discrete region of the iron-sulfur cluster template protein IscU. *J Biol Chem* 277, 27353–27359.
73 Dutkiewicz, R., Schilke, B., Knieszner, H., Walter, W., Craig, E. A. & Marszalek, J. (2003). Ssq1, a mitochondrial Hsp70 involved in iron-sulfur (Fe/S) center biogenesis: Similarities to and differences from its bacterial counterparts. *J Biol Chem* 278, 29719–29727.
74 Craig, E. A. & Marszalek, J. (2002). A specialized mitochondrial molecular chaperone system: a role in formation of Fe/S centers. *Cell Mol Life Sci.* 59, 1658–65.
75 Schwartz, C. J., Djaman, O., Imlay, J. A. & Kiley, P. J. (2000). The cysteine desulferase, IscS, has a major role in in vivo Fe–S cluster formation in *Escherichia coli. Proc Natl Acad Sci USA* 97, 9009–9014.
76 Muhlenhoff, U., Gerber, J., Richhardt, N. & Lill, R. (2003). Components involved in assembly and dislocation of iron-sulfur clusters on the scaffold protein Isu1p. *EMBO J.* 22, 4815–25.
77 Silberg, J. J., Hoff, K. G. & Vickery, L. E. (1998). The Hsc66-Hsc20 chaperone system in Escherichia coli: chaperone activity and interactions with the DnaK-DnaJ-grpE system. *J Bacteriol* 180, 6617–24.
78 Voisine, C., Schilke, B., Ohlson, M., Beinert, H., Marszalek, J. & Craig, E. A. (2000). Role of the mitochondrial Hsp70s, Ssc1 and Ssq1, in the maturation of Yfh1. *Molecular and Cellular Biology* 10, 3677–3684.
79 Cremona, O. (2001). Live stripping of clathrin-coated vesicles. *Dev Cell* 1, 592–4.
80 Schlossman, D. M., Schmid, S. L., Braell, W. A. & Rothman, J. E. (1984). An enzyme that removes clathrin coats: purification of an uncoating ATPase. *J Cell Biol* 99, 723–33.
81 Barouch, W., Prasad, K., Greene, L. E. & Eisenberg, E. (1994). ATPase activity associated with the uncoating of clathrin baskets by Hsp70. *J Biol Chem* 269, 28563–8.
82 Hannan, L. A., Newmyer, S. L. & Schmid, S. L. (1998). ATP- and cytosol-dependent release of adaptor proteins from clathrin-coated vesicles: A dual role for Hsc70. *Mol Biol Cell* 9, 2217–29.
83 Ahle, S. & Ungewickell, E. (1990). Auxilin, a newly identified clathrin-associated protein in coated vesicles from bovine brain. *J Cell Biol* 111, 19–29.
84 Ungewickell, E., Ungewickell, H., Holstein, S. E., Lindner, R., Prasad, K., Barouch, W., Martin, B.,

GREENE, L. E. & EISENBERG, E. (1995). Role of auxilin in uncoating clathrin-coated vesicles. *Nature* 378, 632–5.

85 KANAOKA, Y., KIMURA, S. H., OKAZAKI, I., IKEDA, M. & NOJIMA, H. (1997). GAK: a cyclin G associated kinase contains a tensin/auxilin-like domain. *FEBS Lett* 402, 73–80.

86 KIMURA, S. H., TSURUGA, H., YABUTA, N., ENDO, Y. & NOJIMA, H. (1997). Structure, expression, and chromosomal localization of human GAK. *Genomics* 44, 179–87.

87 GREENER, T., ZHAO, X., NOJIMA, H., EISENBERG, E. & GREENE, L. E. (2000). Role of cyclin G-associated kinase in uncoating clathrin-coated vesicles from non-neuronal cells. *J Biol Chem* 275, 1365–70.

88 UMEDA, A., MEYERHOLZ, A. & UNGEWICKELL, E. (2000). Identification of the universal cofactor (auxilin 2) in clathrin coat dissociation. *Eur J Cell Biol* 79, 336–42.

89 PISHVAEE, B., COSTAGUTA, G., YEUNG, B. G., RYAZANTSEV, S., GREENER, T., GREENE, L. E., EISENBERG, E., MCCAFFERY, J. M. & PAYNE, G. S. (2000). A yeast DNA J protein required for uncoating of clathrin-coated vesicles in vivo. *Nat Cell Biol* 2, 958–63.

90 GREENER, T., GRANT, B., ZHANG, Y., WU, X., GREENE, L. E., HIRSH, D. & EISENBERG, E. (2001). Caenorhabditis elegans auxilin: a J-domain protein essential for clathrin-mediated endocytosis in vivo. *Nat Cell Biol* 3, 215–9.

91 ZHAO, X., GREENER, T., AL-HASANI, H., CUSHMAN, S. W., EISENBERG, E. & GREENE, L. E. (2001). Expression of auxilin or AP180 inhibits endocytosis by mislocalizing clathrin: evidence for formation of nascent pits containing AP1 or AP2 but not clathrin. *J Cell Sci* 114, 353–65.

92 HOLSTEIN, S. E., UNGEWICKELL, H. & UNGEWICKELL, E. (1996). Mechanism of clathrin basket dissociation: separate functions of protein domains of the DnaJ homologue auxilin. *J Cell Biol* 135, 925–37.

93 GALL, W. E., HIGGINBOTHAM, M. A., CHEN, C., INGRAM, M. F., CYR, D. M. & GRAHAM, T. R. (2000). The auxilin-like phosphoprotein Swa2p is required for clathrin function in yeast. *Curr Biol* 10, 1349–58.

94 LEMMON, S. K. (2001). Clathrin uncoating: Auxilin comes to life. *Curr Biol* 11, R49–52.

95 MORGAN, J. R., PRASAD, K., JIN, S., AUGUSTINE, G. J. & LAFER, E. M. (2001). Uncoating of clathrin-coated vesicles in presynaptic terminals: roles for Hsc70 and auxilin. *Neuron* 32, 289–300.

96 JIANG, J., TAYLOR, A. B., PRASAD, K., ISHIKAWA-BRUSH, Y., HART, P. J., LAFER, E. M. & SOUSA, R. (2003). Structure-function analysis of the auxilin J-domain reveals an extended Hsc70 interaction interface. *Biochemistry* 42, 5748–53.

97 NEWMYER, S. L. & SCHMID, S. L. (2001). Dominant-interfering Hsc70 mutants disrupt multiple stages of the clathrin-coated vesicle cycle in vivo. *J Cell Biol* 152, 607–20.

98 JIANG, R., GAO, B., PRASAD, K., GREENE, L. E. & EISENBERG, E. (2000). Hsc70 chaperones clathrin and primes it to interact with vesicle membranes. *J Biol Chem* 275, 8439–47.

99 NEWMYER, S. L., CHRISTENSEN, A. & SEVER, S. (2003). Auxilin-dynamin interactions link the uncoating ATPase chaperone machinery with vesicle formation. *Dev Cell* 4, 929–40.

100 BRONK, P., WENNIGER, J. J., DAWSON-SCULLY, K., GUO, X., HONG, S., ATWOOD, H. L. & ZINSMAIER, K. E. (2001). Drosophila Hsc70-4 is critical for neurotransmitter exocytosis in vivo. *Neuron* 30, 475–88.

101 ARSENE, F., TOMOYASU, T. & BUKAU, B. (2000). The heat shock response of Escherichia coli. *Int J Food Microbiol* 55, 3–9.

102 LUND, P. A. (2001). Regulation of expression of molecular chaperones. In *Moleuclar Chaperones in the Cell*, pp. 235–256. Oxford University Press Inc., New York.

103 STRAUS, D. B., WALTER, W. A. & GROSS, C. A. (1989). The activity of sigma 32 is reduced under conditions

of excess heat shock protein production in Escherichia coli. *Genes Dev* 3, 2003–10.

104 Morita, M. T., Kanemori, M., Yanagi, H. & Yura, T. (2000). Dynamic interplay between antagonistic pathways controlling the sigma 32 level in Escherichia coli. *Proc Natl Acad Sci USA* 97, 5860–5.

105 Herman, C., Thevenet, D., D'Ari, R. & Bouloc, P. (1995). Degradation of sigma 32, the heat shock regulator in Escherichia coli, is governed by HflB. *Proc Natl Acad Sci USA* 92, 3516–20.

106 Tomoyasu, T., Gamer, J., Bukau, B., Kanemori, M., Mori, H., Rutman, A. J., Oppenheim, A. B., Yura, T., Yamanaka, K., Niki, H. et al. (1995). Escherichia coli FtsH is a membrane-bound, ATP-dependent protease which degrades the heat-shock transcription factor sigma 32. *EMBO J* 14, 2551–60.

107 Kanemori, M., Nishihara, K., Yanagi, H. & Yura, T. (1997). Synergistic roles of HslVU and other ATP-dependent proteases in controlling in vivo turnover of sigma32 and abnormal proteins in Escherichia coli. *J Bacteriol* 179, 7219–25.

108 Tatsuta, T., Tomoyasu, T., Bukau, B., Kitagawa, M., Mori, H., Karata, K. & Ogura, T. (1998). Heat shock regulation in the ftsH null mutant of Escherichia coli: dissection of stability and activity control mechanisms of sigma32 in vivo. *Mol Microbiol* 30, 583–93.

109 Straus, D., Walter, W. & Gross, C. A. (1990). DnaK, DnaJ, and GrpE heat shock proteins negatively regulate heat shock gene expression by controlling the synthesis and stability of sigma 32. *Genes Dev* 4, 2202–9.

110 Bukau, B. (1993). Regulation of the Escherichia coli heat-shock response. *Mol Microbiol* 9, 671–80.

111 Tomoyasu, T., Ogura, T., Tatsuta, T. & Bukau, B. (1998). Levels of DnaK and DnaJ provide tight control of heat shock gene expression and protein repair in Escherichia coli. *Mol Microbiol* 30, 567–81.

112 Tatsuta, T., Joob, D. M., Calendar, R., Akiyama, Y. & Ogura, T. (2000). Evidence for an active role of the DnaK chaperone system in the degradation of sigma(32). *FEBS Lett* 478, 271–5.

113 Chattopadhyay, R. & Roy, S. (2002). DnaK-sigma 32 interaction is temperature-dependent. Implication for the mechanism of heat shock response. *J Biol Chem* 277, 33641–7.

114 Laufen, T., Mayer, M. P., Beisel, C., Klostermeier, D., Mogk, A., Reinstein, J. & Bukau, B. (1999). Mechanism of regulation of hsp70 chaperones by DnaJ cochaperones. *Proc Natl Acad Sci USA* 96, 5452–7.

115 Nakahigashi, K., Yanagi, H. & Yura, T. (1995). Isolation and sequence analysis of rpoH genes encoding sigma 32 homologues from gram negative bacteria: conserved mRNA and protein segments for heat shock regulation. *Nucleic Acids Res* 23, 4383–90.

116 McCarty, J. S., Rudiger, S., Schonfeld, H. J., Schneider-Mergener, J., Nakahigashi, K., Yura, T. & Bukau, B. (1996). Regulatory region C of the E. coli heat shock transcription factor, sigma32, constitutes a DnaK binding site and is conserved among eubacteria. *J Mol Biol* 256, 829–37.

117 Arsene, F., Tomoyasu, T., Mogk, A., Schirra, C., Schulze-Specking, A. & Bukau, B. (1999). Role of region C in regulation of the heat shock gene-specific sigma factor of Escherichia coli, sigma32. *J Bacteriol* 181, 3552–61.

118 Nakahigashi, K., Yanagi, H. & Yura, T. (2001). DnaK chaperone-mediated control of activity of a sigma(32) homologue (RpoH) plays a major role in the heat shock response of Agrobacterium tumefaciens. *J Bacteriol* 183, 5302–10.

119 Wickner, S. H. (1990). Three Escherichia coli heat shock proteins are required for P1 plasmid DNA replication: formation of an active complex between E. coli DnaJ protein and the P1 initiator protein. *Proc Natl Acad Sci USA* 87, 2690–4.

120 Wickner, S., Hoskins, J. & McKenney, K. (1991). Function of DnaJ and

DnaK as chaperones in origin-specific DNA binding by RepA. *Nature* 350, 165–7.
121 Wickner, S., Hoskins, J. & McKenney, K. (1991). Monomerization of RepA dimers by heat shock proteins activates binding to DNA replication origin. *Proc Natl Acad Sci USA* 88, 7903–7.
122 Wickner, S., Skowyra, D., Hoskins, J. & McKenney, K. (1992). DnaJ, DnaK, and GrpE heat shock proteins are required in oriP1 DNA replication solely at the RepA monomerization step. *Proc Natl Acad Sci USA* 89, 10345–9.
123 Tilly, K. & Yarmolinsky, M. (1989). Participation of Escherichia coli heat shock proteins DnaJ, DnaK, and GrpE in P1 plasmid replication. *J Bacteriol* 171, 6025–9.
124 Kim, S. Y., Sharma, S., Hoskins, J. R. & Wickner, S. (2002). Interaction of the DnaK and DnaJ chaperone system with a native substrate, P1 RepA. *J Biol Chem* 277, 44778–83.
125 Konieczny, I. & Liberek, K. (2002). Cooperative action of Escherichia coli ClpB protein and DnaK chaperone in the activation of a replication initiation protein. *J Biol Chem* 277, 18483–8.
126 Giraldo, R. & Diaz-Orejas, R. (2001). Similarities between the DNA replication initiators of Gram-negative bacteria plasmids (RepA) and eukaryotes (Orc4p)/archaea (Cdc6p). *Proc Natl Acad Sci USA* 98, 4938–43.
127 Sullivan, C. S. & Pipas, J. M. (2001). The virus-chaperone connection. *Virology* 287, 1–8.
128 Alfano, C. & McMacken, R. (1989). Heat shock protein-mediated disassembly of nucleoprotein structures is required for the initiation of bacteriophage lambda DNA replication. *J Biol Chem* 264, 10709–18.
129 Liberek, K., Osipiuk, J., Zylicz, M., Ang, D., Skorko, J. & Georgopoulos, C. (1990). Physical interactions between bacteriophage and Escherichia coli proteins required for initiation of lambda DNA replication. *J Biol Chem* 265, 3022–9.
130 Zylicz, M., Ang, D., Liberek, K. & Georgopoulos, C. (1989). Initiation of lambda DNA replication with purified host- and bacteriophage-encoded proteins: the role of the dnaK, dnaJ and grpE heat shock proteins. *EMBO J* 8, 1601–8.
131 Tanguy Le Gac, N. & Boehmer, P. E. (2002). Activation of the herpes simplex virus type-1 origin-binding protein (UL9) by heat shock proteins. *J Biol Chem* 277, 5660–6.
132 Eom, C. Y. & Lehman, I. R. (2002). The human DnaJ protein, hTid-1, enhances binding of a multimer of the herpes simplex virus type 1 UL9 protein to oris, an origin of viral DNA replication. *Proc Natl Acad Sci USA* 99, 1894–8.
133 Liu, J. S., Kuo, S. R., Makhov, A. M., Cyr, D. M., Griffith, J. D., Broker, T. R. & Chow, L. T. (1998). Human Hsp70 and Hsp40 chaperone proteins facilitate human papillomavirus-11 E1 protein binding to the origin and stimulate cell-free DNA replication. *J Biol Chem* 273, 30704–12.
134 Lin, B. Y., Makhov, A. M., Griffith, J. D., Broker, T. R. & Chow, L. T. (2002). Chaperone proteins abrogate inhibition of the human papillomavirus (HPV) E1 replicative helicase by the HPV E2 protein. *Mol Cell Biol* 22, 6592–604.
135 Marchler, G. & Wu, C. (2001). Modulation of Drosophila heat shock transcription factor activity by the molecular chaperone DROJ1. *EMBO J* 20, 499–509.
136 Lopez, N., Aron, R. & Craig, E. A. (2003). Specificity of Class II Hsp40 Sis1 in Maintenance of Yeast Prion [RNQ(+)]. *Mol Biol Cell* 14, 1172–81.
137 Hallstrom, T. C., Katzmann, D. J., Torres, R. J., Sharp, W. J. & Moye-Rowley, W. S. (1998). Regulation of transcription factor Pdr1p function by an Hsp70 protein in Saccharomyces cerevisiae. *Mol Cell Biol* 18, 1147–55.
138 Morishima, N., Nakagawa, K., Yamamoto, E. & Shibata, T. (1990). A subunit of yeast site-specific endonuclease SceI is a mitochondrial

version of the 70-kDa heat shock protein. *J Biol Chem* 265, 15189–97.

139 SHIBATA, T., NAKAGAWA, K. & MORISHIMA, N. (1995). Multi-site-specific endonucleases and the initiation of homologous genetic recombination in yeast. *Adv Biophys* 31, 77–91.

140 MIZUMURA, H., SHIBATA, T. & MORISHIMA, N. (1999). Stable association of 70-kDa heat shock protein induces latent multisite specificity of a unisite-specific endonuclease in yeast mitochondria. *J Biol Chem* 274, 25682–90.

141 MIZUMURA, H., SHIBATA, T. & MORISHIMA, N. (2002). Association of HSP70 with endonucleases allows the expression of otherwise silent mutations. *FEBS Lett* 522, 177–82.

15
Regulation of Hsp70 Chaperones by Co-chaperones

Matthias P. Mayer and Bernd Bukau

15.1
Introduction

The 70-kDa heat shock proteins (Hsp70s) are central components of the chaperone network that assists and controls numerous protein folding processes inside the cell. As part of their quality-control functions, Hsp70s are involved in refolding of stress-denatured proteins. They continuously survey the folding status of cellular proteins and bind to misfolded proteins to prevent their aggregation or, in conjunction with Hsp100 proteins, to disaggregate already-aggregated proteins. Since these functions are especially important under stress conditions, in most organisms at least one Hsp70 encoding gene is under heat stress control. In addition to the quality-control functions, Hsp70 are built-in components of cellular pathways. They are involved in the folding of newly synthesized proteins in the cytosol as well as in the translocation into and the folding within organelles. Hsp70s assist in assembly and disassembly of oligomeric protein structures, for example, the clathrin coat, and they are involved in signal transduction and cell cycle progression by controlling the stability and activity of regulatory proteins such as receptors, protein kinases, and transcription factors. To no other class of chaperones has such a plethora of functions been attributed. Hsp70 proteins are consequently abundant and constitutive cellular components that are, at least in eukaryotic organisms, essential for viability under all conditions. For more details on Hsp70 functions see chapters 14 and 36.

The interaction of Hsp70 proteins with their substrates has features that contribute to their versatility. First, their interaction with substrate proteins is restricted to a short peptide stretch within the substrate polypeptide and is therefore independent of the size and overall structure of the substrate protein. Second, the interaction with substrates is transient and is controlled in a nucleotide-dependent manner. Third, Hsp70 chaperones are influenced by a number of co-chaperones, which regulate binding to substrates and the lifetime of the Hsp70-substrate complex and thereby specify Hsp70-substrate interactions. Finally, Hsp70s cooperate with chaperones of other classes, for example, Hsp90 and Hsp100, to accomplish spe-

Protein Folding Handbook. Part II. Edited by J. Buchner and T. Kiefhaber

ISBN: 3-527-30784-2

cific tasks such as control of regulatory proteins and disaggregation of protein aggregates.

In this chapter we will briefly describe the Hsp70 structure and the functional cycle of Hsp70s, but we focus mainly on the control of Hsp70 function by co-chaperones. Special emphasis will be given to the description of biochemical and biophysical methods used for the analysis of Hsp70 functions.

15.2 Hsp70 Proteins

15.2.1 Structure and Conservation

Hsp70 proteins are composed of an N-terminal ATPase domain of ca. 45 kDa, a substrate-binding domain of ca. 15 kDa, and a C-terminal domain of ca. 10 kDa, the function of which is not completely clear. The ATPase domain, which is structurally homologous to actin, hexokinase, and glycerokinase [1–3], is built of two subdomains (I and II) that are linked to each other via two crossed α-helices. The two subdomains form a deep cleft between each other, at the bottom of which nucleotide is bound in complex with one Mg^{2+} and two K^+ ions (Figure 15.1A).

The substrate-binding domain is made up of a sandwich of two twisted β-sheets with four antiparallel strands each. The substrate-binding pocket, which seems to be tailored for a large hydrophobic residue, preferably leucine, is formed by the upward-bent strands 1 and 2 together with the upward-protruding connecting loops $L_{1,2}$ and $L_{3,4}$. These loops are stabilized by a second layer of loops ($L_{4,5}$ and $L_{5,6}$) and the two α-helices A and B. The distal part of helix B is connected to the outer loops by hydrogen bonds and a salt bridge, thereby forming the so-called latch that is in part responsible for the tight binding to substrates when no nucleotide or ADP is bound to the ATPase domain (Figure 15.1A).

The Hsp70 family of proteins is highly conserved throughout evolution, with around 50% sequence identity between bacteria and human. The overall homology is highest in the ATPase domain and decreases continuously towards the C-terminus. Parallel to this orthologous conservation, a radiative adaptation has taken place, creating paralogs with significantly lower sequence identity; for example, the sequence identity between *Escherichia coli* HscC and *E. coli* DnaK is around 27%. In addition, more distant relatives are found in eukaryotic cells: the Hsp110 and Hsp170 chaperones. Homology between the Hsp70 family and the Hsp110 and Hsp170 proteins is comparatively high (up to 40% sequence identity) in the ATPase domain, and homology modeling yields a very similar structure. Overall sequence identity in the part that is C-terminal to the ATPase domain is much lower and is scattered over a large portion of this domain, which together with secondary structure predictions suggests a homologous three-dimensional structure. It therefore could function as a substrate-binding domain [4]. A biochemical deletion

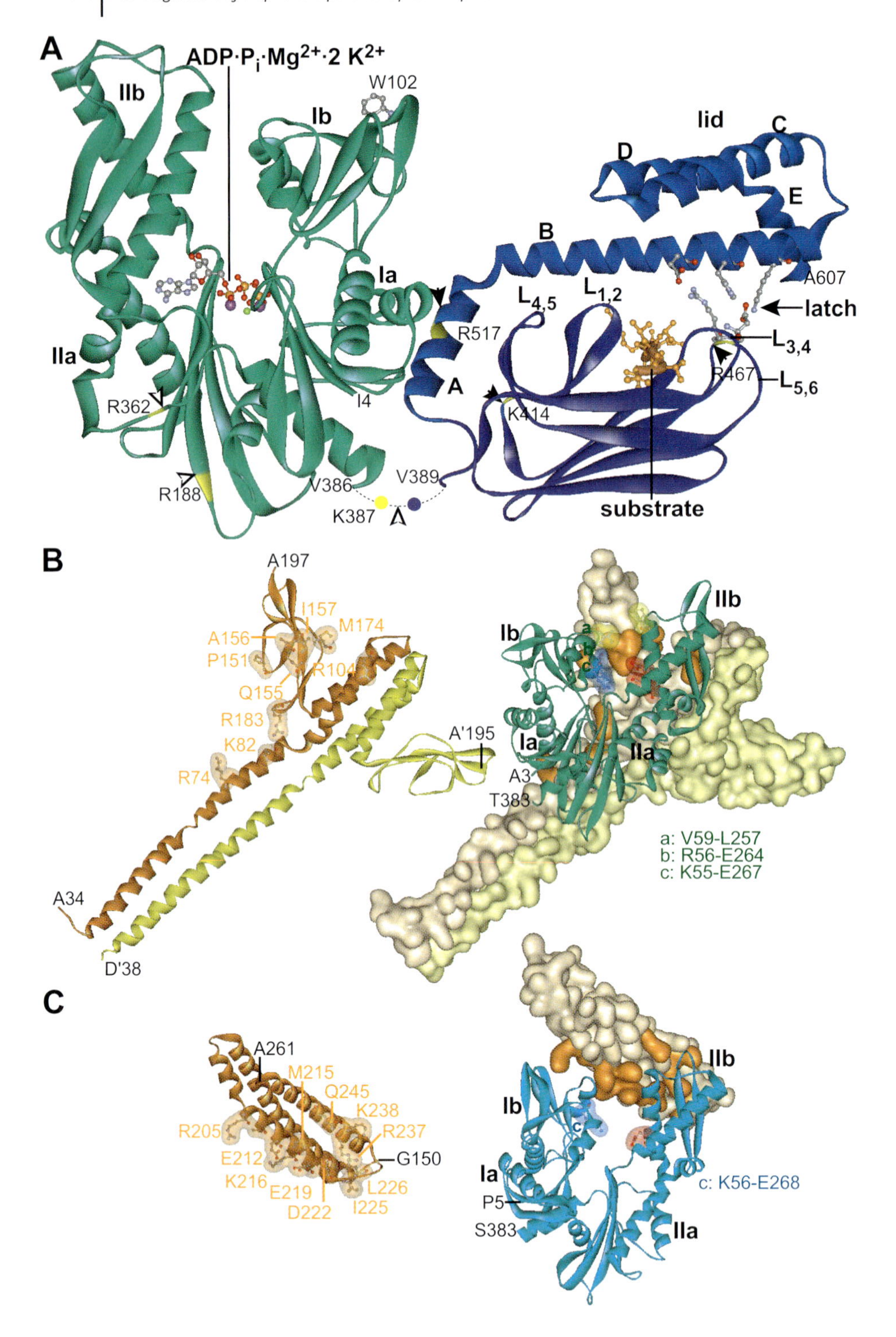

A
ADP·P$_i$·Mg^{2+}·2 K^{2+}
W102
IIb
Ib
lid
C
D
E
B
Ia
A607
latch
$L_{4,5}$
$L_{1,2}$
R517
$L_{3,4}$
IIa
R467
$L_{5,6}$
A
R362
I4
K414
V389
V386
R188
K387
substrate
B
A197
I157
M174
A156
P151
R104
Q155
R183
K82
R74
A'195
A34
D'38
Ib
IIb
Ia
IIa
A3
T383
a: V59-L257
b: R56-E264
c: K55-E267
C
A261
M215
Q245
K238
R237
R205
E212
G150
K216
L226
E219
D222
I225
IIb
Ib
Ia
P5
S383
IIa
c: K56-E268

analysis suggests that this part of the Hsp110 proteins indeed has a substrate-binding function [5]. The Hsp110 and Hsp170 proteins have characteristic insertions of unknown function within their substrate-binding domain. On the other end of the scale, there are Hsp70 homologues with a much-reduced C-terminal domain, including the yeast protein Ssz1 and the STCH protein of higher eukaryotes [6–9].

15.2.2
ATPase Cycle

The basic principles of the Hsp70 ATPase cycle have been elucidated using mainly *E. coli* DnaK, human Hsc70, and hamster BiP as model proteins. Therefore, we will first describe the ATPase cycle for *E. coli* DnaK and will subsequently discuss some variations found in other Hsp70 proteins (Figure 15.2). In the ATP-bound state DnaK has a low affinity for substrates but high substrate association and dissociation rate constants. In contrast, in the ADP-bound state the affinity for substrates is high but substrate association and dissociation rates are low [10–12]. Genetic and biochemical data clearly demonstrate that ATP hydrolysis is essential for DnaK's chaperone function in vitro and in vivo. However, the intrinsic ATP hydrolysis rate of DnaK is very low (ca. 6×10^{-4} s^{-1}; [13–16]) and is considered much

Fig. 15.1. Structure of the ATPase domain and substrate-binding domain of Hsp70 and of nucleotide-exchange factors. (A) (Left) Ribbon model of the ATPase domain of DnaK modeled onto the crystal structure of bovine Hsc70 in complex with Mg^{2+}·ADP·P_i, and two K^+ ions (1BUP; SWISS-MODEL [116, 216–218]). Indicated are first and last residues, the unique tryptophan (W102), and the trypsin cleavage sites (yellow). (Right) Ribbon model of the substrate-binding domain of DnaK in complex with a substrate peptide (orange) (1DKX) [93]. Indicated are the substrate-enclosing loops $L_{1,2}$ to $L_{5,6}$, helices A–E, and the lid. The latch-forming residues are shown as ball-and-stick models. The two missing amino acids between the C-terminus of the ATPase domain and the N-terminus of the substrate-binding domain are represented by a dashed line and two dots. The trypsin cleavage sites are marked in yellow. Filled arrowheads indicate residues that are more accessible for trypsin cleavage in the ATP-bound state, and open arrowheads indicate residues that become protected from trypsin cleavage in the ATP-bound state. (B) (Left) Ribbon representation of the asymmetric GrpE dimer from the co-crystal structure of GrpE with the ATPase domain of DnaK (1DKG) [110]. The DnaK-interacting residues are shown in ball-and-stick representation and are enveloped in the solvent-accessible surface. (Right) Surface representation of the GrpE dimer in complex with the ATPase domain of DnaK in ribbon representation. Indicated in orange are the DnaK-interacting parts of GrpE. The hydrophobic patches and the two pairs of charged residues that bridge the nucleotide-binding cleft in the closed conformation are shown in ball-and-stick and surface representation and are marked with the letters a–c. (C) (Left) Ribbon representation of the Bag domain of Bag-1 from the co-crystal structure of the Bag domain with the ATPase domain of bovine Hsc70 (3HSC) [120]. The Hsc70-interacting residues are shown in ball-and-stick representation and are enveloped in the solvent-accessible surface. (Right) Surface representation of the Bag domain in complex with the ATPase domain of Hsc70 in ribbon representation. Indicated in orange are the Hsc70-interacting parts of the Bag domain. The pair of charged residues that bridge the nucleotide-binding cleft in the closed conformation are shown in ball-and-stick and surface representation and marked with the letter c.

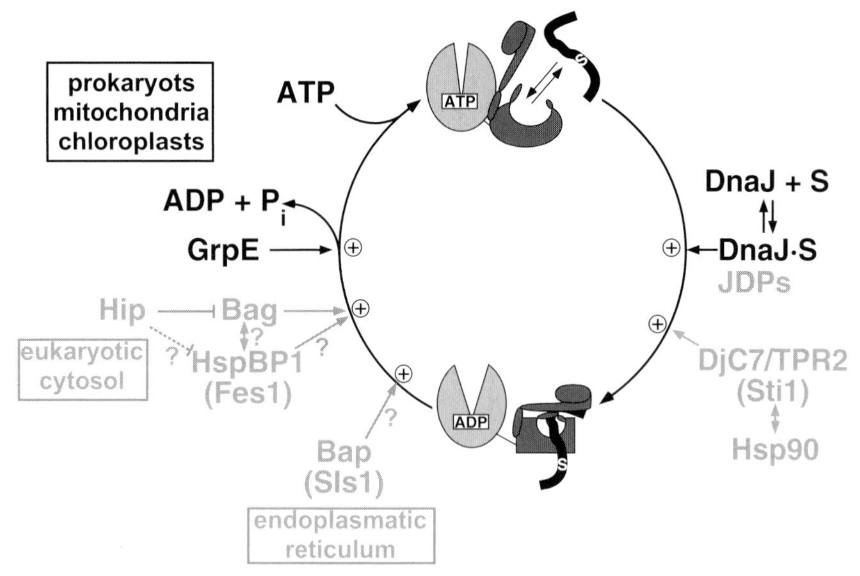

Fig. 15.2. Functional cycle of Hsp70 chaperones. In black is shown the basal cycle as described in Section 15.2.2 for *E. coli* DnaK and in gray are shown variations of the cycle found in eukaryotes. Protein names for mammalian cells are given with yeast homologues and analogues, respectively, in parenthesis.

too low to be of any physiological relevance. Similar basal ATPase rates, in the range of 3×10^{-4} s^{-1} to 1.5×10^{-2} s^{-1}, have been reported for all Hsp70s so far investigated [17–23]. This intrinsic ATPase rate, however, is stimulated by substrates by a factor of 2 to 10 and by the DnaJ co-chaperone by a factor of 5 at physiological concentrations (see Section 15.3) [13, 23, 24]. If both DnaJ and substrate are present at the same time, the ATP hydrolysis rate is stimulated synergistically several thousand-fold [25–29]. From these data it was concluded that substrates associate primarily to DnaK·ATP with high association rates and are subsequently trapped by the substrate and DnaJ-induced ATP hydrolysis, which leads to the transition from the low-affinity to the high-affinity state [25–27, 30, 31]. The ADP dissociation rate is approximately one order of magnitude greater than substrate dissociation rates in the high-affinity state. Under physiological conditions, i.e., high ATP concentrations, ADP will dissociate and ATP will rebind to DnaK, triggering the transition to the low-affinity state prior to substrate release. Nucleotide exchange therefore determines the lifetime of the DnaK-substrate complex. For DnaK, ADP dissociation is accelerated by the nucleotide-exchange factor GrpE several thousand-fold [24, 32]. ATP binds to the nucleotide-free DnaK with high association rates (1.2×10^5 M^{-1} s^{-1}; [33, 34]) and high affinity (1 nM; [34]), leading to the transition to the low-affinity state of DnaK and substrate release. The cycle then restarts with association of a new substrate. Under optimal refolding conditions for the model substrate luciferase (with 80 nM luciferase, 800 nM DnaK, 160 nM

DnaJ, and 400 nM GrpE), one cycle takes about 1 s. Since reactivation of 50% of the luciferase molecules takes about 5 min, many cycles are on average necessary to achieve refolding.

Although the basic scheme of the ATPase cycle seems to be conserved, a number of variations of this theme have been observed within the Hsp70 family. First, nucleotide dissociation rates vary dramatically between Hsp70 proteins. Human Hsc70 and *E. coli* HscA have a 20- and 700-fold higher dissociation rate, respectively, as compared to DnaK [35]. The structural reason for these differences was found in two salt bridges and an exposed loop in the ATPase domain and was used to classify the Hsp70 proteins in three subfamilies, the DnaK-type, the Hsc70-type, and the HscA-type Hsp70 chaperones [35].

Second, variations have also been found in the difference between the high- and low-affinity states for substrates, which for DnaK is a factor of 400–2500 in substrate dissociation rate constants (k_{diss}) and 10–20 in substrate dissociation equilibrium constants (K_d). For example, for the *E. coli* Hsp70 homologue HscC, identical K_d values were measured for the HscC-peptide complex in both nucleotide states. The measured value was thereby similar to the K_d of the corresponding DnaK-peptide complex in the ADP state. However, k_{diss} was stimulated 20-fold by ATP, meaning that dissociation and association rates change to the same extent in response to ATP binding [19]. Therefore, the affinity of HscC for substrates is affected not by ATP but by the turnover rate.

15.2.3
Structural Investigations

ATP binding and hydrolysis in Hsp70 proteins are coupled to conformational changes, not only within the ATPase domain but also within the substrate-binding domain. As described above, ATP binding to the ATPase domain leads to the transition of the high-affinity state of the substrate-binding domain to the low-affinity state, while ATP hydrolysis causes the reverse transformation. Since crystallization of a full-length Hsp70 protein has failed so far and since all available structures from ATPase domain and substrate-binding domain represent the high-affinity state, a number of different techniques have been employed to probe nucleotide-induced conformational changes, including partial proteolysis by trypsin, tryptophan fluorescence, Fourier transform infrared spectroscopy, small angle X-ray scattering, and, most recently, amide hydrogen exchange [36–42] (C. Graf, W. Rist, B. Bukau, and M. P. Mayer, unpublished results).

Comparison of tryptic digestion patterns of the nucleotide-free, ADP-, and ATP-bound states of *E. coli* DnaK revealed that digestion sites within the ATPase domain (R188, R362) and the region connecting the ATPase domain with the substrate-binding domain, the so-called linker region (K387), become less accessible upon ATP binding, while sites within the substrate-binding domain (K414, R467, R517) become more accessible (Figure 15.1A). These observations demonstrate that ATP binding leads to a tighter conformational packing of the ATPase domain and to a more open conformation of the substrate-binding domain [39].

The digestion pattern of eukaryotic Hsp70s (e.g., hamster BiP and yeast Ssa1) also shows characteristic changes upon ATP binding. The ATP-dependent appearance of a 60-kDa fragment, which contains the entire ATPase domain and part of the substrate-binding domain, indicates a stabilization of the ATPase domain and a decrease in accessibility of the linker region [40, 41, 43]. On the basis of these results, it was suggested that the ATPase domain and the substrate-binding domain are more tightly linked in the ATP state than in the ADP or nucleotide-free states. Such a close-up of the ATPase domain and substrate-binding domain was also indicated by small angle X-ray scattering data [42]. Upon ATP binding the radius of gyration R_g decreased by 3.6 Å, and the pair distribution function $P(r)$ (as defined in Ref. [42]) showed a shift from an elongated molecule in the presence of ADP to a more compact shape in the presence of ATP. These data, however, relate only to a truncated version of the Hsp70 protein, the 60-kDa fragment, because small angle X-ray scattering data of full-length Hsp70 cannot easily be interpreted due to the tendency for self-assembly into heterogeneous oligomers.

Tryptophan fluorescence was also used to monitor nucleotide-dependent conformational changes in Hsp70 proteins [33, 36, 39, 44]. *E. coli* DnaK has a single tryptophan in the ATPase domain (position 102) (Figure 15.1A). Upon binding of ATP – but not AMP, ADP, ADP + inorganic phosphate or non-hydrolyzable ATP analogues (AMPPCP, AMPPNP, ATPγS) – the maximum of fluorescence shifts by 3–4 nm towards shorter wavelength (blue shift) and decreases by about 15% [33, 36, 39]. Quench and blue shift are almost entirely due to conformational changes in the substrate-binding domain and to movement of both domains relative to each other, since the isolated ATPase domain does not show any blue shift and shows a quench of only about 5% [39]. Measurements of the fluorescence lifetime and quenching of the fluorescence with polar quenchers indicate that accessibility of the tryptophan is drastically reduced upon ATP binding [45]. Due to the observed ATP-dependent quench, tryptophan fluorescence was used to determine the kinetics of ATP binding and hydrolysis in *E. coli* DnaK [46, 47] and bovine Hsc70 [44].

15.2.4 Interactions With Substrates

On the one hand, Hsp70 chaperones interact promiscuously with virtually all unfolded proteins but generally do not bind their native counterparts. On the other hand, they recognize certain folded proteins with high specificity. Therefore, an important question is how Hsp70s can combine in their substrate specificity both seemingly contradictory properties. A number of different approaches have been used to elucidate the substrate specificity of Hsp70s. One approach used the f1 phage peptide display library method, selecting high-affinity binding phages out of a library of phages exposing a stretch of 6–12 residues with random sequence [48, 49]. The advantage of this method is that a significant part, although not all, of the theoretical possible sequence space can be accessed (with a stretch of nine residues, the complexity of the library would have to be at least 4^{27} ($\approx 10^{16}$), equal to 100 L phage lysate with a titer of 10^{11} pfu mL^{-1}). The drawback of this method is that the number of clones, which have to be sequenced to elucidate the binding motif, increases exponentially with the degenerateness of the binding motif. In one study a phage library of 6×10^9 clones was incubated with immobilized DnaK, unbound phages were washed away, and bound phages were eluted by addition of ATP. Forty-eight clones out of the selected pool, and, for comparison, 44 clones out of the original pool, were sequenced. DnaK was found to prefer positively charged and hydrophobic residues, whereby the hydrophobic residues are more favorable in a central position within the peptide and negatively charged amino acids are disfavored [49]. Using the same phage display technique, it was found for the Hsp70 homologue of the endoplasmic reticulum, BiP, that binding peptides are enriched in aromatic and hydrophobic amino acids in alternating positions, suggesting binding of the substrates to BiP in an extended conformation [48]. Another approach screened synthetic peptides, scanning the sequence of a natural Hsp70 substrate for binding to Hsp70 and for ability to stimulate Hsp70's ATPase activity [50–52]. In this way the potential binding sites of Hsp70 within the tested protein sequence could be identified, but no general motif was derived.

The most extensive analysis of the substrate specificity of an Hsp70 protein used a library of cellulose-bound 13mer peptides scanning the sequences of natural proteins with an overlap of 10 amino acids [53]. Although this method, due to the limited number of peptides (< 5000 in the case of Ref. [53]), does not allow one to exhaust the theoretically possible sequences, it provides a relatively large basis for a statistical analysis. In addition, the use of protein sequences guarantees that the tested peptides represent biologically relevant sequences. The peptides are attached over a β-alanine linker via their C-terminus to the cellulose membrane. The Hsp70 protein (0.1–1 μM) is incubated with the cellulose membrane at room temperature. The membranes are washed and bound proteins are transferred by fractionated electroblotting onto PVDF membranes for immunodetection using a visualization method with a high dynamic range, such as chemifluorescence (ECF™, Amersham Bioscience). Quantification of all spots results in values that correspond to the affinity of the peptides for the Hsp70 protein tested, revealing directly the potential binding sites of Hsp70 in the protein, the sequence of which was used in the peptide library. To elucidate the binding motif, a statistical analysis of a sufficient number of peptides has to be performed. First, the contribution of each amino acid to the binding affinity can be determined by comparing the relative abundance of each residue in high-affinity peptides with the average occurrence in the library (compare Ref. [19]). Second, since protein sequences are scanned with 13mer peptides and an overlap of 10 residues, in general, three or four peptides in a row show significant affinity to the Hsp70 protein. These peptides can therefore be aligned, and the overlapping part corresponds to the region that most likely is bound in the substrate-binding pocket of the Hsp70. The aligned regions can be rooted at a position where a residue is found that contributes most to binding affinity, and the preference for neighboring amino acids can be analyzed. Using this method on a library with more than 4000 peptides, the binding motif for DnaK was elucidated. It consists of a core of five amino acids enriched in hydrophobic residues, flanked on both sides by a region where positively charged residues are

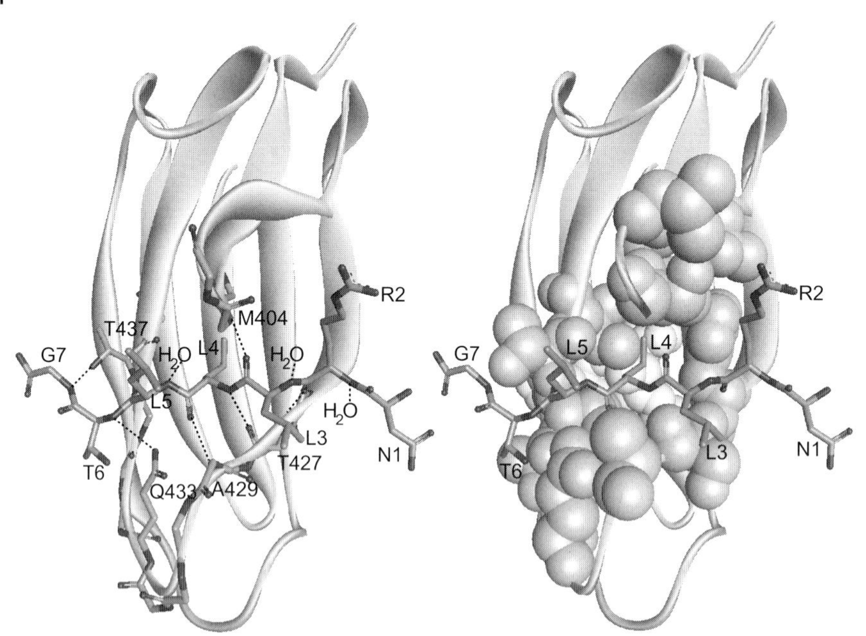

Fig. 15.3. DnaK-substrate interactions. The left panel shows the hydrogen bonding between the substrate-binding domain of DnaK and the backbone of the co-crystallized substrate peptide (NRLLLTG). The substrate is given in stick representation and the substrate-binding domain without α-helices is shown as a ribbon model with the interacting backbone and side chain residues (Q433, T437) in stick representation. Hydrogen bonds are indicated as dashed lines. Two of the seven hydrogen bonds between DnaK and the substrate are not, or are only barely, visible (R2:O to DnaK-T427:N; L5:O to DnaK-T437:N). The right panel emphasizes the hydrophobic interaction between the substrate-binding domain of DnaK and the side chains of the substrate peptide. The hydrophobic cleft with the deep pocket for the central substrate side chain (L4) is clearly visible. For clarity, methionine 404, which arches over the substrate backbone, is left out.

preferred. This binding motif is fully consistent with the crystal structure of the substrate-binding domain of DnaK in complex with a substrate peptide (Figure 15.3).

Two problems can occur in such an analysis. First, the size of the library could be too small to elucidate the binding motif or to determine whether rare (e.g., Met, Trp) or highly abundant (e.g., Leu) residues contribute to binding affinity. Second, since natural sequences are used, some residues may not be randomly distributed. For example, in the library of 557 peptides used for the analysis of the binding specificity of *E. coli* HscC, isoleucine was significantly more frequent in negatively charged peptides than in positively charged peptides. This was even more prominent when peptides were analyzed that contained two or more isoleucines. The case for methionine was similar. Since HscC was strongly biased against negatively charged amino acids, isoleucine and methionine were found with a lower frequency in HscC binding peptides than in the average of the library. From this finding, however, it cannot be concluded that the presence of isoleucine or methionine in a peptide disfavors binding of HscC. In contrast, glycine was much more abundant in positively charged peptides than in negatively charged peptides. Since positively charged residues are strongly favored by HscC, it should be expected that glycine is enriched in peptides with high affinity for HscC. The result that glycine was not enriched in HscC-binding peptides, therefore, has to be interpreted such that the presence of glycine in a peptide is not favorable for binding to HscC.

The binding motif of DnaK is frequent in protein sequences. In the native state these sites are generally buried in the hydrophobic core of the protein. This explains the promiscuous binding of DnaK to unfolded polypeptides. Comparison of the binding preferences of DnaK with *E. coli* HscA and HscC revealed that substrate specificity can vary substantially within the Hsp70 family, probably as a manifestation of adaptive specialization [19, 54] (unpublished data). The structural reason for these differences is found in the substrate-binding cavity, in particular in the arch-forming amino acids that, in contrast to the other substrate-contacting residues, show a significantly lower degree of evolutionary conservation [55].

To understand the chaperone activity of Hsp70 proteins it is necessary not only to know the binding specificity but also to analyze the kinetics of association and dissociation of substrates. These parameters were investigated for peptide substrates using fluorescence intensity/anisotropy and surface plasmon resonance spectroscopy [10–12, 27, 56–60]. For *E. coli* DnaK, a number of different peptides labeled at N- or C-terminal cysteines with fluorescent dyes such as 2-(4′-(iodoacetamido)anilino) naphthalene-6-sulfonic acid (IAANS; Molecular Probes) or 6-acryloyl-2-dimethylaminonaphthalene (acrylodan; Molecular Probes) gave a significant change in fluorescence emission upon binding that was used to determine association and dissociation rate constants and the dissociation equilibrium constant to DnaK in the ADP and ATP states [10, 11, 58]. Such changes in fluorescence emission, however, were not observed when these peptides were incubated with C-terminally truncated DnaK variants [11, 61], *E. coli* HscA, or human Hsc70 (unpublished results), despite the fact that the peptides bound with high affinity to these Hsp70 proteins. More recently a 5-dimethylaminonaphthalene-1-sulfonyl (dansyl) modified peptide was found to be suitable for fluorescence intensity measurements with human Hsc70, giving a much improved signal-to-noise ratio [62]. Together these findings show that Hsp70 proteins differ with respect to the interaction with fluorescent labels attached to binding peptides. As an alternative approach, the association and dissociation rates can be determined by fluorescence anisotropy as shown for human Hsc70 and a fluorescein-labeled peptide [60].

The affinity of DnaK in the nucleotide-free or ADP-bound state for peptide substrates varies dramatically, with K_d values between 50 nM and >100 μM [57]. In the ATP state the K_d values are around 10- to 20-fold higher. When the K_d value for the DnaK-peptide complexes in the ATP state was blotted against the corresponding K_d value in the ADP state, a linear relationship was observed [11]. This was interpreted such that peptides bind to the substrate-binding domain of DnaK in the ADP and ATP states in a similar way. The alteration of affinity of DnaK for peptides with the nucleotide bound to the ATPase domain was illustrated by measurements of the association and dissociation rates. While the association rates of peptides increased by up to two orders of magnitude upon ATP binding to the ATPase domain, the dissociation rates changed by up to three orders of magnitude [10, 11, 58]. This led to the model that in the ATP state the substrate-binding domain is in an open conformation, where association and dissociation of substrates occur with high rates. In the ADP state, in contrast, the substrate-binding domain is in a closed conformation, where binding and release of substrates are slow. The elucidation of the crystal structure of the substrate-binding domain of DnaK in complex with a substrate peptide made clear that a closed conformation as found in the crystal structure would not allow association or dissociation of the substrate. In this conformation the substrate is completely enclosed by the substrate-binding cavity, with seven hydrogen bonds to the backbone of the substrate peptide and numerous hydrophobic contacts to side chains of the substrate. However, since association and dissociation of substrates do occur even when DnaK is in the ADP-state, it was concluded that both the open and closed conformations of the substrate-binding domain exist in the ADP state with continuous interconversion [11]. This hypothesis is supported by measurements of the association rates of peptides to DnaK in the ADP state, which followed biphasic kinetics indicating a two-step process, the first being limited by the opening rate of the substrate-binding domain [59]. Similarly, it was hypothesized that in the ATP state the substrate-binding domain is not always in the open conformation but also cycles between the open and closed states, only with a much higher frequency than in the ADP state [11]. This hypothesis was based on the findings that the association rates of peptides to DnaK·ATP are three orders of magnitude lower than expected from a diffusion-controlled process.

Surface plasmon resonance spectroscopy was also used to determine binding affinities to substrates. Peptides or proteins were thereby immobilized via a thiol-linkage through a cysteine residue or by chemical cross-linking using EDC/NHS, and the flow cell was perfused with the Hsp70 protein (yeast Kar2 and *E. coli* DnaK) at increasing concentrations [27, 56]. This method allowed determination of the dissociation equilibrium constants. The association and dissociation rate constants, however, could not be determined due to the complexity of the observed kinetics, which was probably due to oligomerization of the Hsp70 protein (for a detailed discussion, see Section 15.3.3).

15.3 J-domain Protein Family

15.3.1 Structure and Conservation

The J-domain proteins (JDP) (Hsp40s, DnaJ proteins) comprise a large family of multi-domain proteins that are characterized by a highly conserved stretch of 70 amino acids referred to as the J-domain [63–65]. The prototype for this family,

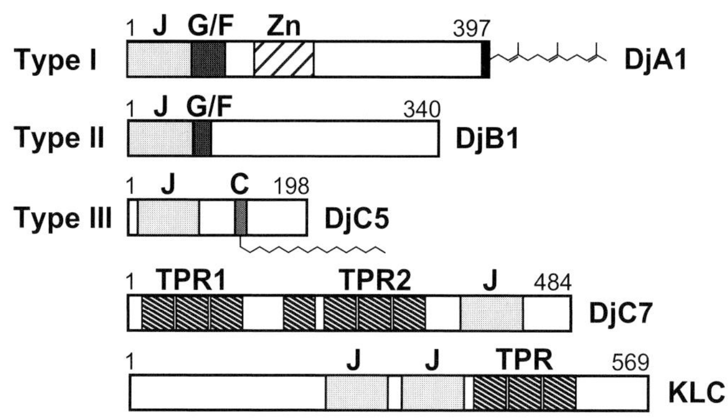

Fig. 15.4. Domain structure of J-domain proteins. For type I and type II JDPs, the human proteins DjA1/Hdj2 and DjB1/Hdj1 are given as examples, and for type III JDPs, the human proteins DjC5/cyteine string protein, DjC7/TPR2, and kinesin light chain (KLC1) are given as examples. For DjA1 and DjC5, farnesylation of the C-terminal CaaX-box and palmitoylation of the cysteine string (C), respectively, are indicated.

the *E. coli* protein DnaJ, has, in addition to its N-terminal J-domain, a glycine/phenylalanine-rich region (G/F), a Zn^{2+}-binding domain, and a C-terminal β-sheet domain. According to the degree of homology to DnaJ, the JDP family is divided into three subfamilies. The type I JDPs share significant homology to all four domains of DnaJ. The type II JDPs show homology to the C-terminal domain of DnaJ and generally contain a G/F-rich region, but do not have a Zn^{2+}-binding domain. The type III JDPs have only the J-domain in common with DnaJ (Figure 15.4). A unifying nomenclature for all mammalian JDPs was proposed by Ohtsuka and Hata [66], consisting of the acronym Dj for DnaJ-like protein, followed by the letter A, B, or C, for type I, II, or III, and a number. Type I and type II JDPs always have their J-domain at the N-terminus, while in type III JDPs the J-domain can be at any position. Type III JDPs can have a large variety of additional domains and protein motifs, including transmembrane anchors (e.g., *E. coli* DjlA, human DjC9/hSec63, yeast mitochondrial Pam18), tetratricopeptide repeat (TPR) motifs (e.g., mouse DjC2/Zrf1/Mida1, human DjC3/hp58, and DjC7/hTpr2), and cysteine-rich domains (mammalian DjC5/cysteine string protein) (Figure 15.4). A number of JDPs of all three types contain at their C-terminus a CaaX box (C, cysteine; a, aliphatic amino acid; X, any amino acid) as farnesylation signal (e.g., yeast Ydj1, mammalian DjA1/Hdj-2, mammalian DjC11/Mdg1). The farnesylation signals found in JDPs are usually not optimal, and the farnesylation ratio in vivo is usually well below 100%. However, a removal of the farnesylation signal in yeast Ydj1 leads to a temperature-sensitive phenotype, indicating its importance [67]. The number of JDPs per organism has increased dramatically in the course of evolution. Genome sequencing revealed that in *E. coli* six homologues exist, in *S. cerevisiae*, 19; in *Caenorhabditis elegans*, 29; in *Drosophila melanogaster*, 38; in humans, 44; and in *Arabidopsis thaliana*, 94 [68].

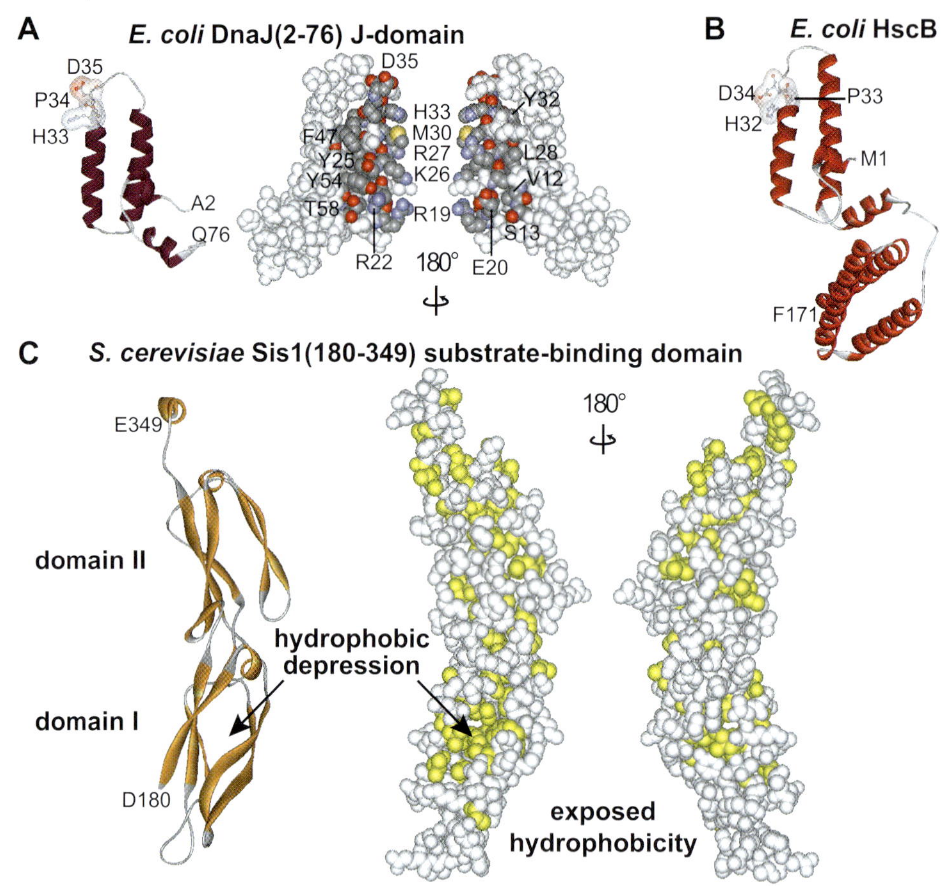

Fig. 15.5. Structures of J-domain proteins. (A) NMR structure of the J-domain of *E. coli* DnaK (2–76) (1XBL) [70]. Left, ribbon model with the invariable HPD motif indicated in ball-and-stick and surface representation. Middle and right, two faces of the J-domain in space-filling representation with the DnaK-interacting residues as elucidated by chemical shift perturbation and line broadening in NMR experiments in atomic colors. Not labeled are residues of the hydrophobic core. (B) Ribbon model of the crystal structure of *E. coli* HscB (1FPO) [82]. (C) Ribbon (left) and space-filling (right) models of the crystal structure of the substrate-binding domain of *S. cerevisiae* Sis1 (residues 180–349) (1C3G) [75]. Indicated is the hydrophobic depression that was proposed to be the substrate-binding site. Hydrophobic residues are shown in yellow.

Structural information is available for the J-domains of *E. coli* DnaJ, human DjB1/Hdj-1, bovine auxilin, polyomavirus T-antigen [69–73], the Zn^{2+}-binding domain of *E. coli* DnaJ (residues 121–209; [74]), the C-terminal domain of yeast Sis1 (residues 171–352) [75], and the Zn^{2+}-binding domain plus C-terminal domain of yeast Ydj1, which has a high homology to the C-terminal domain of DnaJ [76] (Figure 15.5, see p. 528/529). All structures are unique to DnaJ proteins and have not been found in other structures listed in databases.

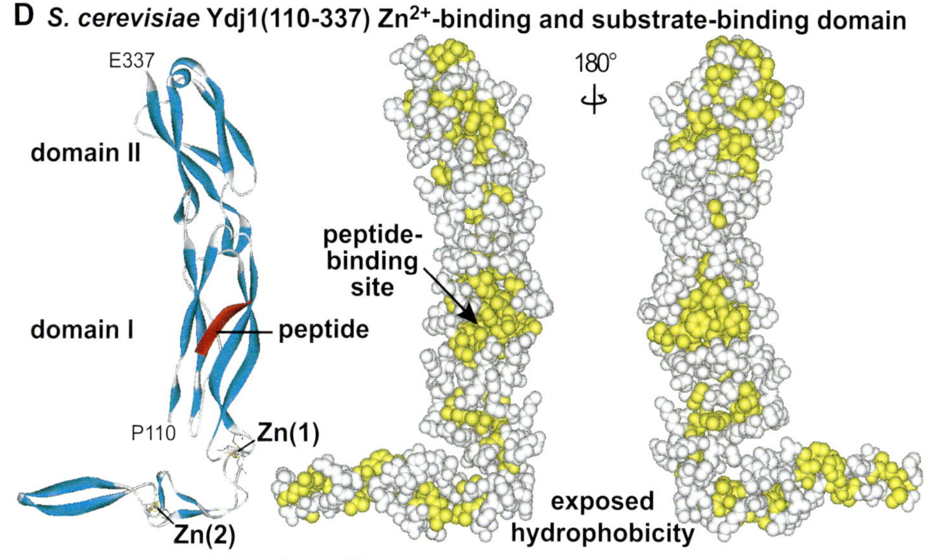

Fig. 15.5. (D) Ribbon (left) and space-filling (right) models of the crystal structure of the substrate-binding domain of *S. cerevisiae* Ydj1 (residues 110–337) (1NLT) [76]. The co-crystallized peptide is shown in red. Hydrophobic residues are shown in yellow.

The J-domain is essential for the JDP function to stimulate the ATPase activity of Hsp70 proteins [13, 24–27, 77, 78] (Figure 15.5A). It consists of four helices. Helices II and III form an antiparallel helical coiled-coil. Helix I runs from the end of helix III to the middle of helix II, approximately parallel to the plane formed by helices II and III. Helix IV starts near the C-terminal end of helix III on the same side of the coiled-coil as helix I, but it is oriented approximately perpendicular to the plane of the helices II and III. Helices I through III form a hydrophobic core between them. The loop connecting helices II and III is of variable length and contains the highly conserved and functionally essential histidine–proline–aspartic acid (HPD) motif. The surface of the J-domain is mainly positively charged due to a number of conserved basic residues, in particular in helix II.

The Zn^{2+}-binding domain consists of a segment of 76 amino acids characterized by four repeats of the motif C-X-X-C-X-G-X-G [79]. NMR spectroscopy revealed a V-shaped extended hairpin topology consisting of three pairs of antiparallel β-strands separated by the two Zn^{2+}-binding sites [74]. Due to this topology, the N- and C-termini are on the same side of the structure, and the first and the last C-X-X-C-X-G-X-G motif of the sequence form the first Zn^{2+}-binding site, while the two middle C-X-X-C-X-G-X-G motifs form the second Zn^{2+}-binding site. Removal of the Zn^{2+} ions causes unfolding of the structure, demonstrating the importance of the Zn^{2+} ions for the structural stability of the domain. The functional role of the Zn^{2+}-binding domain remains obscure.

Most structural information on how a JDP may bind substrates comes from the 2.7-Å crystal structures of the C-terminal substrate-binding domain (residues 171–

352) of the type II homologue, Sis1 of *S. cerevisiae* [75], and the Zn^{2+}-binding and C-terminal domains (102–350) of the type I homologue, Ydj1 of *S. cerevisiae* [76]. The Sis1 substrate-binding domain consists of two highly similar domains, each formed by a sandwich of two β-sheets and a short α-helix (Figure 15.5C). The C-terminus of the second domain extends into a short α-helix, which is involved in the dimerization of Sis1 in the crystals. Although Sis1 and other DnaJ homologues form dimers and higher-order oligomers in solution, it is not clear whether a dimer is the active form. The structure of the Sis1 fragment did not reveal an obvious substrate-binding cavity. However, the authors proposed a small hydrophobic depression in domain I as peptide-binding site for two reasons. First, the analogous hydrophobic depression in domain II is occupied by the aromatic side chain of a phenylalanine residue. Second, in the crystal packing the side chain of a proline residue from an adjacent Sis1 molecule was inserted into this hydrophobic depression [75]. Alteration of hydrophobic residues within this pocket affected the ability of Sis1 to bind substrates in vitro [80].

The Ydj1 structure is L-shaped and consists of three domains, whereby domains I and II are superimposable to the structure of Sis1 and the Zn^{2+}-binding domain is very similar to the NMR structure of the DnaJ Zn^{2+}-binding domain [76] (Figure 15.5D). Ydj1 was crystallized with the peptide GWLYEIS, which was bound in a small hydrophobic groove in domain I of Ydj1. This location is homologous to the hydrophobic depression in Sis1 that was proposed to serve as substrate-binding site. The peptide was bound in an extended conformation, forming an antiparallel β-strand to strand 2 of domain I of Ydj1. The central Leu of the peptide is completely buried in a hydrophobic pocket, while the aromatic side chains of the peptide do not make contact with Ydj1 in the crystal structure. The apparent strong contribution of hydrogen bonds to the backbone of the peptide and the missing contacts of the aromatic residues are difficult to reconcile with the substrate-binding specificity of the type I homologue *E. coli* DnaJ ([81]; discussed below in Section 15.3.3). Although the Ydj1 fragment was crystallized as a monomer, the authors argue that in vivo it forms a dimer similar to Sis1. The reason for the monomer structure was suggested to be due to an alteration of Phe335, the homologous residue of which in Sis1 is involved in dimerization. This Phe335 to Asp alteration was introduced because the authors could not obtain suitable crystals of the wild-type Ydj1 fragment.

The crystal structure of the *E. coli* type III protein HscB represents the only full-length structure of a JDP (Figure 15.5B). The N-terminal J-domain, which is very similar to the J-domain structures solved by NMR despite the very limited degree of sequence identity, is connected to a C-terminal three-helix bundle by a flexible linker [82]. The structure did not convey any idea of where and how HscB might interact with substrates.

15.3.2
Interaction With Hsp70s

The physical interaction of full-length JDP with Hsp70s was first investigated by size-exclusion chromatography and an ELISA assay using *E. coli* DnaJ and *E. coli*

DnaK as model proteins [83]. With these assays a complex between the two proteins could be detected only when ATP was present. Mutational alteration of the HPD motif in the J-domain completely abolished the interaction, indicating that the J-domain is involved. C-terminal truncation of DnaJ also reduced the interaction significantly, suggesting that the C-terminal domains also participate in the interaction [78, 84]. A more detailed analysis using surface plasmon resonance spectroscopy with immobilized DnaJ demonstrated that the thus observed interactions are very similar if not identical to the DnaJ-mediated interaction of Hsp70 proteins with substrates [56, 85]. This conclusion is based on the findings that the interaction depends on DnaJ-stimulated ATP hydrolysis, an intact inter-domain communication between ATPase and the substrate-binding domain of DnaK, and a high-affinity interaction with the substrate-binding cavity of DnaK. Interactions of the J-domain alone with full-length DnaK, as well as interactions of full-length DnaJ with the ATPase domain of DnaK, could not be detected by this method. The DnaJ interaction with the substrate-binding domain of DnaK was further investigated using a photoactivatable heterobifunctional cross-linker attached to cysteines in close proximity to the substrate-binding cavity of DnaK [26]. This approach demonstrated that DnaJ could be cross-linked to DnaK in a nucleotide-modulated manner. This cross-link was abolished in the presence of a good DnaK-binding peptide but not in the presence of a peptide that binds only to DnaJ and not to DnaK. These results indicate that DnaJ binds to DnaK in close proximity to, if not within the substrate-binding cavity of, DnaK [26]. The physiological significance of this interaction is still unclear. It was proposed that this interaction is an intermediate during the transfer of substrates from DnaJ to DnaK [85]. Investigations using the J-domain of the SV40 large T-antigen revealed that Hsp70 proteins bind to and stabilize almost all proteins to which a J-domain is added [86]. In this reaction, multiple Hsp70 molecules were observed to bind to the same protein. Similar results were obtained when the J-domain of DnaJ was fused to biotin carrier protein and the J-domain of Sec63 was fused to glutathione-S-transferase [85, 87]. The J-domain may therefore signify to Hsp70 proteins the close proximity of a substrate, whereby Hsp70 binds to DnaJ itself if DnaJ has no substrate bound.

The interaction site of the J-domain of DnaJ within DnaK was identified by two distinct approaches. The *dnaK*-R167H mutant was isolated as an allele-specific extragenic suppressor of the *dnaJ*-D35N mutant allele [88]. Independently, the crystal structure of the ATPase domain of Hsc70 was searched for highly conserved surface-exposed residues that are in close proximity to the nucleotide-binding cleft. The identified residues were exchanged by alanine-scanning mutagenesis and the interaction with DnaJ was analyzed. All thus identified residues are localized within a conserved channel in the ATPase domain [89].

To study the physical interaction of the J-domain with Hsp70 on the J-domain site, NMR was employed with ^{15}N-labeled J-domain of *E. coli* DnaJ and either full-length DnaK or the ATPase domain of DnaK [90]. Residues that are directly involved in the interaction, or indirectly affected through a conformational change during binding, were hereby revealed by quenching of the NMR signal and by chemical shift perturbation upon addition of DnaK (Figure 15.5A). Most affected by the interaction are residues in helix II and the loop between helices II and III

[165, 166]. By competing with Bag for binding to the ATPase domain of Hsc70, Hip prevents accelerated ADP dissociation and thereby slows down the ATP-mediated substrate release. Three cellular processes have been described where Hip seems to be involved: (1) the activation of nuclear receptors, (2) the translocation of proteins across the lysosomal membrane, and (3) the internalization of G-protein-coupled receptors [159, 167, 168]. In all three processes Hip seems to contribute to the efficient binding of Hsc70 to its substrates, which would be consistent with its role to counteract the effects of Bag proteins. These observations taken together support the notion that the co-chaperones Hip and Bag fine-tune the Hsp70 chaperone machinery for certain substrates or at specific locations.

15.5.2
Hop

The 60-kDa Hsp70-Hsp90 organizing protein (Hop/Sti1/p60) has been identified in a genetic screen for proteins involved in the regulation of the heat shock response in yeast [169] and was subsequently found to be a component of the progesterone receptor complex [170]. The Hop protein is characterized by three TPR domains with three TPR motifs each and it forms a dimer in solution. The TPR domains bind to the EEVD motif present at the C-termini of most eukaryotic cytosolic Hsp70s and Hsp90s. The recent elucidation of the crystal structures of two TPR domains of Hop, in complex with a peptide containing the EEVD motifs of Hsp70 and Hsp90, respectively, and biochemical data show that these motifs allow specific binding to the TPR domains [171–173].

A Hop protein preparation from rabbit reticulocyte lysate was shown to stimulate the ATPase and chaperone activity of Hsp70 in vitro. This stimulation was proposed to result from Hop's acceleration of ADP release and ATP binding [174] and the thereby accelerated substrate release. However, the ATP association rate and ADP and substrate dissociation rates in the presence of Hop were not determined, and it was therefore not clear how Hsp70 chaperone function was enhanced by Hop. More recently, it was shown that the yeast homologue Sti1 stimulates the steady-state ATPase activity of its cognate Hsp70 protein, Ssa1, up to 200-fold even in the absence of a J-domain protein [175]. This stimulation could not be explained with an acceleration of the nucleotide exchange since γ-phosphate cleavage was rate-limiting under the conditions tested. Sti1 therefore acted in these experiments like a JDP. In the same study the mammalian homologue Hop had no effect on the ATPase activity of its cognate Hsp70 protein, and the question was put forward whether the yeast and mammalian Hsp70/Hsp90 systems act through a similar mechanism. However, a recent study on DjC7/hTPR2 showed that this protein is similar to Sti1 in yeast, stimulating the ATPase activity of Hsp70 with its J-domain and interacting with Hsp70 and Hsp90 with its TPR domains [176].

Hop is an important, although not strictly essential, component of the Hsp70-Hsp90 chaperone machinery that activates nuclear receptors, protein kinases, and other regulatory proteins [177–180]. In contrast to its action on Hsp70, Hop inhibits the ATPase activity of Hsp90, thereby holding it in an open conformation, ready to bind to substrates [181, 182]. Thus, Hop coordinates the action of Hsp70 and Hsp90 during the activation of client proteins. For more detail on Hop, see Chapter 23.

15.5.3 Chip

The 35-kDa C-terminus of Hsc70-interacting protein (Chip) consists of an N-terminal TPR motif domain, a region with a high number of positively and negatively charged residues (mixed charge region), and a C-terminal RING-finger-type E3 ubiquitin-protein isopeptide ligase motif, a so-called U box [183, 184]. Based on these criteria, Chip proteins can be identified in a variety of eukaryotic organisms ranging from *C. elegans* to man. *A. thaliana* Chip shows a significant degree of identity to only the U box of human Chip. In contrast, the homology of this protein to the TPR motifs of human Chip is very weak, and it does not contain a mixed charge region. It nevertheless seems to function as a bona fide Chip protein [185]. No Chip homologues exist in *S. cerevisiae* or *Schizosaccharomyces pombe*. Chip dimerizes through its middle domain, which is predicted to form a coiled-coil, and dimerization is essential for its function [186].

The N-terminal TPR motifs are involved in the interaction of Chip with the C-terminal EEVD domain of Hsp70 and Hsp90, whereby Chip competes with other TPR domain proteins such as Hop and immunophilins. Chip inhibits the in vitro chaperone activity of Hsc70 and Hsp70, as measured by preventing the aggregation of denatured rhodanese and refolding of denatured luciferase. Chip inhibits the ATPase activity of Hsc70 without influencing ADP dissociation [184]. How Chip acts on ATPase activity and whether this inhibitory activity is responsible for the effects of Chip on the chaperone activity of Hsc70 and Hsp70 are unclear.

Chip was recently identified as an E3 ubiquitin-protein isopeptide ligase by several laboratories based on the criteria of (1) self-ubiquitination; (2) ubiquitination of several native substrates, including Hsp70, Raf-1 kinase, and unidentified proteins of a bacterial extract, and denatured substrate proteins such as heat-denatured luciferase; and (3) a physical and/or functional interaction of Chip with E2 enzymes such as UbcH5A that act specifically in the degradation of misfolded and aberrant proteins [187–190]. It is therefore believed that Chip takes part in the triage decision by ubiquitin-tagging of hopelessly misfolded proteins for degradation by the proteasome [191, 192]. In vivo and in vitro, increasing concentrations of Chip induce the ubiquitination of Hsc70 and Hsp90 substrates, including the glucocorticoid receptor, the dioxin receptor, the cystic fibrosis transmembrane conductance regulator (CFTR), and the receptor tyrosine kinase Erb2, and promote their degradation by the proteasome [193–197]. Chip also promoted the ubiquitination and degradation of a misfolded mutant form of αB-crystalline, suggesting a general function in the clearance of aberrant proteins and thereby slowing down the accumulation of large aggregates [198]. In addition, Chip was able to polyubiquitinate Bag-1 in vivo and in vitro without influencing the half-life of Bag-1 [199]. This ubiquitination was proposed to enhance the affinity of Bag-1 to the proteasome. Chip therefore not only shares with Bag-1 the potential to connect the activity of Hsp70 chaperones with proteolysis but also seems to act synergistically with Bag-1.

There is, however, also evidence for a seemingly degradation-independent function of Chip. The trafficking of the endothelial NO synthase to the plasmalemma was influenced by overexpression of Chip, and the activity of the NO synthase was thereby downregulated. In contrast, the half-life of the NO synthase was not decreased by Chip [200]. The heat shock factor HSF1 was found to be activated by the interaction with Chip and translocated into the nucleus in complex with Chip and Hsp70 to activate transcription of heat shock genes [201]. Chip may therefore have more functions to be discovered, and ubiquitination by Chip may not always doom to degradation as has been shown for other E3 ubiquitin ligases.

15.6 Concluding Remarks

In the course of evolution, the Hsp70 chaperone machinery has adapted to a large variety of protein-folding tasks by (1) variation of the Hsp70 protein itself, (2) cooperation with the modular J-domain proteins that are ideally suited as targeting factors due to the large variety of protein-protein interaction domains, and (3) an increasing number of auxiliary co-chaperones that fine-tune the action of the Hsp70-J-domain protein team or link it to other chaperones such as Hsp90. As novel co-chaperones for the Hsp70 chaperones are found, novel protein-folding tasks for the Hsp70 folding machine may be discovered.

15.7 Experimental Protocols

All purification steps are performed on ice or at 4 °C. In particular, ammonium sulfate precipitation is carried out on ice as described in Ref. [202], using fine powdered ammonium sulfate. For buffer changes, samples are generally dialyzed overnight against 100 sample volumes of the respective buffer. Purified proteins are aliquoted, quick frozen in liquid N_2, and stored at −80 °C.

15.7.1 Hsp70s

Many of the classical Hsp70 proteins can be purified by a combination of anion-exchange chromatography and affinity chromatography over an ATP-agarose column [203, 204], as described in the following for the *E. coli* homologue DnaK. DnaK-overproducing cells are lysed in 50 mM TRIS/HCl pH 7.6, 18 mM Spermidin, 10 mM $(NH_4)_2SO_4$, 5 mM EDTA, 5 mM DTT, 10% sucrose, 1 mM phenylmethylsulfonyl fluoride using a French press or sonication. After clarifying centrifugations (20 min, 30 000 g; 2 h, 100 000 g), the cell extract is subjected to a differential ammonium sulfate precipitation (30% and 60% saturation). The second DnaK-containing ammonium sulfate pellet is resuspended in buffer A (25 mM HEPES/KOH pH 7.6, 50 mM KCl, 5 mM $MgCl_2$, 10 mM β-mercaptoethanol, 1 mM EDTA), dialyzed overnight against buffer A, and applied to anion-exchange chromatography over a DEAE Sepharose column equilibrated in the same buffer. DnaK is eluted using a linear KCl gradient of 0–500 mM. The DnaK-containing fractions are pooled and incubated in a batch procedure with an appropriate amount of ATP-agarose (ATP linked over C8; 1 mL per 10 mg DnaK, swollen in water, washed in 10 bed volumes of buffer A, 10 bed volumes of buffer B, and again 10 bed volumes of buffer A) rotating end over end at 4 °C for 30 min. The slurry is poured into a plastic, disposable 10-mL column (BioRad), and unspecific contaminants are removed with subsequent washes of five column volumes buffer A and five column volumes buffer A plus 1 M KCl and re-equilibrated to buffer A. For the elution of DnaK, the ATP-agarose is gently resuspended in the column in one column volume buffer A plus ATP (5 mM) and incubated at 4 °C for 45 min. After collecting the eluate, the column is eluted with a second column volume of ATP-containing buffer A. To remove bound substrates, the pooled and concentrated DnaK-containing fractions of the ATP-agarose are further purified by size-exclusion chromatography in the presence of ATP (1 mM) and finally concentrated by chromatography over a strong anion-exchange column (e.g., resource Q; Amersham Bioscience). Alternative purification methods for DnaK, not including an ATP-agarose step, have also been used successfully [33].

The protocol described here works well for many Hsp70 proteins from overproducing sources, whereas for some proteins the salt concentrations of the anion-exchange chromatography have to be adjusted. The protocol can even be adapted to the purification from sources with a natural Hsp70 content (e.g., Ref. [205]). However, not all Hsp70 proteins can be purified by this protocol because the affinity for nucleotide varies substantially between the different Hsp70s; consequently, not all Hsp70s bind tightly enough to ATP-agarose (see, e.g., Refs. [19, 35, 206]). Purification of Hsp70 proteins using a tag (e.g., His_6-tag) is not advisable because tags on both ends influence the ATPase activity and interaction with co-chaperones.

15.7.2
J-Domain Proteins

E. coli DnaJ can be purified by taking advantage of its high isoelectric point (pI $\approx$ 8) using chromatography on a cation-exchange resin and hydroxyapatite [100, 207]. One of the major obstacles in DnaJ purification is the hydrophobicity of the protein and its tendency to aggregate and to associate with membranes. To overcome these obstacles, urea (2 M) and the detergent Brij 58 (0.6%) is used.

DnaJ-overproducing *E. coli* cells, resuspended in lysis buffer (50 mM TRIS/HCl, pH 8, 10 mM DTT, 0.6% (w/v) Brij 58, 1 mM phenylmethylsulfonyl fluoride, 0.8 mg mL^{-1} lysozyme), are lysed using a French press or sonication, and cell debris is removed by centrifugation (20 min, 30 000 g; 2 h, 100 000 g). The supernatant is diluted by addition of one volume of buffer A (50 mM sodium phosphate buffer pH 7, 5 mM DTT, 1 mM EDTA, 0.1% (w/v) Brij 58), and DnaJ is precipitated by addition of $(NH_4)_2SO_4$ to a final concentration of 65%. The ammonium sulfate pellet is dissolved in buffer B (50 mM sodium phosphate buffer pH 7, 5 mM DTT, 1 mM EDTA, 0.1% (w/v) Brij 58, 2 M urea) and dialyzed against buffer B. DnaJ is loaded onto an equilibrated strong cation-exchange column (e.g., Poros SP20, Applied Biosystems), washed with buffer B, and eluted with a linear gradient of 0–666 mM KCl (15 column volumes). DnaJ-containing fractions are pooled and dialyzed against buffer C (50 mM TRIS/HCl, pH 7.5, 2 M urea, 0.1% (w/v) Brij 58, 5 mM DTT, 50 mM KCl). The dialyzed sample is loaded onto a Bio-Gel HT hydroxyapatite column equilibrated in buffer C and washed with one column volume buffer C + 1 M KCl and two column volumes buffer C. DnaJ is eluted with a linear gradient of one column volume 0–50% buffer D (50 mM TRIS/HCl, pH 7.5, 2 M urea, 0.1% (w/v) Brij 58, 5 mM DTT, 50 mM KCl, 600 mM KH_2PO_4) and two column volumes of 50% buffer D. The DnaJ-containing fractions are pooled and dialyzed against buffer E (50 mM TRIS/HCl, pH 7.7, 100 mM KCl). This protocol works equally well for the human J-domain proteins DjB1/Hdj-1 and DjA1/Hdj-2 [62].

The J-domain of DnaJ can be purified according to a protocol from Karzai and McMacken [25]. The cleared extract from a J-domain-overproducing *E. coli* strain in buffer A (50 mM HEPES/NaOH pH 7.6, 2 mM DTT, 1 M NaCl, 2 mM $MgCl_2$) is subjected to ammonium sulfate precipitation by addition of solid ammonium sulfate to 75% saturation and centrifugation (1 h, 30 000 g). The J-domain-containing supernatant is dialyzed extensively against buffer B (50 mM HEPES/NaOH, pH 7.6, 2 mM DTT, 25 mM NaCl, 10% glycerol), applied onto an equilibrated cation-exchange column, washed with buffer B, and eluted with a linear gradient of 0.025–0.7 M NaCl in buffer B. The J-domain-containing fractions are applied to a Bio-Gel HT hydroxyapatite column equilibrated in buffer C (50 mM potassium phosphate buffer pH 6.8, 2 mM DTT, 0.15 M KCl, 10% glycerol), washed with buffer C, and eluted with a linear gradient of 0.05–0.5 M potassium phosphate in buffer C. The purified J-domain is dialyzed against buffer D (25 mM HEPES/KOH, pH 7.6, 100 mM KCl, 2 mM DTT).

15.7.3
GrpE

GrpE can be purified by affinity chromatography on a DnaK column [114] or, alternatively, by a combination of anion-exchange chromatography, hydroxyapatite chromatography, and size-exclusion chromatography [109]. In the first protocol cells of the *E. coli* strain C600 dnaK103 [208] are resuspended in 5 mL per gram cell paste buffer A (50 mM TRIS/HCl pH 8, 0.06 μg mL^{-1} spermidine, 6.6 mM EDTA, 6.6 mM DTT, 200 mM ammonium sulfate, 10% sucrose, 1 mM phenylmethylsulfonyl fluoride)and lysed using a French press or sonication, and cell debris is removed by centrifugation (30 min 100 000 g). Ammonium sulfate is added to the supernatant to a final concentration of 0.35 g mL^{-1}. The GrpE-containing precipitate is resuspended in 2 mL buffer B (25 mM HEPES/KOH pH 8, 1 mM EDTA, 10 mM 2-mercaptoethanol, 10% glycerol) and dialyzed against three times 1 L of buffer B. The dialyzed sample is diluted with buffer B to a final volume of 20 mL and applied to a 20-mL equilibrated DnaK column previously prepared by cross-linking purified DnaK to an activated chromatographic material (e.g., Affi-Gel 10; BioRad) according to the manufacturer's recommendations. The column is washed successively with 200 mL of 50-mM KCl in buffer B, 70 mL of 0.5-M KCl in buffer B, and 70 mL of 2-M KCl in buffer B. The column is re-equilibrated with 100 mL of 50-mM KCl in buffer B, followed by elution of GrpE with 50 mL of 50-mM KCl, 20 mM $MgCl_2$, and 10 mM ATP in buffer B (pH adjusted to 7.6). The GrpE fraction is dialyzed three times against 2 L buffer C (25 mM HEPES/KOH pH 8, 50 mM KCl, 0.1 mM EDTA, 10 mM 2-mercaptoethanol, 20% glycerol). The number and duration of final dialysis steps are particularly important to remove all traces of ATP.

In the second protocol GrpE is purified from overproducing cells of a ΔdnaK strain (e.g., BB1553) [209]. The cell pellet is resuspended in lysis buffer (50 mM TRIS/HCl, pH 7.5, 100 mM KCl, 2 mM DTT, 3 mM EDTA, 1 mM phenylmethylsulfonyl fluoride) at 5 mL per gram cell paste, the cells are lysed, and cell debris is removed by centrifugation (20 min, 30 000 g; 90 min, 100 000 g). Ammonium sulfate is added to the supernatant to a final concentration of 0.35 g mL^{-1}, and the precipitate is dissolved in buffer A (50 mM TRIS/HCl, pH 7.5, 100 mM KCl, 1 mM DTT, 1 mM EDTA, 10% glycerol) and dialyzed twice against buffer A. The dialyzed sample is applied onto a DEAE-Sepharose column equilibrated with buffer A, and GrpE is eluted with a linear gradient of one column volume 0–30% buffer B (50 mM TRIS/HCl, pH 7.5, 1 M KCl, 1 mM DTT, 1 mM EDTA, 10% glycerol). The pooled GrpE-containing fractions are dialyzed against buffer C (10 mM potassium phosphate buffer, pH 6.8, 1 mM DTT, 10% glycerol), applied to an equilibrated Bio-Gel HT hydroxyapatite column, and eluted by stepwise increase of the phosphate concentration (4%, 8%, 15%, 20%, and 100% buffer D; 0.5 M potassium phosphate buffer, pH 6.8, 1 mM DTT, 10% glycerol). The pooled GrpE-containing fractions are applied to a Superdex 200 gel-filtration column equilibrated in buffer A and afterwards are concentrated on a strong anion-exchange column (e.g., Resource Q, Amersham Bioscience) using a steep gradient of buffer B. The GrpE fraction is dialyzed against buffer E (25 mM HEPES/KOH pH 8, 50 mM KCl, 0.1 mM EDTA, 10 mM 2-mercaptoethanol, 20% glycerol). If GrpE is purified from a wild-type *E. coli* strain, co-purifying DnaK can be removed by extensive washing of the Bio-Gel HT hydroxyapatite-bound GrpE with buffer C containing 10 mM ATP before stepwise elution. In addition, the size-exclusion chromatography is carried out in the presence of 1 mM ATP.

15.7.4 Bag-1

Bag-1 is purified by a combination of anion exchange, hydroxyapatite, and size-exclusion chromatography according to a published protocol [62]. The Bag-1-overproducing *E. coli* cells are lysed in lysis buffer (50 mM TRIS/HCl pH 7.6, 18 mM spermidine, 100 mM $(NH_4)_2SO_4$, 5 mM EDTA, 5 mM DTT, 10% sucrose, 1 mM phenylmethylsulfonyl fluoride) using a French press or sonication, and cell debris is removed by centrifugation (20 min, 30 000 g; 2 h, 100 000 g). The supernatant is subjected to a fractionated ammonium sulfate precipitation on ice by adding first 226 g L^{-1}, removing the precipitated proteins by centrifugation and then adding again 187 g L^{-1} to the supernatant to reach a final ammonium sulfate concentration of 70% saturation. The Bag-1-containing precipitate is resuspended in buffer A (20 mM MOPS/KOH pH 7.2, 50 mM KCl, 1 mM EDTA, 1 mM 2-mercaptoethanol), dialyzed extensively against buffer A, and applied to an anion-exchange chromatography column (DEAE-Sepharose) equilibrated in buffer A. The column is washed with four column volumes of buffer A and Bag-1 is eluted with a linear gradient of 50–500 mM KCl over 10 column volumes. The Bag-1-containing fractions are pooled, dialyzed against buffer B (50 mM potassium phosphate buffer, pH 7, 1 mM 2-mercaptoethanol), and applied onto an equilibrated Bio-Gel HT hydroxyapatite (BioRad) column. The column is washed with buffer B, Bag-1 is eluted using a linear gradient of 50–500 mM potassium phosphate in buffer B over five column volumes, and the Bag-1-containing fractions are dialyzed against buffer C (20 mM MOPS/KOH pH 7.2, 20 mM KCl, 0.5 mM EDTA, 1 mM 2-mercaptoethanol). Contaminating ATPases are removed by incubation of the dialyzed Bag-1 sample with 0.5 mL ATP-agarose (swollen and equilibrated as described above in Section 15.7.1) rotating end over end at 4 °C for 30 min. The agarose is pelleted and the supernatant is transferred to a new tube. Bag-1 is precipitated by addition of ammonium sulfate to a final concentration of 80%, and the precipitate is resuspended in a minimal volume of buffer C and applied to size-exclusion chromatography on a Superdex 200 column (Amersham Bioscience). The Bag-1-containing fractions are concentrated by chromatography on a strong anion-exchange column (MonoQ HR5/5; Amersham Bioscience) and eluted with a linear gradient of 100–500 mM KCl over 10 column volumes. Bag-1 is dialyzed against buffer C.

15.7.5 Hip

Hip is purified by a combination of anion-exchange, hydroxyapatite, and size-exclusion chromatography following a modified protocol by Höhfeld and Jentsch [119]. The Hip-overproducing *E. coli* cells are lysed in lysis buffer (50 mM TRIS/HCl pH 7.6, 18 mM spermidine, 100 mM $(NH_4)_2SO_4$, 5 mM EDTA, 5 mM DTT, 10% sucrose, 1 mM phenylmethylsulfonyl fluoride) using a French press or sonication, and cell debris is removed by centrifugation (20 min, 30 000 g; 2 h, 100 000 g). Ammonium sulfate is added to the supernatant to a final concentration of 0.29 g mL^{-1} (50% saturation), and the Hip-containing precipitate is dissolved in buffer A (20 mM MOPS/KOH pH 7.2, 50 mM KCl, 1 mM EDTA, 1 mM 2-mercaptoethanol), dialyzed extensively against buffer A, and applied to an anion-exchange chromatography column (DEAE-Sepharose) equilibrated in buffer A. The column is washed with four column volumes of buffer A and Hip is eluted with a linear gradient of 50–500 mM KCl over 10 column volumes. The Hip-containing fractions are pooled, dialyzed against buffer B (50 mM potassium phosphate buffer, pH 7, 1 mM 2-mercaptoethanol), and applied onto an equilibrated Bio-Gel HT hydroxyapatite (BioRad) column. The column is washed with buffer B, Hip is eluted using a linear gradient of 50–500 mM potassium phosphate buffer over five column volumes, and the Hip-containing fractions are pooled and precipitated by addition of ammonium sulfate to a final concentration of 80%. The precipitate is resuspended in a minimal volume of buffer C (20 mM MOPS/KOH pH 7.2, 20 mM KCl, 0.5 mM EDTA, 1 mM 2-mercaptoethanol) and applied to size-exclusion chromatography on a Superdex 200 column (Amersham Bioscience) equilibrated in the same buffer. The Hip-containing fractions are concentrated by chromatography on a strong anion-exchange column (MonoQ HR5/5; Amersham Bioscience) and eluted with a linear gradient of 100–1000 mM KCl over 10 column volumes. The purified Hip is dialyzed against buffer C.

15.7.6 Hop

Hop is purified according to a protocol by Buchner and coworkers [210] using anion-exchange chromatography, hydroxyapatite, and size-exclusion chromatography. The cell pellet from a human Hop-overproducing *E. coli* strain is resuspended in 3 mL of 10 mM TRIS/HCl, pH 7.5 per gram of wet cell pellet containing 1 mM phenylmethylsulfonyl fluoride, *p*-aminobenzoic acid (PABA 520 μg mL^{-1}), and pepstatin A (1 μg mL^{-1}) as protease inhibitors. The cells are lysed and the suspension is clarified by centrifugation at 40 000 g for 60 min at 4 °C. The supernatant is loaded onto a 40-mL Q-Sepharose column (Amersham Bioscience) equilibrated with lysis buffer. Bound protein is eluted from the column with a linear gradient of 0–15 M NaCl over 10 column volumes. Hop elutes around 0.13 M NaCl. Hop-containing fractions are pooled and concentrated by ultrafiltration. The concentrate is dialyzed overnight against 10 mM potassium phosphate buffer, pH 7.5. The Hop pool is then loaded onto a pre-equilibrated 2-mL FPLC hydroxyapatite column (BioRad) and eluted with a linear potassium phosphate buffer gradient ranging from 10 to 300 mM over 15 column volumes (pH 7.5). Hop elutes between 80 and 220 mM phosphate. The Hop-containing fractions are pooled, concentrated as described above, and loaded onto a 120-mL Superdex 200 preparative-grade gel-filtration column (Amersham Bioscience) equilibrated with two column volumes of 10 mM TRIS/HCl, pH 7.5, 400 mM NaCl. Hop-containing fractions are pooled, concentrated as described above to 5 mg mL^{-1}, and finally dialyzed against 40 mM HEPES-KOH, pH 7.5.

15.7.7 Chip

Wild-type untagged human Chip is purified from overproducing *E. coli* cells using three steps: anion-exchange, cation-exchange, and size-exclusion chromatography according to a published protocol [186]. The cell pellet from the Chip-producing

E. coli strain is resuspended in buffer A (50 mM HEPES, pH 7.0, 50 mM KCl, 5 mM DTT, 10% glycerol, 1 mM phenylmethylsulfonyl fluoride) containing 0.4 mg mL^{-1} lysozyme, lysed by sonification, and clarified by ultracentrifugation at 100 000 g for 1 h. The supernatant is precipitated by addition of ammonium sulfate to a final saturation of 40%. The pellet is resuspended in buffer A, dialyzed against the same buffer, and loaded onto a DEAE-Sepharose column (Sigma) that is equilibrated with the same buffer. The column is washed with two column volumes of buffer A, and bound proteins are eluted with a linear gradient from 0% to 100% buffer B (50 mM HEPES/KOH, pH 7.0, 1 M KCl, 5 mM DTT, 10% glycerol, 1 mM phenylmethylsulfonyl fluoride) over two column volumes. Chip-containing fractions are pooled and dialyzed against buffer A and loaded onto an SP Sepharose column (Amersham Bioscience). The column is washed with two column volumes of buffer A, and bound proteins are eluted with a linear gradient of 0–100% buffer B over five column volumes. The peak fractions are pooled, concentrated, and applied to a Superdex 200 high-load gel-filtration column (Amersham Biosciences) equilibrated with buffer A. Highly pure Chip-containing fractions are pooled, frozen in liquid nitrogen, and stored at −80 °C.

References

1 R. J. Fletterick, D. J. Bates, and T. A. Steitz. The structure of a yeast hexokinase monomer and its complexes with substrates at 2.7-A resolution. *Proc. Natl. Acad. Sci. USA* **1975**, *72*, 38–42.

2 W. Kabsch, H. G. Mannherz, D. Suck, E. F. Pai, and K. C. Holmes. Atomic structure of the actin:DNase I complex. *Nature* **1990**, *347*, 37–44.

3 J. H. Hurley, H. R. Faber, D. Worthylake, N. D. Meadow, S. Roseman, D. W. Pettigrew, and S. J. Remington. Structure of the regulatory complex of *E. coli* IIIGlc with glycerol kinase. *Science* **1993**, *259*, 673–677.

4 D. P. Easton, Y. Kaneko, and J. R. Subjeck. The Hsp110 and Grp170 stress proteins: newly recognized relatives of the Hsp70s. *Cell Stress & Chaperones* **2000**, *5*, 276–290.

5 H. J. Oh, D. Easton, M. Murawski, Y. Kaneko, and J. R. Subjeck. The chaperoning activity of hsp110. Identification of functional domains by use of targeted deletions. *J. Biol. Chem.* **1999**, *274*, 15712–15718.

6 M. Gautschi, H. Lilie, U. Fünfschilling, A. Mun, S. Ross, T. Lithgow, P. Rucknagel, and S. Rospert. RAC, a stable ribosome-associated complex in yeast formed by the DnaK-DnaJ homologs Ssz1p and zuotin. *Proc. Natl. Acad. Sci. USA* **2001**, *98*, 3762–3767.

7 H. Hundley, H. Eisenman, W. Walter, T. Evans, Y. Hotokezaka, M. Wiedmann, and E. Craig. The in vivo function of the ribosome-associated Hsp70, Ssz1, does not require its putative peptide-binding domain. *Proc Natl Acad Sci USA* **2002**, *99*, 4203–4208.

8 G. A. Otterson, G. C. Flynn, R. A. Kratzke, A. Coxon, P. G. Johnston, and F. J. Kaye. Stch encodes the 'ATPase core' of a microsomal stress 70 protein. *Embo J* **1994**, *13*, 1216–1225.

9 G. A. Otterson, and F. J. Kaye. A 'core ATPase', Hsp70-like structure is conserved in human, rat, and C. elegans STCH proteins. *Gene* **1997**, *199*, 287–292.

10 D. Schmid, A. Baici, H. Gehring,

and P. Christen. Kinetics of molecular chaperone action. *Science* **1994**, *263*, 971–973.

11 M. P. Mayer, H. Schröder, S. Rüdiger, K. Paal, T. Laufen, and B. Bukau. Multistep mechanism of substrate binding determines chaperone activity of Hsp70. *Nature Struct. Biol.* **2000**, *7*, 586–593.

12 S. M. Gisler, E. V. Pierpaoli, and P. Christen. Catapult Mechanism Renders the Chaperone Action of Hsp70 Unidirectional. *J. Mol. Biol.* **1998**, *279*, 833–840.

13 J. S. McCarty, A. Buchberger, J. Reinstein, and B. Bukau. The role of ATP in the functional cycle of the DnaK chaperone system. *J. Mol. Biol.* **1995**, *249*, 126–137.

14 M. Zylicz, J. H. LeBowitz, R. McMacken, and C. Georgopoulos. The dnaK protein of *Escherichia coli* possesses an ATPase and autophosphorylating activity and is essential in an in vitro DNA replication system. *Proc. Natl. Acad. Sci. USA* **1983**, *80*, 6431–6435.

15 R. Jordan, and R. McMacken. Modulation of the ATPase activity of the molecular chaperone DnaK by peptides and the DnaJ and GrpE heat shock proteins. *J. Biol. Chem.* **1995**, *270*, 4563–4569.

16 D. R. Palleros, K. L. Reid, L. Shi, W. J. Welch, and A. L. Fink. ATP-induced protein-Hsp70 complex dissociation requires K^+ but not ATP hydrolysis. *Nature* **1993**, *365*, 664–666.

17 J.-H. Ha, E. R. Johnson, D. B. McKay, M. C. Sousa, S. Takeda, and S. M. Wilbanks. Structure and mechanism of Hsp70 proteins. In *Molecular Chaperones and Folding Catalysts. Regulation, Cellular Function and Mechanism* (B. Bukau, ed), pp. 573–607, Harwood Academic Publishers, Amsterdam, **1999**.

18 J. J. Silberg, and L. E. Vickery. Kinetic characterization of the ATPase cycle of the molecular chaperone Hsc66 from *Escherichia coli*. *J. Biol. Chem.* **2000**, *275*, 7779–7786.

19 C. Kluck, H. Pazelt, P. Genevaux, D. Brehmer, W. Rist, J. Schneider-Mergener, B. Bukau, and M. P. Mayer. Structure-function analysis of HscC, the *Escherichia coli* member of a novel subfamily of specialized Hsp70 chaperones. *J. Biol. Chem.* **2002**, *277*, 41060–41069.

20 J.-H. Ha, and D. B. McKay. ATPase kinetics of recombinant bovine 70 kDa heat shock cognate protein and its amino-terminal ATPase domain. *Biochemistry* **1994**, *33*, 14625–14635.

21 P. Lopez-Buesa, C. Pfund, and E. A. Craig. The biochemical properties of the ATPase activity of a 70-kDa heat shock protein (Hsp70) are governed by the C-terminal domains. *Proc. Natl. Acad. Sci. USA* **1998**, *95*, 15253–15258.

22 D. Klostermeier, R. Seidel, and J. Reinstein. The functional cycle and regulation of the *Thermus thermophilus* DnaK chaperone system. *J. Mol. Biol.* **1999**, *287*, 511–525.

23 G. C. Flynn, T. G. Chappell, and J. E. Rothman. Peptide binding and release by proteins implicated as catalysts of protein assembly. *Science* **1989**, *245*, 385–390.

24 K. Liberek, J. Marszalek, D. Ang, C. Georgopoulos, and M. Zylicz. *Escherichia coli* DnaJ and GrpE heat shock proteins jointly stimulate ATPase activity of DnaK. *Proc. Natl. Acad. Sci. USA* **1991**, *88*, 2874–2878.

25 A. W. Karzai, and R. McMacken. A bipartite signaling mechanism involved in DnaJ-mediated activation of the *Escherichia coli* DnaK protein. *J. Biol. Chem.* **1996**, *271*, 11236–11246.

26 T. Laufen, M. P. Mayer, C. Beisel, D. Klostermeier, J. Reinstein, and B. Bukau. Mechanism of regulation of Hsp70 chaperones by DnaJ co-chaperones. *Proc. Natl. Acad. Sci. USA* **1999**, *96*, 5452–5457.

27 B. Misselwitz, O. Staeck, and T. A. Rapoport. J proteins catalytically activate Hsp70 molecules to trap a wide range of peptide sequences. *Mol Cell* **1998**, *2*, 593–603.

28 K. G. Hoff, J. J. Silberg, and L. E. Vickery. Interaction of the iron-sulfur cluster assembly protein IscU with the Hsc66/Hsc20 molecular chaperone

system of *Escherichia coli*. *Proc. Natl. Acad. Sci. USA* **2000**, *97*, 7790–7795.

29 W. Barouch, K. Prasad, L. Greene, and E. Eisenberg. Auxilin-induced interaction of the molecular chaperone Hsc70 with clathrin baskets. *Biochemistry* **1997**, *36*, 4303–4308.

30 B. Bukau, and A. L. Horwich. The Hsp70 and Hsp60 chaperone machines. *Cell* **1998**, *92*, 351–366.

31 A. Szabo, T. Langer, H. Schröder, J. Flanagan, B. Bukau, and F. U. Hartl. The ATP hydrolysis-dependent reaction cycle of the *Escherichia coli* Hsp70 system-DnaK, DnaJ and GrpE. *Proc. Natl. Acad. Sci. USA* **1994**, *91*, 10345–10349.

32 L. Packschies, H. Theyssen, A. Buchberger, B. Bukau, R. S. Goody, and J. Reinstein. GrpE accelerates nucleotide exchange of the molecular chaperone DnaK with an associative displacement mechanism. *Biochemistry* **1997**, *36*, 3417–3422.

33 H. Theyssen, H.-P. Schuster, B. Bukau, and J. Reinstein. The second step of ATP binding to DnaK induces peptide release. *J. Mol. Biol.* **1996**, *263*, 657–670.

34 R. Russell, R. Jordan, and R. McMacken. Kinetic Characterization of the ATPase Cycle of the DnaK Molecular Chaperone. *Biochemistry* **1998**, *37*, 596–607.

35 D. Brehmer, S. Rüdiger, C. S. Gässler, D. Klostermeier, L. Packschies, J. Reinstein, M. P. Mayer, and B. Bukau. Tuning of chaperone activity of Hsp70 proteins by modulation of nucleotide exchange. *Nat Struct Biol.* **2001**, *8*, 427–432.

36 D. R. Palleros, K. L. Reid, J. S. McCarty, G. C. Walker, and A. L. Fink. DnaK, hsp73, and their molten globules. Two different ways heat shock proteins respond to heat. *J. Biol. Chem.* **1992**, *267*, 5279–5285.

37 B. Banecki, M. Zylicz, E. Bertoli, and F. Tanfani. Structural and functional relationships in DnaK and DnaK756 heat-shock proteins from *Escherichia coli*. *J. Biol. Chem.* **1992**, *267*, 25051–25058.

38 K. Liberek, D. Skowyra, M. Zylicz, C. Johnson, and C. Georgopoulos. The *Escherichia coli* DnaK chaperone, the 70-kDa heat shock protein eukaryotic equivalent, changes conformation upon ATP hydrolysis, thus triggering its dissociation from a bound target protein. *J. Biol. Chem.* **1991**, *266*, 14491–14496.

39 A. Buchberger, H. Theyssen, H. Schröder, J. S. McCarty, G. Virgallita, P. Milkereit, J. Reinstein, and B. Bukau. Nucleotide-induced conformational changes in the ATPase and substrate binding domains of the DnaK chaperone provide evidence for interdomain communication. *J. Biol. Chem.* **1995**, *270*, 16903–16910.

40 J. Wei, J. R. Gaut, and L. M. Hendershot. in vitro Dissociation of BiP-Peptide Complexes Requires a Conformational Change in BiP after ATP Binding but Does Not Require ATP Hydrolysis. *J. Biol. Chem.* **1995**, *270*, 26677–26682.

41 J. Wei, and L. M. Hendershot. Characterization of the Nucleotide Binding Properties and ATPase Activity of Recombinant Hamster BiP Purified from Bacteria. *J. Biol. Chem.* **1995**, *270*, 26670–26676.

42 S. M. Wilbanks, L. Chen, H. Tsuruta, K. O. Hodgson, and D. B. McKay. Solution small-angle X-ray scattering study of the molecular chaperone hsc70 and its subfragments. *Biochem.* **1995**, *34*, 12095–12106.

43 K. L. Fung, L. Hilgenberg, N. M. Wang, and W. J. Chirico. Conformations of the Nucleotide and Polypeptide Binding Domains of a Cytosolic Hsp70 Molecular Chaperone Are Coupled. *J. Biol. Chem.* **1996**, *271*, 21559–21565.

44 J.-H. Ha, and D. B. McKay. Kinetics of Nucleotide-Induced Changes in the Tryptophane Fluorescence of the Molecular Chaperone Hsc70 and Its Subfragments Suggest the ATP-Induced Conformational Change Follows Initial ATP Binding. *Biochemistry* **1995**, *34*, 11635–11644.

45 F. Moro, V. Fernandez, and A. Muga. Interdomain interaction through helices A and B of DnaK peptide binding domain. *FEBS Lett* **2003**, *533*, 119–123.
46 B. Banecki, and M. Zylicz. Real Time Kinetics of the DnaK/DnaJ/GrpE Molecular Chaperone Machine Action. *J. Biol. Chem.* **1996**, *271*, 6137–6143.
47 S. V. Slepenkov, and S. N. Witt. Kinetics of the Reactions of the *Escherichia coli* Molecular Chaperone DnaK with ATP: Evidence that a Three-Step Reaction Precedes ATP Hydrolysis. *Biochemistry* **1998**, *37*, 1015–1024.
48 S. Blond-Elguindi, S. E. Cwirla, W. J. Dower, R. J. Lipshutz, S. R. Sprang, J. F. Sambrook, and M.-J. H. Gething. Affinity panning of a library of peptides displayed on bacteriophages reveals the binding specificity of BiP. *Cell* **1993**, *75*, 717–728.
49 A. Gragerov, L. Zeng, X. Zhao, W. Burkholder, and M. E. Gottesman. Specificity of DnaK-peptide binding. *J. Mol. Biol.* **1994**, *235*, 848–854.
50 A. M. Fourie, J. F. Sambrook, and M. J. Gething. Common and divergent peptide binding specificities of Hsp70 molecular chaperones. *J. Biol. Chem.* **1994**, *269*, 30470–30478.
51 A. M. Fourie, T. R. Hupp, D. P. Lane, B. C. Sang, M. S. Barbosa, J. F. Sambrook, and M. J. Gething. HSP70 binding sites in the tumor suppressor protein p53. *J Biol Chem* **1997**, *272*, 19471–19479.
52 G. Knarr, M. J. Gething, S. Modrow, and J. Buchner. BiP binding sequences in antibodies. *J Biol Chem* **1995**, *270*, 27589–27594.
53 S. Rüdiger, L. Germeroth, J. Schneider-Mergener, and B. Bukau. Substrate specificity of the DnaK chaperone determined by screening cellulose-bound peptide libraries. *EMBO J.* **1997**, *16*, 1501–1507.
54 K. G. Hoff, D. T. Ta, T. L. Tapley, J. J. Silberg, and L. E. Vickery. Hsc66 substrate specificity is directed toward a discrete region of the iron-sulfur cluster template protein IscU. *J Biol Chem* **2002**, *277*, 27353–27359.
55 S. Rüdiger, M. P. Mayer, J. Schneider-Mergener, and B. Bukau. Modulation of the specificity of the Hsp70 chaperone DnaK by altering a hydrophobic arch. *J. Mol. Biol.* **2000**, *304*, 245–251.
56 M. P. Mayer, T. Laufen, K. Paal, J. S. McCarty, and B. Bukau. Investigation of the interaction between DnaK and DnaJ by surface plasmon resonance spectroscopy. *J. Mol. Biol.* **1999**, *289*, 1131–1144.
57 J. S. McCarty, S. Rüdiger, H.-J. Schönfeld, J. Schneider-Mergener, K. Nakahigashi, T. Yura, and B. Bukau. Regulatory region C of the *E. coli* heat shock transcription factor, σ^{32}, constitutes a DnaK binding site and is conserved among eubacteria. *J. Mol. Biol.* **1996**, *256*, 829–837.
58 E. V. Pierpaoli, E. Sandmeier, A. Baici, H.-J. Schönfeld, S. Gisler, and P. Christen. The power stroke of the DnaK/DnaJ/GrpE molecular chaperone system. *J. Mol. Biol.* **1997**, *269*, 757–768.
59 E. V. Pierpaoli, S. M. Gisler, and P. Christen. Sequence-specific rates of interaction of target peptides with the molecular chaperones DnaK and DnaJ. *Biochemistry* **1998**, *37*, 16741–16748.
60 S. Takeda, and D. B. McKay. Kinetics of Peptide Binding to the Bovine 70 kDa Heat Shock Cognate Protein, a Molecular Chaperone. *Biochemistry* **1996**, *35*, 4636–4644.
61 M. Pellecchia, D. L. Montgomery, S. Y. Stevens, C. W. Vander Kooi, H. Feng, L. M. Gierasch, and E. R. P. Zuiderweg. Structural insights into substrate binding by the molecular chaperone DnaK. *Nat Struct Biol.* **2000**, *7*, 298–303.
62 C. S. Gässler, T. Wiederkehr, D. Brehmer, B. Bukau, and M. P. Mayer. Bag-1M accelerates nucleotide release for human Hsc70 and Hsp70 and can act concentration-dependent as positive and negative cofactor. *J Biol Chem* **2001**, *276*, 32538–32544.
63 M. I. E. Cheetham, and A. J. Caplan.

Structure, function and evolution of DnaJ: conservation and adaptation of chaperone function. *Cell Stress Chap.* **1998**, *3*, 28–36.

64 W. L. Kelley. The J-domain family and the recruitment of chaperone power. *Trends Biochem Sci* **1998**, *23*, 222–227.

65 T. Laufen, U. Zuber, A. Buchberger, and B. Bukau. DnaJ proteins. In *Molecular chaperones in proteins: Structure, function, and mode of action* (A. L. Fink, and Y. Goto, eds), pp. 241–274, Marcel Dekker, New York, **1998**.

66 K. Ohtsuka, and M. Hata. Mammalian HSP40/DNAJ homologs: cloning of novel cDNAs and a proposal for their classification and nomenclature. *Cell Stress & Chaperones* **2000**, *5*, 98–112.

67 A. J. Caplan, J. Tsai, P. J. Casey, and M. G. Douglas. Farnesylation of YDJ1p is required for function at elevated growth temperatures in Saccharomyces cerevisiae. *J. Biol. Chem.* **1992**, *267*, 18890–18895.

68 J. C. Venter, M. D. Adams, E. W. Myers, et al. The sequence of the human genome. *Science* **2001**, *291*, 1304–1351.

69 J. Jiang, A. B. Taylor, K. Prasad, Y. Ishikawa-Brush, P. J. Hart, E. M. Lafer, and R. Sousa. Structure-function analysis of the auxilin J-domain reveals an extended Hsc70 interaction interface. *Biochemistry* **2003**, *42*, 5748–5753.

70 M. Pellecchia, T. Szyperski, D. Wall, C. Georgopoulos, and K. Wüthrich. NMR structure of the J-domain and the Gly/Phe-rich region of the *Escherichia coli* DnaJ chaperone. *J. Mol. Biol.* **1996**, *260*, 236–250.

71 Y. Q. Qian, D. Patel, F. U. Hartl, and D. J. McColl. Nuclear magnetic resonance solution structure of the human Hsp40 (HDJ-1) J-domain. *J. Mol. Biol.* **1996**, *260*, 224–235.

72 T. Szyperski, M. Pellecchia, D. Wall, C. Georgopoulos, and K. Wüthrich. NMR structure determination of the *Escherichia coli* DnaJ molelcular chaperone: secondary structure and backbone fold of the N-terminal region (residues 2–108) containing the highly conserved J domain. *Proc. Natl. Acad. Sci. USA* **1994**, *91*, 11343–11347.

73 M. V. Berjanskii, M. I. Riley, A. Xie, V. Semenchenko, W. R. Folk, and S. R. Van Doren. NMR structure of the N-terminal J domain of murine polyomavirus T antigens. Implications for DnaJ-like domains and for mutations of T antigens. *J. Biol. Chem.* **2000**, *275*, 36094–36103.

74 M. Martinez-Yamout, G. B. Legge, O. Zhang, P. E. Wright, and H. J. Dyson. Solution structure of the cysteine-rich domain of the *Escherichia coli* chaperone protein DnaJ. *J. Mol. Biol.* **2000**, *300*, 805–818.

75 B. Sha, S. Lee, and D. M. Cyr. The crystal structure of the peptide-binding fragment from the yeast Hsp40 protein Sis1. *Structure* **2000**, *15*, 799–807.

76 J. Li, X. Qian, and B. Sha. The crystal structure of the yeast Hsp40 Ydj1 complexed with its peptide substrate. *Structure (Camb)* **2003**, *11*, 1475–1483.

77 R. Russell, A. Wali Karzai, A. F. Mehl, and R. McMacken. DnaJ dramatically stimulates ATP hydrolysis by DnaK: Insight into targeting of Hsp70 proteins to polypeptide substrates. *Biochemistry* **1999**, *38*, 4165–4176.

78 D. Wall, M. Zylicz, and C. Georgopoulos. The NH_2-terminal 108 amino acids of the *Escherichia coli* DnaJ protein stimulate the ATPase activity of DnaK and are sufficient for λ replication. *J. Biol. Chem.* **1994**, *269*, 5446–5451.

79 J. C. Bardwell, K. Tilly, E. Craig, J. King, M. Zylicz, and C. Georgopoulos. The nucleotide sequence of the Escherichia coli K12 dnaJ+ gene. A gene that encodes a heat shock protein. *J. Biol. Chem.* **1986**, *261*, 1782–1785.

80 S. Lee, C. Y. Fan, J. M. Younger, H. Ren, and D. M. Cyr. Identification of essential residues in the type II Hsp40 Sis1 that function in polypeptide

binding. *J Biol Chem* **2002**, *277*, 21675–21682.

81 S. Rüdiger, J. Schneider-Mergener, and B. Bukau. Its substrate specificity characterizes the DnaJ chaperone as scanning factor for the DnaK chaperone. *EMBO J.* **2001**, *20*, 1–9.

82 J. R. Cupp-Vickery, and L. E. Vickery. Crystal structure of Hsc20, a J-type Co-chaperone from Escherichia coli. *J Mol Biol* **2000**, *304*, 835–845.

83 A. Wawrzynów, and M. Zylicz. Divergent effects of ATP on the binding of the DnaK and DnaJ chaperones to each other, or to their various native and denatured protein substrates. *J. Biol. Chem.* **1995**, *270*, 19300–19306.

84 D. Wall, M. Zylicz, and C. Georgopoulos. The conserved G/F motif of the DnaJ chaperone is necessary for the activation of the substrate binding properties of the DnaK chaperone. *J. Biol. Chem.* **1995**, *270*, 2139–2144.

85 W.-C. Suh, C. Z. Lu, and C. A. Gross. Structural features required for the interaction of the Hsp70 molecular chaperone DnaK with its cochaperone DnaJ. *J. Biol. Chem.* **1999**, *274*, 30534–30539.

86 R. Schirmbeck, M. Kwissa, N. Fissolo, S. Elkholy, P. Riedl, and J. Reimann. Priming polyvalent immunity by DNA vaccines expressing chimeric antigens with a stress protein-capturing, viral J-domain. *Faseb J* **2002**, *16*, 1108–1110.

87 B. Misselwitz, O. Staeck, K. E. S. Matlack, and T. A. Rapoport. Interaction of BiP with the J-domain of the Sec63p component of the endoplasmic reticulum protein translocation complex. *J. Biol. Chem.* **1999**, *274*, 20110–20115.

88 W.-C. Suh, W. F. Burkholder, C. Z. Lu, X. Zhao, M. E. Gottesman, and C. A. Gross. Interactions of the Hsp70 Molecular Chaperone, DnaK, with its Cochaperone DnaJ. *Proc. Natl. Acad. Sci. USA* **1998**, *95*, 15223–15228.

89 C. S. Gässler, A. Buchberger, T. Laufen, M. P. Mayer, H. Schröder, A. Valencia, and B. Bukau. Mutations in the DnaK chaperone affecting interaction with the DnaJ co-chaperone. *Proc. Natl. Acad. Sci. USA* **1998**, *95*, 15229–15234.

90 M. K. Greene, K. Maskos, and S. J. Landry. Role of the J-domain in the cooperation of Hsp40 with Hsp70. *Proc. Nat. Acad. Sci. USA* **1998**, *95*, 6108–6113.

91 S. J. Landry. Structure and Energetics of an Allele-Specific Genetic Interaction between dnaJ and dnaK: Correlation of Nuclear Magnetic Resonance Chemical Shift Perturbations in the J-Domain of Hsp40/DnaJ with Binding Affinity for the ATPase Domain of Hsp70/DnaK. *Biochemistry* **2003**, *42*, 4926–4936.

92 P. Genevaux, F. Schwager, C. Georgopoulos, and W. L. Kelley. Scanning mutagenesis identifies amino acid residues essential for the in vivo activity of the Escherichia coli DnaJ (Hsp40) J-domain. *Genetics* **2002**, *162*, 1045–1053.

93 X. Zhu, X. Zhao, W. F. Burkholder, A. Gragerov, C. M. Ogata, M. Gottesman, and W. A. Hendrickson. Structural analysis of substrate binding by the molecular chaperone DnaK. *Science* **1996**, *272*, 1606–1614.

94 B. Feifel, H.-J. Schönfeld, and P. Christen. D-peptide ligands for the co-chaperone DnaJ. *J. Biol. Chem.* **1998**, *273*, 11999–12002.

95 B. Banecki, K. Liberek, D. Wall, A. Wawrzynów, C. Georgopoulos, E. Bertoli, F. Tanfani, and M. Zylicz. Structure-function analysis of the zinc finger region of the DnaJ molecular chaperone. *J. Biol. Chem.* **1996**, *271*, 14840–14848.

96 A. Szabo, R. Korzun, F. U. Hartl, and J. Flanagan. A zinc finger-like domain of the molecular chaperone DnaJ is involved in binding to denatured protein substrates. *EMBO J.* **1996**, *15*, 408–417.

97 K. Linke, T. Wolfram, J. Bussemer, and U. Jakob. The roles of the two zinc binding sites in DnaJ. *J Biol Chem* **2003**, *278*, 44457–44466.

98 C. Y. Fan, S. Lee, H. Y. Ren, and D. M. Cyr. Exchangeable chaperone modules contribute to specification of Type I and Type II Hsp40 cellular function. *Mol Biol Cell* **2004**, *15*, 761–773.

99 J. Gamer, G. Multhaup, T. Tomoyasu, J. S. McCarty, S. Rudiger, H. J. Schonfeld, C. Schirra, H. Bujard, and B. Bukau. A cycle of binding and release of the DnaK, DnaJ and GrpE chaperones regulates activity of the Escherichia coli heat shock transcription factor sigma32. *Embo J* **1996**, *15*, 607–617.

100 H.-J. Schönfeld, D. Schmidt, and M. Zulauf. Investigation of the molecular chaperone DnaJ by analytical ultracentrifugation. *Progr Colloid Polym Sci* **1995**, *99*, 7–10.

101 W. Yan, B. Schilke, C. Pfund, W. Walter, S. Kim, and E. A. Craig. Zuotin, a ribosome-associated DnaJ molecular chaperone. *EMBO J.* **1998**, *17*, 4809–4817.

102 D. Feldheim, J. Rothblatt, and R. Schekman. Topology and functional domains of Sec63p, an endoplasmic reticulum membrane protein required for secretory protein translocation. *Mol Cell Biol.* **1992**, *12*, 3288–3296.

103 K. N. Truscott, W. Voos, A. E. Frazier, et al. A J-protein is an essential subunit of the presequence translocase-associated protein import motor of mitochondria. *J. Cell Biol.* **2003**, *163*, 707–713.

104 D. Mokranjac, M. Sichting, W. Neupert, and K. Hell. Tim14, a novel key component of the import motor of the TIM23 protein translocase of mitochondria. *Embo J* **2003**, *22*, 4945–4956.

105 P. D. D'Silva, B. Schilke, W. Walter, A. Andrew, and E. A. Craig. J protein cochaperone of the mitochondrial inner membrane required for protein import into the mitochondrial matrix. *Proc Natl Acad Sci USA* **2003**, *100*, 13839–13844.

106 G. J. Evans, A. Morgan, and R. D. Burgoyne. Tying everything together: the multiple roles of cysteine string protein (CSP) in regulated exocytosis. *Traffic* **2003**, *4*, 653–659.

107 L. H. Chamberlain, and R. D. Burgoyne. Cysteine-string protein: the chaperone at the synapse. *J Neurochem* **2000**, *74*, 1781–1789.

108 P. Wittung-Stafshede, J. Guidry, B. E. Horne, and S. J. Landry. The J-domain of Hsp40 couples ATP hydrolysis to substrate capture in Hsp70. *Biochemistry* **2003**, *42*, 4937–4944.

109 H.-J. Schönfeld, D. Schmidt, H. Schröder, and B. Bukau. The DnaK chaperone system of *Escherichia coli*: quaternary structures and interactions of the DnaK and GrpE components. *J. Biol. Chem.* **1995**, *270*, 2183–2189.

110 C. J. Harrison, M. Hayer-Hartl, M. Di Liberto, F.-U. Hartl, and J. Kuriyan. Crystal structure of the nucleotide exchange factor GrpE bound to the ATPase domain of the molecular chaperone DnaK. *Science* **1997**, *276*, 431–435.

111 D. Ang, G. N. Chandrasekhar, M. Zylicz, and C. Georgopoulos. Escherichia coli grpE gene codes for heat shock protein B25.3, essential for both lambda DNA replication at all temperatures and host growth at high temperature. *J. Bacteriol.* **1986**, *167*, 25–29.

112 B. Wu, D. Ang, M. Snavely, and C. Georgopoulos. Isolation and Characterization of Point Mutations in the *Escherichia coli grpE* Heat Shock Gene. *J. Bacteriol.* **1994**, *176*, 6965–6973.

113 A. Buchberger, H. Schröder, M. Büttner, A. Valencia, and B. Bukau. A conserved loop in the ATPase domain of the DnaK chaperone is essential for stable binding of GrpE. *Nat Struct Biol.* **1994**, *1*, 95–101.

114 M. Zylicz, D. Ang, and C. Georgopoulos. The grpE protein of *Escherichia coli*. Purification and properties. *J. Biol. Chem.* **1987**, *262*, 17437–17442.

115 L. S. Chesnokova, S. V. Slepenkov, I. I. Protasevich, M. G. Sehorn, C. G. Brouillette, and S. N. Witt. Deletion of DnaK's lid strengthens

binding to the nucleotide exchange factor, GrpE: a kinetic and thermodynamic analysis. *Biochemistry* **2003**, *42*, 9028–9040.

116 K. M. Flaherty, C. DeLuca-Flaherty, and D. B. McKay. Three-dimensional structure of the ATPase fragment of a 70K heat-shock cognate protein. *Nature* **1990**, *346*, 623–628.

117 A. Mally, and S. N. Witt. GrpE accelarates peptide binding and release from the high affinity state of DnaK. *Nat Struct Biol.* **2001**, *8*, 254–257.

118 W. Han, and P. Christen. Interdomain communication in the molecular chaperone DnaK. *Biochem J* **2003**, *369*, 627–634.

119 J. Höhfeld, and S. Jentsch. GrpE-like regulation of the Hsc70 chaperone by the anti-apoptotic protein BAG-1. *EMBO J.* **1997**, *16*, 6209–6216.

120 H. Sondermann, C. Scheufler, C. Scheider, J. Höhfeld, F.-U. Hartl, and I. Moarefi. Structure of a Bag/Hsc70 Complex: Convergent functional evolution of Hsp70 nucleotide exchange factors. *Science* **2001**, *291*, 1553–1557.

121 K. Briknarova, S. Takayama, L. Brive, et al. Structural analysis of BAG1 cochaperone and its interactions with Hsc70 heat shock protein. *Nat Struct Biol* **2001**, *8*, 349–352.

122 S. Takayama, T. Sato, S. Krajewski, K. Kochel, S. Irie, J. A. Millan, and J. C. Reed. Cloning and Functional Analysis of BAG-1: A Novel Bcl-2-Binding Protein with Anti-Cell Death Activity. *Cell* **1995**, *80*, 279–284.

123 H.-G. Wang, S. Takayam, U. R. Rapp, and J. C. Reed. Bcl-2 interacting protein, BAG-1, binds to and activates the kinase Raf-1. *Proc. Natl. Acad. Sci. USA* **1996**, *93*, 7063–7068.

124 M. Zeiner, and U. Gehring. A protein that interacts with members of the nuclear hormone receptor family: Identification and cDNA cloning. *Proc. Natl. Acad. Sci. USA* **1995**, *92*, 11465–11469.

125 A. Bardelli, P. Longati, D. Alberto, S. Goruppi, C. Schneider, C. Ponzetto, and P. M. Comoglio. HGF receptor associates with the anti-apoptotic protein BAG-1 and prevents cell death. *EMBO J.* **1996**, *15*, 6205–6212.

126 C. V. Clevenger, K. Thickman, W. Ngo, W. P. Chang, S. Takayama, and J. C. Reed. Role of Bag-1 in the survival and proliferation of the cytokine-dependent lymphocyte lines, Ba/F3 ad Nb2. *Mol Endocrinol* **1997**, *11*, 608–618.

127 M. Zeiner, M. Gebauer, and U. Gehring. Mammalian protein RAP46: an interaction partner and modulator of 70 kDa heat shock proteins. *EMBO J.* **1997**, *16*, 5483–5490.

128 J. Schneikert, S. Hubner, E. Martin, and A. C. Cato. A nuclear action of the eukaryotic cochaperone RAP46 in downregulation of glucocorticoid receptor activity. *J Cell Biol* **1999**, *146*, 929–940.

129 J. Lüders, J. Demand, and J. Höhfeld. The ubiquitin-related BAG-1 provides a link between the molecular chaperones Hsc70/Hsp70 and the proteasome. *J. Biol. Chem.* **2000**, *275*, 4613–4617.

130 S. Takayama, and J. C. Reed. Molecular chaperone targeting and regulation by BAG family proteins. *Nat Cell Biol* **2001**, *3*, E237–241.

131 S. Takayama, Z. Xie, and J. C. Reed. An evolutionarily conserved family of Hsp70/Hsc70 molecular chaperone regulators. *J Biol Chem* **1999**, *274*, 781–786.

132 K. Thress, J. Song, R. I. Morimoto, and S. Kornbluth. Reversible inhibition of Hsp70 chaperone function by Scythe and Reaper. *EMBO J.* **2001**, *20*, 1033–1041.

133 X. Yang, G. Chernenko, Y. Hao, Z. Ding, M. M. Pater, A. Pater, and S. C. Tang. Human BAG-1/RAP46 protein is generated as four isoforms by alternative translation initiation and overexpressed in cancer cells. *Oncogene* **1998**, *17*, 981–989.

134 S. Takayama, S. Krajewski, M. Krajewski, et al. Expression and

location of Hsp70/Hsc-binding anti-apoptotic protein BAG-1 and its variants in normal tissues and tumor cell lines. *Cancer Res.* **1998**, *58*, 3116–3131.

135 U. Schmidt, G. M. Wochnik, M. C. Rosenhagen, J. C. Young, F. U. Hartl, F. Holsboer, and T. Rein. Essential Role of the Unusual DNA-binding Motif of BAG-1 for Inhibition of the Glucocorticoid Receptor. *J Biol Chem* **2003**, *278*, 4926–4931.

136 M. Zeiner, Y. Niyaz, and U. Gehring. The hsp70-associating protein Hap46 binds to DNA and stimulates transcription. *Proc Natl Acad Sci USA* **1999**, *96*, 10194–10199.

137 A. Crocoll, J. Schneikert, S. Hubner, E. Martin, and A. C. Cato. BAG-1M: a potential specificity determinant of corticosteroid receptor action. *Kidney Int* **2000**, *57*, 1265–1269.

138 D. Bimston, J. Song, D. Winchester, S. Takayama, J. C. Reed, and R. I. Morimoto. BAG-1, a negative regulator of Hsp70 chaperone activity, uncouples nucleotide hydrolysis from substrate release. *EMBO J.* **1998**, *17*, 6871–6878.

139 J. Lüders, J. Demand, S. Schönfelder, M. Frien, R. Zimmermann, and J. Höhfeld. Cofactor-Induced Modulation of the Functional Specificity of the Molecular Chaperone Hsc70. *Biol. Chem.* **1998**, *379*, 1217–1226.

140 E. A. A. Nollen, J. F. Brunsting, J. Song, H. H. Kampinga, and R. I. Morimoto. Bag1 functions in vivo as a negative regulator of Hsp70 chaperone activity. *Mol Cell Biol.* **2000**, *20*, 1083–1088.

141 S. Takayama, D. N. Bimston, S.-i. Matsuzawa, B. C. Freeman, C. Aime-Sempe, Z. Xie, R. I. Morimoto, and J. C. Reed. BAG-1 modulates the chaperone activity of Hsp70/Hsc70. *EMBO J.* **1997**, *16*, 4887–4896.

142 K. C. Kanelakis, Y. Morishima, K. D. Dittmar, M. d. Galigniana, S. Takayama, J. C. Reed, and W. B. Pratt. Differential effects of the hsp70-binding protein BAG-1 on glucocorticoid receptor folding by the hsp90-based chaperone machinery. *J. Biol. Chem.* **1999**, *274*, 34134–34140.

143 M. Kabani, J. M. Beckerich, and C. Gaillardin. Sls1p stimulates Sec63p-mediated activation of Kar2p in a conformation-dependent manner in the yeast endoplasmic reticulum. *Mol Cell Biol* **2000**, *20*, 6923–6934.

144 K. T. Chung, Y. Shen, and L. M. Hendershot. BAP, a mammalian BiP-associated protein, is a nucleotide exchange factor that regulates the ATPase activity of BiP. *J Biol Chem* **2002**, *277*, 47557–47563.

145 M. Kabani, J. M. Beckerich, and J. L. Brodsky. Nucleotide exchange factor for the yeast Hsp70 molecular chaperone Ssa1p. *Mol Cell Biol* **2002**, *22*, 4677–4689.

146 M. Kabani, C. McLellan, D. A. Raynes, V. Guerriero, and J. L. Brodsky. HspBP1, a homologue of the yeast Fes1 and Sls1 proteins, is an Hsc70 nucleotide exchange factor. *FEBS Lett* **2002**, *531*, 339–342.

147 J. P. Grimshaw, I. Jelesarov, H. J. Schönfeld, and P. Christen. Reversible thermal transition in GrpE, the nucleotide exchange factor of the DnaK heat-shock system. *J. Biol. Chem.* **2001**, *276*, 6098–6104.

148 J. P. Grimshaw, I. Jelesarov, R. K. Siegenthaler, and P. Christen. Thermosensor Action of GrpE: The DnaK chaperone system at heat shock temperatures. *J Biol Chem* **2003**, *278*, 19048–19053.

149 Y. Groemping, and J. Reinstein. Folding properties of the nucleotide exchange factor GrpE from Thermus thermophilus: GrpE is a thermosensor that mediates heat shock response. *J Mol Biol* **2001**, *314*, 167–178.

150 A. D. Gelinas, K. Langsetmo, J. Toth, K. A. Bethoney, W. F. Stafford, and C. J. Harrison. A structure-based interpretation of E. coli GrpE thermodynamic properties. *J Mol Biol* **2002**, *323*, 131–142.

151 A. D. Gelinas, J. Toth, K. A. Bethoney, K. Langsetmo, W. F. Stafford, and C. J. Harrison. Thermodynamic linkage in the GrpE

nucleotide exchange factor, a molecular thermosensor. *Biochemistry* **2003**, *42*, 9050–9059.

152 L. D. D'Andrea, and L. Regan. TPR proteins: the versatile helix. *Trends Biochem Sci* **2003**, *28*, 655–662.

153 S. Chen, and D. F. Smith. Hop as an Adaptor in the Heat Shock Protein 70 (Hsp70) and Hsp90 Chaperone Machinery. *J. Biol. Chem.* **1998**, *273*, 35194–35200.

154 S. Tobaben, P. Thakur, R. Fernandez-Chacon, T. C. Sudhof, J. Rettig, and B. Stahl. A trimeric protein complex functions as a synaptic chaperone machine. *Neuron* **2001**, *31*, 987–999.

155 J. C. Young, N. J. Hoogenraad, and F. U. Hartl. Molecular chaperones Hsp90 and Hsp70 deliver preproteins to the mitochondrial import receptor Tom70. *Cell* **2003**, *112*, 41–50.

156 J. C. Young, J. M. Barral, and F. Ulrich Hartl. More than folding: localized functions of cytosolic chaperones. *Trends Biochem Sci* **2003**, *28*, 541–547.

157 J. Höhfeld, Y. Minami, and F. U. Hartl. Hip, a novel cochaperone involved in the eukaryotic Hsc70/Hsp40 reaction cycle. *Cell* **1995**, *83*, 589–598.

158 J. Frydman, and J. Hohfeld. Chaperones get in touch: the Hip-Hop connection. *Trends Biochem Sci* **1997**, *22*, 87–92.

159 V. Prapapanich, S. Chen, S. Nair, R. Rimerman, and D. Smith. Molecular cloning of human p48, a transient component of progesterone receptor complexes and an Hsp70-binding protein. *Mol Endocrinol.* **1996**, *10*, 420–431.

160 H. Irmer, and J. Höhfeld. Characterization of Functional Domains of the Eukaryotic Co-chaperone Hip. *J. Biol. Chem.* **1997**, *272*, 2230–2235.

161 F. Vignols, N. Mouaheb, D. Thomas, and Y. Meyer. Redox control of Hsp70-Co-chaperone interaction revealed by expression of a thioredoxin-like Arabidopsis protein. *J Biol Chem* **2003**, *278*, 4516–4523.

162 M. A. Webb, J. M. Cavaletto, P. Klanrit, and G. A. Thompson. Orthologs in Arabidopsis thaliana of the Hsp70 interacting protein Hip. *Cell Stress Chaperones* **2001**, *6*, 247–255.

163 M. Velten, B. O. Villoutreix, and M. M. Ladjimi. Quaternary structure of HSC70 cochaperone HIP. *Biochemistry* **2000**, *39*, 307–315.

164 B. D. Bruce, and J. Churchich. Characterization of the molecular-chaperone function of the heat-shock-cognate-70-interacting protein. *Eur. J. Biochem.* **1997**, *245*, 738–744.

165 K. C. Kanelakis, P. J. Murphy, M. D. Galigniana, Y. Morishima, S. Takayama, J. C. Reed, D. O. Toft, and W. B. Pratt. hsp70 interacting protein Hip does not affect glucocorticoid receptor folding by the hsp90-based chaperone machinery except to oppose the effect of BAG-1. *Biochemistry* **2000**, *39*, 14314–14321.

166 E. A. Nollen, A. E. Kabakov, J. F. Brunsting, B. Kanon, J. Hohfeld, and H. H. Kampinga. Modulation of in vivo HSP70 chaperone activity by Hip and Bag-1. *J Biol Chem* **2001**, *276*, 4677–4682.

167 F. A. Agarraberes, and J. F. Dice. A molecular chaperone complex at the lysosomal membrane is required for protein translocation. *J Cell Sci* **2001**, *114*, 2491–2499.

168 G. H. Fan, W. Yang, J. Sai, and A. Richmond. Hsc/Hsp70 interacting protein (hip) associates with CXCR2 and regulates the receptor signaling and trafficking. *J Biol Chem* **2002**, *277*, 6590–6597.

169 C. Nicolet, and E. Craig. Isolation and characterization of STI1, a stress-inducible gene from *Saccharomyces cerevisiae*. *Mol Cell Biol.* **1989**, *9*, 3638–3646.

170 D. F. Smith, W. P. Sullivan, T. N. Marion, K. Zaitsu, B. Madden, D. J. McCormick, and D. O. Toft. Identification of a 60-Kilodalton Stress-Related Protein, p60, which interacts with hsp90 and hsp70. *Mol Cell Biol.* **1993**, *13*, 869–876.

171 C. Scheufler, A. Brinker, G.

Bourenkov, S. Pegoraro, L. Moroder, H. Bartunik, F. U. Hartl, and I. Moarefi. Struture of TPR domain-peptide complexes: critical elements in the assembly of the Hsp70-Hsp90 multichaperone machine. *Cell* **2000**, *101*, 199–210.

172 A. Brinker, C. Scheufler, F. Von Der Mulbe, B. Fleckenstein, C. Herrmann, G. Jung, I. Moarefi, and F. U. Hartl. Ligand discrimination by TPR domains. Relevance and selectivity of EEVD-recognition in Hsp70 × Hop × Hsp90 complexes. *J Biol Chem* **2002**, *277*, 19265–19275.

173 O. O. Odunuga, J. A. Hornby, C. Bies, R. Zimmermann, D. J. Pugh, and G. L. Blatch. Tetratricopeptide repeat motif-mediated Hsc70-mSTI1 interaction. Molecular characterization of the critical contacts for successful binding and specificity. *J Biol Chem* **2003**, *278*, 6896–6904.

174 M. Gross, and S. Hessefort. Purification and Characterization of a 66-kDa Protein from Rabbit Reticulocyte Lysate which Promotes the Recycling of Hsp70. *J. Biol. Chem.* **1996**, *271*, 16833–16841.

175 H. Wegele, M. Haslbeck, J. Reinstein, and J. Buchner. Sti1 is a novel activator of the Ssa proteins. *J Biol Chem* **2003**, *278*, 25970–25976.

176 A. Brychzy, T. Rein, K. F. Winklhofer, F. U. Hartl, J. C. Young, and W. M. Obermann. Cofactor Tpr2 combines two TPR domains and a J domain to regulate the Hsp70/Hsp90 chaperone system. *Embo J* **2003**, *22*, 3613–3623.

177 Y. Morishima, K. C. Kanelakis, A. M. Silverstein, K. D. Dittmar, L. Estrada, and W. B. Pratt. The Hsp organizer protein Hop enhances the rate of but is not essential for glucocorticoid receptor folding by the multiprotein Hsp90-based chaperone system. *J. Biol. Chem.* **2000**, *275*, 6894–6900.

178 P. J. Murphy, K. C. Kanelakis, M. D. Galigniana, Y. Morishima, and W. B. Pratt. Stoichiometry, abundance, and functional significance of the hsp90/hsp70-based multiprotein chaperone machinery in reticulocyte lysate. *J Biol Chem* **2001**, *276*, 30092–30098.

179 K. Richter, and J. Buchner. Hsp90: chaperoning signal transduction. *J Cell Physiol* **2001**, *188*, 281–290.

180 J. C. Young, I. Moarefi, and F. U. Hartl. Hsp90: a specialized but essential protein-folding tool. *J Cell Biol* **2001**, *154*, 267–273.

181 K. Richter, P. Muschler, O. Hainzl, J. Reinstein, and J. Buchner. Sti1 is a non-competitive inhibitor of the Hsp90 ATPase. Binding prevents the N-terminal dimerization reaction during the atpase cycle. *J Biol Chem* **2003**, *278*, 10328–10333.

182 C. Prodromou, G. Siligardi, R. O'Brien, D. N. Woolfson, L. Regan, B. Panaretou, J. E. Ladbury, P. W. Piper, and L. H. Pearl. Regulation of Hsp90 ATPase activity by tetratricopeptide repeat (TPR)-domain co-chaperones. *EMBO J.* **1999**, *18*, 754–762.

183 L. Aravind, and E. V. Koonin. The U box is a modified RING finger – a common domain in ubiquitination. *Curr Biol* **2000**, *10*, R132–134.

184 C. A. Ballinger, P. Connell, Y. Wu, Z. Hu, L. J. Thompson, L. Y. Yin, and C. Patterson. Identification of CHIP, a novel tetratricopeptide repeat-containing protein that interacts with heat shock proteins and negatively regulates chaperone functions. *Mol Cell Biol* **1999**, *19*, 4535–4545.

185 J. Yan, J. Wang, Q. Li, J. R. Hwang, C. Patterson, and H. Zhang. AtCHIP, a U-box-containing E3 ubiquitin ligase, plays a critical role in temperature stress tolerance in Arabidopsis. *Plant Physiol* **2003**, *132*, 861–869.

186 R. Nikolay, T. Wiederkehr, W. Rist, G. Kramer, M. P. Mayer, and B. Bukau. Dimerization of the human E3 ligase, CHIP, via a coiled-coil domain is essential for its activity. *J Biol Chem* **2004**, *279*, 2673–2678.

187 J. Demand, S. Alberti, C. Patterson, and J. Hohfeld. Cooperation of a ubiquitin domain protein and an E3 ubiquitin ligase during chaperone/

proteasome coupling. *Curr Biol* **2001**, *11*, 1569–1577.
188 S. Hatakeyama, M. Yada, M. Matsumoto, N. Ishida, and K. I. Nakayama. U box proteins as a new family of ubiquitin-protein ligases. *J Biol Chem* **2001**, *276*, 33111–33120.
189 J. Jiang, C. A. Ballinger, Y. Wu, Q. Dai, D. M. Cyr, J. Hohfeld, and C. Patterson. CHIP is a U-box-dependent E3 ubiquitin ligase: Identification of Hsc70 as a target for ubiquitylation. *J Biol Chem* **2001**, *13*, 13.
190 S. Murata, Y. Minami, M. Minami, T. Chiba, and K. Tanaka. CHIP is a chaperone-dependent E3 ligase that ubiquitylates unfolded protein. *EMBO Rep* **2001**, *2*, 1133–1138.
191 D. M. Cyr, J. Hohfeld, and C. Patterson. Protein quality control: U-box-containing E3 ubiquitin ligases join the fold. *Trends Biochem Sci* **2002**, *27*, 368–375.
192 T. Wiederkehr, B. Bukau, and A. Buchberger. Protein turnover: a CHIP programmed for proteolysis. *Curr Biol* **2002**, *12*, R26–28.
193 P. Connell, C. A. Ballinger, J. Jiang, Y. Wu, L. J. Thompson, J. Hohfeld, and C. Patterson. The co-chaperone CHIP regulates protein triage decisions mediated by heat-shock proteins. *Nat Cell Biol* **2001**, *3*, 93–96.
194 G. C. Meacham, C. Patterson, W. Zhang, J. M. Younger, and D. M. Cyr. The Hsc70 co-chaperone CHIP targets immature CFTR for proteasomal degradation. *Nat Cell Biol.* **2001**, *3*, 100–105.
195 W. Xu, M. Marcu, X. Yuan, E. Mimnaugh, C. Patterson, and L. Neckers. Chaperone-dependent E3 ubiquitin ligase CHIP mediates a degradative pathway for c-ErbB2/Neu. *Proc Natl Acad Sci USA* **2002**, *99*, 12847–12852.
196 P. Zhou, N. Fernandes, I. L. Dodge, et al. ErbB2 degradation mediated by the co-chaperone protein CHIP. *J Biol Chem* **2003**, *278*, 13829–13837.
197 M. J. Lees, D. J. Peet, and M. L. Whitelaw. Defining the role for XAP2 in stabilization of the dioxin receptor. *J Biol Chem* **2003**, *278*, 35878–35888.
198 A. T. Chavez Zobel, A. Loranger, N. Marceau, J. R. Theriault, H. Lambert, and J. Landry. Distinct chaperone mechanisms can delay the formation of aggresomes by the myopathy-causing R120G alphaB-crystallin mutant. *Hum Mol Genet* **2003**, *12*, 1609–1620.
199 S. Alberti, J. Demand, C. Esser, N. Emmerich, H. Schild, and J. Hohfeld. Ubiquitylation of BAG-1 suggests a novel regulatory mechanism during the sorting of chaperone substrates to the proteasome. *J Biol Chem* **2002**, *277*, 45920–45927.
200 J. Jiang, D. Cyr, R. W. Babbitt, W. C. Sessa, and C. Patterson. Chaperone-dependent regulation of endothelial nitric-oxide synthase intracellular trafficking by the co-chaperone/ubiquitin ligase CHIP. *J Biol Chem* **2003**, *278*, 49332–49341.
201 Q. Dai, C. Zhang, Y. Wu, et al. CHIP activates HSF1 and confers protection against apoptosis and cellular stress. *EMBO J* **2003**, *22*, 5446–5458.
202 M. P. Deutscher (ed) *Guide to protein purification* Vol. 182. Methods in Enzymology. Edited by J. N. Abelson, and M. I. Simon, Academic Press. Inc., San Diego, **1990**.
203 W. J. Welch, and J. R. Feramisco. Rapid Purification of Mammalian 70,000-Dalton Stress Proteins: Affinity of the Proteins for Nucleotides. *Mol. Cell. Biol.* **1985**, *5*, 1229–1237.
204 D. M. Schlossman, S. L. Schmid, W. A. Braell, and J. E. Rothman. An Enzyme that Removes Clathrin Coats: Purification of an Uncoating ATPase. *J. Cell Biol.* **1984**, *99*, 723–733.
205 J. V. Anderson, D. W. Haskell, and C. L. Guy. Differential influence of ATP on native spinach 70-kilodalton heat-shock cognates. *Plant Physiol* **1994**, *104*, 1371–1380.
206 L. E. S. Vickery, Jonathan J. and Ta, Dennis T. Hsc66 and Hsc20, a new

heat shock cognate molecular chaperone system from *Escherichia coli*. *Protein Sci.* **1997**, *6*, 1047–1056.

207 M. Zylicz, D. Ang, K. Liberek, and C. Georgopoulos. Initiation of λ DNA replication with purified host- and bacteriophage-encoded proteins: the role of the DnaK, DnaJ and GrpE heat shock proteins. *EMBO J.* **1989**, *8*, 1601–1608.

208 J. Spence, A. Cegielska, and C. Georgopoulos. Role of *Escherichia coli* heat shock proteins DnaK and HtpG (C62.5) in response to nutritional deprivation. *J. Bacteriol* **1990**, *172*, 7157–7166.

209 B. Bukau, and G. Walker. Mutations altering heat shock specific subunit of RNA polymerase suppress major cellular defects of *E. coli* mutants lacking the DnaK chaperone. *EMBO J.* **1990**, *9*, 4027–4036.

210 J. Buchner, T. Weikl, H. Bugl, F. Pirkl, and S. Bose. Purification of Hsp90 partner proteins Hop/p60, p23, and FKBP52. *Methods Enzymol* **1998**, *290*, 418–429.

211 B. Miao, J. E. Davis, and E. A. Craig. Mge1 functions as a nucleotide release factor for Ssc1, a mitochondrial Hsp70 of *Saccharomyces cerevisiae*. *J. Mol. Biol.* **1997**, *265*, 541–552.

212 H. C. Schneider, B. Westermann, W. Neupert, and M. Brunner. The nucleotide exchange factor MGE exerts a key function in the ATP-dependent cycle of mt-Hsp70-Tim44 interaction driving mitochondrial protein import. *Embo J* **1996**, *15*, 5796–5803.

213 S. Schmidt, A. Strub, K. Rottgers, N. Zufall, and W. Voos. The two mitochondrial heat shock proteins 70, Ssc1 and Ssq1, compete for the cochaperone Mge1. *J Mol Biol* **2001**, *313*, 13–26.

214 D. A. Raynes, and V. Guerriero, Jr. Inhibition of Hsp70 ATPase Activity and Protein Renaturation by a Novel Hsp70-binding Protein. *J. Biol. Chem.* **1998**, *273*, 32883–32888.

215 A. Boisramé, M. Kabani, J. M. Beckerich, E. Hartmann, and C. Gaillardin. Interaction of Kar2p and Sls1p is required for efficient co-translational translocation of secreted proteins in the yeast Yarrowia lipolytica. *J Biol Chem* **1998**, *273*, 30903–30908.

216 N. Guex, and M. C. Peitsch. SWISS-MODEL and the Swiss-PdbViewer: An environment for comparative protein modelling. *Electrophoresis* **1997**, *18*, 2714–2723.

217 M. C. Peitsch. Protein modeling by E-mail. *Bio/Technology* **1995**, *13*, 658–660.

218 M. C. Peitsch. ProMod and Swiss-Model: Internet-based tools for automated comparative protein modelling. *Biochem. Soc. Trans.* **1996**, *24*, 274–279.

219 D. Brehmer, C. Gässler, W. Rist, M. P. Mayer, and B. Bukau. Influence of GrpE on DnaK-substrate interactions. *J Biol Chem* **2004**, *279*, 27957–27964.

16
Protein Folding in the Endoplasmic Reticulum Via the Hsp70 Family

Ying Shen, Kyung Tae Chung, and Linda M. Hendershot

16.1
Introduction

The endoplasmic reticulum (ER) is a membrane-enclosed organelle that is found in all eukaryotic organisms and that represents the entry site or origin of the secretory pathway. As such, the ER is a major site of protein folding and assembly. In some highly specialized secretory cells – such as immunoglobulin-producing plasma cells, serum-protein-producing liver cells, or insulin-producing pancreatic cells – the ER is highly developed and becomes the major site of protein biosynthesis for the cell. Proteins that are destined for the secretory pathway are synthesized in the cytosol on ER-associated ribosomes. A hydrophobic signal sequence present on the nascent polypeptide chain directs it to the translocon, which is a proteinaceous channel that traverses the ER membrane [108], allowing the protein to be translocated into the lumen of the ER, in many cases as it is being synthesized. The elongating nascent chain first passes through a channel in the ribosome and then through the translocon. This requires that ~70 amino acids of the nascent chain be translated before the N-terminus can enter the ER lumen [51]. It appears that the growing polypeptide chain remains unfolded during its transit through the ribosome and translocon [123, 216]. After it enters the ER, N-terminal signal sequences are often removed by a signal peptidase that is positioned at the lumenal side of the translocon. Once the site is ~14 amino acids into the lumen, *N*-linked glycans are added co-translationally by oligosaccharyl transferase (OST) to asparagine residues that are followed by a second amino acid, which can be anything but a proline, and then a serine or a threonine (N-X-S/T) [216]. The OST complex is also associated with the translocon. Inside the ER, the polypeptide chain begins folding co-translationally [7, 38], and in some cases, subunit assembly can occur before the individual chains are completely translated and inside the ER [8].

The modification of secretory pathway proteins with *N*-linked glycans serves in part to limit the ways the nascent protein can fold. In addition, the ER environment itself causes further constraints on and benefits to protein folding and assembly. The calcium required for many signal transduction pathways is stored here. Thus, proteins that are synthesized in this organelle have evolved to fold

Protein Folding Handbook. Part II. Edited by J. Buchner and T. Kiefhaber

ISBN: 3-527-30784-2

of proteins with BiP and a block in their secretion [55]. However, it does not appear that BiP release is the only requirement for all proteins to fold properly. While the in vitro release of BiP from Ig light chains with ATP can lead to folding and disulfide bond formation [119], a similar release of BiP from Ig heavy chains results in the formation of large aggregates [207]. Thus, individual properties of proteins dictate whether they will be able to fold upon BiP release. Given estimates on molecular crowding inside the cell and the subsequent effects crowding can have on protein folding [142], it is reasonable to assume that BiP release might be a tightly regulated process in vivo. Data to support this idea came from studies to measure the rate of BiP cycling on and off unassembled Ig heavy chains. Both wild-type and ATPase-defective hamster BiP were co-expressed with Ig heavy chains under conditions where heavy chains bound equally to both wild-type and mutant BiP during a short pulse [207]. The mutant BiP was expected to act as a kinetic trap by binding to nascent heavy chains but not leasing them, whereas the wild-type BiP should release during its normal course of cycling. Each time the wild-type BiP releases, it should be replaced half the time by another wild-type BiP and the other half by a mutant. Eventually, and under steady-state conditions, all the heavy chains should be bound to mutant BiP. Instead, the study revealed that heavy chains remained bound to both wild-type and mutant BiP and that the ratio of the two types of BiP did not change over a long chase period [207]. However, introduction of light chains rapidly displaced wild-type but not mutant BiP, suggesting that the ATPase cycle of BiP is "stalled" when heavy chains are expressed alone and that assembly with light chains either releases a repressive factor or allows access of a stimulatory factor. This further suggests that the substrate itself can play a role in contributing to the release of BiP or the rate at which the ATPase cycle turns.

Although peptide-binding preferences for BiP have been determined, relatively little is known about actual BiP-binding sites on proteins. Early studies to identify the BiP-binding site on unassembled Ig heavy chains revealed that for all isotypes, the C_H1 domain, which pairs with Ig light chains, possesses the "stable" BiP-binding site [95]. A more recent study revealed that, unlike other domains on the heavy chain, the C_H1 domain remains unfolded until it assembles with light chains [131], thus providing a long-term BiP-binding site. An algorithm based on the binding of BiP to peptides displayed on bacterial phages [14] was applied to Ig domains, where it was found that each Ig domain possessed multiple potential BiP-binding sites [118]. Peptides corresponding to these sequences did indeed stimulate the ATPase activity of BiP, demonstrating that they were capable of binding to the chaperone. The more rapid folding of other domains probably accounts for the fact that BiP associates only transiently with them [112] or, in some cases, that BiP fails to bind altogether [94]. The fact that multiple potential BiP-binding sites were identified on the C_H1 domain [118], even though BiP binds in a 1:1 stoichiometry to this domain in vivo [149], suggests either that the binding of BiP to one site precludes its binding to others or that the algorithm does not accurately identify BiP-binding sites that are used on proteins in vivo. Further studies are needed to determine which is correct.

In addition to BiP, both yeast and mammals possess a second ER-localized

Hsp70 family member, Lhs1p [48] and GRP170 [57], respectively, that is less well characterized than BiP. Lhs1p (also known as Cer1p and Ssi1p) is an Hsp70-related protein that appears to be important for the translocation of a subset of proteins into the ER of *S. cerevisiae* [81, 206]. Lhs1p/Cer1p possesses an ATP-binding site like Kar2p and other members of the Hsp70 family [82]. The *CER1* RNA levels increase during UPR activation [48] and at lower temperatures, which is in keeping with the more severe defects in folding and translocation that are observed in the Δ*cer* strain at low temperatures [82]. These results suggest that Lhs1p/Cer1p provides an additional chaperoning activity in processes known to require Kar2p. In mammalian cells, GRP170 has been identified as an ATP-binding Hsp70 homologue [57] that is induced by ER stress [138] and hypoxia, which has led to its alternative designation as oxygen-regulated protein ORP150 [114, 166]. GRP170 can be co-precipitated with unassembled Ig heavy chains [138], although it is not clear whether it binds to them directly or as a component of the multi-chaperone complex that was recently identified in the ER [149]. GRP170 shows sequence similarity to Lhs1p and may also play a role in the translocation of nascent polypeptides into the ER lumen [53].

16.3
ER-localized DnaJ Homologues

The ATPase activity of Hsp70 proteins is stimulated by interactions with DnaJ proteins. The first DnaJ was discovered in *Escherichia coli*, where it cooperates with DnaK to aid in lambda phage replication [222]. It is a 43-kDa protein containing four domains, beginning with the N-terminal, ~73-amino-acid J-domain, which is present in all DnaJ family members and contains the hallmark tripeptide HPD motif (His-Pro-Asp) [36, 116]. A flexible Gly/Phe-rich domain links the J-domain to a cysteine-rich Zn^{2+}-binding domain [4, 195]. Distal to the Cys-rich domain is a poorly conserved region that accounts for nearly half of the DnaJ molecule and that may contain the substrate-binding domain [98]. Presently, more than 100 DnaJ family members have been identified, which can be found in all species and organelles. They can be divided into three subgroups based upon the degree of domain conservation with *E. coli* DnaJ [98]. Type I DnaJ-like proteins have the highest domain homology with *E. coli* DnaJ and possess all four domains. Type II DnaJ proteins have an N-terminal J-domain and the Gly/Phe-rich linker but lack the Zn^{2+}-binding domain. Type III proteins possess only a J-domain, which can occur anywhere in the protein.

Organelle-specific DnaJs work as cofactors to cooperate with their specific Hsp70 partners, and, unlike Hsp70s, when multiple DnaJs are present in an organelle, each one often regulates a different function. The yeast ER contains three DnaJ-like proteins: Sec63p [182, 183], Jem1p [160], and Scj1p [186]. Sec63p and Jem1p are both type III DnaJ proteins [59, 160], whereas Scj1p is a type I protein [186]. Sec63p is an integral membrane protein possessing three transmembrane domains, with its J-domain in the ER lumen and a large C-terminal domain exposed to the cytosol [59]. As one subunit of the ER translocon, Sec63p interacts with Sec61p, Sec62p, Sec71p, and Sec72p and recruits yeast BiP, Kar2p, to the lumenal side of the translocation apparatus [45]. Sec63p assists Kar2p in pulling nascent proteins into the ER lumen during both co-translational and posttranslational translocation [23, 145, 225]. Sec63p also appears to be a component of the retrograde translocon, which plays a role in removing terminally misfolded proteins from the ER for degradation by the 26S proteasome [171]. Unlike Sec63p, which is a membrane protein required for growth, Jem1p and Scj1p are both soluble ER lumenal proteins that are not essential for cell viability under normal growth conditions. However, a double disruption of the *JEM1* and *SCJ1* genes causes growth arrest at elevated temperatures. Jem1p interacts with Kar2p to mediate nuclear membrane fusion or karyogamy during mating [160], while Scj1p cooperates with Kar2p to fold and assemble proteins in the ER lumen [186]. Recent studies show that both Scj1p and Jem1p may also assist Kar2p in maintaining lumenal ERAD substrates in a retrotranslocation-competent state [161].

Recently, five mammalian ER DnaJ homologues have been identified. They appear to include two Sec63 homologues, one Scj1 homologue, and two novel eukaryotic family members (Table 16.1). However, no JEM homologues have been identified at this time. According to their order of discovery, we propose they be named ERdj1–5 for ER-associated DnaJ proteins. They are ERdj1/Mtj1 [22, 41], ERdj2/hSec63 [191, 205], ERdj3/HEDJ [127, 149, 226], ERdj4/Mdg1 [176, 188], and ERdj5/JPDI [52, 102]. All of the ERdjs contain a J-domain, which interacts with

Tab. 16.1. Identification and characterization of mammalian ER DnaJ proteins

ERdj protein	*M.W.*	*Yeast homol.*	*Topology*	*UPR induction*
ERdj1/Mtj1	63 kD	Sec63	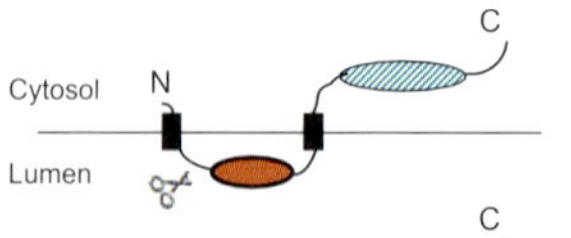	No
ERdj2/hSec63	85 kD	Sec63	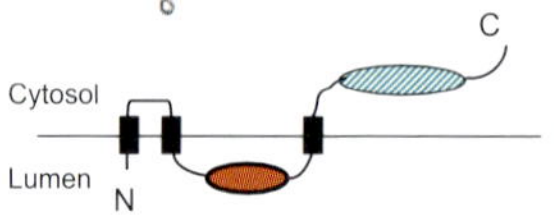	No
ERdj3/HEDJ	43 kD	Scj1	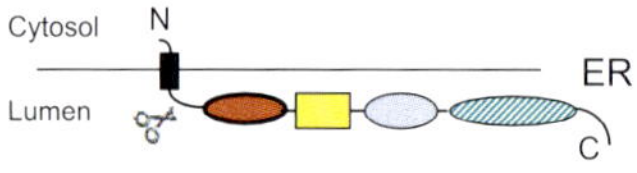	Yes
ERdj4/Mdg1	25 kD	none	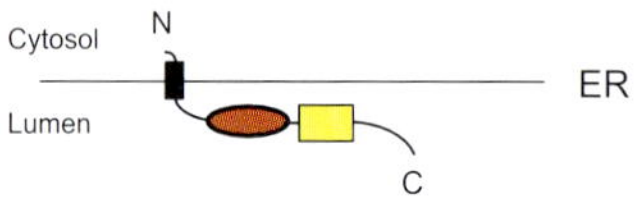	Yes
ERdj5/JPOI	96 kD	none	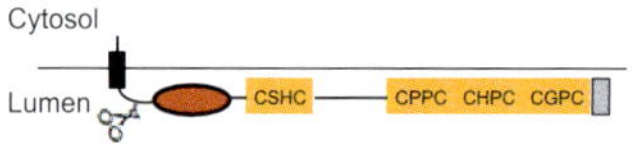	Yes

Signal peptide or TM; Gly/Phe domain; Cys-rich domain; CXXC thioredoxin, PDI-like domain; J domain; KDEL; C-term domain

BiP and stimulates its ATPase activity in vitro. Since mammalian BiP has multiple functions in the ER, it is possible that these ERdjs each regulate different BiP functions in vivo.

Like yeast Sec63p, ERdj1/Mtj1 (66 kDa) and ERdj2/hSec63 (97 kDa) are type III DnaJ proteins associated with the mammalian Sec61p complex in the ER membrane [56, 150, 205]. The N-terminal 190 amino acids of ERdj1/Mtj1 are oriented in the ER lumen and contain the J-domain, which could serve to recruit BiP to the translocon. A large C-terminal cytosolic domain is in close contact with SRP in the cytosol and appears to modulate translation [56]. Inspection of gene databases reveals that ERdj1/Mtj1 is highly conserved in mammals, with homologues present in *Drosophila melanogaster*, *Ciona intestinalis*, and *Anopheles gambiae*. ERdj2/hSec63 displays similar membrane topology and shares ~ 44% amino acid sequence similarity (~26% identity) with yeast Sec63p [191]. Like ERdj1/Mtj1, ERdj2/hSec63 is associated with Sec61 and Sec62, where it may act with BiP to translocate nascent polypeptides into the ER [150, 205]. ERdj2/hSec63 is expressed at relatively higher levels (1.98 μM) in pancreatic microsomes than is ERdj1/Mtj1 (0.36 μM) [56]. ERdj2/hSec63 homologues can also be found in organisms as diverse as *Arabidopsis thaliana*, *Caenorhabditis elegans*, *Neurospora crassa*, and *A. gambiae*.

ERdj3/HEDJ is a 43-kDa type I DnaJ protein that shows ~46% sequence similarity (~35% identity) to yeast Scj1p. Despite the absence of a KDEL retention signal, it appears to be a resident ER protein, although there is some controversy as to its intracellular localization [12, 127, 149, 226] and whether it is a membrane-anchored or soluble protein [127, 149, 226]. ERdj3/HEDJ was identified in association with Shiga toxin [226], which is taken up by the cell at the plasma membrane and passes through the ER before being retrotranslocated to the cytosol, where it performs its pathogenic function [107, 185]. The fact that ERdj3/HEDJ interacts with long-lived, BiP-bound, unassembled Ig heavy chains in vivo [149] suggests that it is likely to play a role in keeping nascent proteins from aggregating and in aiding protein folding and assembly. Alternatively, ERdj3 could act to target ERAD substrates to the retrotranslocon for degradation. Either function would be compatible with its association with Shiga toxin [226]. ERdj3 appears to be widely expressed, with highly homologous orthologs present in *D. melanogaster*, *C. elegans*, *Danio rerio*, *A. gambiae*, *A. thaliana*, and *Zea mays* in addition to yeast.

ERdj4/Mdg1 (26 kDa) is the first type II DnaJ homologue to be found in the ER of any organism [188]. It appears to be largely restricted to vertebrates, since no potential ERdj4 homologues could be found in the *D. melanogaster*, *C. elegans*, or *S. cerevisiae* genomic databases. ERdj4/Mdj1 is anchored in the ER membrane through its uncleaved signal peptide, thus orienting its J-domain and the rest of the protein inside the ER where it interacts with BiP [188]. However, another study using an N-terminal-tagged version of ERdj4/Mdg2 reported nuclear localization after heat shock and cytosolic expression before heat shock [176]. ERdj4 is expressed at the highest levels in secretory tissues and is highly induced by ER stress, suggesting that ERdj4/Mdj1 may play a role in either protein folding or ERAD [188]. Overexpression of ERdj4/Mdg1 inhibited cell death induced by ER stress [125], suggesting that it plays some protective role during the UPR.

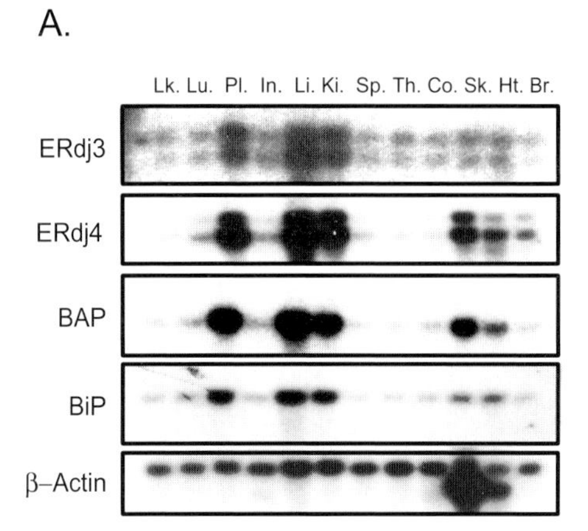

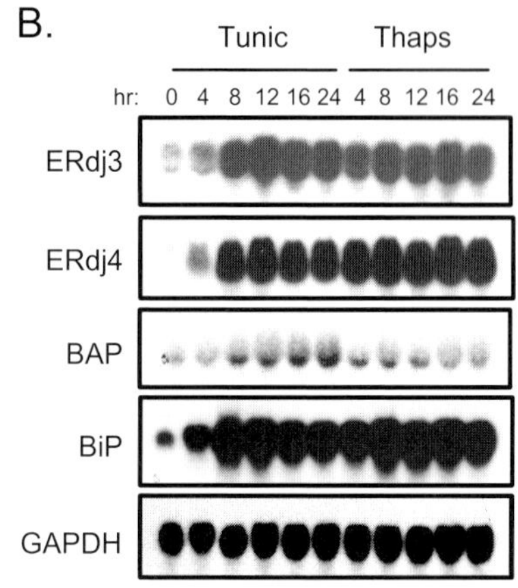

Fig. 16.1. Tissue distribution and ER stress inducibility of BiP and its cofactors. (A) A human multi-tissue Northern blot was hybridized with probes corresponding to the coding regions of human ERdj3, mouse ERdj4, human BAP, hamster BiP, and human β-actin genes. Lane 1: peripheral blood leukocytes; lane 2: lung; lane 3: placenta; lane 4: small intestine; lane 5: liver; lane 6: kidney; lane 7: spleen; lane 8: thymus; lane 9: colon; lane 10: skeletal muscle; lane 11: heart; lane 12: brain. (B) NIH3T3 fibroblasts were incubated with the ER stress-inducing agents tunicamycin or thapsigargin. RNA was isolated at the indicated times and subjected to Northern blot analyses using the indicated probes.

The last mammalian DnaJ homologue to be discovered, ERdj5/JPDI (87 kDa), contains a J-domain at its N-terminus, a PDI-like domain, a thioredoxin domain, and a KDEL sequence at its C-terminus [102]. This latter feature suggests that it is likely to be a soluble ER lumenal protein, although this has not been formally demonstrated. The presence of a thioredoxin-like domain suggests that ERdj5 may be involved in assisting protein folding and disulfide bond formation in the ER. Homologues of ERdj5/JPDI can also be found in *C. elegans*, *D. melanogaster*, *Ciona intestinalis*, and *A. gambiae*, but not in yeast or plants.

Like BiP and other ER chaperones, some of the ERdjs are also upregulated in response to ER stress. ERdj3 and ERdj4 transcripts are dramatically upregulated in response to UPR activation [188]. Combined with the fact that both are expressed at the highest levels in secretory tissues (Figure 16.1), it is likely that they play roles in either the refolding of unfolded proteins or the retrotranslocation of misfolded proteins, both of which diminish the accumulation of unfolded proteins that occurs in the ER during stress conditions. Inspection of genomic sequences upstream of these two genes reveals that *ERdj3/HEDJ* contains a putative ERSE element in its promoter, which could serve to regulate ERdj3 expression, but no potential ERSE can be found in the ERdj4 promoter. ERdj5/JPDI has been reported to contain one ERSE in its promoter and to respond to ER stress, albeit less dramatically than either ERdj3 or ERdj4 [52]. Unlike yeast Sec63, ERdj2/hSec63 is not upregulated during ER stress [188]. This is compatible with the fact that translation and translocation of newly synthesized peptides slow down when mammalian cells encounter ER stress and that, unlike yeast, mammalian cells do not respond to ER stress by dramatically expanding the ER membranes [46, 203].

Because BiP has multiple functions in vivo and because a second Hsp70 homologue, GRP170/ORP150 [138], also exists in the mammalian ER, we anticipate that various ER DnaJ homologues will be specific to the different functions of the mammalian ER Hsp70s and that more ERdjs are therefore likely to be discovered. At present, little functional data are available for most of the mammalian ER DnaJs, and it is not known whether any of them interact with GRP170 in vivo.

16.4 ER-localized Nucleotide-exchange/releasing Factors

In order for Hsp70 proteins to be efficiently released from unfolded substrate proteins, ATP must replace ADP in the nucleotide-binding cleft. In bacteria, this feat is accomplished by the nucleotide-exchange factor GrpE, which binds to the ATPase domain of DnaK (the bacterial Hsp70 ortholog) and promotes the exchange of ADP to ATP, consequently releasing the unfolded substrate [136, 196]. In the mammalian cytosol, a number of both positive and negative regulators of nucleotide exchange have been identified. Hip is an Hsc70-interacting protein that binds to the ATPase domain of Hsc70 and stabilizes it in the ADP-bound state [101, 175]. The Bcl-2-binding anti-apoptotic factor BAG-1 was found to enhance nucleotide exchange from Hsp70 [13, 100, 193], but interestingly this does not always lead to enhanced folding activity. The third cofactor for Hsp70, HspBP1, promotes nucleotide exchange from Hsp70 in vitro, which actually inhibits its chaperoning activity for some proteins [178]. A homologue of HspBP1, Fes1p, was recently identified in *S. cerevisiae*, where it appears to be a much less efficient nucleotide-exchange factor than HspBP1 [110]. Interestingly, none of these Hsp70 cofactors share any homology with GrpE.

The first potential nuclear-exchange factor for BiP came from a genetic screen in the yeast *Yarrowia lipolytica* to identify genes that interacted with the signal recognition particle in co-translational translocation. This screen identified *SLS1*, a resident ER protein [15]. A homologue of Sls1p, ScSls1p, was identified in *S. cerevisiae*, which was independently identified as Per100p [203] and Sil1p [206]. Further studies revealed that Sls1p binds directly to the ATPase domain of Kar2p and, in the presence of the J-domain of Sec63p, enhances ATP hydrolysis [110, 206]. Binding studies revealed that Sls1p prefers Kar2p when it is in the ADP-bound form, suggesting that it could act as a nucleotide-exchange factor [16]. *SLS1/SIL1* is a nonessential gene, but its overexpression can suppress the lethal phenotype observed in yeasts that lack both Ire1p, the kinase that signals the unfolded protein response in yeast, and Lhs1p, a second resident ER Hsp70 family member [206]. Recently, a mammalian nucleotide-releasing factor for BiP was identified in a yeast two-hybrid screen that used the ATPase domain of a BiP mutant as bait [43]. BAP (BiP-associated protein) shares low sequence homology with both Sls1p and HspBP1. BAP is an ER-localized glycoprotein that is ubiquitously expressed but which shows the highest levels of expression in tissues with a well-developed secretory pathway. BAP stimulates the ATPase activity of BiP by accelerating nucleotide release from BiP. Surprisingly, both ATP and ADP can be released by BAP in vitro, but, like Sls1p [16], BAP appears to bind better or more stably to the ADP-bound state of BiP [43]. This property may be critical in allowing BAP to drive the ATPase cycle of BiP forward. In contrast to Sls1p, whose expression level is increased by ER stress [206], BAP transcripts are not induced. In fact, BAP protein levels actually appear to decline [43], suggesting that mammalian cells might be able to regulate BiP release from substrates by controlling the ratio of BiP and BAP. It is not known whether BAP can also serve as a nucleotide-releasing factor for GRP170 or if another specific release factor exists. It has been suggested that nucleotide-exchange/releasing factors are not necessary for some Hsp70s including BiP, since nucleotide hydrolysis, not exchange, appears to be the rate-limiting step in the reaction [69, 143]. However, it is important to point out that these assays are done in the presence of only ATP and not with a combination of ATP and ADP as would be expected to occur in vivo, although currently there are no measurements of the relative ratio of the two nucleotides. The rate-limiting step in the ATPase cycle of DnaK is also ATP hydrolysis [144, 201], but it is clear that the nucleotide-exchange factor GrpE plays an important role in the DnaK cycle in vivo [173].

16.5 Organization and Relative Levels of Chaperones in the ER

Data obtained from a number of studies have demonstrated that multiple ER chaperones can associate with a given nascent protein. Sitia and coworkers demonstrated that unoxidized Ig light chains form disulfide bonds transiently with both PDI and ERp72 and suggested that these proteins may form a kind of affinity matrix in the ER that impedes the transport of unoxidized nascent proteins [179]. Similarly, both thyroglobulin [126] and HCGβ [60] can be cross-linked to BiP, GRP94, and ERp72 during their maturation, and the influenza hemagglutinin protein binds to a number of ER proteins including BiP, GRP94, calreticulin, and calnexin when cross-linking agents are added to the cells [200]. However, it was not clear from these studies whether the chaperones were binding as a complex or whether the individual chaperones were binding to distinct unfolded regions on these proteins. Recently, membrane-permeable cross-linking studies demonstrated the existence of a large ER-localized multi-protein complex bound to unassembled Ig heavy chains that is comprised of the molecular chaperones BiP, GRP94, CaBP1, PDI, ERdj3, cyclophilin B, ERp72, GRP170, UDP-glucosyltransferase (UDP-GT), and SDF2-L1 [149]. Except for ERdj3, and to a lesser extent PDI, this complex also forms in the absence of nascent protein synthesis and is found in a variety of cell types, suggesting that this subset of ER chaperones forms an ER network that can bind to unfolded protein substrates instead of existing as free pools that assemble onto substrate proteins. It is notable that most of the components of the calnexin/calreticulin system, which include some of the most abundant chaperones inside the ER, either are not detected in this complex or are only very poorly represented [149]. Further support for this type of sub-organellar organization of chaperones comes from a study in which fluorescence microscopy was used. The precursor of the human asialoglycoprotein receptor, H2a, and the free heavy chains of MHC class I molecules accumulated in a compartment containing calnexin and calreticulin, but not BiP, PDI, or UDP-GT, when proteasomal degradation was inhibited [113]. Together these studies suggest a spatial separation of the two chaperone systems that may account for the temporal interactions observed in other studies [84, 117].

16.6 Regulation of ER Chaperone Levels

Changes in the normal physical environment of the cell (e.g., decreases in pH, energy, oxygen, glucose, or other nutrients) can dramatically affect the normal biosynthesis of proteins in the ER and can result in the accumulation of unfolded proteins. Under these conditions, a signal transduction pathway, termed the unfolded protein response (UPR), is activated to protect the cell by preventing the formation of insoluble protein aggregates. The hallmark of the ER stress response, and perhaps the only component of the response that is truly ER-specific, is the coordinate transcriptional upregulation of most ER chaperones and folding enzymes [128]. A second characteristic of the UPR, and one unique to metazoans, is the inhibition of protein synthesis, which serves to limit the accumulation of unfolded proteins in the ER. This occurs via phosphorylation of the α-subunit of eukaryotic translation initiation factor 2 (eIF2α) at Ser^{51}, which reduces the frequency of translation initiation and thereby inhibits new protein synthesis [169].

The UPR pathway was first delineated in yeast. The identification of an unfolded protein response element (UPRE) in the yeast Kar2 (BiP) promoter [120] allowed investigators to use genetic approaches to characterize the signaling pathway. Ire1/Ern1, the first component to be identified, is a transmembrane, ER-localized kinase that possesses an N-terminal stress-sensing domain in the lumen of the ER and cytoplasmic kinase domain [46, 154]. In response to ER stress, Ire1 dimerizes and is phosphorylated, which serves to activate a unique endonuclease activity at its C-terminus [190]. The target of this activity is a precursor mRNA that encodes the Hac1 transcription factor. After cleavage by Ire1p and religation by Rlg1, Hac1p is synthesized and regulates the expression of UPR target genes by binding to the UPRE in their promoter [47]. This single signaling cascade is responsible for activating the UPR in yeast.

In higher eukaryotes, the elements of this signaling pathway are conserved but greatly expanded. Two Ire1 homologues exist in mammalian cells; Ire1α, which is ubiquitously expressed [202], and Ire1β, whose expression is restricted to gut epithelium [209]. Both proteins possess a lumen stress-sensing domain, a cytosolic kinase domain, and the unique endonuclease domain found in yeast Ire1p. The target of Ire1's endonuclease activity is XBP1 mRNA, which alters the C-terminus of this transcription factor so that it encodes a protein with both a DNA-binding domain and a strong transactivation domain [31, 224]. Although XBP1 was first identified in a screen to identify proteins that bind to ER stress-regulated elements in mammalian chaperone promoters [223], it does not appear that either Ire1 or XBP1 [9, 130] is required for chaperone upregulation during ER stress, since mouse embryonic fibroblasts that do not express these proteins are still capable of inducing the chaperones. Instead, it is possible that the ATF6 transcription factor, which is synthesized as an ER-localized transmembrane protein [91] is responsible for their induction [130]. Activation of the UPR in mammalian cells leads to transport of ATF6 to the Golgi, where it is cleaved by the S1P and S2P proteases [221], thus liberating the cytosolic transcription factor domain from the membranes. Cleaved ATF6 can then enter the nucleus, bind to conserved ER stress elements (ERSE) that are found in multiple copies in the promoters of most ER chaperones and folding enzymes, and presumably upregulate their transcription during ER stress [223]. The latter point remains to be formally demonstrated either by examining chaperone induction in cells that are null for ATF6 or by showing that endogenous ATF6 binds to the chaperone promoters in a stress-inducible manner. A third arm of the mammalian UPR is represented by PERK/PEK, an eIF-2α kinase that is responsible for the transient inhibition of protein synthesis that occurs during ER stress [89, 189]. Although PERK null cells can still upregulate ER chaperones and folding enzymes during ER stress, the magnitude of the response is not as high as in wild-type cells [88], suggesting that something downstream of PERK contributes to their transcriptional upregulation.

The mechanism by which the ER stress signal is transduced has been recently determined. Earlier studies revealed that the initial signal for activating the UPR was the accumulation of unfolded proteins in the ER [124] and that all the agents that induced the UPR would be expected to dramatically affect protein folding in this organelle [129]. However, not all unfolded proteins are able to activate the response. Apparently, those that bind to BiP do [124, 133], whereas those that bind to other chaperones do not [77]. Since only BiP-binding proteins appeared to activate the response, these studies suggested that levels of free BiP might be monitored by the cell to judge changes in the folding environment of the ER. Identification of the UPR transducer proteins allowed this to be examined directly. Studies revealed that both Ire1 and PERK were associated with BiP during normal physiological conditions, which kept them in a monomeric, non-activated state. Activation of the UPR with thapsigargin or DTT led to the rapid loss of BiP from the lumenal domain of these proteins and a concomitant oligomerization and activation of the two signal transducers [10]. Similar results have been obtained with yeast Ire1p [165]. Another study showed that the lumenal domain of ATF6 was also associated with BiP prior to stress, which in this case served to retain ATF6 in the ER [187]. Activation of the stress response leads to BiP release and transport of ATF6 to the Golgi, where the cytosolic transcription factor domain is liberated by the S1P and S2P proteases that reside there [221]. Thus, BiP directly regulates the UPR by controlling the activation status of the three transducers. It is likely that as stress conditions are alleviated and the pool of free BiP increases, BiP also plays a role in shutting down the response. In keeping with a central role of BiP in regulating the response, a recent study found that BiP is not readily translated early in the stress response even though BiP transcripts begin to increase very early [139].

16.7 Disposal of BiP-associated Proteins That Fail to Fold or Assemble

Proteins that have ultimately failed ER quality control are degraded to prevent their accumulation in the ER, which might either titrate out the components of the chaperone systems or form large insoluble aggregates that would be toxic to the cell. This turnover mechanism is termed ER-associated protein degradation (ERAD), which is conserved from yeast to mammals [87, 214]. The final steps of this ERAD process have been best characterized in yeast. Both malfolded proteins and excess subunits of multimeric proteins are retrotranslocated or dislocated back into the cytosol through a structure, which appears to be similar to the translocon used by nascent polypeptide chains to enter the ER lumen [25]. This retrotranslocation process is usually coupled with ubiquitination, which occurs at the cytosolic surface of the ER membrane. Ubiquitin (Ub) is a highly conserved small protein that is universally expressed in eukaryotic cells. Ubiquitination of substrates is a multi-step process that is dependent on a Ub-activating enzyme (E1), a Ub-conjugating enzyme (E2), and a Ub ligase (E3) enzyme [67, 99]. E1 binds to Ub, adenylates its C-terminus, and then binds to an E2 to transfer Ub to its catalytic subunit. E3, which is usually substrate-specific, will bring the substrate to an E2 and mediate the polyubiquitination process. E2 enzymes, such as Ubc6p and Ubc7p in yeast, are recruited by Cue1p or E3 enzymes such as Hrd1p/Der3p to the ER membrane and positioned near the translocon to directly facilitate the ERAD process [11, 75, 218]. The ubiquitinated ERAD substrates are then degraded by the cytosolic 26S proteasome [162]. This appears to be an important process in maintaining ER homeostasis during normal physiological conditions, since interfering with this process by either expressing mutants of the ERAD pathway [68, 158, 203] or using proteasomal inhibitors [29] results in activation of the ER stress pathway.

However, the upstream ERAD signals that help cells select malfolded proteins and feed them into the downstream degradation mechanism remain fairly elusive. One mechanism for identifying malfolded glycoproteins for ERAD involves the ER chaperones calnexin, calreticulin, and calmegin, which constitute a machinery called the "CNX cycle" [30, 92]. Glycoproteins with nine mannoses are allowed to bind to and are then released from calnexin by alternating actions of glucosidase II and UDP-GT. The incorrectly folded proteins are allowed multiple rounds of association and dissociation to acquire the correct conformation until the outermost unit of mannose from the middle branch of the sugar is cleaved by ER mannosidase I. Glycoproteins tagged with Man8-glycans now have a lower affinity for UDP-GT [168] but a higher affinity for EDEM [103]. Unlike CNX, EDEM is a UPR-inducible ER membrane protein with homology to α-mannosidase but lacks mannosidase activity. EDEM will then extract malfolded ERAD substrates from the CNX cycle and feed them into the downstream ERAD machinery via mechanisms that are still unclear [152, 163]. Calnexin is required for degradation of ERAD substrates in an in vitro system in which ER membranes from yeast are used [25]. Kar2p was shown to be important in keeping the ERAD substrates in a soluble and retrotranslocation-competent state in the yeast system [161]. Recent data on ERAD in mammalian cells have suggested that calnexin and BiP play sequential roles in identifying and targeting ERAD substrates for degradation [153].

16.8 Other Roles of BiP in the ER

In addition to their role in folding nascent proteins, both calnexin and BiP appear to aid the translocation of nascent polypeptide chains into the ER. BiP has been shown to "plug" the translocon during early stages of protein translocation to maintain the permeability barrier between the ER and cytosol [83]. This puts BiP in an ideal place to bind nascent chains as they enter the ER, and indeed a number of studies have shown that Kar2p together with Sec63p is required for the translocation of nascent proteins into the yeast ER [24, 184, 208]. However, there are currently no data to show a similar role for mammalian BiP. The recent identification of two mammalian homologues of Sec63 [22, 191] is certainly compatible with such a role. A final function of the ER is to house the calcium stores that are essential for many intracellular signaling pathways. Along with a number of other ER chaperones, BiP is a calcium-binding protein that contributes significantly to the calcium stores of the ER [137].

16.9 Concluding Comments

In conclusion, the ER is the site of most secretory protein synthesis, where aqueous channels must be opened to allow the nascent polypeptide chains to enter the ER lumen. Care must be taken to ensure that the permeability barrier between the ER and cytosol are preserved in order to maintain the unique environment of the ER. Once inside the ER lumen, the protein must fold and assemble into its mature tertiary and quaternary form. Proteins that fail to do so must be identified and targeted for retrotranslocation into the cytosol, where they become substrates for the 26S proteasome. Changes in the physiological environment of the ER that could affect protein maturation must be monitored and responded to by increasing ER chaperones and presumably proteins involved in maintaining and restoring the ER environment. BiP has been shown to be involved in all of these functions and to directly regulate activation of the UPR that maintains ER homeostasis. As such BiP can be considered a master sensor and regulator of ER function. All of the ER functions, with the exception of calcium storage, require BiP's ATPase activity and as a result are likely to involve BiP regulators including the ERdjs and BAP. Thus, we believe that functions for the regulators in these processes will be revealed in the future and that additional family members are likely to be discovered.

16.10 Experimental Protocols

16.10.1 Production of Recombinant ER Proteins

16.10.1.1 General Concerns

Production of large amounts of biologically active recombinant ER proteins suffers from the same difficulties as encountered with any other protein as well as a number of unique ones. The common ones like codon usage, size limitations, internal start sites, and inefficient translation are dealt with in other chapters, and thus there is no need to repeat them here. Instead we will focus on those problems unique to ER proteins and systems for dealing with them. Due to the oxidizing and calcium-rich environment of the ER and to the presence of an apparatus for assembling and adding *N*-linked glycans to nascent chains, many ER-localized proteins have evolved to fold properly in the presence of calcium and only when they have been glycosylated and have formed disulfide bonds. Thus, the expression of recombinant proteins in the bacterial cytosol can have deleterious effects on the folding of ER proteins. This said, large quantities of enzymatically active BiP protein (which has no internal disulfide bonds and is not glycosylated) have been produced by this method [213]. The targeting of ER proteins to the periplasmic space of bacteria can support the formation of disulfide bonds in proteins, which in some cases is sufficient to allow their proper folding [54, 197]. However, as bacteria do not possess the enzymes for *N*-linked glycosylation, proteins that require this modification for solubility or function must be produced in a eukaryotic system. In the end, the choice of expression systems often relies on empirical trials. The second problem encountered is where and how to tag an ER protein for purification purposes. The presence of an N-terminal-targeting sequence, which will be present in all cases except bacterial cytosolic expression, makes the choice and placement of a tag more difficult. If it is clear where cleavage of the signal sequence will occur and if any downstream sequence is required, then it is possible to add the tag sequence just C-terminal of the downstream sequence. Otherwise, tags need to be added to the C-terminus of the recombinant protein, which can interfere with ER retention mechanisms [156]. Tags of choice include both N- and C-terminal hexahistidine (His_6) and glutathione S-transferase (GST), which allows the recombinant protein to be purified on Ni-agarose or glutathione beads, respectively. The insertion of a cleavage sequence (i.e., Factor Xa, thrombin, TEV, or enterokinase) makes it possible to remove the tag after purification, although in the case of recombinant BiP protein, an N-terminal His_6 did not interfere with enzyme activity [213].

16.10.1.2 **Bacterial Expression**

In spite of the limitations for ER proteins, bacterial expression systems remain very popular due to their simplicity, speed, and the high production levels that can be obtained. Two sites of production are possible: cytosolic and periplasmic. If aggregation is a major problem with either expression site, it is possible to minimize this by using shorter times of IPTG induction, lower IPTG concentrations, or lower temperatures for induction. In all cases, lower expression levels occur, but higher yields of soluble, biologically active protein can be obtained [19, 43].

A: Cytosolic expression Secretory pathway proteins are targeted to the ER membrane by a stretch of 11–20 hydrophobic amino acids, which is often present at their N-termini and which may be removed after they are translocated into the ER lumen. Removal of this sequence from the cDNA ensures cytosolic expression, produces a protein more similar to the mature protein, and in some cases may increase the solubility of the recombinant protein. Algorithms are available for predicting the site of signal sequence cleavage [44]. This type of expression has been used successfully to produce recombinant wild-type and mutant BiP proteins [212, 213], the J-domains of ERdj proteins [41, 127, 188, 205], and BAP [43].

Materials and solutions for producing His-tagged proteins:

LB broth: 10 g L^{-1} tryptone, 5 g L^{-1} yeast extract, 10 g L^{-1} NaCl

IPTG (1 M): filter-sterilized and stored in aliquots at −20 °C

Ampicillin stock solution: 100 mg mL^{-1} in H_2O, filter-sterilized and stored in aliquots at −20 °C

Kanamycin stock solution: 25 mg mL^{-1} in H_2O, filter-sterilized and stored in aliquots at −20 °C

Lysis buffer: 50 mM NaH_2PO_4, 300 mM NaCl, pH 8.0, containing 10 mM imidazole

Wash buffer: 50 mM NaH_2PO_4, 300 mM NaCl, pH 8.0, containing 20 mM imidazole

Elution buffers: 50 mM NaH_2PO_4, 300 mM NaCl, containing 20 mM imidazole (at pH 7.0, 6.0, and 5.0).

Procedure:

1. Inoculate LB broth with M15 *E. coli* that have been transformed with the appropriate construct and incubate until OD_{600} reaches 0.7–0.9, then add 1 mM IPTG to induce protein expression.
2. After 1–3 h, pellet cells and resuspend in 8 mL lysis buffer containing 1 mg mL^{-1} lysozyme (add lysozyme fresh each time).
3. Transfer to a 30-mL Corex tube and incubate on ice for 5 min.
4. Sonicate on ice for total of 3 min and centrifuge the lysate at 8000 *g* for 30 min at 4 °C.
5. Apply the supernatant to a 1-mL bed-size Ni-NTA column that is equilibrated with lysis buffer and allow the sample to run through the column by gravity. Wash the column with 15 mL of wash buffer.
6. Elute the histidine-tagged protein by using a stepwise decrease in the pH of the elution buffer: pH 7.0 (3 mL), pH 6.0 (3 mL), and pH 5.0 (5 mL). Collect 1-mL fractions.
7. Run each fraction on an SDS-PAGE gel to find the histidine-tagged protein.
8. Pool the protein fractions. Keep the protein at 4 °C for immediate use; otherwise, concentrate the protein by a Centricon filter device, add glycerol to 50% (v/v), and store at −20 °C.

Alternative purification strategies:

1. Instead of eluting the Ni-agarose–bound protein with low pH, it is possible to use imidazole. In this case, 250 mM imidazole is added to the lysing buffer solution (50 mM NaH_2PO_4, 300 mM NaCl, pH 8.0). Binding and washing are done as above. Imidazole absorbs at 280 nm and therefore must be considered when monitoring elution and determining protein concentration.
2. Denaturants (either 6 M GuHCl or 8 M urea) can be added to aid in the solubilization of proteins. In this case the denaturant is added to a lysis buffer (100 mM NaH_2PO_4, 10 mM Tris, pH 8.0), which is used to bind the proteins to the column. After washing with the same buffer at pH 6.3, samples are eluted with the same buffer at pH 5.9 or pH 4.5. Solutions containing urea cannot be autoclaved and the pH must be adjusted just prior to their use. Solutions containing GuHCl precipitate in the presence of SDS, so samples must be TCA-precipitated before they can be analyzed on SDS gels. Refolding assays are the same as for other proteins that are obtained under denaturing conditions.
3. A GST-fusion system is widely used for purification of recombinant proteins because of its simplicity. The basic steps of this method are similar to those of the His_6-tagged system. The glutathione S-transferase sequence is added in frame to the target protein. Often, a cleavage site (i.e., Xa, thrombin, etc.) is added between the GST and the target protein in order to remove the GST moiety after purification. This method requires only PBS as the buffer to resuspend cells and to equilibrate and/or wash glutathione Sepharose beads. Bacteria are lysed by sonication followed by adding Triton X-100 to a final concentration of 1%. The recombinant protein can be eluted with glutathione-eluting solution

(100 mM reduced glutathione in sterile distilled water, aliquoted and stored at −20 °C).

B: Periplasmic expression The expression of an ER protein in the periplasmic space of bacteria, which supports the formation of disulfide bonds, can remedy the problems of insolubility or inactivity that are often encountered when secretory pathway proteins are expressed in cytosol. This targeting is achieved by fusing a bacterial signal sequence from ompT [40], ompA [19], pelB [174], or alkaline phosphatase [164] to the N-terminus of the protein. Although periplasmic expression allows a more simplified purification by osmotic shock [157], the yield of the desired protein is often lower than that achieved with cytoplasmic expression. Tags for purification can be engineered on the expressed protein either at the C-terminus or following the signal sequence cleavage site [197]. It is important to note that targeting secretory pathway proteins to the periplasmic space is not always sufficient to remedy insolubility problems, as the overexpressed recombinant proteins can still form inclusion bodies in this organelle [19]. This method was used to produce mouse BiP, which is free from contamination by DnaK [40].

Merits:

- The *Escherichia coli* host system provides a rapid and easily manipulable method for protein production.
- Very high expression rates can be obtained.
- A variety of expression systems are available.
- It is economical and easy to maintain the bacterial system compared to other expression systems

Limitations:

- Most ER-specific posttranslational modifications of the recombinant protein are not achieved.
- Eukaryotic proteins expressed intracellularly in *E. coli* often form inclusion bodies, which require denaturation, renaturation, and refolding processes that are not always successful.
- Codon usage for eukaryotic genes is different, which can result in lower protein expression.

16.10.1.3 Yeast Expression

Yeast systems for recombinant protein expression have proven attractive for the following reasons: (1) they allow production of soluble proteins from many different eukaryotes including mammalian species, (2) they support the production of proteins with the appropriate posttranslational modifications including *N*-linked and *O*-linked glycosylation, and (3) secretion of the protein to the extracellular medium can occur, making purification easier. Two different species of yeast are generally used to produce foreign proteins: *Saccharomyces cerevisiae* and *Pichia pastoris*. *S. cerevisiae* provides well-developed expression vectors and host strains, and many genetically modified strains are available. The advantages encountered in using *P. pastoris* are the fact that (1) 10- to 100-fold higher expression levels of recombinant proteins can be achieved compared to those in *S. cerevisiae* and (2) *N*-linked glycosylation occurs with a shorter oligosaccharide chain that lacks the terminal α1–3 glycan linkage found in *S. cerevisiae* but which is not found in mammalian proteins [34, 50]. Protein expression vectors for *S. cerevisiae* contain yeast signal sequences derived from various secreted proteins including invertase, α-factor, and acid phosphatase, which are fused in-frame to the desired protein. At present it is necessary to engineer a yeast signal sequence in front of the protein of interest when using *P. pastoris*, and studies have shown that codon usage and yeast poly A sequences can be important for synthesis of mammalian proteins in this system [177]. Protocols for using either organism can be found in *Current Protocols in Molecular Biology* and *Protein Science* and in the Invitrogen manual.

16.10.1.4 **Baculovirus**

The baculovirus-based system utilizes an insect virus to transfer genes to insect cell lines, such as Sf9, Sf21, and High Five cells. This eukaryotic expression method supports processing events and posttranslational modifications, such as phosphorylation, myristoylation, and palmitoylation, similar to those found in higher eukaryotes, and the majority of the overexpressed protein often remains soluble in insect cells, in contrast to the inclusion bodies that often form in bacteria. In addition, the machinery for targeting proteins to the ER is highly conserved between insect and vertebrate cells, so it is not necessary to replace the native ER-targeting sequence. However, there are data to indicate that some posttranslational processing events appear to be different from those found in vertebrate cells [170]. In addition, cleavage of signal sequences, removal of hormonal prosequences, and cleavage of polyproteins do not always occur properly in the expressed proteins [93]. Glycosylation is generally similar between insect and vertebrate cells, except that the *N*-linked oligosaccharides in insect cells remain in the high-mannose form and are not processed to the complex form containing fucose, galactose, and sialic acid. *O*-linked glycosylation has been reported, but this is less well-characterized in insect cells [106]. Protocols of producing recombinant proteins are available in books [1] or on websites and manuals of several companies that provide the vectors and transfection reagents.

Procedure:

1. If protein must be tagged for detection or isolation, it is necessary to add the tag either at the C-terminus or just after the signal sequence cleavage site.
2. Clone cDNA of interest into pFASTBAC and transform MAX efficiency DH10 BAC-competent cells (both from Gibco BRL-Inuitrogen) to produce recombinant bacmid for transfection. Spread serial dilutions of bacteria (10^{-1}, 10^{-2}, and 10^{-3}) on agar plates and incubate for 24–48 h at 37 °C.
3. Pick white colonies that contain the recombinant bacmid and inoculate overnight at 37 °C (option: streak to fresh plates to confirm the color).

4. Isolate bacmid DNA with the CONCERT High Purity Plasmid Isolation System (GibcoBRL-Invitrogen) or other methods developed for isolating large plasmids (>100 kb). For detailed steps, see the Bac-to-Bac Baculovirus Expression Systems instruction manual, GibcoBRL-Invitrogen.
5. Produce viruses by tranfecting insect cells with bacmid DNA and CELLFECTIN reagent according to instructions. A viral titer of 1×10^7 to 4×10^7 pfu mL^{-1} can be expected from the initial transfection. For the method to determine the viral titer, see Bac-to-Bac Baculovirus Expression Systems instruction manual, GibcoBRL-Invitrogen.
6. Amplify virus stock by infecting a log-phase suspension or monolayer Sf9 culture at a multiplicity of infection (MOI) of 0.01 to 0.1.
7. Harvest virus 48 h later by pelleting cells at 1000 g for 20 min, and store supernatant containing virus at 4 °C, protected from light.
8. Culture Sf9 cells in spinner flasks at 27 °C, and infect log-phase cells with virus at an MOI of 0.01 to 0.1.
9. After 48 h (or optional time for maximum protein expression), collect an aliquot ($0.5–1.0 \times 10^6$) of infected cells and control cells to monitor protein production. Lyse cells with lysing buffer (50 mM NaH_2PO_4 pH 8.0, 300 mM NaCl, 1% of NP-40, 1 mM PMSF, and 10 mM imidazole) and run cell extract on a SDS-PAGE gel. The protein expression levels obtained with the baculovirus system are usually not as high as those achieved through bacterial expression methods. Thus, a Western blot instead of coomassie staining is usually employed to detect the protein.
10. Harvest the remaining cells by centrifugation for 3 min at 500 g and wash twice with PBS (media residue may affect the purification efficiency).
11. Freeze and thaw cells twice to break the cell membrane and resuspend in lysing buffer (as described above).
12. Centrifuge the lysate at 10 000 g for 10 min at 4 °C.
13. Purify the recombinant protein as described above.

This method was developed by researchers at Monsanto. It is faster and more efficient than the traditional methods, which take weeks to pick a positive recombinant virus clone and achieve final high-titer stocks. However, the vector is not yet available with a C-terminal His_6 tag. Therefore, an additional step is required to tag the gene of interest with either a C-terminal tag or an N-terminal tag following the signal sequence.

Merits:

- It is a eukaryotic expression system using modification, processing, and transport systems similar to those found in vertebrate cells.
- Adaptation for growth in suspension cultures makes it possible to obtain moderate quantities of recombinant protein with relative ease.
- The majority of the proteins produced are soluble and possess the correct post-translational modifications.

- Adaptation to non-serum cultures cuts the costs and makes purification of the secreted proteins easier.
- It is safe and not hazardous to humans unlike viral expression systems.
- The virus can be preserved at 4 °C for long periods of time.

Limitations:

- Different modifications other than those observed in vertebrate species may occur.
- It is more time-consuming than bacterial expression methods.
- Insect cells recover slowly from frozen stocks.
- Cells grow fast but cannot be cultured at very low concentrations: frequent splitting at a 2:5 ratio every 2–3 days leads to high serum and medium costs.
- Additional tagging steps are required to isolate ER proteins.

16.10.1.5 **Mammalian Cells**

Mammalian cells are good hosts for expressing properly folded glycoproteins that possess all the correct posttranslational modifications that control their physiological functions as native proteins. There are a number of choices for expressing proteins in mammalian cells as described below, including both transient and stable methods.

A: Plasmid-based transfection When plasmid DNA is transferred into mammalian host cells, the majority of DNA remains extra-chromosomal and can support expression. The cells used for expression often express the SV40 large T antigen, which will replicate vectors containing the SV40 origin of replication (ori). Popular cell lines include COS and 293T cells. The vectors used for expression of the protein of interest often use strong viral promoters such as the cytomegalovirus (CMV) or adenovirus promoters, as well as the SV40 ori. DNA is introduced into the cells by Ca^{2+} precipitation, DEAE-dextran, or lipofection. These methods are quick and can provide a one-time production of from one to a few hundred micrograms of purified protein [1, 219]. On the other hand, a small percent of the DNA can integrate into a transcriptionally active chromosomal locus and express the protein on a permanent basis from generation to generation if an appropriate selectable marker is included in the transfection. This stable transfection allows the selection of clones that express high levels of the recombinant proteins. Popular host cell lines for stable expression include CHO (Chinese hamster ovary) cells, 293T (transformed human embryonic kidney) cells, myeloma cells, and BHK-21 (baby hamster kidney) cells. The DNA can be transfected into these host cells by using Ca^{2+} precipitation, DEAE-dextran, lipofection, or electroporation. Detailed protocols are described in Refs. [1, 2]. Finally, it is possible to obtain stable cell lines that can express toxic proteins by using regulated expression systems (i.e., tetracycline [76], heavy metal [215], or hormone-regulated systems [220, 227]).

Merits:

- The proteins produced are identical to the native proteins with good solubility and proper structures and functions.
- Transient expression is timesaving compared to the baculovirus system, if a single transient expression can produce enough protein.
- For stable lines, clones can be frozen and thawed for later repeat use.
- Regulated expression makes long-term production of toxic protein possible.
- There is no exposure to viruses and therefore no significant safety issue.

Limitations:

- It is time-consuming to screen a high-production, stable line.
- The product yield is often low compared to the baculovirus system; this may be improved by using suspension cultures that are grown in spinner flasks (1–50 mg L^{-1}).
- The types of host cells to choose from are limited.
- It is relatively expensive due to the high costs for medium and serum.

B: Viral infection Retrovirus infection is a useful technique for efficiently producing stable cell lines that express a heterologous protein. The non-viral gene is cloned into a retrovirus vector and transduced into a packaging cell line to produce viruses that carry the gene of interest. The virus can then be used to infect host cells as long as they are dividing and carry the appropriate retroviral receptors on their surface. After integration into the host chromosome, the retrovirus will stably produce a single copy of the viral genome including the gene of interest from the viral LTR. In order to lower the risk of this system, replication-incompetent retroviral vectors are derived from proviruses by deleting some or all of the genes encoding virion structural proteins. These vectors need to be transfected into a packaging cell line that provides these genes in *trans*. By using packaging lines that produce an ecotropic murine retrovirus instead of an amphotropic one, it is possible to further decrease the risk to humans. Although the efficiency of viral infection can be as high as 100%, retroviral infection is not commonly used to produce recombinant proteins because the LTR is not a strong promoter, and therefore the yield of recombinant protein is usually fairly low. However, retroviral infection is widely used to study the protein of interest in mammalian cells, due to the high efficiency and broad type of host cell lines that can be used.

Materials and reagents:

293T cells and NIH3T3 cells, transfection reagents (it is recommended to use Ca^{2+} precipitation or FuGENE6 (Boehringer-Mannheim))

0.45-μM filter bottle

Polybrene 2 mg mL^{-1} in water, aliquoted and stored at −20 °C

Procedures:

1. For virus production, seed 3×10^6 293T cells/100-mm dish and incubate overnight.
2. Transfect 20 μg total DNA including retrovirus vector containing the gene of interest along with helper vector, and incubate overnight.
3. In the morning, remove media, wash cells, and add 10 mL fresh media.
4. Incubate 4–5 h, aspirate media, and add 4 mL of fresh media to cells.
5. Begin harvesting the media containing retroviral particles by removing and adding fresh media every 4–6 h for 2 days. Combine all the media and filter through a 0.45-μM filter. Aliquot virus stock and save at −80 °C.
6. For recombinant protein expression, seed 2×10^5 NIH3T3 cells/100-mm dish and incubate overnight.
7. Aspirate all the media from the dishes, add enough virus stock containing 8–10 μg mL^{-1} polybrene to cover the cells, and then add additional virus every 3 h, at least 3 times.
8. Harvest cells ~ 48 h after infection or treat cells with proper agent (i.e., G418) to select infected cells.

A second viral system that is commonly used for transient expression of a protein is the DNA-based adenovirus system [219]. Very high levels of expression can be obtained that can reach up to 10–20% of total protein. Adenovirus particles can target both dividing and non-dividing cells from a majority of human and many non-human cell types and produce multiple copies of the gene of interest [39, 71]. This wide variety of target cells combined with high-level gene expression levels makes the adenovirus system ideal for a number of research applications, including recombinant protein production, gene therapy, gene function analyses, antisense strategies, vaccine development, and transgenic animal studies. The disadvantages of adenoviral delivery include safety concerns due to the fact that they can infect human cells and the rather complex and time-consuming methods required to produce the virus. The lentivirus (an HIV-related retrovirus) represents another expression system that has been used to constitutively express proteins in a wide range of mammalian cells [192]. An advantage is that the host cell does not need to be replicating, but for recombinant protein production, this is rarely a concern. Again, their wide target range represents both an advantage and a disadvantage. Detailed protocols for using viral strategies for producing recombinant proteins can be found in the ViraPower manual under the section on adenoviral/lentiviral expression systems and in the Invitrogen manual. Methods for purifying recombinant proteins from mammalian cells infected with these viruses are similar to those for baculovirus-infected insect cells, as described above.

Merits:

- These viral delivery systems all use mammalian cells to produce proteins and should therefore be identical to the native proteins with good solubility and proper structures and functions.

- A broad range of host cells can be infected, including in some cases non-dividing cells.
- The high efficiency of infection makes it unnecessary to sort positive cells.
- Very high levels of expression can be obtained with adenovirus.

Limitations:

- Large fragments of DNA (>8 kb) are difficult to transduce with this approach.
- The protein production levels achieved with retroviruses are relatively low.
- Adenovirus and retroviruses can infect human cells; therefore, extra caution must be used when handling them.
- Methods for producing and infecting with viruses are more cumbersome than simple transfection with other vectors.

16.10.2
Yeast Two-hybrid Screen for Identifying Interacting Partners of ER Proteins

The identification of interacting proteins can often provide insights into the regulation and function of a given protein. The yeast two-hybrid system has proven extremely useful for identifying interacting proteins, but it is best suited for cytosolic or nuclear proteins due to the underlying principles of the procedure. The bait and target proteins must both be expressed in the cytosol and transported to the nucleus to drive transcription of a reporter gene [141]. Thus, to find proteins that interact with an ER lumenal protein, the target cDNAs must have lost their ER-targeting sequence when the library is made, both the bait and target must fold properly in the cytosol in the absence of an oxidizing environment and *N*-linked glycosylation, and the interaction between the two proteins must be supported by the reducing, low-calcium environment of the cytosol and nucleus. In spite of these limitations, it has been possible to identify proteins that interact with secretory proteins by this method [43]. The problems encountered with finding interacting proteins for ER chaperones become even larger. Because chaperones interact with many unfolded proteins, it is reasonable to expect that by expressing the secretory pathway in the wrong environment, many "false positives" would be found. In fact, Hsps are often identified as false positives in screens with various bait proteins (http://www.fccc.edu/research/labs/golemis/main_false.html). However, a two-hybrid screen with cytosolic Hsc70 was successfully employed to identify HiP, a protein that interacts with the ATPase domain of Hsc70 and stabilizes its binding to ADP [101]. In order to minimize false positives, the peptide-binding domain of Hsc70 was removed. Similarly, a screen with the ATPase domain of human Hsp70 identified HspBP1, a protein that is abundant in heart and skeletal tissues and that regulates nucleotide release from Hsp70 [111, 178]. To identify BiP-interacting protein(s), the ATPase domain of a BiP mutant (T229G) that is unable to hydrolyze ATP [212] was used as bait in the screen. Since BiP is an ER-resident protein, the ER-targeting signal sequence was removed to prevent the Gal4-mutant ATPase domain fusion protein from being targeted to the ER [43]. Interestingly, none of the regulatory proteins that interact with cytosolic or nuclear Hsp70s were obtained in this screen.

There are several different commercially available yeast two-hybrid systems (BD Sciences, Invitrogen, and Stratagene) and a number of comprehensive descriptions of them [3, 61]. Briefly, the yeast transcription factor GAL4 contains an N-terminal DNA-binding domain (DNA-BD) and a C-terminal activation domain (AD). These two domains are functionally separable but must be brought together to initiate transcription. A known protein (the bait) is fused to the GAL4 DNA-BD. A cDNA library is expressed as fusions to GAL4 AD. When the bait and library fusion proteins interact, the DNA-BD and AD are brought into close proximity to allow the two components of GAL4 to come together and drive the transcription of a reporter gene (i.e., *lacZ*). Yeast colonies are then screened for β-galactosidase activity.

Materials and solutions:

Strain: HF7c transformed with the pAS (bait) vector. The transformed strain must be negative in an X-gal assay. In our case, a bait vector contained the ATPase domain of a BiP mutant without the signal sequence fused to Gal4 (pAS(T229G)44K) [43].

cDNA library: a human liver cDNA library (pACTII base) from Clontech (now BD Sciences). The cDNA library was amplified according to the manufacturer's instruction, and the amplified cDNA library was checked for its quality by a PCR reaction for known ER proteins and by restriction enzyme digestion of the library to detect the smear of inserted cDNAs.

SD synthetic minimal media: SD/dropout (DO) medium with DO supplements (Clontech): to prepare SD/–Leu/–Trp agar, you will need to combine SD minimal agar with –Leu/–Trp DO supplement (#8608-1).

1 M 3-AT (3-amino-1,2,4-triazole; Sigma #A-8056): prepared in deionized H_2O and filter-sterilized. Store at 4 °C. Plates containing 3-AT can be stored at 4 °C for up to 2 months.

X-gal (20 mg mL^{-1} in DMF): dissolve 5-bromo-4-chloro-3-indolyl-b-D-galactopyranoside in *N,N*-dimethylformamide. Store in the dark at −20 °C.

LiSORB: 100 mM lithium acetate, 10 mM Tris pH 8.0, 1 mM EDTA, 1 M sorbitol.

Procedure:

1. Incubate the recipient strain (HF7c) carrying pAS-bait vector in 2 mL of SD-Trp overnight at 30 °C.
2. Transfer the overnight culture to 100 mL of SD-Trp in a 500-mL flask and incubate overnight at 30 °C.
3. Dilute with YPD broth to obtain $OD_{600} = 0.3 \sim 0.4$.
4. Incubate the above culture in a 1000-mL flask for 3–4 h until $OD_{600} = 0.5 \sim 0.8$.

The tissue or cells must first be made into single-cell suspensions and homogenized to break the cells into the various organelles as described in Refs. [207, 210].

From rat liver:

1. Remove liver from a freshly sacrificed rat that has fasted overnight. Keep the liver on ice as much as possible, weigh the liver, and then mince it coarsely with scissors or a razor blade.
2. Homogenize the minced liver at 20% (w/v) in homogenization medium (0.25–0.3 M sucrose) using five up-down strokes at 1700 rpm with a high-torque, motor-driven pestle.
3. Filter the homogenate through four layers of cheesecloth.

From cell lines:

1. Collect about 20–70 million cells, wash twice with PBS, and resuspend in 1 mL homogenization buffer (0.25 M sucrose, 10 mM HEPES pH 7.5).
2. Transfer to the Dounce homogenizer and homogenize 8–15 strokes, depending on the cells and the homogenizer.

Centrifugation to separate microsomes:

1. In both cases, spin the homogenate at 600 g in a microcentrifuge for 10 min at 4 °C to remove nuclei and unbroken cells.
2. Remove supernatant and centrifuge at 25 000 g using a SW50.1 rotor for 10 min at 4 °C to remove mitochondria.
3. Remove supernatant and either centrifuge at 124 000 g using a SW50.1 rotor for 1 h at 4 °C to pellet all remaining membranous organelles or layer over a stepwise sucrose gradient (0.6 M, 1.0 M, 1.3 M, and 2.0 M sucrose) to separate the rough ER, smooth ER, Golgi, and plasma membranes from the cytosol [17].
4. The purity of each fraction can be assayed by Western blotting an aliquot of each fraction for the various for organelle markers (Table 16.2).

A simplified method for isolating microsomes has been described [207, 210]. Briefly, cells are resuspended in homogenization buffer (25 mM HEPES-KOH, 125 mM KCl, pH 7.2) and broken by Dounce homogenization. The crude homogenates are centrifuged at 600 g to remove cell debris and nuclei. The supernatant containing ER microsomes and cytosol is centrifuged at 10 000 g to pellet the microsomes, which remain contaminated with mitochondria.

16.10.3.4 Determination of Topology

A: Protease protection assays Proteins or protein domains located inside the ER are protected by the impermeable membrane from digestion with proteases. Resistance to treatment with protease K can reveal whether the entire protein is located inside the ER or if a portion of the protein is located inside the ER. When the membranes are treated with detergent and disrupted, the microsomes no longer protect the protein or domains unless they themselves are protease-resistant. Thus, detergent treatment of vesicles before protease should be included as a control. Either microsomes extracted from cell lines or tissues can be used if antibodies are available for multiple epitopes or at least for known epitopes. Alternatively, the full-length cDNA can be used in coupled in vitro transcription/translation assays that have dog pancreatic microsomes added to the reaction mixture. Because the cDNA is the primary protein translated, antibodies are not required to detect the protein. Known ER-resident proteins are recommended as controls (e.g., BiP and calnexin) to ensure that the membranes remain intact and the proteinase K behaves appropriately.

Procedure:

1. The microsomes are divided into three aliquots: one is left untreated, one is digested with 150 μg mL^{-1} proteinase K, and one is made 1% NP40 prior to treatment with protease and all are incubated for 1 h at 37 °C.
2. The samples are treated with 1 mM PMSF for 15 min to neutralize the protease.
3. SDS sample buffer is added to each reaction and the samples are analyzed by SDS-PAGE. If the samples are translated and labeled in vitro, they can be visualized by autoradiography; if not, the proteins must be detected by Western blotting.

B: Integral membrane vs. soluble protein Inspection of the primary sequence can often predict whether a protein is likely to be a membrane protein if it contains a second hydrophobic stretch of amino acids that can serve as a transmembrane domain. However, a single transmembrane domain can also act to anchor a protein in the membrane, if it remains uncleaved. Additionally, hydrophobic sequences that are a bit short or that contain charged amino acids can make it hard to predict whether or not they are transmembrane domains. Three methods are used to determine whether a protein is anchored to the membranes or soluble in the ER. The first relies on the ability of low concentrations of mild detergents to allow soluble proteins to leak out of ER membrane vesicles without completely destroying the membrane or solubilizing integral proteins.

Procedure:

1. Microsomes are pelleted and resuspended in 100 μL of cold PBS buffer alone or PBS containing either 0.1–0.2% digitonin or 1% in deoxycholic acid (DOC).
2. After rocking at 4 °C for 1 h, samples are centrifuged at 10 000 g for 5 min to sediment residual membranes [226].
3. The supernatant and pellet are separated and prepared for SDS-PAGE and Western blotting. To avoid cross-contamination, the pellet should be rinsed once with cold PBS.

Transmembrane proteins remain in the pellet fraction, whereas soluble lumenal proteins are released into the supernatant in the presence of low concentrations of digitonin. However, both protein types are found in the supernatant fraction when the microsomes are treated with 1% DOC. A good control for membrane-associated proteins is calnexin, and BiP can be used for soluble lumenal proteins.

The second method utilizes the fact that treatment of ER vesicles with high-pH buffers transforms the vesicles into open sheets, which release soluble lumenal contents but retain integral membrane proteins. The membrane sheets can be pelleted by centrifugation to separate membrane and soluble proteins [159], which can be analyzed as above. The third method takes advantage of preferential detergent binding to hydrophobic regions of proteins and of the fact that at a given temperature, detergents reach a "cloud point" that allows them to be precipitated from solutions. The detergent TritonX-114 is particularly useful for this application, since it precipitates at 20 °C and goes back into solution at 0 °C [18]. Unlike the characteristics of many detergents, these temperatures are not denaturing to proteins. Briefly, labeled cells or ER vesicles are resuspended in 200 μL separation buffer (10 mM Tris-HCl, pH 7.4, 150 mM NaCl), containing 0.5–1.0% Triton X-114 at 0 °C, and then overlaid on a sucrose cushion (the same buffer containing 6% sucrose, 0.06% Triton X-114). The sample is incubated at 30 °C for 3 min, and the clouded sample is centrifuged for 3 min at 300 *g* at 30 °C. The membrane proteins pellet to the bottom with the detergent phase, and soluble proteins remain in the upper aqueous phase. After separating the phases, detergent should be re-added to the aqueous fraction, and the detergent fraction should be resuspended in separation buffer. Both fractions should be cooled to 0 °C and re-fractionated. This can be repeated several times to increase the purity of the fractions [18].

16.10.3.5 *N*-linked Glycosylation

The covalent addition of oligosaccharides to translocating proteins is one of the major biosynthetic functions of the ER. Most secretory proteins that are made in the ER are glycoproteins. Conversely, very few proteins in the cytosol or nucleus are glycosylated, and those that are receive only a single trisaccharide addition that is not *N*-linked [90]. Oligosaccharides are specifically added co-translationally to asparagine residues in the sequence Asn-X-Ser/Thr, where X is any amino acid except proline. In addition, a simpler sugar modification can be added to the –OH group of serine, threonine, or hydroxylysine residues in the Golgi. Asparagine-linked or *N*-linked glycosylation is a stepwise procedure of oligosaccharide addition and removal that begins with the addition of a high-mannose dolichol intermediate. Glucose and mannose residues are trimmed while the protein is still in the ER, and these events serve as recognition structures for calnexin and calreticulin. The presence of *N*-linked glycosylation on a protein is a clear indication that it was synthesized in the ER. Tunicamycin, a fungal metabolite that inhibits the addition of the dolichol intermediate to nascent chains [132], is widely used to determine whether a protein is glycosylated. However, this approach is useful only in examining newly synthesized proteins and works best in biosynthetic assays. The *N*-linked glycans are further processed as the protein is transported through the *cis*, medial, and *trans* Golgi. A number of glycosidases that have specificity for different glycan processing intermediates have been purified from bacteria and fungus and can be used to distinguish the various forms. Endoglycosidase H (Endo H) recognizes only the immature *N*-glycans found on proteins that are still in the ER [204], not those on proteins that have been transported to the Golgi [122]. These Endo H–resistant glycoproteins can be digested with either Endo D, which specifically recognizes and cleaves the processed complex *N*-linked sugars [121], or *N*-glycosidase F (PNGase F), which removes the *N*-linked oligosaccharides from both high-mannose ER forms and processed post-ER forms of glycoproteins [172, 198]. Thus, sensitivity of a protein to these various endoglycosidases can provide information on the subcellular localization of a glycoprotein. Because these treatments are done on cell lysates or immunoprecipitated proteins, they can be used in both biosynthetic and Western blotting assays.

A: Tunicamycin treatment

1. Pretreat cells with 1 μg mL^{-1} (effective concentration varies from 0.15–10 μg mL^{-1}) tunicamycin for ~1 h before labeling the cells with ^{35}S methionine. If Western blotting will be used to detect the protein of interest, longer treatment is required, depending on the synthetic and turnover rate of the protein. This treatment induces an unfolded protein response in cells, which can lead to lower protein synthetic rates and more rapid protein turnover.
2. Lyse cells and immunoprecipitate protein with the appropriate antibody.
3. Subject protein samples to SDS-PAGE analyses and autoradiography. Glycosylation slows down the protein mobility on the gel; thus, a tunicamycin-treated, non-glycosylated protein runs faster than the non-treated protein on SDS gels. One oligosaccharide adds ~2 kDa to the apparent molecular weight of the protein.

B: Endo H digestion

1. Obtain proteins either from in vitro translation reactions performed in the presence of microsomes or from immunoprecipitates of cell lysates. If in vitro–translated proteins are used, it is highly recommended to pellet the membrane fraction and discard the reticulocyte lysate before lysing to eliminate contamination with untranslocated, non-glycosylated forms of the protein.
2. After washing immunoprecipitated material three times, wash one additional time with reaction buffer (0.1 M sodium citrate, pH 5.5).
3. Resuspend in 49 μL of reaction buffer containing PMSF and add 3 mU of Endo H.
4. Incubate at 30–37 °C overnight.
2. Stop reaction by adding SDS sample buffer and heat to 95 °C for 5 min.
3. Subject to SDS-PAGE gel to detect the mobility changes by autoradiography or Western blotting.

In general, 50–250 mU of Endo H is sufficient to deglycosylate up to 1 mg high-mannose glycoprotein when incubated overnight. The pH optimum of Endo H is ~5–6. For glycans that are not readily removed, it is sometimes possible to increase their accessibility to the enzyme by denaturing the protein [1]. This is accomplished by adding SDS to 0.25% and mercaptoethanol (2-ME) to 0.5% and heating the sample to 95 °C for 5 min. The denatured sample should then be diluted ~1:3, and PMSF (1 mM) should be added to protect the protein and to prevent the inactivation of Endo H.

C: PNGase F digestion Obtain protein samples from in vitro translation reactions done in the presence of microsomes or from material immunoprecipitated from cell lysates. If in vitro–translated proteins are used, it is highly recommended to pellet the membrane fraction and discard the reticulocyte lysate to eliminate the untranslocated, non-glycosylated proteins.

1. Denature proteins (75–100 μg in ≤25 μL) by adding 25 μL freshly made denaturing buffer (0.5% SDS, 1% 2-ME) and heat to 95 °C for 5–15 min.
2. Add the following in order (50 μL total): 25 μL 0.5 M TrisCl, pH 8.0; 10 μL 0.1 M 1,10-phenanthroline; 10 μL 10% nonionic detergent; 5 μL 200 to 250 mU mL^{-1} PNGase F.
3. Incubate overnight at 30 °C.

The optimum working pH for PNGase F is from 7 to 9, although enzyme has some activity between pH 5 and 7.

D: Manipulation of glycosylation sites Not all ER proteins are glycosylated. The localization and topology of non-glycosylated ER protein can be determined by the protease-protection and detergent-release experiments described above. In addition, chimera proteins can be made in which the protein of interest is fused in-frame to a known ER glycoprotein that has had its signal sequence removed. This strategy was recently used to demonstrate that ERdj4 is a type II protein with an uncleaved signal sequence serving as the membrane anchor and with its J-domain oriented inside the ER lumen [188]. This method can also be used to confirm orientation of transmembrane proteins. Alternatively, *N*-linked glycosylation sites (NXS/T) can often be introduced into the protein of interest by replacing a single amino acid. In most cases the introduced site will be glycosylated, which can be monitored with tunicamycin treatment or Endo H digestion.

16.10.4
Nucleotide Binding, Hydrolysis, and Exchange Assays

16.10.4.1 Nucleotide-binding Assays

All Hsp70 family members bind and hydrolyze ATP, which regulates the ability of their C-terminal domain to bind and release unfolded protein substrates [26, 74].

The amino-terminal ~44-kDa domain of Hsp70 proteins encodes the nucleotide-binding site [35], and crystal structures for the ATP- and ADP-bound forms of bovine and rat Hsc70 have been determined [62, 63, 146]. The ATPase domain is comprised of two lobes with a deep cleft between them with nucleotide binding occurring at the base of the cleft. Due to the high degree of sequence similarly among family members, all are likely to form a similar structure. The differences between the ATP-bound and ADP-bound states of this isolated domain are limited to minor rearrangements of solvent-assessable side chains and those present near the scissile bond. Conversely, the two nucleotides dramatically alter the conformation of the full-length molecule as detected by protease-sensitivity assays [35, 115, 145, 212] and changes in far-UV CD spectra [42]. Currently there are no structures available for any full-length Hsp70s to determine how nucleotide binding alters the overall structure the Hsp70s. The ability to bind nucleotide can be altered by mutations [145, 212], and the nucleotide-bound state of an Hsp70 can be regulated by proteins that prevent exchange [101] or proteins that promote exchange [5, 19, 78, 110, 111]. When a potential cofactor for an Hsp70 protein is identified, an initial step is to assay its effects on the binding of nucleotide to the client Hsp70. Also, it is important to check whether the cofactor itself binds nucleotides. There is a critical difference between cofactors that bind and transfer nucleotides to target proteins and those that interact with the target protein and alter its ability to bind nucleotides. To date all of the known Hsp70 cofactors that have been identified are of the second class: non-nucleotide-binding. Following are two simple techniques to obtain qualitative data as to whether a protein binds nucleotides [43, 109] and one quantitative technique [69, 70]. The two simpler techniques worked equally well when characterizing BAP, which serves as a nucleotide-releasing factor for BiP.

A: Nitrocellulose membrane method

Procedure:

1. Mix 6 μg of each purified protein, such as BSA (negative control), recombinant BiP (positive control), or recombinant BAP, with 50 μCi [α-^{32}P]-ATP, add 2× buffer A (40 mM HEPES pH 7.2, 100 mM KCl and 10 mM $MgCl_2$, and 20 mM DTT) to make final reaction mix 1× at a total volume of 50 μL, and incubate for 10 min at 30 °C.
2. During the incubation, wet a piece of nitrocellulose membrane with chromatography buffer (20 mM Tris-HCl, pH 7.5, 20 mM NaCl, and 1 mM DTT). Remove extra buffer B, but avoid drying the nitrocellulose membrane.
3. Spot 2 μL of protein samples onto the wet nitrocellulose membrane and let it dry until spot marks disappear.
4. Rinse the nitrocellulose membrane with 10 mL buffer B twice for 10 s each to remove free ATP, which does not bind to the filter.
5. Air-dry the nitrocellulose membrane and expose to X-ray film.

B: Micro spin-column chromatography

Materials and solutions:

MicroSpin G-50 columns (Amersham Bioscience): In our experience, approximately 2/3 of the applied protein was recovered from a column. Do not apply more than 10 μL on a column; otherwise, free [α-^{32}P]ATP will appear in flow-through.

Thin-layer chromatography plate: polyethyleneimine cellulose sheets (Sigma)

Developing solution: 0.5 M formic acid and 0.5 M LiCl

Developing chamber

A desktop centrifuge with refrigerating function

Procedure:

1. Mix 3 μg of each protein with 50 μCi [α-^{32}P]-ATP with 2× buffer A to a final volume of 10 μL and incubate for 10 min at 30 °C.
2. During the incubation period, prepare MicroSpin G-50 columns at 4 °C according to manufacture's instruction and keep columns at 4 °C.
3. After incubation, transfer 10 μL of each protein sample onto a MicroSpin G-50 column and centrifuge for 2 min at 3000 rpm, 4 °C.
4. Analyze 2 μL of the flow-through solution (which contains the protein and bound nucleotide but not the free nucleotide) by thin-layer chromatography on a polyethyleneimine cellulose sheet using 0.5 M formic acid and 0.5 M LiCl.
   (Option: If separation of hydrolyzed products is not required, a nitrocellulose membrane can be substituted for the polyethyleneimine cellulose sheet and analyzed as above.)
5. Air-dry the TLC plate and expose to X-ray film.

C: Equilibrium dialysis More lengthy equilibrium dialysis assays are useful for determining actual binding constants for nucleotides and have been described previously [69, 70].

16.10.4.2 ATP Hydrolysis Assays

In addition to binding nucleotide, all Hsp70 proteins have a very weak intrinsic ATPase activity [79, 213, 228], and both potassium and magnesium are required for full activity [115, 167, 217]. This intrinsic ATPase activity can be regulated by positive and negative cofactors [13, 43, 72, 100, 111, 151]. The procedure introduced here employs use of a radioactive isotope of ATP, which can be either [γ-^{32}P]ATP or [α-^{32}P]ATP. In terms of separation, free [γ-^{32}P] gives slightly better resolution from unhydrolyzed ATP than does [α-^{32}P]ADP.

Materials and solutions:

Purified proteins should be dialyzed against the ATPase assay buffer.

[γ-^{32}P]ATP (3000 Ci mmol^{-1}, Amersham Bioscience)

Unlabeled ATP stock: 100 mM ATP in H_2O, pH 7.0. Store in small aliquots at −80 °C

Mixture of [γ-^{32}P]ATP and unlabeled ATP: 2 μL [γ-^{32}P]ATP (3000 Ci mmol^{-1}), 5 μL of unlabeled ATP (10 mM), and 43 μL of ATPase assay buffer. Final concentrations: 20 μCi and 1 mM ATP

ATPase assay buffer (freshly made): 20 mM HEPES, pH 7.2, 50 mM KCl, 5 mM $MgCl_2$, and 10 mM DTT

TLC plate: polyethyleneimine cellulose sheets (Sigma)

Developing solution: 0.5 M formic acid and 0.5 M LiCl

Procedure:

1. Pre-incubate the reaction mixture (0.5 μM BiP with and without the desired concentration of a cofactor (or combinations of cofactors) in ATPase assay buffer at a total volume of 45 μL) for 5 min at room temperature. Also, set up a negative control omitting BiP and adding bovine serum albumin.
2. Start the reaction by adding 5 μL of the mixture of [γ-^{32}P]ATP and unlabeled ATP to the pre-incubated reaction mixture.
3. Incubate at 30 °C.
4. At desired time points, remove a 2-μL aliquot of the reaction mixture and spot on a TLC plate immediately.
5. Once all time points have been spotted and dried, transfer plate to a developing chamber that has been saturated with the developing solution.
6. Allow the solvent front to migrate to approximately 15 cm from the spotted line, dry immediately, and expose the TLC plate to X-ray film.
7. Quantify ATP hydrolysis by a PhosphoImager (Molecular Dynamics).

16.10.4.3 Nucleotide Exchange Assays

In contrast to GTP exchange factors, all of the Hsp70 nucleotide-exchange or nucleotide-releasing factors that have been identified thus far do not bind to nucleotides themselves. Instead, the binding of these cofactors causes a conformational change in the Hsp70 protein, which decreases its affinity for ADP and/or ATP. The technique currently employed for nucleotide-exchange assays utilizes size-exclusive mini spin-column chromatography performed on a desktop centrifuge [43, 100, 151]. Spin-column chromatography is very rapid and convenient, allowing the immediate separation of protein-associated nucleotides from free nucleotides [136]. The assay consists of two parts. The first part is to prepare and isolate a complex of [α-^{32}P]ATP-bound Hsc70 (here, BiP). The second part is to exchange the bound hot nucleotides with cold ones and then to recover the Hsp70. Therefore, the radioactive signal decreases more quickly if a nucleotide-exchange cofactor is present.

Materials and solutions:

Purified proteins should be dialyzed against ATPase assay buffer prior to assaying.

MicroSpin G-50 columns (Amersham Bioscience)

[α-^{32}P]ATP (3000 Ci mmol^{-1}, Amersham Bioscience)

Unlabeled ATP stock: 100 mM ATP in H_2O, pH 7.0. Store in small aliquots at −80 °C

Mixture of 100 μCi [α-^{32}P]ATP and 250 μM unlabeled ATP in ATPase assay buffer

ATPase assay buffer (freshly made): 20 mM HEPES, pH 7.2, 50 mM KCl, 5 mM $MgCl_2$, and 10 mM DTT

TLC plate: polyethyleneimine cellulose sheets (Sigma)

Developing solution: 0.5 M formic acid and 0.5 M LiCl

Procedure:

1. Incubate 2.5 μM BiP with 50 μM [α-^{32}P]ATP in 50 μL ATPase assay buffer for 5 min at room temperature and for 5 min on ice.
2. Apply 10 μL of reaction mixture per a cold MicroSpin G-50 column to separate the [α-^{32}P]ATP-BiP complex from free nucleotide, which is retained on the column.
3. Transfer approximately 0.5 μM [α-^{32}P]ATP-BiP complex to ATPase assay buffer containing 100 μM cold ATP and any regulator (e.g., 1 μM ERdj4 or 1 μM ERdj4 plus 1 μM BAP). Keep the final volume at 50 μL.
4. Incubate the reaction mixture at room temperature and remove 12-μL aliquots at 1, 3, 5, and 10 min. Freeze the removed aliquots immediately in an ethanol–dry ice bath.
5. Thaw aliquots one at a time at room temperature, and as soon as ice particles disappear, put 10 μL onto a cold MicroSpin G-50 column and spin.
6. Keep the flow-through on ice and repeat 5–6 steps for all other aliquots.
7. Remove 2 μL of the flow-through and spot on TLC plate.
8. Dry immediately and transfer to a developing chamber that has been saturated with developing solution.
9. Allow samples to migrate ~15 cm from the origin, dry immediately, and analyze the bound nucleotide by autoradiography.
10. Quantify signal by a PhosphoImager (Molecular Dynamics) if necessary.

Note: Unlike with Hsc70, we experienced difficulty in maintaining BiP in the ATP-bound form after the second micro-spin column. A significant portion of the nucleotide was consistently hydrolyzed to ADP. This might be due to biochemical differences between BiP and Hsc70; for example, the k_{cat} value of BiP ($k_{cat} = 0.40$ min^{-1}) is three times higher than that of Hsp70 ($k_{ca} = 0.14\ min^{-1}$) [13, 213]. Therefore, it is strongly recommended to do the assay as rapidly at possible and to keep everything at 4 °C to maintain as much [α-^{32}P]ATP-BiP as possible.

16.10.5
Assays for Protein–Protein Interactions in Vitro/in Vivo

ER chaperones form complexes with unfolded proteins that are essential to their functions of aiding and monitoring protein folding. These complexes can be detected using both in vitro and in vivo assays. Commonly used methods include in vitro GST pull-down assays, co-immunoprecipitation, covalent cross-linking agents, and yeast two-hybrid interactions. It is highly recommended to use a combination of several methods to avoid false positives or false negatives that can result from limitations in the experimental assays.

16.10.5.1 In Vitro GST Pull-down Assay

The in vitro GST pull-down assay is a convenient way to detect protein-protein interactions. One of the proteins of interest can be produced as a recombinant protein by the methods described above or can be in vitro translated. The other protein is produced as a GST fusion protein. The two are allowed to interact, and the GST-fusion protein is isolated by binding to glutathione-conjugated beads. Samples are separated by SDS-PAGE. Detection of the second interacting protein can be done by autoradiography, Western blotting, or simple Coomassie blue staining of the gel. This procedure allows the investigator to alter binding conditions, such as pH, divalent cations, or nucleotides. Some chaperone–co-chaperone interactions (i.e., Hsp70-DnaJ or BiP-BAP) are nucleotide-dependent [16, 43, 211].

Methods:

1. Produce one of the proteins of interest with a GST tag as described [226]. The other protein of interest should not be GST-tagged but can possess another tag for purification purposes (i.e., His_6, HA, or myc tag). The concerns about tagging ER proteins described above may have less of an effect in this type of in vitro assay.
2. Wash the glutathione beads with the appropriate assay buffer three times at 4 °C.
3. Gently rock the GST-fusion protein with the glutathione beads for 20 min at 4 °C.
4. Pellet beads and remove unbound GST-fusion proteins by washing with assay buffer.
5. Add the second protein to the GST-fusion protein–bound beads in a 100-μL reaction mix. Generally, for detecting BiP's interaction with its cofactors, use 1 mM ATP or ADP in standard ATPase buffer.
6. Rock the reaction system gently for 1 h at 4 °C.

7. Wash the beads with assay buffer in the absence or presence of nucleotide as indicated.
8. Separate bound proteins on SDS gels and stain with Coomassie blue or transfer for Western blotting.

16.10.5.2 **Co-immunoprecipitation**

The in vivo association between two proteins can be detected by immunoprecipitating one protein and examining the proteins that co-precipitate. It can be carried out under a variety of conditions, including the use of covalent cross-linking agents or mild detergents to stabilize complexes. In a typical experiment, cells are disrupted and a whole-cell extract is prepared under non-denaturing conditions. The protein of interest is precipitated from the lysate by using an appropriate antibody together with protein A/G beads to isolate the target protein along with any interacting proteins. The precipitate can be analyzed for the presence of other proteins directly by autoradiography or after ionic detergent disruption of the complex followed by re-precipitation with specific antisera, if the proteins are labeled, or by Western blotting or peptide sequencing, if the proteins are not labeled. This approach can be used for both native and epitope-tagged exogenous proteins. Negative controls to test the specificity of the interactions are crucial.

Methods:

1. Prepare whole-cell lysates by lysing with a nonionic detergent like DOC or NP40 (i.e., 50 mM Tris, pH 7.5, 150 mM NaCl, 0.5% DOC, and 0.5% NP40).
2. Clear the cell lysates of nuclei and other cellular debris by centrifugation at full speed in a microcentrifuge for 10 min at 4 °C.
3. Remove supernatant to at least two Eppendorf tubes and add corresponding antibody to one tube to a final concentration of ~10 μg mL^{-1} and leave second tube as is.
4. Rock the tubes gently for 1.5 h at 4 °C.
5. Add 50 μL protein A (for most antibody isotypes) or protein G (for IgM antibodies) beads (resuspended at a 1:1 volume of beads to lysing buffer).
6. Rock gently for 0.5 h at 4 °C.
7. Wash samples three times with lysing buffer containing 400 mM NaCl to reduce nonspecific binding.
8. Perform SDS-PAGE gel followed by autoradiography or Western blotting analysis.

This method can also be modified for using recombinant proteins or proteins translated in vitro.

16.10.5.3 **Chemical Cross-linking**

The detergents used to solubilize cellular membranes can often disrupt protein-protein interactions. Membrane-permeable, thiol-cleavable, covalent cross-linkers, such as dithiobis (succinimidyl propionate) (DSP) [33], have been widely used to stabilize ER chaperone–client protein complexes for investigation [147–149, 181,

188]. After cross-linking, cell extracts can be immunoprecipitated and complexes can be analyzed by one- or two-dimensional gel electrophoresis that combines non-reducing and reducing electrophoretic separation in perpendicular directions.

Protocol:

1. Wash cells three times in PBS to remove serum protein.
2. Resuspend cells in cross-linking buffer (25 mM HEPES-KOH, 125 mM KCl, pH 8.3).
3. Prepare a fresh 5-mg mL^{-1} solution of DSP in dimethyl sulfoxide and add to the cells to achieve a final concentration of 150 μg mL^{-1}; control sample is treated with DMSO containing no cross-linker.
4. Incubate on ice for 1 h with occasional shaking, and then quench with glycine (100 mM) for an additional 15 min on ice.
5. Lyse the cells and immunoprecipitate with appropriate antibody (see co-immunoprecipitation section for the detailed procedure).
6. For one-dimensional gel separation, load samples under reducing conditions to separate components of the complex.
7. For two-dimensional gel separation, electrophorese the sample under non-reducing conditions to separate different cross-linked complexes that may be present. Cut the gel strip corresponding to a single sample lane from the first gel and equilibrate in 5 mL of reducing SDS sample buffer for 40 min at room temperature on a rocker to reduce DSP and to liberate the various proteins in the complex. Then place the gel strip on the top of a second gel and run at a 90° angle to the first.
8. After electrophoresis, either stain the gel with Coomassie blue or silver nitrate or transfer for Western blotting.

Other membrane-permeable and -reversible cross-linkers include ethyleneglycolbis (succinimidylsuccinate) (EGS) [6], m-maleimidobenzoyl-*N*-hydroxysuccinimide ester (MBS) [27], dimethyl adipimidate (DMA) [28], and dimethyl suberimidate (DMS) [28].

16.10.5.4 **Yeast Two-hybrid System**

The interaction of proteins can also be assayed in vivo through the use of a modified yeast two-hybrid system. Briefly, one protein is fused to the GAL4 DNA-binding domain and the other to the GAL4 transactivation domain. The vectors encoding them are used to transform yeast cells, and the resulting colonies are screened for interaction by plating on selective media as described above.

16.10.6 In Vivo Folding, Assembly, and Chaperone-binding Assays

16.10.6.1 **Monitoring Oxidation of Intrachain Disulfide Bonds**

The folding of proteins in the ER is often stabilized by the formation of disulfide bonds between cysteine residues that are juxtaposed in the folded protein. These bonds can restrain the conformation of a denatured protein, which can result in an increased mobility of the protein on SDS gels as compared to the same protein with the disulfide bonds broken by a reducing agent such as DTT or 2-ME. This property has been used to monitor the oxidation status of ER proteins, as an indication of their folded state [20, 21, 96, 131]. Because most ER proteins fold co-translationally [7, 38], a method was developed to maintain a pool of unoxidized protein in the cell with the reversible reducing agent DTT [20]. These proteins appear to be incompletely or unstably folded, as they are bound to ER chaperones during this time [20, 96, 131]. Removal of DTT from the culture allows the ER to reestablish an oxidizing environment and allows the pool of unfolded proteins to fold and form disulfide bonds, all of which is followed by monitoring their migration on non-reducing SDS gels. NEM, an alkylating agent, is added at the time of lysis to prevent post-lysis oxidation of free cysteine residues. Assembly of multimeric proteins that are stabilized by disulfide bonds can be monitored similarly on non-reducing gels, a technique that was originally used to demonstrate that BiP binds to incompletely assembled Ig intermediates but not to completely assembled H_2L_2 molecules [17].

16.10.6.2 **Detection of Chaperone Binding**

In many cases the binding of chaperones to nascent proteins is rapid and can even occur on proteins before they have reached their full length, making it very difficult or often impossible to catch these transient associations. For mutant proteins, or proteins that require subunit assembly to complete their folding, this can be easier. In most cases, pulse-chase experiments are preferable, as they allow the investigator to specifically monitor newly synthesized proteins that are more likely to be the targets of chaperone binding. Cells are disrupted with gentle nonionic detergents (e.g., NP40, DOC, digitonin, 3-[(3-cholamidopropyl)dimethylammonio]propanesulfonate (CHAPS), dodecylmaltoside (DDM), and Triton X-100) to aid in the preservation of protein-protein interactions. The chaperone-protein complexes are detected by co-immunoprecipitation assays, which can often be done in both directions [96]. Antibodies to the chaperone co-precipitate the client protein, which can be analyzed under non-reducing conditions to monitor its oxidation status, and antibodies to the client protein can be used to identify the chaperone(s) that co-precipitate. When the interaction of BiP with proteins is being monitored, it is often important to add apyrase to the lysing buffer. This hydrolyzes ATP to AMP and ensures that the cytosolic pools of ATP do not induce post-lysis release of BiP. A second method that has proven useful is to co-express BiP ATPase mutants with the protein of interest using a transient transfection method. The ATPase mutants still interact with most client proteins but are not released either in vivo or in vitro, so the associations are easier to detect [49, 96, 155]. These types of co-immunoprecipitation experiments have been used to demonstrate that BiP binds to those regions of proteins that remain unoxidized [96, 131, 140].

For other ER chaperones (e.g., PDI, GRP94, ERdj3, and ERdj4) that do not bind ATP, the use of apyrase is not useful in stabilizing complexes, and mutants have not been identified that bind more stably to client proteins in vivo. In addition, the binding of some of these chaperones to client proteins is sensitive to the detergents that are used to disrupt membranes and gain access to ER proteins [149]. In these cases, membrane-permeable chemical cross-linking agents, such as DSP, have been very useful in stabilizing complexes for co-precipitation experiments, allowing investigators to monitor the binding of these chaperones to unfolded proteins [147–149, 181, 188]. After immunoprecipitating the client protein or the individual chaperone, the complex is dissociated with SDS sample buffer containing reducing agents. This liberates the various components, which can then be separated by electrophoresis on SDS gels. However, because client proteins and chaperone complexes are disrupted with 2-ME or DTT, it is not possible to examine the oxidation status of the associated protein.

Acknowledgements

We wish to thank Ms. Melissa Mann for help in preparing and editing the manuscript.

References

1 (1995). *Current Protocols in Protein Science* John Wiley & Sons, Inc.
2 Ausubel, F. M., Brent, R., Kingston, R. E., Moore, D. D., Seidman, J. G., Smith, J. A., & Struhl, K. (1989). *Current Protocols in Molecular Biology* John Wiley and sons, New York, NY.
3 Bai, C. & Elledge, S. J. (1996). Gene identification using the yeast two-hybrid system. *Methods Enzymol.* **273**, 331–347.
4 Banecki, B., Liberek, K., Wall, D., Wawrzynow, A., Georgopoulos, C., Bertoli, E., Tanfani, F., & Zylicz, M. (1996). Structure-function analysis of the zinc finger region of the DnaJ molecular chaperone. *J. Biol. Chem.* **271**, 14840–14848.
5 Banecki, B. & Zylicz, M. (1996). Real time kinetics of the DnaK/DnaJ/GrpE molecular chaperone machine action. *J. Biol. Chem.* **271**, 6137–6143.
6 Baskin, L. S. & Yang, C. S. (1980). Cross-linking studies of cytochrome P-450 and reduced nicotinamide adenine dinucleotide phosphate-cytochrome P-450 reductase. *Biochem.* **19**, 2260–2264.
7 Bergman, L. W. & Kuehl, W. M. (1979). Formation of an intrachain disulfide bond on nascent immunoglobulin light chains. *J. Biol. Chem.* **254**, 8869–8876.
8 Bergman, L. W. & Kuehl, W. M. (1979). Formation of intermolecular disulfide bonds on nascent immunoglobulin polypeptides. *J. Biol. Chem.* **254**, 5690–5694.
9 Bertolotti, A., Wang, X., Novoa, I., Jungreis, R., Schlessinger, K., Cho, J. H., West, A. B., & Ron, D. (2001). Increased sensitivity to dextran sodium sulfate colitis in IRE1beta-deficient mice. *J. Clin. Invest* **107**, 585–593.
10 Bertolotti, A., Zhang, Y., Hendershot, L. M., Harding, H. P., & Ron, D. (2000). Dynamic interaction of BiP and ER stress transducers in the unfolded-protein response. *Nat. Cell Biol.* **2**, 326–332.
11 Biederer, T., Volkwein, C., & Sommer, T. (1997). Role of Cue1p in ubiquitination and degradation at the ER surface. *Science* **278**, 1806–1809.
12 Bies, C., Guth, S., Janoschek, K.,

Nastainczyk, W., Volkmer, J., & Zimmermann, R. (1999). A Scj1p homolog and folding catalysts present in dog pancreas microsomes. *Biol. Chem.* **380**, 1175–1182.

13 Bimston, D., Song, J., Winchester, D., Takayama, S., Reed, J. C., & Morimoto, R. I. (1998). BAG-1, a negative regulator of Hsp70 chaperone activity, uncouples nucleotide hydrolysis from substrate release. *EMBO Journal* **17**, 6871–6878.

14 Blond-Elguindi, S., Cwirla, S. E., Dower, W. J., Lipshutz, R. J., Sprang, S. R., Sambrook, J. F., & Gething, M. J. (1993). Affinity panning of a library of peptides displayed on bacteriophages reveals the binding specificity of BiP. *Cell* **75**, 717–728.

15 Boisrame, A., Beckerich, J. M., & Gaillardin, C. (1996). Sls1p, an endoplasmic reticulum component, is involved in the protein translocation process in the yeast Yarrowia lipolytica. *J. Biol. Chem.* **271**, 11668–11675.

16 Boisrame, A., Kabani, M., Beckerich, J. M., Hartmann, E., & Gaillardin, C. (1998). Interaction of Kar2p and Sls1p is required for efficient co-translational translocation of secreted proteins in the yeast Yarrowia lipolytica. *J. Biol. Chem.* **273**, 30903–30908.

17 Bole, D. G., Hendershot, L. M., & Kearney, J. F. (1986). Posttranslational association of immunoglobulin heavy chain binding protein with nascent heavy chains in nonsecreting and secreting hybridomas. *J. Cell. Biol.* **102**, 1558–1566.

18 Bordier, C. (1981). Phase separation of integral membrane proteins in Triton X-114 solution. *J. Biol. Chem.* **256**, 1604–1607.

19 Bowden, G. A. & Georgiou, G. (1990). Folding and aggregation of beta-lactamase in the periplasmic space of Escherichia coli. *J. Biol. Chem.* **265**, 16760–16766.

20 Braakman, I., Helenius, J., & Helenius, A. (1992). Manipulating disulfide bond formation and protein folding in the endoplasmic reticulum. *EMBO. J.* **11**, 1717–1722.

21 Braakman, I., Hoover Litty, H., Wagner, K. R., & Helenius, A. (1991). Folding of influenza hemagglutinin in the endoplasmic reticulum. *J. Cell. Biol.* **114**, 401–411.

22 Brightman, S. E., Blatch, G. L., & Zetter, B. R. (1995). Isolation of a mouse cDNA encoding MTJ1, a new murine member of the DnaJ family of proteins. *Gene* **153**, 249–254.

23 Brodsky, J. L., Goeckeler, J., & Schekman, R. (1995). BiP and Sec63p are required for both co- and posttranslational protein translocation into the yeast endoplasmic reticulum. *Proc. Natl. Acad. Sci. U.S.A.* **92**, 9643–9646.

24 Brodsky, J. L. & Schekman, R. (1993). A Sec63p-BiP complex from yeast is required for protein translocation in a reconstituted proteoliposome. *J. Cell Biol.* **123**, 1355–1363.

25 Brodsky, J. L., Werner, E. D., Dubas, M. E., Goeckeler, J. L., Kruse, K. B., & McCracken, A. A. (1999). The requirement for molecular chaperones during endoplasmic reticulum-associated protein degradation demonstrates that protein export and import are mechanistically distinct. *J. Biol. Chem.* **274**, 3453–3460.

26 Bukau, B. & Horwich, A. L. (1998). The Hsp70 and Hsp60 chaperone machines. *Cell* **92**, 351–366.

27 Burgess, A. J., Matthews, I., Grimes, E. A., Mata, A. M., Munkonge, F. M., Lee, A. G., & East, J. M. (1991). Chemical cross-linking and enzyme kinetics provide no evidence for a regulatory role for the 53 kDa glycoprotein of sarcoplasmic reticulum in calcium transport. *Biochim. Biophys. Acta* **1064**, 139–147.

28 Burt, H. M. & Jackson, J. K. (1990). Role of membrane proteins in monosodium urate crystal-membrane interactions. II. Effect of pretreatments of erythrocyte membranes with membrane permeable and impermeable protein cross-linking agents. *J. Rheumatol.* **17**, 1359–1363.

29 Bush, K. T., Goldberg, A. L., & Nigam, S. K. (1997). Proteasome inhibition leads to a heat-shock response, induction of endoplasmic reticulum chaperones, and thermotolerance. *J. Biol. Chem.* **272**, 9086–9092.

30 Cabral, C. M., Liu, Y., & Sifers, R. N. (2001). Dissecting glycoprotein quality control in the secretory pathway. *Trends Biochem. Sci.* **26**, 619–624.

31 Calfon, M., Zeng, H., Urano, F., Till, J. H., Hubbard, S. R., Harding, H. P., Clask, S. G., & Ron, D. (2002). IRE1 couples endoplasmic reticulum load to secretory capacity by processing the *XBP-1* mRNA. *Nature* **415**, 92–96.

32 Cao, X., Zhou, Y., & Lee, A. S. (1995). Requirement of tyrosine- and serine/threonine kinases in the transcriptional activation of the mammalian grp78/BiP promoter by thapsigargin. *J. Biol. Chem.* **270**, 494–502.

33 Carlsson, J., Drevin, H., & Axen, R. (1978). Protein thiolation and reversible protein-protein conjugation. N-Succinimidyl 3-(2-pyridyldithio)propionate, a new heterobifunctional reagent. *Biochem. J.* **173**, 723–737.

34 Cereghino, G. P., Cereghino, J. L., Ilgen, C., & Cregg, J. M. (2002). Production of recombinant proteins in fermenter cultures of the yeast Pichia pastoris. *Curr. Opin. Biotechnol.* **13**, 329–332.

35 Chappell, T. G., Konforti, B. B., Schmid, S. L., & Rothman, J. E. (1987). The ATPase core of a clathrin uncoating protein. *J. Biol. Chem.* **262**, 746–751.

36 Cheetham, M. E. & Caplan, A. J. (1998). Structure, function and evolution of DnaJ: conservation and adaptation of chaperone function. *Cell Stress & Chaperones* **3**, 28–36.

37 Chen, S. & Smith, D. F. (1998). Hop as an adaptor in the heat shock protein 70 (Hsp70) and hsp90 chaperone machinery. *J. Biol. Chem.* **273**, 35194–35200.

38 Chen, W., Helenius, J., Braakman, I., & Helenius, A. (1995). Cotranslational folding and calnexin binding during glycoprotein synthesis. *Proc. Natl. Acad. Sci. U.S.A* **92**, 6229–6233.

39 Chengalvala, M. V., Lubeck, M. D., Selling, B. J., Natuk, R. J., Hsu, K. H., Mason, B. B., Chanda, P. K., Bhat, R. A., Bhat, B. M., Mizutani, S., et al. (1991). Adenovirus vectors for gene expression. *Curr. Opin. Biotechnol.* **2**, 718–722.

40 Chevalier, M., King, L., & Blond, S. (1998). Purification and properties of BiP. *Methods Enzymol.* **290**, 384–409.

41 Chevalier, M., Rhee, H., Elguindi, E. C., & Blond, S. Y. (2000). Interaction of murine BiP/GRP78 with the DnaJ homologue MTJ1. *J. Biol. Chem.* **275**, 19620–19627.

42 Chirico, W. J., Markey, M. L., & Fink, A. L. (1998). Conformational changes of an Hsp70 molecular chaperone induced by nucleotides, polypeptides, and N-ethylmaleimide. *Biochem.* **37**, 13862–13870.

43 Chung, K. T., Shen, Y., & Hendershot, L. M. (2002). BAP, a mammalian BiP associated protein, is a nucleotide exchange factor that regulates the ATPase activity of BiP. *J. Biol. Chem.* **277**, 47557–47563.

44 Claros, M. G., Brunak, S., & von Heijne, G. (1997). Prediction of N-terminal protein sorting signals. *Curr. Opin. Struct. Biol.* **7**, 394–398.

45 Corsi, A. K. & Schekman, R. (1997). The lumenal domain of Sec63p stimulates the ATPase activity of BiP and mediates BiP recruitment to the translocon in Saccharomyces cerevisiae. *J. Cell Biol.* **137**, 1483–1493.

46 Cox, J. S., Shamu, C. E., & Walter, P. (1993). Transcriptional induction of genes encoding endoplasmic reticulum resident proteins requires a transmembrane protein kinase. *Cell* **73**, 1197–1206.

47 Cox, J. S. & Walter, P. (1996). A novel mechanism for regulating activity of a transcription factor that controls the unfolded protein response [see comments]. *Cell* **87**, 391–404.

48 Craven, R. A., Egerton, M., & Stirling, C. J. (1996). A novel Hsp70

of the yeast ER lumen is required for the efficient translocation of a number of protein precursors. *EMBO J.* **15**, 2640–2650.

49 Creemers, J. W., van de Loo, J. W., Plets, E., Hendershot, L. M., & Van De Ven, W. J. (2000). Binding of BiP to the processing enzyme lymphoma proprotein convertase prevents aggregation, but slows down maturation. *J. Biol. Chem.* **275**, 38842–38847.

50 Cregg, J. M., Vedvick, T. S., & Raschke, W. C. (1993). Recent advances in the expression of foreign genes in Pichia pastoris. *Biotechnology (N. Y.)* **11**, 905–910.

51 Crowley, K. S., Liao, S., Worrell, V. E., Reinhart, G. D., & Johnson, A. E. (1994). Secretory proteins move through the endoplasmic reticulum membrane via an aqueous, gated pore. *Cell* **78**, 461–471.

52 Cunnea, P. M., Miranda-Vizuete, A., Bertoli, G., Simmen, T., Damdimopoulos, A. E., Hermann, S., Leinonen, S., Huikko, M. P., Gustafsson, J. A., Sitia, R., & Spyrou, G. (2003). ERdj5, an endoplasmic reticulum (ER)-resident protein containing DnaJ and thioredoxin domains, is expressed in secretory cells or following ER stress. *J. Biol. Chem.* **278**, 1059–1066.

53 Dierks, T., Volkmer, J., Schlenstedt, G., Jung, C., Sandholzer, U., Zachmann, K., Schlotterhose, P., Neifer, K., Schmidt, B., & Zimmermann, R. (1996). A microsomal ATP-binding protein involved in efficient protein transport into the mammalian endoplasmic reticulum. *EMBO J.* **15**, 6931–6942.

54 Digilio, F. A., Morra, R., Pedone, E., Bartolucci, S., & Rossi, M. (2003). High-level expression of Aliciclobacillus acidocaldarius thioredoxin in Pichia pastoris and Bacillus subtilis. *Protein Expr. Purif.* **30**, 179–184.

55 Dorner, A. J., Wasley, L. C., & Kaufman, R. J. (1990). Protein dissociation from GRP78 and secretion are blocked by depletion of cellular ATP levels. *Proc. Natl. Acad. Sci. U.S.A.* **87**, 7429–7432.

56 Dudek, J., Volkmer, J., Bies, C., Guth, S., Muller, A., Lerner, M., Feick, P., Schafer, K. H., Morgenstern, E., Hennessy, F., Blatch, G. L., Janoscheck, K., Heim, N., Scholtes, P., Frien, M., Nastainczyk, W., & Zimmermann, R. (2002). A novel type of co-chaperone mediates transmembrane recruitment of DnaK-like chaperones to ribosomes. *EMBO J.* **21**, 2958–2967.

57 Easton, D. P., Kaneko, Y., & Subjeck, J. R. (2000). The hsp110 and Grp170 stress proteins: newly recognized relatives of the Hsp70s. *Cell Stress. Chaperones.* **5**, 276–290.

58 Ellgaard, L., Molinari, M., & Helenius, A. (1999). Setting the standards: quality control in the secretory pathway. *Science* **286**, 1882–1888.

59 Feldheim, D., Rothblatt, J., & Schekman, R. (1992). Topology and functional domains of Sec63p, an endoplasmic reticulum membrane protein required for secretory protein translocation. *Mol. Cell. Biol.* **12**, 3288–3296.

60 Feng, W., Matzuk, M. M., Mountjoy, K., Bedows, E., Ruddon, R. W., & Boime, I. (1995). The asparagine-linked oligosaccharides of the human chorionic gonadotropin beta subunit facilitate correct disulfide bond pairing. *J. Biol. Chem.* **270**, 11851–11859.

61 Fields, S. & Song, O. (1989). A novel genetic system to detect protein-protein interactions. *Nature* **340**, 245–246.

62 Flaherty, K. M., DeLuca Flaherty, C., & McKay, D. B. (1990). Three-dimensional structure of the ATPase fragment of a 70K heat-shock cognate protein. *Nature* **346**, 623–628.

63 Flaherty, K. M., Wilbanks, S. M., DeLuca Flaherty, C., & McKay, D. B. (1994). Structural basis of the 70-kilodalton heat shock cognate protein ATP hydrolytic activity. II. Structure of the active site with ADP or ATP bound to wild type and mutant ATPase

fragment. *J. Biol. Chem.* **269**, 12899–12907.

64 Fleischer, S. & Kervina, M. (1974). Subcellular fractionation of rat liver. *Methods Enzymol.* **31**, 6–41.

65 Flynn, G. C., Chappell, T. G., & Rothman, J. E. (1989). Peptide binding and release by proteins implicated as catalysts of protein assembly. *Science* **245**, 385–390.

66 Flynn, G. C., Pohl, J., Flocco, M. T., & Rothman, J. E. (1991). Peptide-binding specificity of the molecular chaperone BiP. *Nature* **353**, 726–730.

67 Freiman, R. N. & Tjian, R. (2003). Regulating the regulators: lysine modifications make their mark. *Cell* **112**, 11–17.

68 Friedlander, R., Jarosch, E., Urban, J., Volkwein, C., & Sommer, T. (2000). A regulatory link between ER-associated protein degradation and the unfolded-protein response. *Nat. Cell Biol.* **2**, 379–384.

69 Gao, B., Emoto, Y., Greene, L., & Eisenberg, E. (1993). Nucleotide binding properties of bovine brain uncoating ATPase. *J. Biol. Chem.* **268**, 8507–8513.

70 Gao, B., Greene, L., & Eisenberg, E. (1994). Characterization of nucleotide-free uncoating ATPase and its binding to ATP, ADP, and ATP analogues. *Biochem.* **33**, 2048–2054.

71 Garnier, A., Cote, J., Nadeau, I., Kamen, A., & Massie, B. (1994). Scale-up of the adenovirus expression system for the production of recombinant protein in human 293S cells. *Cytotechnology* **15**, 145–155.

72 Gassler, C. S., Wiederkehr, T., Brehmer, D., Bukau, B., & Mayer, M. P. (2001). Bag-1M accelerates nucleotide release for human Hsc70 and Hsp70 and can act concentration-dependent as positive and negative cofactor. *J. Biol. Chem.* **276**, 32538–32544.

73 Gaut, J. R. & Hendershot, L. M. (1993). Mutations within the nucleotide binding site of immunoglobulin-binding protein inhibit ATPase activity and interfere with release of immunoglobulin heavy chain. *J. Biol. Chem.* **268**, 7248–7255.

74 Gething, M. J. & Sambrook, J. (1992). Protein folding in the cell. *Nature* **355**, 33–45.

75 Gilon, T., Chomsky, O., & Kulka, R. G. (2000). Degradation signals recognized by the Ubc6p–Ubc7p ubiquitin-conjugating enzyme pair. *Mol. Cell Biol.* **20**, 7214–7219.

76 Gossen, M. & Bujard, H. (1992). Tight control of gene expression in mammalian cells by tetracycline-responsive promoters. *Proc. Natl. Acad. Sci. U.S.A.* **89**, 5547–5551.

77 Graham, K. S., Le, A., & Sifers, R. N. (1990). Accumulation of the insoluble PiZ variant of human alpha 1-antitrypsin within the hepatic endoplasmic reticulum does not elevate the steady-state level of grp78/BiP. *J. Biol. Chem.* **265**, 20463–20468.

78 Gross, M. & Hessefort, S. (1996). Purification and characterization of a 66-kDa protein from rabbit reticulocyte lysate which promotes the recycling of hsp 70. *J. Biol. Chem.* **271**, 16833–16841.

79 Ha, J. H. & McKay, D. B. (1994). ATPase kinetics of recombinant bovine 70 kDa heat shock cognate protein and its amino-terminal ATPase domain. *Biochemistry.* **33**, 14625–14635.

80 Haas, I. G. & Wabl, M. (1983). Immunoglobulin heavy chain binding protein. *Nature* **306**, 387–389.

81 Hamilton, T. G. & Flynn, G. C. (1996). Cer1p, a novel Hsp70-related protein required for posttranslational endoplasmic reticulum translocation in yeast. *J. Biol. Chem.* **271**, 30610–30613.

82 Hamilton, T. G., Norris, T. B., Tsuruda, P. R., & Flynn, G. C. (1999). Cer1p functions as a molecular chaperone in the endoplasmic reticulum of Saccharomyces cerevisiae. *Mol. Cell Biol.* **19**, 5298–5307.

83 Hamman, B. D., Hendershot, L. M., & Johnson, A. E. (1998). BiP maintains the permeability barrier of the ER membrane by sealing the lumenal end of the translocon pore

before and early in translocation. *Cell* **92**, 747–758.

84 Hammond, C. & Helenius, A. (1994). Folding of VSV G protein: sequential interaction with BiP and calnexin. *Science* **266**, 456–458.

85 Hammond, C. & Helenius, A. (1995). Quality control in the secretory pathway. *Curr. Opin. Cell Biol.* **7**, 523–529.

86 Hampton, R. Y. (2002). ER-associated degradation in protein quality control and cellular regulation. *Curr. Opin. Cell Biol. 2002. Aug. ;14(4):476. -82.* **14**, 476–482.

87 Hampton, R. Y. (2002). ER-associated degradation in protein quality control and cellular regulation. *Curr. Opin. Cell Biol.* **14**, 476–482.

88 Harding, H. P., Novoa, I., Zhang, Y., Zeng, H., Wek, R., Schapira, M., & Ron, D. (2000). Regulated translation initiation controls stress-induced gene expression in mammalian cells. *Mol. Cell* **6**, 1099–1108.

89 Harding, H. P., Zhang, Y., & Ron, D. (1999). Protein translation and folding are coupled by an endo-plasmic-reticulum-resident kinase. *Nature* **397**, 271–274.

90 Hart, G. W. (1997). Dynamic O-linked glycosylation of nuclear and cytoskeletal proteins. *Annu. Rev. Biochem.* **66**, 315–335.

91 Haze, K., Yoshida, H., Yanagi, H., Yura, T., & Mori, K. (1999). Mammalian transcription factor ATF6 is synthesized as a transmembrane protein and activated by proteolysis in response to endoplasmic reticulum stress. *Mol. Biol. Cell* **10**, 3787–3799.

92 Helenius, A. & Aebi, M. (2001). Intracellular functions of N-linked glycans. *Science* **291**, 2364–2369.

93 Hellers, M., Gunne, H., & Steiner, H. (1991). Expression of posttranslational processing of preprocecropin A using a baculovirus vector. *Eur. J. Biochem.* **199**, 435–439.

94 Hellman, R., Vanhove, M., Lejeune, A., Stevens, F. J., & Hendershot, L. M. (1999). The in vivo association of BiP with newly synthesized proteins is dependent on their rate and stability of folding and simply on the presence of sequences that can bind to BiP. *J. Cell. Biol.* **144**, 21–30.

95 Hendershot, L., Bole, D., Kohler, G., & Kearney, J. F. (1987). Assembly and secretion of heavy chains that do not associate posttranslationally with immunoglobulin heavy chain-binding protein. *J. Cell. Biol.* **104**, 761–767.

96 Hendershot, L., Wei, J., Gaut, J., Melnick, J., Aviel, S., & Argon, Y. (1996). Inhibition of immunoglobulin folding and secretion by dominant negative BiP ATPase mutants. *Proc. Natl. Acad. Sci. U.S.A.* **93**, 5269–5274.

97 Hendershot, L. M., Wei, J.-Y., Gaut, J. R., Lawson, B., Freiden, P. J., & Murti, K. G. (1995). In vivo expression of mammalian BiP ATPase mutants causes disruption of the endoplasmic reticulum. *Mol. Biol. Cell.* **6**, 283–296.

98 Hennessy, F., Cheetham, M. E., Dirr, H. W., & Blatch, G. L. (2000). Analysis of the levels of conservation of the J-domain among the various types of DnaJ-like proteins. *Cell Stress. Chaperones.* **5**, 347–358.

99 Hershko, A. & Ciechanover, A. (1998). The ubiquitin system. *Annu. Rev. Biochem.* **67**, 425–479.

100 Hohfeld, J. & Jentsch, S. (1997). GrpE-like regulation of the hsc70 chaperone by the anti-apoptotic protein BAG-1. *EMBO J.* **16**, 6209–6216.

101 Hohfeld, J., Minami, Y., & Hartl, F. U. (1995). Hip, a novel cochaperone involved in the eukaryotic Hsc70/Hsp40 reaction cycle. *Cell* **83**, 589–598.

102 Hosoda, A., Kimata, Y., Tsuru, A., & Kohno, K. (2003). JPDI, a novel endoplasmic reticulum-resident protein containing both a BiP-interacting J-domain and thioredoxin-like motifs. *J. Biol. Chem.* **278**, 2669–2676.

103 Hosokawa, N., Wada, I., Hasegawa, K., Yorihuzi, T., Tremblay, L. O., Herscovics, A., & Nagata, K. (2001). A novel ER alpha-mannosidase-like protein accelerates ER-associated degradation. *EMBO Rep.* **2**, 415–422.

104 Hwang, C., Sinskey, A. J., & Lodish, H. F. (1992). Oxidized redox state of glutathione in the endoplasmic reticulum. *Science* **257**, 1496–1502.

105 Jarosch, E., Lenk, U., & Sommer, T. (2003). Endoplasmic reticulum-associated protein degradation. *Int. Rev. Cytol. 2003. ;223. :39. -81.* **223**, 39–81.

106 Jarvis DL and Summers MD (1992). *Baculovirus expression vectors.* In Recombinant DNA vaccines: Rationale and Strategies, pp. 265–291, Marcel Dekker, New York.

107 Johannes, L. & Goud, B. (2000). Facing inward from compartment shores: how many pathways were we looking for? *Traffic.* **1**, 119–123.

108 Johnson, A. E. & van Waes, M. A. (1999). The translocon: a dynamic gateway at the ER membrane. *Annu. Rev. Cell Dev. Biol.* **15**, 799–842.

109 Jones, S., Litt, R. J., Richardson, C. J., & Segev, N. (1995). Requirement of nucleotide exchange factor for Ypt1 GTPase mediated protein transport. *J. Cell Biol.* **130**, 1051–1061.

110 Kabani, M., Beckerich, J. M., & Gaillardin, C. (2000). Sls1p stimulates Sec63p-mediated activation of Kar2p in a conformation-dependent manner in the yeast endoplasmic reticulum. *Mol. Cell Biol.* **20**, 6923–6934.

111 Kabani, M., McLellan, C., Raynes, D. A., Guerriero, V., & Brodsky, J. L. (2002). HspBP1, a homologue of the yeast Fes1 and Sls1 proteins, is an Hsc70 nucleotide exchange factor. *FEBS Lett.* **531**, 339–342.

112 Kaloff, C. R. & Haas, I. G. (1995). Coordination of immunoglobulin chain folding and immunoglobulin chain assembly is essential for the formation of functional IgG. *Immunity* **2**, 629–637.

113 Kamhi-Nesher, S., Shenkman, M., Tolchinsky, S., Fromm, S. V., Ehrlich, R., & Lederkremer, G. Z. (2001). A novel quality control compartment derived from the endoplasmic reticulum. *Mol. Biol. Cell* **12**, 1711–1723.

114 Kaneda, S., Yura, T., & Yanagi, H. (2000). Production of three distinct mRNAs of 150 kDa oxygen-regulated protein (ORP150) by alternative promoters: preferential induction of one species under stress conditions. *J. Biochem. (Tokyo)* **128**, 529–538.

115 Kassenbrock, C. K. & Kelly, R. B. (1989). Interaction of heavy chain binding protein (BiP/GRP78) with adenine nucleotides. *EMBO. J.* **8**, 1461–1467.

116 Kelley, W. L. (1998). The J-domain family and the recruitment of chaperone power. *Trends Biochem. Sci.* **23**, 222–227.

117 Kim, P. S. & Arvan, P. (1995). Calnexin and BiP act as sequential molecular chaperones during thyroglobulin folding in the endoplasmic reticulum. *J. Cell Biol.* **128**, 29–38.

118 Knarr, G., Gething, M. J., Modrow, S., & Buchner, J. (1995). BiP binding sequences in antibodies. *J. Biol. Chem.* **270**, 27589–27594.

119 Knittler, M. R., Dirks, S., & Haas, I. G. (1995). Molecular chaperones involved in protein degradation in the endoplasmic reticulum: Quantitative interaction of the heat shock cognate protein BiP with partially folded immunoglobulin light chains that are degraded in the endoplasmic reticulum. *Proc. Natl. Acad. Sci. U.S.A.* **92**, 1764–1768.

120 Kohno, K., Normington, K., Sambrook, J., Gething, M. J., & Mori, K. (1993). The promoter region of the yeast KAR2 (BiP) gene contains a regulatory domain that responds to the presence of unfolded proteins in the endoplasmic reticulum. *Mol. Cell. Biol.* **13**, 877–890.

121 Koide, N. & Muramatsu, T. (1975). Specific inhibition of endo-beta-N-acetylglucosaminidase D by mannose and simple mannosides. *Biochem. Biophys. Res. Commun.* **66**, 411–416.

122 Kornfeld, R. & Kornfeld, S. (1985). Assembly of asparagine-linked oligosaccharides. *Annu. Rev. Biochem* **54**, 631–664.

123 Kowarik, M., Kung, S., Martoglio, B., & Helenius, A. (2002). Protein

folding during cotranslational translocation in the endoplasmic reticulum. *Mol. Cell* **10**, 769–778.

124 Kozutsumi, Y., Segal, M., Normington, K., Gething, M. J., & Sambrook, J. (1988). The presence of malfolded proteins in the endoplasmic reticulum signals the induction of glucose-regulated proteins. *Nature* **332**, 462–464.

125 Kurisu, J., Honma, A., Miyajima, H., Kondo, S., Okumura, M., & Imaizumi, K. (2003). MDG1/ERdj4, an ER-resident DnaJ family member, suppresses cell death induced by ER stress. *Genes Cells* **8**, 189–202.

126 Kuznetsov, G., Chen, L. B., & Nigam, S. K. (1997). Multiple molecular chaperones complex with misfolded large oligomeric glycoproteins in the endoplasmic reticulum. *J. Biol. Chem.* **272**, 3057–3063.

127 Lau, P. P., Villanueva, H., Kobayashi, K., Nakamuta, M., Chang, B. H., & Chan, L. (2001). A DnaJ protein, apobec-1-binding protein-2, modulates apolipoprotein B mRNA editing. *J. Biol. Chem.* **276**, 46445–46452.

128 Lee, A. S. (1987). Coordinated regulation of a set of genes by glucose and calcium ionophores in mammalian cells. *Trends. Biochem Sci.* **12**, 20–23.

129 Lee, A. S. (1992). Mammalian stress response: induction of the glucose-regulated protein family. *Curr. Opin. Cell Biol.* **4**, 267–273.

130 Lee, K., Tirasophon, W., Shen, X., Michalak, M., Prywes, R., Okada, T., Yoshida, H., Mori, K., & Kaufman, R. J. (2002). IRE1-mediated unconventional mRNA splicing and S2P-mediated ATF6 cleavage merge to regulate XBP1 in signaling the unfolded protein response. *Genes Dev.* **16**, 452–466.

131 Lee, Y.-K., Brewer, J. W., Hellman, R., & Hendershot, L. M. (1999). BiP and Ig light chain cooperate to control the folding of heavy chain and ensure the fidelity of immunoglobulin assembly. *Mol. Biol. Cell.* **10**, 2209–2219.

132 Lehle, L. & Tanner, W. (1976). The specific site of tunicamycin inhibition in the formation of dolichol-bound N-acetylglucosamine derivatives. *FEBS Lett.* **72**, 167–170.

133 Lenny, N. & Green, M. (1991). Regulation of endoplasmic reticulum stress proteins in COS cells transfected with immunoglobulin mu heavy chain cDNA. *J. Biol. Chem.* **266**, 20532–20537.

134 Lewis, M. J. & Pelham, H. R. (1992). Sequence of a second human KDEL receptor. *J. Mol. Biol.* **226**, 913–916.

135 Lewis, M. J., Sweet, D. J., & Pelham, H. R. (1990). The ERD2 gene determines the specificity of the luminal ER protein retention system. *Cell* **61**, 1359–1363.

136 Liberek, K., Marszalek, J., Ang, D., Georgopoulos, C., & Zylicz, M. (1991). Escherichia coli DnaJ and GrpE heat shock proteins jointly stimulate ATPase activity of DnaK. *Proc. Natl. Acad. Sci. U.S.A.* **88**, 2874–2878.

137 Lievremont, J. P., Rizzuto, R., Hendershot, L., & Meldolesi, J. (1997). BiP, a major chaperone protein of the endoplasmic reticulum lumen, plays a direct and important role in the storage of the rapidly exchanging pool of Ca2+. *J. Biol. Chem.* **272**, 30873–30879.

138 Lin, H. Y., Masso-Welch, P., Di, Y. P., Cai, J. W., Shen, J. W., & Subjeck, J. R. (1993). The 170-kDa glucose-regulated stress protein is an endoplasmic reticulum protein that binds immunoglobulin. *Mol. Biol. Cell.* **4**, 1109–1119.

139 Ma, Y. & Hendershot, L. M. (2003). Delineation of the negative feedback regulatory loop that controls protein translation during ER stress. *J. Biol. Chem.* **278**, 34864–34873.

140 Machamer, C. E., Doms, R. W., Bole, D. G., Helenius, A., & Rose, J. K. (1990). Heavy chain binding protein recognizes incompletely disulfide-bonded forms of vesicular stomatitis virus G protein. *J. Biol. Chem.* **265**, 6879–6883.

141 Malby, R. L., Caldwell, J. B., Gruen,

L. C., Harley, V. R., Ivancic, N., Kortt, A. A., Lilley, G. G., Power, B. E., Webster, R. G., Colman, P. M., et al. (1993). Recombinant antineuraminidase single chain antibody: expression, characterization, and crystallization in complex with antigen. *Proteins* **16**, 57–63.

142 Martin, J. & Hartl, F. U. (1997). The effect of macromolecular crowding on chaperonin-mediated protein folding. *Proc. Natl. Acad. Sci. U.S.A* **94**, 1107–1112.

143 Mayer, M., Reinstein, J., & Buchner, J. (2003). Modulation of the ATPase cycle of BiP by peptides and proteins. *J. Mol. Biol.* **330**, 137–144.

144 McCarty, J. S., Buchberger, A., Reinstein, J., & Bukau, B. (1995). The role of ATP in the functional cycle of the DnaK chaperone system. *J. Mol. Biol.* **249**, 126–137.

145 McClellan, A. J., Endres, J. B., Vogel, J. P., Palazzi, D., Rose, M. D., & Brodsky, J. L. (1998). Specific molecular chaperone interactions and an ATP-dependent conformational change are required during post-translational protein translocation into the yeast ER. *Mol. Biol. Cell* **9**, 3533–3545.

146 McKay, D. B. (1993). Structure and mechanism of 70-kDa heat-shock-related proteins. *Adv. Protein. Chem.* **44**, 67–98.

147 Melnick, J., Aviel, S., & Argon, Y. (1992). The endoplasmic reticulum stress protein GRP94, in addition to BiP, associates with unassembled immunoglobulin chains. *J. Biol. Chem.* **267**, 21303–21306.

148 Melnick, J., Dul, J. L., & Argon, Y. (1994). Sequential interaction of the chaperones BiP and Grp94 with immunoglobulin chains in the endoplasmic reticulum. *Nature* **370**, 373–375.

149 Meunier, L., Usherwood, Y. K., Chung, K. T., & Hendershot, L. M. (2002). A subset of chaperones and folding enzymes form multiprotein complexes in endoplasmic reticulum to bind nascent proteins. *Mol. Biol. Cell* **13**, 4456–4469.

150 Meyer, H. A., Grau, H., Kraft, R., Kostka, S., Prehn, S., Kalies, K. U., & Hartmann, E. (2000). Mammalian Sec61 is associated with Sec62 and Sec63. *J. Biol. Chem.* **275**, 14550–14557.

151 Minami, Y., Hohfeld, J., Ohtsuka, K., & Hartl, F. U. (1996). Regulation of the heat-shock protein 70 reaction cycle by the mammalian DnaJ homolog, Hsp40. *J. Biol. Chem.* **271**, 19617–19624.

152 Molinari, M., Calanca, V., Galli, C., Lucca, P., & Paganetti, P. (2003). Role of EDEM in the release of misfolded glycoproteins from the calnexin cycle. *Science* **299**, 1397–1400.

153 Molinari, M., Galli, C., Piccaluga, V., Pieren, M., & Paganetti, P. (2002). Sequential assistance of molecular chaperones and transient formation of covalent complexes during protein degradation from the ER. *J. Cell Biol. 2002. Jul. 22. ;158. (2):247. -57.* **158**, 247–257.

154 Mori, K., Ma, W., Gething, M. J., & Sambrook, J. (1993). A transmembrane protein with a cdc2+/CDC28-related kinase activity is required for signalling from the ER to the nucleus. *Cell* **74**, 743–756.

155 Morris, J. A., Dorner, A. J., Edwards, C. A., Hendershot, L. M., & Kaufman, R. J. (1997). Immunoglobulin binding protein (BiP) function is required to protect cells from endoplasmic reticulum stress but is not required for the secretion of selective proteins. *J. Biol. Chem.* **272**, 4327–4334.

156 Munro, S. & Pelham, H. R. (1987). A C-terminal signal prevents secretion of luminal ER proteins. *Cell* **48**, 899–907.

157 Neu, H. C. & Heppel, L. A. (1965). The release of enzymes from Escherichia coli by osmotic shock and during the formation of spheroplasts. *J. Biol. Chem.* **240**, 3685–3692.

158 Ng, D. T., Spear, E. D., & Walter, P. (2000). The unfolded protein response regulates multiple aspects of secretory and membrane protein biogenesis and endoplasmic reticulum quality control. *J. Cell Biol.* **150**, 77–88.

159 Nicchitta, C. V. & Blobel, G. (1993). Lumenal proteins of the mammalian endoplasmic reticulum are required to complete protein translocation. *Cell* **73**, 989–998.

160 Nishikawa, S. & Endo, T. (1997). The yeast JEM1p is a DnaJ-like protein of the endoplasmic reticulum membrane required for nuclear fusion. *J. Biol. Chem.* **272**, 12889–12892.

161 Nishikawa, S. I., Fewell, S. W., Kato, Y., Brodsky, J. L., & Endo, T. (2001). Molecular chaperones in the yeast endoplasmic reticulum maintain the solubility of proteins for retro-translocation and degradation. *J. Cell Biol.* **153**, 1061–1070.

162 Oberdorf, J., Carlson, E. J., & Skach, W. R. (2001). Redundancy of mammalian proteasome beta subunit function during endoplasmic reticulum associated degradation. *Biochem.* **40**, 13397–13405.

163 Oda, Y., Hosokawa, N., Wada, I., & Nagata, K. (2003). EDEM as an acceptor of terminally misfolded glycoproteins released from calnexin. *Science* **299**, 1394–1397.

164 Oka, T., Sakamoto, S., Miyoshi, K., Fuwa, T., Yoda, K., Yamasaki, M., Tamura, G., & Miyake, T. (1985). Synthesis and secretion of human epidermal growth factor by Escherichia coli. *Proc. Natl. Acad. Sci. U.S.A* **82**, 7212–7216.

165 Okamura, K., Kimata, Y., Higashio, H., Tsuru, A., & Kohno, K. (2000). Dissociation of Kar2p/BiP from an ER sensory molecule, Ire1p, triggers the unfolded protein response in yeast. *Biochem. Biophys. Res. Commun.* **279**, 445–450.

166 Ozawa, K., Tsukamoto, Y., Hori, O., Kitao, Y., Yanagi, H., Stern, D. M., & Ogawa, S. (2001). Regulation of tumor angiogenesis by oxygen-regulated protein 150, an inducible endoplasmic reticulum chaperone. *Cancer Res.* **61**, 4206–4213.

167 Palleros, D. R., Reid, K. L., Shi, L., Welch, W. J., & Fink, A. L. (1993). ATP-induced protein-Hsp70 complex dissociation requires K+ but not ATP hydrolysis. *Nature* **365**, 664–666.

168 Parodi, A. J. (2000). Role of N-oligosaccharide endoplasmic reticulum processing reactions in glycoprotein folding and degradation. *Biochem. J.* **348 Pt 1**, 1–13.

169 Pathak, V. K., Schindler, D., & Hershey, J. W. (1988). Generation of a mutant form of protein synthesis initiation factor eIF-2 lacking the site of phosphorylation by eIF-2 kinases. *Mol. Cell. Biol.* **8**, 993–995.

170 Piwnica-Worms, H., Williams, N. G., Cheng, S. H., & Roberts, T. M. (1990). Regulation of pp60c-src and its interaction with polyomavirus middle T antigen in insect cells. *J. Virol.* **64**, 61–68.

171 Plemper, R. K., Bohmler, S., Bordallo, J., Sommer, T., & Wolf, D. H. (1997). Mutant analysis links the translocon and BiP to retrograde protein transport for ER degradation. *Nature* **388**, 891–895.

172 Plummer, T. H., Jr., Elder, J. H., Alexander, S., Phelan, A. W., & Tarentino, A. L. (1984). Demonstration of peptide: N-glycosidase F activity in endo-beta-N-acetylglucosaminidase F preparations. *J. Biol. Chem.* **259**, 10700–10704.

173 Polissi, A., Goffin, L., & Georgopoulos, C. (1995). The Escherichia coli heat shock response and bacteriophage lambda development. *FEMS Microbiol. Rev.* **17**, 159–169.

174 Power, B. E., Ivancic, N., Harley, V. R., Webster, R. G., Kortt, A. A., Irving, R. A., & Hudson, P. J. (1992). High-level temperature-induced synthesis of an antibody VH-domain in Escherichia coli using the PelB secretion signal. *Gene* **113**, 95–99.

175 Prapapanich, V., Chen, S., Toran, E. J., Rimerman, R. A., & Smith, D. F. (1996). Mutational analysis of the hsp70-interacting protein Hip. *Mol. Cell Biol.* **16**, 6200–6207.

176 Prols, F., Mayer, M. P., Renner, O., Czarnecki, P. G., Ast, M., Gassler, C., Wilting, J., Kurz, H., & Christ, B. (2001). Upregulation of the cochaperone Mdg1 in endothelial cells is induced by stress and during in

vitro angiogenesis. *Exp. Cell Res.* **269**, 42–53.

177 Qian, P., Li, X., Tong, G., & Chen, H. (2003). High-level Expression of the ORF6 Gene of Porcine Reproductive and Respiratory Syndrome Virus (PRRSV) in Pichia pastoris. *Virus Genes* **27**, 189–196.

178 Raynes, D. A. & Guerriero, V., Jr. (1998). Inhibition of Hsp70 ATPase activity and protein renaturation by a novel Hsp70-binding protein. *J. Biol. Chem.* **273**, 32883–32888.

179 Reddy, P., Sparvoli, A., Fagioli, C., Fassina, G., & Sitia, R. (1996). Formation of reversible disulfide bonds with the protein matrix of the endoplasmic reticulum correlates with the retention of unassembled Ig light chains. *EMBO. J.* **15**, 2077–2085.

180 Rose, M. D., Misra, L. M., & Vogel, J. P. (1989). KAR2, a karyogamy gene, is the yeast homolog of the mammalian BiP/GRP78 gene [published erratum appears in Cell 1989 Aug 25; 58(4): following 801]. *Cell* **57**, 1211–1221.

181 Roth, R. A. & Pierce, S. B. (1987). In vivo cross-linking of protein disulfide isomerase to immunoglobulins. *Biochem.* **26**, 4179–4182.

182 Rothblatt, J. A., Deshaies, R. J., Sanders, S. L., Daum, G., & Schekman, R. (1989). Multiple genes are required for proper insertion of secretory proteins into the endoplasmic reticulum in yeast. *J. Cell Biol.* **109**, 2641–2652.

183 Sadler, I., Chiang, A., Kurihara, T., Rothblatt, J., Way, J., & Silver, P. (1989). A yeast gene important for protein assembly into the endoplasmic reticulum and the nucleus has homology to DnaJ, an Escherichia coli heat shock protein. *J. Cell Biol.* **109**, 2665–2675.

184 Sanders, S. L., Whitfield, K. M., Vogel, J. P., Rose, M. D., & Schekman, R. W. (1992). Sec61p and BiP directly facilitate polypeptide translocation into the ER. *Cell* **69**, 353–365.

185 Sandvig, K., Garred, O., Prydz, K., Kozlov, J. V., Hansen, S. H., & van Deurs, B. (1992). Retrograde transport of endocytosed Shiga toxin to the endoplasmic reticulum. *Nature* **358**, 510–512.

186 Schlenstedt, G., Harris, S., Risse, B., Lill, R., & Silver, P. A. (1995). A yeast DnaJ homologue, Scj1p, can function in the endoplasmic reticulum with BiP/Kar2p via a conserved domain that specifies interactions with Hsp70s. *J. Cell Biol.* **129**, 979–988.

187 Shen, J., Chen, X., Hendershot, L., & Prywes, R. (2002). ER stress regulation of ATF6 localization by dissociation of BiP/GRP78 binding and unmasking of Golgi localization signals. *Dev. Cell 2002. Jul. ;3(1):99. -111.* **3**, 99–111.

188 Shen, Y., Meunier, L., & Hendershot, L. M. (2002). Identification and characterization of a novel endoplasmic reticulum (ER) DnaJ homologue, which stimulates ATPase activity of BiP in vitro and is induced by ER stress. *J. Biol. Chem. 2002. May. 3;277. (18.):15947. -56.* **277**, 15947–15956.

189 Shi, Y., Vattem, K. M., Sood, R., An, J., Liang, J., Stramm, L., & Wek, R. C. (1998). Identification and characterization of pancreatic eukaryotic initiation factor 2 alpha-subunit kinase, PEK, involved in translational control. *Mol. Cell Biol.* **18**, 7499–7509.

190 Sidrauski, C. & Walter, P. (1997). The transmembrane kinase Ire1p is a site-specific endonuclease that initiates mRNA splicing in the unfolded protein response. *Cell* **90**, 1031–1039.

191 Skowronek, M. H., Rotter, M., & Haas, I. G. (1999). Molecular characterization of a novel mammalian DnaJ-like Sec63p homolog. *Biol. Chem.* **380**, 1133–1138.

192 Stebbins, J. & Debouck, C. (1994). Expression systems for retroviral proteases. *Methods Enzymol.* **241**, 3–16.

193 Stuart, J. K., Myszka, D. G., Joss, L., Mitchell, R. S., McDonald, S. M., Xie, Z., Takayama, S., Reed, J. C., & Ely, K. R. (1998). Characterization of interactions between the anti-apoptotic

protein BAG-1 and Hsc70 molecular chaperones. *J. Biol. Chem.* **273**, 22506–22514.

194 Suzuki, C. K., Bonifacino, J. S., Lin, A. Y., Davis, M. M., & Klausner, R. D. (1991). Regulating the retention of T-cell receptor alpha chain variants within the endoplasmic reticulum: Ca(2+)-dependent association with BiP. *J. Cell. Biol.* **114**, 189–205.

195 Szabo, A., Korszun, R., Hartl, F. U., & Flanagan, J. (1996). A zinc finger-like domain of the molecular chaperone DnaJ is involved in binding to denatured protein substrates. *EMBO J.* **15**, 408–417.

196 Szabo, A., Langer, T., Schroder, H., Flanagan, J., Bukau, B., & Hartl, F. U. (1994). The ATP hydrolysis-dependent reaction cycle of the Escherichia coli Hsp70 system DnaK, DnaJ, and GrpE. *Proc. Natl. Acad. Sci. U.S.A.* **91**, 10345–10349.

197 Tachibana, H., Takekoshi, M., Cheng, X. J., Maeda, F., Aotsuka, S., & Ihara, S. (1999). Bacterial expression of a neutralizing mouse monoclonal antibody Fab fragment to a 150-kilodalton surface antigen of Entamoeba histolytica. *Am. J. Trop. Med. Hyg.* **60**, 35–40.

198 Takahashi, N. (1977). Demonstration of a new amidase acting on glycopeptides. *Biochem. Biophys. Res. Commun.* **76**, 1194–1201.

199 Takayama, S., Bimston, D. N., Matsuzawa, S., Freeman, B. C., Aime-Sempe, C., Xie, Z., Morimoto, R. I., & Reed, J. C. (1997). BAG-1 modulates the chaperone activity of Hsp70/Hsc70. *EMBO Journal.* **16**, 4887–4896.

200 Tatu, U. & Helenius, A. (1997). Interactions between newly synthesized glycoproteins, calnexin and a network of resident chaperones in the endoplasmic reticulum. *J. Cell Biol.* **136**, 555–565.

201 Theyssen, H., Schuster, H. P., Packschies, L., Bukau, B., & Reinstein, J. (1996). The second step of ATP binding to DnaK induces peptide release. *J. Mol. Biol.* **263**, 657–670.

202 Tirasophon, W., Welihinda, A. A., & Kaufman, R. J. (1998). A stress response pathway from the endoplasmic reticulum to the nucleus requires a novel bifunctional protein kinase/endoribonuclease (Ire1p) in mammalian cells. *Genes. Dev.* **12**, 1812–1824.

203 Travers, K. J., Patil, C. K., Wodicka, L., Lockhart, D. J., Weissman, J. S., & Walter, P. (2000). Functional and genomic analyses reveal an essential coordination between the unfolded protein response and ER-associated degradation. *Cell* **101**, 249–258.

204 Trimble, R. B., Tarentino, A. L., Plummer, T. H., Jr., & Maley, F. (1978). Asparaginyl glycopeptides with a low mannose content are hydrolyzed by endo-beta-N-acetylglucosaminidase H. *J. Biol. Chem.* **253**, 4508–4511.

205 Tyedmers, J., Lerner, M., Bies, C., Dudek, J., Skowronek, M. H., Haas, I. G., Heim, N., Nastainczyk, W., Volkmer, J., & Zimmermann, R. (2000). Homologs of the yeast Sec complex subunits Sec62p and Sec63p are abundant proteins in dog pancreas microsomes. *Proc. Natl. Acad. Sci. U.S.A* **97**, 7214–7219.

206 Tyson, J. R. & Stirling, C. J. (2000). LHS1 and SIL1 provide a lumenal function that is essential for protein translocation into the endoplasmic reticulum. *EMBO J.* **19**, 6440–6452.

207 Vanhove, M., Usherwood, Y.-K., & Hendershot, L. M. (2001). Unassembled Ig heavy chains do not cycle from BiP in vivo, but require light chains to trigger their release. *Immunity* **15**, 105–114.

208 Vogel, J. P., Misra, L. M., & Rose, M. D. (1990). Loss of BiP/GRP78 function blocks translocation of secretory proteins in yeast. *J. Cell. Biol.* **110**, 1885–1895.

209 Wang, X.-Z., Harding, H. P., Zhang, Y., Jolicoeur, E. M., Kuroda, M., & Ron, D. (1998). Cloning of mammalian Ire1 reveals diversity in the ER stress responses. *EMBO. J.* **17**, 5708–5717.

210 Watkins, J. D., Hermanowski, A. L., & Balch, W. E. (1993). Oligomerization of immunoglobulin G heavy and light chains in vitro. A cell-free assay to study the assembly of the endoplasmic reticulum. *J. Biol. Chem.* **268**, 5182–5192.

211 Wawrzynow, A. & Zylicz, M. (1995). Divergent effects of ATP on the binding of the DnaK and DnaJ chaperones to each other, or to their various native and denatured protein substrates. *Journal. of. Biological. Chemistry.* **270**, 19300–19306.

212 Wei, J.-Y., Gaut, J. R., & Hendershot, L. M. (1995). In vitro dissociation of BiP: peptide complexes requires a conformational change in BiP after ATP binding but does not require ATP hydrolysis. *J. Biol. Chem.* **270**, 26677–26682.

213 Wei, J.-Y. & Hendershot, L. M. (1995). Characterization of the nucleotide binding properties and ATPase activity of recombinant hamster BiP purified from bacteria. *J. Biol. Chem.* **270**, 26670–26677.

214 Werner, E. D., Brodsky, J. L., & McCracken, A. A. (1996). Proteasome-dependent endoplasmic reticulum-associated protein degradation: an unconventional route to a familiar fate. *Proc. Natl. Acad. Sci. U.S.A.* **93**, 13797–13801.

215 Westin, G., Gerster, T., Muller, M. M., Schaffner, G., & Schaffner, W. (1987). OVEC, a versatile system to study transcription in mammalian cells and cell-free extracts. *Nucleic. Acids. Res.* **15**, 6787–6798.

216 Whitley, P., Nilsson, I. M., & von Heijne, G. (1996). A nascent secretory protein may traverse the ribosome/endoplasmic reticulum translocase complex as an extended chain. *J. Biol. Chem.* **271**, 6241–6244.

217 Wilbanks, S. M., DeLuca Flaherty, C., & McKay, D. B. (1994). Structural basis of the 70-kilodalton heat shock cognate protein ATP hydrolytic activity. I. Kinetic analyses of active site mutants. *J. Biol. Chem.* **269**, 12893–12898.

218 Wilhovsky, S., Gardner, R., & Hampton, R. (2000). HRD gene dependence of endoplasmic reticulum-associated degradation. *Mol. Biol. Cell* **11**, 1697–1708.

219 Wurm, F. & Bernard, A. (1999). Large-scale transient expression in mammalian cells for recombinant protein production. *Curr. Opin. Biotechnol.* **10**, 156–159.

220 Xu, R., Kume, A., Matsuda, K. M., Ueda, Y., Kodaira, H., Ogasawara, Y., Urabe, M., Kato, I., Hasegawa, M., & Ozawa, K. (1999). A selective amplifier gene for tamoxifen-inducible expansion of hematopoietic cells. *J. Gene Med.* **1**, 236–244.

221 Ye, J., Rawson, R. B., Komuro, R., Chen, X., Dave, U. P., Prywes, R., Brown, M. S., & Goldstein, J. L. (2000). ER stress induces cleavage of membrane-bound ATF6 by the same proteases that process SREBPs. *Mol. Cell* **6**, 1355–1364.

222 Yochem, J., Uchida, H., Sunshine, M., Saito, H., Georgopoulos, C. P., & Feiss, M. (1978). Genetic analysis of two genes, dnaJ and dnaK, necessary for Escherichia coli and bacteriophage lambda DNA replication. *Mol. Gen. Genet.* **164**, 9–14.

223 Yoshida, H., Haze, K., Yanagi, H., Yura, T., & Mori, K. (1998). Identification of the cis-acting endoplasmic reticulum stress response element responsible for transcriptional induction of mammalian glucose-regulated proteins. Involvement of basic leucine zipper transcription factors. *J. Biol. Chem.* **273**, 33741–33749.

224 Yoshida, H., Matsui, T., Yamamoto, A., Okada, T., & Mori, K. (2001). XBP1 mRNA is induced by ATF6 and spliced by IRE1 in response to ER stress to produce a highly active transcription factor. *Cell* **107**, 881–891.

225 Young, B. P., Craven, R. A., Reid, P. J., Willer, M., & Stirling, C. J. (2001). Sec63p and Kar2p are required for the translocation of SRP-dependent precursors into the yeast endoplasmic reticulum in vivo. *EMBO J.* **20**, 262–271.

226 Yu, M., Haslam, R. H., & Haslam, D. B. (2000). HEDJ, an Hsp40 co-chaperone localized to the endoplasmic reticulum of human cells. *J. Biol. Chem.* **275**, 24984–24992.

227 Zerby, D., Sakhuja, K., Reddy, P. S., Zimmerman, H., Kayda, D., Ganesh, S., Pattison, S., Brann, T., Kadan, M. J., Kaleko, M., & Connelly, S. (2003). In vivo ligand-inducible regulation of gene expression in a gutless adenoviral vector system. *Hum. Gene Ther.* **14**, 749–761.

228 Zylicz, M., LeBowitz, J. H., McMacken, R., & Georgopoulos, C. (1983). The dnaK protein of Escherichia coli possesses an ATPase and autophosphorylating activity and is essential in an in vitro DNA replication system. *Proc. Natl. Acad. Sci. U.S.A.* **80**, 6431–6435.

17
Quality Control In Glycoprotein Folding

E. Sergio Trombetta and Armando J. Parodi

17.1
Introduction

The concept of quality control of protein folding in the secretory pathway emerged in the late 1970s and early 1980s when it was noticed that not in all cases did insertion of proteins in the endoplasmic reticulum (ER) result in their appearance at the expected final destination, intra- or extracellular. Several experimental results showed that cells displayed mechanisms that ensured that only proteins in their native conformations could be produced by the secretory pathway. Those mechanisms received the collective denomination of "quality control" [1].

Protein folding in living cells is a complex, error-prone process. Numerous mechanisms are in place to ensure that newly synthesized proteins reach their folded functional form. One such mechanism is the addition of glycans occurring in the ER lumen. Covalently linked *N*-glycans affect protein folding in cell-free assays, as they provide bulky, highly hydrophilic substituents that maintain molecules in solution while protein moieties successively adopt a variety of different conformations before reaching their final structures. In addition, the highly hydrophilic nature of *N*-glycans forces the asparagine units to which they are linked and neighboring amino acids to be in or close to the water-protein interphase. This chapter will not deal with those effects, which certainly also occur in vivo, but with folding-efficiency enhancement and ER retention of folding intermediates and irreparably misfolded species mediated by the interaction of a specific glycan structure (monoglucosylated polymannose-type compounds) with the ER lectins calnexin (CNX) and calreticulin (CRT). Recent evidence suggesting a role for a specific putative lectin (EDEM/Htm1p/Man1p) on the disposal of irreparably misfolded glycoproteins will be discussed also.

17.2
ER *N*-glycan Processing Reactions

Glycosylation of asparagine units in eukaryotic cells (*N*-glycosylation) involves the transfer of a glycan ($Glc_3Man_9GlcNAc_2$ in most cells; Figure 17.1) from dolichol-P-

Protein Folding Handbook. Part II. Edited by J. Buchner and T. Kiefhaber

ISBN: 3-527-30784-2

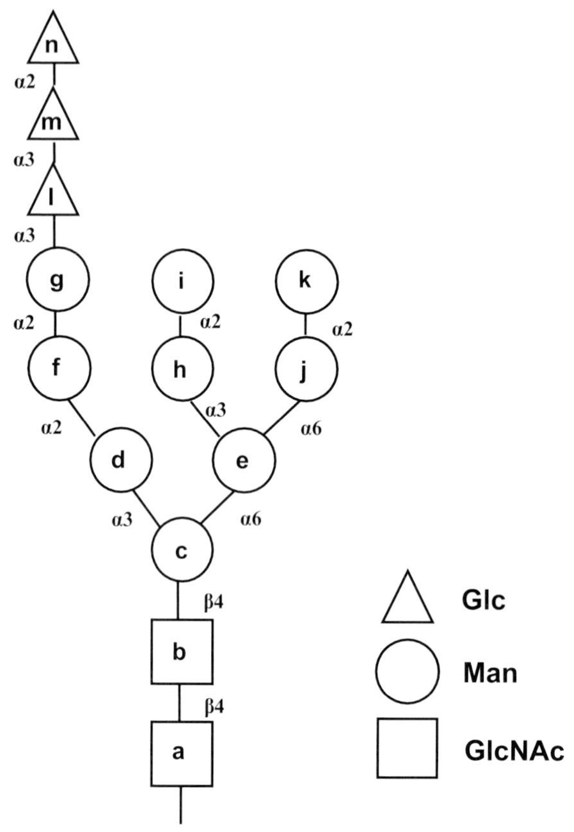

Fig. 17.1. *N*-glycan structures. Lettering (*a–n*) follows the order of addition of monosaccharides in the synthesis of $Glc_3Man_9GlcNAc_2$-P-P-Dol. Squares, circles, and triangles represent GlcNAc, Man, and Glc residues, respectively. GI removes residue *n* and GII removes residues *l* and *m*. GT adds residue *l* to residue *g*. The $Man_8GlcNAc_2$ isomer formed by mammalian cell or *S. cerevisiae* ER α-mannosidase I (M8B) lacks residues *i* and *l–n*, and the isomer formed by mammalian cell ER α-mannosidase II (M8C) is devoid of residues *k* and *l–n*. $Man_8GlcNAc_2$ isomer A (M8A) lacks residues *l–n* and *g*.

P (Dol-P-P) derivatives to consensus sequences (Asn-X-Ser/Thr, where X may be any amino acid except for Pro) in nascent polypeptide chains on the lumen of the ER [2]. Except for trypanosomatid protozoa (see below), the presence of the three glucose (Glc) units is required for efficient transfer of the glycan [3]. On the other hand, the consensus sequence is a necessary but insufficient condition for *N*-glycosylation. Processing of glycans (removal and addition of monosaccharide units) starts immediately after their transfer to proteins with the co-translational removal of the α(1,2)-linked external Glc unit (residue *n*; Figure 17.1) by glucosidase I (GI) (Figures 17.1 and 17.2). Further removal of the two remaining α(1,3)-linked Glc units is catalyzed by glucosidase II (GII). Additional *N*-glycan processing reactions such as removal of mannose (Man) units (residue *i* by mammalian or

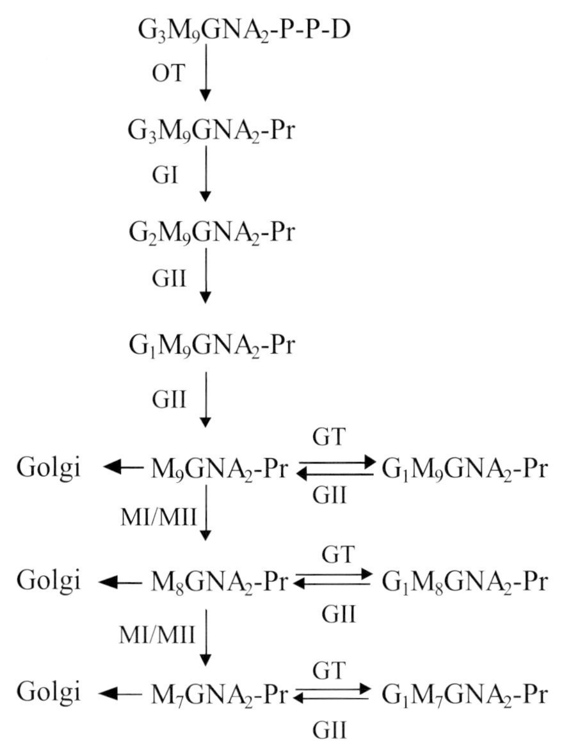

Fig. 17.2. Glycan processing in the mammalian cell ER. G, M, GNA, D, and Pr stand for Glc, Man, GlcNAc, Dol, and protein, respectively. OT, GI, GII, GT, MI and MII stand for oligosaccharyltransferase, glucosidase I, glucosidase II, UDP-Glc:glycoprotein glucosyltransferase, mannosidase I, and mannosidase II, respectively.

Saccharomyces cerevisiae ER α-mannosidase I and/or residue *k* by mammalian ER α-mannosidase II) do not necessarily occur for all glycoproteins [4]. Both the structure of the glycan transferred and the processing reactions occurring in the ER present only slight variations among protozoan, fungal, plant, and mammalian cells.

17.3
The UDP-Glc:Glycoprotein Glucosyltransferase

Beyond Glc and Man trimming, an additional processing reaction occurring in the ER is the transient reglucosylation of Glc-free protein-linked glycans ($Man_{7-9}GlcNAc_2$) (Figure 17.2) catalyzed by the UDP-Glc:glycoprotein glucosyltransferase (GT) [5–7]. The reaction products are the respective monoglucosylated derivatives that were previously deglucosylated by GII [8]. The single Glc unit is always added to Man g (Figure 17.1) in an α(1,3) linkage. That is, the compound formed by partial deglucosylation of the protein-bound glycan (removal of residues *m* and *n*; Figure 17.1) is re-created by GT. The monoglucosylated species have a transient existence in vivo, as GII readily removes the re-added Glc residue. As will be further described below, a continuous interconversion between monoglucosylated and unglucosylated structures, catalyzed by the opposing activities of GT and GII, is then established. This energy-consuming, apparently futile cycle lasts until glycoproteins acquire their final native tertiary structure or, alternatively, until irreparably misfolded species are diverted to proteasomal degradation in the cytosol. GT, which is the only soluble glycosyltransferase occurring in the secretory pathway, has been detected in the ER and in the ER-Golgi intermediate compartment (ERGIC) [8, 9]. Except for that described in the protozoon *Trypanosoma cruzi*, all known GTs display KDEL-related sequences at their C-termini [6, 10]. GTs are large proteins (about 1500 amino acids) that require millimolar Ca^{2+} concentrations for activity. The relatively high Ca^{2+} requirement agrees with their subcellular location. GTs specifically use UDP-Glc as a sugar donor and an antiport transport system by which entrance of UDP-Glc into the mammalian cell ER lumen is coupled to exit of UMP [11]. Moreover, two UDPase/GDPase activities have been described in the same subcellular location [12, 13]. Except for glucuronosyltransferases that occur in the liver of higher organisms, the ubiquitous GT is the only other nucleotide sugar-dependent glycosyltransferase described to date in the ER lumen. The budding yeast *S. cerevisiae* is the only eukaryote known so far to lack an enzymatically active GT [14]. Removal of Man units by ER mannosidases I and/or II in the acceptor glycoprotein decreases glucosylation rates. Man removal from monoglucosylated glycans also decreases GII-mediated Glc excision [15, 16].

The most remarkable GT property is that it does not glucosylate properly folded or completely unfolded glycoproteins [8, 15]. Apparently, the enzyme has to somehow detect exposure of hydrophobic amino acid patches in collapsed, molten globule-like conformers to catalyze transfer of Glc units [17]. Glycans linked to peptide moieties displaying random coil conformations, such as ribonuclease B molecules in which disulfide bonds had been reduced and further alkylated, were only poorly glucosylated in spite of having hydrophobic amino acid-rich stretches [17, 18]. Similarly, addition of dithiothreitol to live *Schizosaccharomyces pombe* cells did not enhance GT-mediated protein glucosylation [19]. This result probably reflects the inability of most glycoproteins, which normally have disulfide bonds, to reach a compact, molten globule-like conformation when formation of those bonds is prevented. Further, GT only glucosylated *N*-glycans present in the vicinity of protein structural perturbations that result in the exposure of the above-mentioned patches [20].

GT is composed of at least two domains [6]. The N-terminal domain comprises 80% of the molecule, has no homology to other known proteins, and is probably involved in nonnative conformer recognition. The C-terminal or catalytic domain binds [β-^{32}P]5N_3UDP-Glc and displays a similar size and significant similarity to glycosyltransferase family 8 members [21]. All GT C-terminal domains from different species share a significant similarity (65–70%), but no such similarity occurs between N-terminal domains [6]. For instance, *Rattus norvegicus* and *Drosophila melanogaster* GT N-terminal domains share a 32.6% similarity between them, but they show only a respective 15.5% and 16.3% similarity with the same portion of *S. pombe* GT. Nevertheless, the N-terminal domains of the fly and yeast enzymes were found to be mutually interchangeable, thus showing that they probably share common structural and functional features [22]. The notion that the N-terminal domain is responsible for recognition of nonnative conformers is supported by the enzymatic activities of chimeras constructed with N- and C-terminal domains of human cell GT homologues [23]. These cells express two GT homologues, only one of which is able to glucosylate misfolded glycoproteins. A chimera containing the catalytic domain of the inactive enzyme plus the N-terminal portion of the active one was found to glucosylate misfolded conformers. The notion of the N-terminal domain being the conformation sensor region agrees with the poor similarity found between those domains in different GTs. It should not be expected that the wide variety of hydrophobic amino acid patches exposed by different glycoproteins during their folding processes would require the corresponding sensor to display a stringently defined structure. Although the junction between both N- and C-terminal domains is extremely sensitive to proteolysis, both domains in the cleaved molecules could not be separated by a number of analytical procedures without losing enzymatic activity [22]. Further, the presence of the N-terminal domain appeared to be required for proper folding of the C-terminal one [22].

17.4 Protein Folding in the ER

The ER lumen has certain features that differentiate this subcellular compartment from others that also support protein folding, such as the cytosol or the mitochondria. The ER lumen is particularly rich in Ca^{2+}, which is required by several chaperones and folding-facilitating enzymes for activity, and it displays an oxidizing redox potential. Proteins following the secretory pathway are rich in cysteine residues, and several enzymes belonging to the protein disulfide isomerase family that facilitate proper formation of disulfide bridges have been described in the ER. Enzymes that catalyze *cis-trans* proline isomerization and several classical chaperones (Grp78/BiP, Grp94, Grp170) are also present in the ER lumen. Further, addition of *N*-glycans, an ER-exclusive process, is required for proper folding of most glycoproteins [5, 24].

17.5 Unconventional Chaperones (Lectins) Are Present in the ER Lumen

Two unconventional chaperones, CNX and CRT, are also present in the ER lumen [25–28]. As will be further described below, they do not directly interact with protein moieties of folding intermediates, as classical chaperones do, but, as lectins, they specifically recognize monoglucosylated glycans either generated by GT-mediated reglucosylation or formed by partial deglucosylation of $Glc_3Man_9GlcNAc_2$ (Figure 17.1).

Mammalian CNX is a type I, approximately 572-amino-acid long transmembrane protein. The cytosolic portion has an ER-retrieval sequence at its C-terminus (RKPRRE). The single transmembrane stretch is followed by the lumenal domain. The middle region of the latter, called the P or proline-rich domain, contains two motifs, each of which is tandemly repeated four times. The lumenal portion is composed of a globular lectin domain and a long hairpin loop composed by an antiparallel arrangement of the repeated motifs. CRT is an approximately 400-amino-acid long soluble protein that has a consensus KDEL-related ER retrieval signal at its C-terminus. The middle or P-domain is highly similar to the respective CNX portion, but its two motifs are repeated only three times instead of four as in CNX. The lectin domains in CNX and CRT are structurally similar. In the ER lumen, both CNX and CRT P-domains are associated with ERp57, a protein belonging to the protein disulfide isomerase family that promotes correct disulfide bonding in monoglucosylated glycoproteins [29, 30]. Chemical cross-linking studies conducted on intact cells showed that CNX and CRT form part of a large, weakly associated, heterogeneous protein network including Grp78/BiP, Grp94, and other ER resident proteins [31].

In vitro studies have confirmed results previously obtained in vivo (see below) showing that both CNX and CRT behave as monovalent lectins specific for monoglucosylated *N*-glycans. These were the sole compounds retained by CNX- or CRT-immobilized columns when a mixture of labeled polymannose-type compounds was applied to them. Optimal binding was observed with $Glc_1Man_9GlcNAc_2$. Compounds having a lower Man content showed a diminished binding capacity, but $Glc_1Man_5GlcNAc_2$ (residues *a–g* and *l*; Figure 17.1) still had about 65% of the binding capacity of $Glc_1Man_9GlcNAc_2$ [32, 33]. On the other hand, $Glc_1Man_4GlcNAc_2$ (residues *a–d*, *f*, *g*, and *l*; Figure 17.1) was not retained by CNX or CRT. This indicated that the $\alpha(1,6)$ branch (Figure 17.1) is essential for recognition. It is worth mentioning that both Glc$\alpha(1,3)$Man and Glc$\alpha(1,3)$Glc inhibited binding of $Glc_1Man_9GlcNAc_2$ to the lectins to a similar degree. The former disaccharide is present at the non-reducing end of the glycan probe, whereas the latter one is that present in $Glc_2Man_9GlcNAc_2$, which is not retained by the lectins. This result emphasizes the importance of the polymannose core in glycan-CNX/CRT interaction, as non-relevance of the core would have probably resulted in exclusive inhibition by the former disaccharide. Isothermal titration calorimetry studies confirmed that CRT recognizes the entire Glc$\alpha(1,3)$Man$\alpha(1,2)$Man$\alpha(1,2)$Man branch in the *N*-glycan (Figure 17.1) and that the lectin has a single carbohydrate-binding site per molecule [34]. Moreover, no differences between the glycan-binding properties of CNX and CRT were found. Glycan-CNX/CRT binding required Ca^{2+}, and the presence of adenosine nucleotides or peptides linked to the glycan moiety had little or no effect on it. Monoglucosylated chicken IgG-CRT interaction, studied by surface plasmon resonance, yielded a micromolar association constant. Whereas free

$Glc_1Man_9GlcNAc_2$ inhibited the interaction, no effect of $Man_9GlcNAc_2$ was observed, attesting to its exquisite specificity [35].

17.6 In Vivo Glycoprotein-CNX/CRT Interaction

Numerous in vivo experiments involving pulse-chase labeling with [^{35}S]Met+Cys followed by cell lysis and co-immunoprecipitation with CNX/CRT antisera showed that many newly synthesized glycoproteins transiently interacted with CNX and/or CRT, irrespective of their soluble or membrane-bound status or of their final destination. Interaction was dependent on the presence of monoglucosylated *N*-glycans for the following reasons. (1) No glycoprotein-CNX/CRT interaction was detected in mammalian cell mutants lacking GI or GII activities in which either tri- or diglucosylated *N*-glycans accumulated, respectively, as a result of blocking the formation of protein-bound monoglucosylated glycans (Figure 17.2) [5, 6, 36, 37]. (2) The same effect was observed upon addition of cell-permeable GI/GII inhibitors (deoxynojirimycin or its *N*-methyl or *N*-butyl derivatives, castanospermine). (3) Addition of those same inhibitors actually significantly enhanced glycoprotein-CRT interaction in the human pathogen *T. cruzi* (trypanosomatids lack CNX) [38]. These parasitic protozoa are the only wild-type cells known so far to transfer unglucosylated glycans from Dol-P-P derivatives to nascent proteins ($Man_6GlcNAc_2$, $Man_7GlcNAc_2$, or $Man_9GlcNAc_2$, dependent on the species; the last compound is transferred in *T. cruzi*), and monoglucosylated *N*-glycans are exclusively formed in them by GT-mediated glucosylation. As expected, addition of GII inhibitors in these microorganisms leads to the accumulation of monoglucosylated species (Figure 17.2).

Although the above-mentioned in vitro studies showed no differences between CNX and CRT specificities for *N*-glycans, the pattern of glycoproteins interacting in vivo with both lectins only partially overlapped [5, 6]. This is apparently due to the CNX and CRT respective status of membrane-bound and soluble proteins, since a truncated version of CNX, displaying only its lumenal portion, and CRT recognized the same glycoproteins. The same recognition pattern was also observed in cells expressing an artificially membrane-anchored CRT and CNX. A detailed study using membrane-bound influenza virus hemagglutinin (HA) as a model glycoprotein substrate showed that CRT preferentially interacted with *N*-glycans in the top/hinge domain of the molecule, i.e., those more lumenally oriented, whereas CNX, although being more promiscuous, mainly recognized *N*-glycans in the proximity of the ER inner membrane [39]. An interesting case is that of the assembly of the human class I major histocompatibility complex (MHC). The heavy chain (a membrane glycoprotein) interacted first with CNX, but this interaction ceased upon association of the former protein with β_2 microglobulin (β_2m). CRT then associated with the heavy chain, and this interaction persisted during the rest of the assembly process, which involves transient interaction with the transporter associated with antigen processing (TAP, a trimeric complex formed by TAP1, TAP2, and tapasin or TAP-A) and permanent association with a short, 8–10-amino-acid, peptide [40, 41]. The association first with CNX and then with CRT implies that the heavy-chain single *N*-glycan is first located close to the ER membrane but that a change in the heavy-chain conformation resulting from β_2m binding would make it more accessible to soluble ER proteins such as CRT.

17.7 Effect of CNX/CRT Binding on Glycoprotein Folding and ER Retention

One consequence of glycoprotein-CNX/CRT interaction is the ER retention of folding intermediates or of irreparably misfolded glycoproteins. Thus, the half-times of secretion of different glycoproteins synthesized in human hepatome cells, as well as those of different isotypic class I MHC molecules, correlated with their half-times of interaction with CNX [42, 43]. In addition, whereas thermosensitive G-protein of vesicular stomatitis virus ts045 mutants interacted with CNX at the non-permissive temperature, the interaction was lost and the glycoprotein correctly folded and secreted upon lowering the temperature [44]. Similarly, although both the wild-type and ΔF508 mutants of the chloride channel or transmembrane conductance regulator interacted with CNX, only the former protein was able to escape this interaction and migrate to the plasma membrane [45]. The indicated mutation is present in most cystic fibrosis patients. Furthermore, retention of GT-added Glc units in cruzipain caused by addition of GII inhibitors delayed arrival of this *T. cruzi* cysteine proteinase to lysosomes due to its prolonged interaction with CRT [46].

A model for the CNX-CRT-mediated quality control of glycoprotein folding, as initially proposed by A. Helenius and coworkers, is depicted in Figure 17.3 [47]. According to this model, a monoglucosylated glycoprotein molecule generated either by partial deglucosylation of the transferred glycan or by GT-mediated glucosylation would first interact with CNX/CRT. This interaction would be followed by a shuttle between glucosylated (CNX/CRT-bound) and deglucosylated (CNX/CRT-free) forms of the glycoprotein catalyzed by the opposing activities of GT and GII. Upon acquiring its proper tertiary structure, the glycoprotein would be deglucosylated by GII but not reglucosylated by GT and thus liberated from the CNX/CRT anchor. It follows that irreparably misfolded molecules would continue to interact with the lectins. The element sensing the conformational status of glycoproteins in this retaining mechanism is GT, an enzyme that tags folding intermediates and irreparably misfolded conformers with a Glc unit, not the unconventional chaperones (CNX/CRT). On the other hand, both the low affinity of CNX/CRT for glycoproteins and the CNX 3-D structure suggest that GII-mediated removal of Glc units occurs not on lectin-bound but on lectin-free species [35, 48].

This is not the only ER retention mechanism of incompletely folded glycoproteins. For instance, it was observed that less than 2% of truncated misfolded HA molecules were secreted by CHO cells infected with influenza virus in the presence of suboptimal puromycin concentrations that resulted in the synthesis of

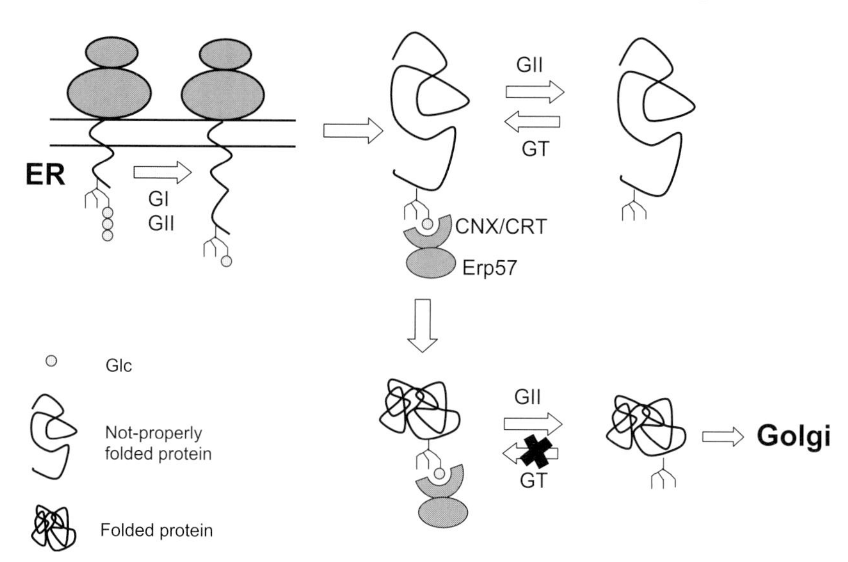

Fig. 17.3. Model proposed for the quality control of glycoprotein folding. Protein-linked $Glc_3Man_9GlcNAc_2$ is partially deglucosylated to the monoglucosylated derivative by GI and GII, and this structure is recognized by CNX/CRT. $Man_9GlcNAc_2$ is glucosylated by GT if complete deglucosylation occurs before lectin binding. The glycoprotein is liberated from the CNX/CRT anchor by GII and reglucosylated by GT only if it is not properly folded. On adoption of the native structure, the glycoprotein is released from CNX/CRT by GII and is not reglucosylated by GT. For further explanation, see text. ERp57 is a CNX/CRT-associated protein of the protein disulfide isomerase family that promotes formation of native disulfide bonds in monoglucosylated glycoproteins.

mostly C-terminal-truncated polypeptides [49]. The bulk of the molecules were found intracellularly in association with CNX/CRT but not with Grp78/BiP. Addition of GI/GII inhibitors increased the percentage of secreted species to 10%, and, as expected, no retained CNX/CRT-associated species were found. Instead and rather surprisingly, part of them was now bound to Grp78/BiP. As mentioned above, a lengthened glycoprotein-CRT interaction caused by the addition of a GII inhibitor to *T. cruzi* cells delayed arrival of a proteinase to lysosomes. Kinetics of arrival of the same glycoprotein in *T. cruzi* GT minus mutant cells was not affected by the inhibitor but was similar to that observed in wild-type cells in the presence of the drug due to a prolonged interaction with Grp78/BiP [10]. Further, synthesis of this last chaperone was upregulated in mutants. It follows that cells have alternative mechanisms applicable to both non-glycosylated proteins and glycoproteins for ER retention of folding intermediates and irreparably misfolded conformers such as binding to classical chaperones or formation of reversible disulfide bonds with ER matrix proteins.

Glycoprotein-CNX/CRT interaction not only prevents ER exit of misfolded conformers but also decreases the folding rate while increasing the folding efficiency of glycoproteins by preventing their premature oligomerization and degradation and by facilitating formation of native disulfide bonds [5, 6, 36, 37]. The role of GT in glycoprotein-folding facilitation is highlighted by the fact that in both *S. pombe* and human cells its synthesis is induced as a consequence of the cell unfolded protein response (see Section 17.8), i.e., under conditions that promote ER accumulation of misfolded conformers (heat shock, inhibition of glycosylation, prevention of disulfide bond formation) [50, 51]. It was reported that simultaneous expression of CNX increased assembly of class I MHC heavy-chain molecules with β_2m in *D. melanogaster* cells [52]. A reduction in the level of aggregates and the use of conformational monoclonal antibodies revealed that this was due to a higher efficiency of heavy-chain folding. Similarly, addition of GI/GII inhibitors increased the folding rate of HA expressed in a rabbit reticulocyte–dog pancreas microsome system, although the overall efficiency decreased due to aggregation, premature trimerization and degradation, and formation of nonnative disulfide bridges [53]. Moreover, the same inhibitors produced similar effects on folding, disulfide bond formation, and productive homodimerization of the human insulin receptor expressed in CHO cells [54]. The effects observed upon glycoprotein-CNX/CRT interaction are not restricted to mammalian cells, as carboxypeptidase Y arrived at a higher rate, but in decreased amounts, at the vacuole (the yeast equivalent of the lysosome) of *S. pombe* GII minus mutants than at those of wild-type cells [55].

Why does glycoprotein-CNX/CRT interaction have such effects on glycoprotein folding? One of the main obstacles for productive folding is aggregation. It may be speculated that upon binding *N*-glycans, CNX/CRT maintain glycoproteins in solution, while the protein moieties acquire their proper tertiary structures helped by the sequential or simultaneous action of classical chaperones and other folding-assisting proteins. Although interaction with CNX/CRT and Grp78/BiP has been reported for several glycoproteins (coagulation factor FVIII, thyroglobulin, VSV-G, Semliki Forest virus [SFV] glycoprotein E1), other glycoproteins such as HA and SFV glycoprotein p62 exclusively interact with CNX/CRT [5, 6]. Analysis of interactions occurring between a number of glycoproteins and CNX/CRT or Grp78/BiP and mutational removal of *N*-glycosylation consensus sequences in HA revealed that *N*-glycans located approximately within the first 50 amino acids of the N-terminus determine the exclusive interaction with CNX/CRT [56]. As Grp78/BiP preferentially recognizes the presence of heptapeptides displaying aliphatic hydrophobic amino acids in alternating positions in extended conformers, steric hindrance caused by *N*-glycan proximity would probably prevent binding of the classical chaperone. It may be speculated that the length of approximately 50 amino acids within which an *N*-glycan apparently impairs Grp78/BiP binding could be related to the primary sequence information required by a polypeptide chain to adopt collapsed molten globule-like conformations that are known to display low or null Grp78/BiP-binding capacity and are suitable substrates for GT-mediated glucosylation [17]. That is, the absence of an *N*-glycan before the polypeptide collapses would allow Grp78/BiP binding, whereas its presence would prevent it. According to its specificity, GT exerts its action during the last steps of protein folding, when a newly synthesized protein accumulates enough sequence information to form long-range hydrophobic clusters characteristic of molten-globule conformations. It may be expected that in previous folding steps other chaperones that have evolved to preferentially recognize extended hydrophobic structures, such as Grp78/BiP, could be more effective. There is evidence in several systems for a sequential intervention of Grp78/BiP and CNX/CRT to assist the entire protein-folding process. The different conformation recognition patterns between Grp78/BiP and GT provide a molecular rationale for such sequential intervention. The action of other chaperones also occurring in the ER (see Section 17.4) in conjunction of CNX/CRT has not been analyzed yet.

ERp57 is an enzyme belonging to the protein disulfide isomerase, intimately interacting with CNX/CRT in vivo as revealed by specific cross-linking with suitable reagents [29, 57–59]. It was found that ERp57 acted exclusively on monoglucosylated glycoproteins; moreover, a direct interaction between the enzyme and monoglucosylated glycoproteins (SFV glycoproteins E1 and p62) could be demonstrated by the isolation of mixed disulfide species [30]. Interaction of the classical protein disulfide isomerase (PDI) with the viral glycoproteins did not occur in ternary complexes with CNX/CRT. The participation of ERp57 in folding facilitation of class I MHC heavy-chain molecules in conjunction with CNX/CRT has also been reported. On the other hand, the effect of ERp57 addition in the proper folding of isolated bovine pancreas monoglucosylated RNase B was enhanced if CNX or CRT was added to the cell-free system [60]. The action of CNX/CRT could be simply to bring together ERp57 and a glycoprotein in order to facilitate formation of native disulfide bonds. Alternatively, the lectins may also facilitate glycoprotein folding by an additional mechanism, as ERp57 perhaps also displays a chaperone activity similar to that ascribed to PDI, a protein with which it shows 29% identity and 56% similarity. It is worth mentioning that ERp57 also interacts with glycoproteins lacking cysteine units through its association with CNX/CRT.

The already-mentioned shuttle between monoglucosylated (CNX/CRT-bound) and deglucosylated (CNX/CRT-free) forms of folding glycoproteins (Figure 17.3) is not necessarily required for increasing folding efficiency. As mentioned above, VSVts045 G-protein associated with CNX/CRT at the non-permissive temperature. Although a higher folding rate was observed upon lowering the temperature in the absence of GI/GII inhibitors (a condition that allowed the shuttle to occur), the overall folding efficiency was the same whether the inhibitors were added or not [44]. Similarly, in a *S. pombe alg6* mutant that synthesizes under-glycosylated glycoproteins, a glycoprotein essential for cell viability required GT-mediated glucosylation but not GII-mediated deglucosylation for proper folding at high temperature (see Section 17.8) [61].

17.8 Glycoprotein-CNX/CRT Interaction Is Not Essential for Unicellular Organisms and Cells in Culture

Folding facilitation and ER retention of folding intermediates and irreparably misfolded species mediated by glycoprotein-CNX/CRT interaction are not required for single-cell viability under normal growth conditions. Mammalian or *S. pombe* cells deficient in GI or GII activities, in which monoglucosylated glycans cannot be formed by partial deglucosylation of $Glc_3Man_9GlcNAc_2$ or by GT-mediated glucosylation, do not present any discernable phenotype [50, 55, 62, 63]. Cells have alternative systems involving different chaperones for helping proteins to acquire their native structures. A deficient system is replaced by an alternative one, and, consequently, a substantial proportion of many glycoproteins may fold correctly in the absence of interaction with CNX/CRT. For instance, as mentioned above, the amount of carboxypeptidase Y that reached the vacuole in GII-deficient *S. pombe* cells after a short pulse with [^{35}S]Met decreased by only 50% with respect to wild-type cells, and about half of HA molecules folded properly when translated in a rabbit reticulocyte–dog pancreas microsome system in the presence of GI/GII inhibitors [53, 55]. Furthermore, accumulation of misfolded glycoproteins in the ER caused by a total or reduced ability of glycoproteins to interact with CNX/CRT triggers an upregulation of chaperones and other folding-assisting proteins (unfolded protein response). This has been observed in mammalian and *S. pombe* cells lacking GI/GII activities as well as in *S. pombe* and *T. cruzi* mutants lacking GT [10, 55, 64]. Further, even though several *T. cruzi* glycoproteins have been identified as essential components of differentiation and mammalian cell invasion processes, total hindering of monoglucosylated *N*-glycan formation caused by disruption of both GT-encoding alleles did not affect cell growth rate of epimastigote-form parasites and only partially affected differentiation and mammalian cell invasion [10].

The dispensable character of glycoprotein-CNX/CRT interaction for single-cell viability may be highlighted by the fact that this folding-facilitating mechanism is probably not operative in *S. cerevisiae* since this yeast lacks GT [14]. Also, contrary to what happens in *S. pombe*, *T. cruzi*, and mammalian cells, no induction of ER chaperones was observed under conditions that prevent formation of monoglucosylated *N*-glycans, thus indicating that the absence of glycoprotein-CNX interaction does not lead to an accumulation of misfolded species in the ER (both *S. cerevisiae* and *S. pombe* lack CRT) [65].

Nonetheless, glycoprotein-CNX/CRT interaction is essential for viability under conditions of excessive ER stress such as those caused by under-glycosylation of glycoproteins and high temperature: *alg6/gpt1 S. pombe* double mutants in which $Man_9GlcNAc_2$ is transferred (inefficiently, see Section 17.2) from lipid derivatives and that are devoid of GT activity grew at 28 °C but not at 37 °C. Growth at high temperature was rescued not only upon transfection with a GT-encoding expression vector but also by 1 M sorbitol addition, thus suggesting that the affected glycoprotein(s) might be involved in cell wall formation [61]. On the other hand, it was reported that GI/GII inhibitors prevented VSV maturation by interfering with G-protein folding, as well as with formation of infectious human immunodeficiency virus (HIV) type I particles, probably due to misfolding of loop V1–V2 in gp120 [66, 67]. In addition, the same inhibitors prevented folding of tyrosinase in melanoma cells and assembly of hepatitis B virus particles by blocking the correct folding of M glycoprotein [68, 69].

The results described above indicate that although not required for cell growth under normal conditions, glycoprotein folding facilitation and irreparably misfolded glycoprotein ER retention mediated by glycoprotein-CNX/CRT interaction are indeed required under special conditions or for proper folding of particular proteins. Whether glycoprotein-CNX/CRT interaction is required for viability of multicellular organisms is presently unknown.

17.9 Diversion of Misfolded Glycoproteins to Proteasomal Degradation

Cells continuously monitor whether newly synthesized glycoproteins are in the process of proper folding or if, alternatively, they are irreparably misfolded. Both folding intermediates and misfolded species may be associated with CNX and CRT and thus retained in the ER. However, as permanent residence of the latter species in the ER is expected to have deleterious effects on cell viability, a mechanism for their recognition and subsequent diversion to degradation has been postulated to be necessarily required [70, 71]. It has been recently proposed that a particular glycan structure ($Man_8GlcNAc_2$ isomer B or M8B; Figure 17.1) could be the signal by which cells recognize that a glycoprotein molecule is unable to reach its native three-dimensional structure. As generation of M8B by ER α-mannosidase I is a slow process when compared to Glc removal by GI or GII, it has been speculated that if a protein folds extremely slowly or if it is irreparably misfolded, it would display the particular M8B structure that in turn would be recognized by an ER lectin and thus diverted to proteasomal degradation, a process known as ERAD (endoplasmic reticulum-associated degradation).

It was initially observed that in both *S. cerevisiae* and mammalian cells, degradation of glycoproteins unable to fold properly was delayed upon addition of ER α-mannosidase I inhibitors (deoxymannojirimycin, DMJ; kifunensin, KFN) to intact cells [72–77]. Furthermore, genetic manipulations that prevented M8B formation led to similar results. On the other hand, neither inhibitor addition nor genetic manipulation affected non-glycosylated protein degradation. Further research indicated the presence in the *S. cerevisiae* ER of an α-mannosidase I homologue (Man1p or Htm1p) that lacked mannosidase activity but for which a role of a lectin that recognized M8B was postulated [78–82]. Biochemical studies in mammalian cells showed that the mouse protein EDEM was a Man1p/Htm1p homologue, that it had no α-mannosidase activity, and that it localized to the ER and was upregulated upon an ER stress that triggered accumulation of misfolded proteins. Overexpression of this protein resulted in the accelerated ERAD of glycoproteins by promoting the release from CNX of irreparably misfolded glycoproteins but not of glycoproteins undergoing productive folding. On the contrary, downregulation of EDEM expression by the RNAi technique decreased misfolded glycoprotein disposal. Addition of KFN prevented the increased degradation that resulted from EDEM overexpression, thus suggesting that M8B was required for EDEM's role in misfolded glycoprotein degradation. EDEM co-immunoprecipitated with the misfolded glycoprotein, but it was not determined whether this represented a protein-

protein or a glycan-protein interaction. Further, EDEM interacted with CNX, but not with CRT, through their transmembrane regions. Both binding of substrates to CNX and their release from CNX were required for ERAD to occur. It was proposed that EDEM somehow prevented GT from reglucosylating irreparably misfolded molecules, thus facilitating their liberation from CNX anchors. Experiments performed with *S. cerevisiae* showed that the yeast EDEM homologue (Man1p/Htm1p) shared many of the mammalian cell protein properties.

It is evident that M8B cannot be per se a signal for degradation in *S. cerevisiae*. Although it is practically the only isomer produced in this yeast, all glycans transferred to nascent polypeptide chains are eventually converted to M8B in the ER in all glycoproteins, even in those that fold properly [83]. It may be speculated, nevertheless, that the putative lectin EDEM/Htm1p/Man1p would have to recognize both the *N*-glycan (M8B) and the misfolded status of the protein moiety to divert glycoproteins to ERAD. Even this model, however, does not explain how the putative lectin would discriminate between irreparably folded glycoproteins from folding intermediates.

Contrary to what happens in *S. cerevisiae*, where a single $Man_8GlcNAc_2$ isomer (M8B) is produced, in mammalian cells at least two isomers are formed (M8B and M8C; Figure 17.1). Whereas mammalian ER α-mannosidase I produces the first of such compounds, ER α-mannosidase II yields the second one [84–87]. ER α-mannosidase I is inhibited by both KFN and DMJ, whereas only the first compound affects ER α-mannosidase II activity. Because present evidence indicates that at least part of the proteins bound for ERAD are first transported to the *cis* Golgi cisternae and then retrieved to the ER, the products yielded by four other α-mannosidases present in the early mammalian cell secretory pathway are worth being described when considering the possible effects of glycan structures on driving misfolded glycoproteins to ERAD [71]. An endomannosidase occurs in the *cis* Golgi and in the ERGIC of high eukaryotes [88]. This enzyme liberates $Glc_{1-3}Man$ from the corresponding $Man_9GlcNAc_2$ derivatives, yielding $Man_8GlcNAc_2$ isomer A (M8A; Figure 17.1), and is not inhibited by either DMJ or KFN. As GT has also been reported to occur in the ERGIC, it is not inconceivable to assume that part of the glycoproteins degraded by proteasomes might display the M8A structure. Endomannosidase degradation of *N*-glycans would stop glucosylation-deglucosylation cycles, as M8A lacks the Man unit to which the Glc is added by GT (residue g; Figure 17.1). Three different *cis* Golgi α-mannosidases (IA, IB, and IC) have been described in mammalian cells. All of them are inhibited by DMJ and KFN and are able to degrade $Man_9GlcNAc_2$ to $Man_5GlcNAc_2$ [89, 90].

Although mammalian cell ER α-mannosidases I and II produce M8B and M8C, respectively, as first degradation products, both isomers may be further degraded to smaller compounds within the same subcellular location: the single *N*-glycan present in 3-hydroxy-3-methylglutaryl-CoA reductase (HMGR), an ER-resident membrane glycoprotein, was found to present the so-called microheterogeneity, as $Man_8GlcNAc_2$, $Man_7GlcNAc_2$, $Man_6GlcNAc_2$, and $Man_5GlcNAc_2$ glycans were detected on the enzyme [91]. M8B and a single $Man_6GlcNAc_2$ isomer were the main species. These results confirm that glycoproteins residing in the mammalian cell

ER for long periods of time, as irreparably misfolded species do, may have substantial amounts of glycans different from M8B and that glycoproteins displaying Man-trimmed glycans are not necessarily bound for degradation. In concurrence with these results, it was recently reported that both *S. cerevisiae* and mammalian ER α-mannosidase I is not as specific as initially thought, as the recombinant species were able to further degrade M8B to smaller glycans in vitro [92].

It was reported that addition of KFN and/or DMJ also delayed degradation of misfolded glycoproteins synthesized in the presence of GI and GII inhibitors [77, 93]. As those compounds prevent removal of Glc units from transferred glycans, this implies that the putative lectin(s) also recognizes $Glc_{1-3}Man_8GlcNAc_2$ in addition to M8B. These results, together with the variety of polymannose glycans present in glycoproteins after an extended ER residence, point not to a restricted but rather to a broad specificity of the putative ER lectin(s) involved in glycoprotein degradation. Further, it was reported that DMJ and KFN delayed ERAD in mutant mammalian cells in which neither M8B nor M8C could be formed, as in them, $Man_5GlcNAc_2$ rather than $Glc_3Man_9GlcNAc_2$ was transferred in protein *N*-glycosylation (residues *a*–*g*; Figure 17.1) [94]. More recent results would suggest that, at least in mammalian cells, *N*-glycans in glycoproteins bound to ERAD are degraded in the ER to $Man_{4-6}GlcNAc_2$ structures lacking Man residue g, i.e., the residue to which the Glc is added by GT [95]. Excision of the Man unit would remove irreparably misfolded glycoproteins from the CNX/CRT cycle, thus allowing their transport to the cytosol. This interpretation is at odds with the effect on ERAD of the addition of both mannosidase and glucosidase inhibitors: if the effect of the former on ERAD were caused by preventing disruption of CNX/CRT-glycoprotein interaction, then no effect of those same inhibitors would have been observed under conditions in which the interaction had been already prevented by retention of 2–3 Glc units in *N*-glycans.

It is clear from the results mentioned above that further studies are required to clarify the role of M8B and other Man-trimmed glycans in diversion of irreparably misfolded glycoproteins to ERAD. Regardless of the specific mechanism involved, the observed effects of KFN and DMJ might not be due to inhibition of α-mannosidase activity; being Man analogues, they could just as well affect lectin–*N*-glycan interactions.

Ubiquitination of misfolded proteins precedes their proteasomal degradation, but it is unclear whether the ubiquitination machinery somehow participates in recognition of irreparably misfolded species because the same *S. cerevisiae* ubiquitination components participate in the regulated degradation of a properly folded membrane glycoprotein such as Hmg2p (the yeast HMGR; see below) and of a soluble lumenal misfolded molecule such as a carboxypeptidase Y mutant [96]. An ubiquitin-ligase complex that specifically recognizes *N*-glycoproteins has been described in neuronal cells. Evidence was presented indicating that the complex was involved in the ERAD of ER glycoproteins, but neither the structure of the glycan moieties nor the folding status of substrate glycoproteins recognized by the complex was characterized [97]. A role of the ligase complex in driving irreparably misfolded species to proteasomal degradation is doubtful, as the interaction of the

ER *N*-glycosylated proteins, by which folding quality control is performed. Monoglucosylated glycans formed by glucosidase I- and glucosidase II (GII)-dependent partial deglucosylation of glycans transferred from lipid pyrophosphate derivatives to proteins ($Glc_3Man_9GlcNAc_2$) mediate glycoprotein recognition by two ER-resident chaperone lectins: calnexin (CNX), a transmembrane protein, and its soluble homologue, calreticulin (CRT). Further deglucosylation of glycans by GII liberates glycoproteins from their CNX/CRT anchors. Glycans may then be reglucosylated by the UDP-Glc:glycoprotein glucosyltransferase (GT), and thus recognized again by CNX/CRT, only when linked to protein moieties that are not yet properly folded. This enzyme behaves, therefore, as a sensor of glycoprotein conformations. Deglucosylation-reglucosylation cycles catalyzed by the opposing activities of GII and GT stop when proper folding is achieved, as glycoproteins then become substrates for GII but not for GT. Permanent liberation from CNX/CRT allows further glycoprotein transit through the secretory pathway. The CNX/CRT-monoglucosylated glycan interaction is one of the alternative mechanisms by which cells retain folding intermediates and irreparably misfolded glycoproteins in the ER; in addition, it enhances folding efficiency by preventing protein aggregation and allowing intervention of additional ER chaperones and folding-facilitating proteins. There is evidence suggesting that Man removal by resident ER mannosidases might be involved in recognition of irreparably misfolded glycoproteins bound for proteasomal degradation.

The mechanism described above constitutes a novel system, different from those of classical molecular chaperones, for retaining incompletely folded conformers and facilitating protein folding and oligomerization. CNX and CRT are unconventional chaperones that apparently do not directly sense the folding status of the substrate proteins as classical chaperones do. This task is reserved for an enzyme, GT, that introduces a covalent modification in glycoproteins not displaying their native conformations. It is this covalent modification that the element recognized by this new kind of chaperone. Although the main features of this system are already well defined, there are particular points that remain obscure and that will undoubtedly be the object of future studies, e.g., the mechanism by which GT recognizes nonnative structures, the precise way by which cells differentiate between folding intermediates from irreparably misfolded species, and the disentangling process of the latter, a process required for their transport through the translocon channel.

17.12
Characterization of *N*-glycans from Glycoproteins

17.12.1
Characterization of *N*-glycans Present in Immunoprecipitated Samples

It is possible to evaluate *N*-glycan structures on immunoprecipitated [^{35}S]-labeled proteins using glycosidases and SDS-PAGE analysis. These methods do not provide detailed carbohydrate structures but do take advantage of the immunoprecipitation technique: multiple samples can be studied simultaneously, the detection limit is very low, few cells are needed, and selected glycoproteins can be studied without the need for purification. When the protein under study is synthesized in large quantities, very short pulses are sufficient to produce an intense signal, and very early events (even co-translational) can be detected. The glycan structures are evaluated by their susceptibility to endo- and exoglycosidases, evidenced as small increases in mobility on SDS-PAGE. Each glycan removed by endo-*β*-*N*-acetylglucosaminidase H (Endo H) results in an increase in mobility of about 2 kDa, but exoglycosidases produce smaller shifts. For large proteins, the shifts are usually very small unless the glycosidases cleave *N*-glycans on several glycosylation sites.

The most common analysis is the evaluation of whether glycoproteins contain Endo H-sensitive glycans. This enzyme does not remove complex-type glycans generated in the Golgi but removes all *N*-glycoforms present in the ER. Therefore, when glycoproteins are sensitive to cleavage by Endo H, they are considered to have *N*-glycan structures similar to those acquired in the ER. Some exceptions to this rule are glycoproteins that traverse the Golgi but carry *N*-glycans that are not modified and thus remain partially or completely sensitive to Endo H. Also, certain yeasts elongate *N*-glycans in the Golgi but generate structures that remain sensitive to Endo H. In these cases, Golgi modifications can be identified by mobility shifts (the Golgi-modified forms run much slower and often diffuse) or with antibodies to *N*-glycans structures produced in the Golgi. To verify that the lack of Endo H cleavage is due to acquisition of complex-type *N*-glycans, parallel samples should be digested with Endo F or with *N*-glycosidase F. These enzymes cleave *N*-glycans irrespective of their structure and thus are insensitive to *N*-glycan modifications acquired in the Golgi.

For those glycoproteins that remain Endo H-sensitive due to their residence in the ER, the presence of glucosylated *N*-glycans can be evaluated by partial resistance to digestion with α-mannosidases displaying only exoglycosidase activity. Eight of the nine Man residues in ER-glycoforms are linked in the α-configuration and can be removed by α-mannosidases. *N*-glycans that carry a terminal Glc residue contain three Man residues protected from α-mannosidases. Therefore, glucosylated *N*-glycans are partially resistant to α-mannosidases, and a mobility difference between completely demannosylated and partially demannosylated proteins can often be detected by high-resolution SDS-PAGE (the magnitude of the shift observed with α-mannosidase digestion is smaller than with Endo H). It is difficult to distinguish between mono-, di-, and triglucosylated *N*-glycans by partial resistance to α-mannosidase. Prctrcatment with GII renders di- and monoglucosylated *N*-glycans fully sensitive to α-mannosidases.

Except for GII, the other glycosidases are commercially available (Roche, Sigma, New England Biolabs, US Biologicals, Calbiochem). When α-mannosidase is obtained as an ammonium sulfate precipitate, it should be dialyzed against 50 mM sodium acetate buffer pH 5.0, and 0.1 mM zinc acetate prior to use. The washed immunoprecipitates are resuspended in 10–50 μL of 10-mM HEPES buffer pH 7.4, 0.5% SDS, 2 mM dithiothreitol and heated at 95 °C for 10 min. After a short centrifugation, 1–5 μL of 10% Nonident-P40 is added to quench the SDS, and sufficient GII, Endo H, Endo F, *N*-glycanase, or α-mannosidase is added in (usually 1 μL). For α-mannosidase digestion, about 2–10 μL of 500-mM sodium acetate buffer pH 5.0, and 0.1 mM zinc acetate should be added. When appropriate, GII digestion can be performed before α-mannosidase to evaluate further the presence of glucosylated *N*-glycans. The samples are resuspended and incubated for 60–120 min at 37 °C. Reactions are stopped by addition of SDS-PAGE sample buffer and analyzed by electrophoresis and autoradiography.

17.12.2
Analysis of Radio-labeled *N*-glycans

The most common way of radio-labeling *N*-glycans is by incubating cells in growth media containing radioactive Glc or Man [108]. To achieve the most efficient labeling, cells need to be depleted from unlabeled intermediates as nucleotide sugars and Dol-P derivatives. This can be done by incubating them in media devoid of those sugars for 15–30 min. If glucosidase and/or mannosidase activities need to be inhibited, suitable compounds may be added at the onset of the starving period at 1–2.5 mM concentration, as those inhibitors penetrate rather slowly into cells. Labeling has to be performed in media containing 2–5 mM Glc, as lower concentrations may result in synthesis of truncated Dol-P-P derivatives and thence to formation of non-physiological protein-linked glycans. When sufficient radioactivity is used ([U-^{14}C]-Glc or [2-^{3}H]-Man), pulses can be kept short (5 min or less), and, conesquently, *N*-glycan processing can be followed during the early stages of glycoprotein biosynthesis. [U-^{14}C]-Glc can be converted into Man and GlcNAc, leading to Glc-, Man-, and GlcNAc-labeled *N*-glycans. Cells also convert Glc rapidly into labeled lipids and amino acids. As a consequence, glycans have to be purified extensively to avoid detection of radioactivity in other molecules. When *N*-glycans are labeled with [2-^{3}H]-Man, radioactivity is confined to Man and fucose residues only, since interconversion of Man into other sugars requires oxidation of the –OH group at position 2; therefore, radio-labeled glycans retain the label exclusively in the Man or fucose. Both Glc and Man internalized by cells can be incorporated into glycoproteins following incorporation into UDP-Glc or GDP-Man and further transferred to Dol-P-Glc or Dol-P-Man [109–111].

17.12.3
Extraction and Analysis of Protein-bound *N*-glycans

After the pulse/chase, cells are extracted by suspension in chloroform/methanol/water (3:2:1) (this is best done in a glass tube, 12 × 75 mm). After vigorous mixing, the sample is centrifuged at 3000 *g* for 5 min. A proteinaceous interphase that forms between two liquid phases is carefully recovered, discarding both the upper and lower liquid phases. The pellet is further washed twice with chloroform/methanol/water (3:2:1), followed by further washes with chloroform/methanol/water (10:10:3) to extract Dol-P-P-glycans (which can be saved for further analysis).

The washed proteinaceous pellet is then digested exhaustively with Pronase (2 mg mL^{-1}, Sigma) for 24 h in 1 mM $CaCl_2$ and 200 mM Tris-HCl buffer, pH 8.0. The digestion converts most of the insoluble pellet into amino acids, short peptides, and glycopeptides. The digest is cleared by centrifugation (10 000 g for 5 min), and the supernatant is desalted on a Sephadex G-10 column (2 × 60 cm), equilibrated, and run in 7% 2-propanol, in which the glycopeptides are excluded and separated from most amino acids and monosaccharides in the hydrolysate. The isolated glycopeptides are dried and resuspended in 50 mM sodium acetate buffer, pH 5.5, and digested with Endo H for 18 h. Tubes are boiled for 5 min and then passed over an Amberlite MB3 acetate column (0.5 × 5 cm) to retain amino acids and other charged contaminants. The neutral glycans (released by Endo H digestion) are recovered in the flow-through of the ion-exchange column and dried. They can be analyzed by a number of chromatographic techniques. The most simple technique, and arguably the one with highest resolving power, is descending paper chromatography in 1-propanol-nitromethane-water (5:2:4) [112]. Another simple alternative is thin-layer chromatography on silica plates [113]. *N*-glycans can also be analyzed by HPLC techniques, requiring dedicated equipment but providing good resolving power for free glycans without modification [114] or after derivatization with 2-aminopyridine [115] or perbenzoylation [116].

17.12.4 GII and GT Assays

17.12.4.1 Assay for GII

GII activity can be measured using two types of substrates: radioactive glycans or artificial substrate analogues. Radioactive [^{14}C]-Glc- or [^{3}H]-Glc-labeled $Glc_1Man_9GlcNAc$ is prepared in vitro by incubating rat liver microsomes (20 mg mL^{-1}) with 10–40 μM UDP-Glc (0.8–3.2 μCi of UDP-linked [^{14}C]-Glc or [^{3}H]-Glc should be added), 1 mM castanospermine or 1-deoxynojirimycin (Sigma, Roche, Calbiochem), 10 mg mL^{-1} denatured thyroglobulin (Sigma) (see Section 17.12.4.2, Assay for GT), 10 mM $CaCl_2$, 1% Triton-X100, 5 mM 2-mercaptoethanol, and 20 mM HEPES buffer, pH 7.4, in a final volume of 200 μL for 60 min at 37 °C. Under these conditions, the radio-labeled Glc is incorporated into *N*-glycans on glycoproteins directly via reglucosylation (mostly on the denatured thyroglobulin added as exogenous acceptor) or, additionally, via the dolichol pathway. At the end of the incubation, samples are extracted with chloroform/methanol/water as described above for isolation of radio-labeled *N*-glycans from cells (see Extraction and Analysis of Protein-bound *N*-glycans). $Glc_1Man_9GlcNAc$ has to be chromatographically purified from $Glc_3Man_9GlcNAc$ and $Glc_2Man_9GlcNAc$ that might have been formed via the Dol-P pathway.

To measure GII activity, radio-labeled glycans are incubated for 5–60 min at 37 °C, with test samples in a final volume of 50–100 μL containing 10 mM HEPES buffer, pH 7.4 (1% Triton-X100 has to be added when microsomal vesicles are used as enzyme source). At the end of the incubation, the released radioactive Glc ([^{14}C]-Glc or [^{3}H]-Glc) can be detected in a number of ways. One possibility is to separate the released Glc from the intact oligosaccharide by paper chromatography [8, 117]. Another approach to separate the released Glc is to add 100 μL of concanavalin A (ConA) (1 mg mL^{-1}, Sigma) to 200 mM Tris-HCl buffer, pH 8.0, and 1 mM $CaCl_2$ to bind all the undegraded glycan, followed by polyethylene glycol addition to precipitate the glycan-lectin complex [118]. The Glc liberated by GII in the assay is not bound by ConA and therefore remains in the supernatant after centrifugation; it is then quantified by liquid scintillation counting.

GII activity can also be measured with the artificial substrate *p*-NO_2-phenyl-α-D-glucopiranoside (Sigma). GI does not cleave this substrate, and therefore GII is the main enzyme capable of cleaving this substrate at pH 8.0, since most lysosomal glycosidases are unstable and/or inactive at such pH. In this case, Glc removal is followed by Abs405 to detect the free *p*-NO_2-phenol released. Assays can be conducted in a final volume of 100 μL containing 1 mM *p*-NO_2-phenyl-α-D-glucopiranoside in 20 mM Tris-HCl buffer, pH 8.0, and 1 mM EDTA in a thermostated cuvette, monitored continuously at 405 nm.

17.12.4.2 Assay for GT

The assay for GT is based on the incorporation of radioactive Glc from the sugar donor (UDP-[^{14}C]-Glc or UDP-[^{3}H]-Glc) into polymannose glycans on unlabeled denatured acceptor glycoproteins. The radio-labeled reaction product (glucosylated glycoproteins) is separated from the radio-labeled substrate (UDP-Glc) by trichloroacetic acid (TCA) precipitation and is quantified by liquid scintillation counting. The acceptor glycoproteins (bovine thyroglobulin; soybean agglutinin, SBA; bovine pancreatic ribonuclease B, RNaseB; available from Sigma, Roche, Worthington) are prepared by chemical denaturation. Glycoproteins are dissolved at high concentrations (20–50 mg mL^{-1}) in 10 mM HEPES buffer, pH 7.4. One gram of solid urea is added per milliliter of protein solution, and the mixture is incubated at 60 °C for 4 h. The samples are then exhaustively dialyzed against 10 mM HEPES buffer, pH 7.4. The assay for GT is very specific, especially when solubilized extracts and subsequently purified fractions are used as source of enzyme. When microsomes are used, incorporation of radioactive Glc into proteins may potentially arise from the Dol-P pathway [119]. This involves the formation of Dol-P-Glc from UDP-Glc and endogenous Dol-P, leading to the formation of $Glc_3Man_9GlcNAc_2$-P-P-Dol followed by transfer of the entire glycans to Asn residues on vacant glycosylation sites on the denatured glycoproteins used as substrates [120]. The enzymes involved in this cascade of reactions are integral membrane proteins and are poorly extracted in the conditions utilized to solubilize GT [8]. As a consequence, radio-labeling of the acceptor glycoproteins via the Dol-P pathway occurs only when crude microsomal membranes are used and does not occur once GT is solubilized from microsomes. In some systems (such as those derived from yeasts or plants), incurporation of radioactive Glc into trichloroacetic acid (TCA)-insoluble polysaccharides may also be observed when crude extracts are used. Reactions are conducted in 50–100 μL final volume, containning 10 mM HEPES buffer, pH 7.4, 10 mM $CaCl_2$, 5 mM 2-mercaptoethanol, and 2.5 μM UDP-Glc (about 0.01–0.05 μCi of UDP-linked [^{14}C]-Glc or [^{3}H]-Glc should be added). Reactions are initiated by addition of the test sample and incubated for 5–30 min at 37 °C and stopped by addition of 1 mL of 10% TCA. After boiling the stopped reactions for 5 min to allow complete protein insolubilization, the precipitated proteins are recovered by low-speed centrifugation (5 min at 2000 g) and washed three times with 1 mL of 10% TCA. The washed pellets are resuspended in 100 μL of 1-N KOH in methanol or other commercial solubilizers, diluted with 3 mL of scintillation cocktail, and quantified by liquid scintillation counting.

17.12.5
Purification of GII and GT from Rat Liver

Both GII and GT are soluble proteins of the lumen of the ER and therefore are soluble in the absence of detergents. Low concentrations of detergents may nevertheless be used to release the soluble content of the microsomes at the start of the purification, but detergents are not used in the subsequent purification steps. GT and GII are minor components of the ER, and typically less than 1 mg of GT or GII is obtained from 100–200 g of liver, with yields below 10%. Both enzymes are highly susceptible to proteases, and therefore it is critical to include protease inhibitors in the homogenization buffers, to maintain the pH above 7.0, and to process the microsomal extracts rapidly. Microsomes can be kept frozen for a few weeks, but it is important to go from the microsomes to the final step with minimal delays. Both enzymes are relatively stable after purification. GT is composed of a single polypeptide that runs at approximately 160 kDa in sodium dodecyl sulfate–polyacrylamide gel electrophoresis (SDS-PAGE) [121]. GII is composed of two different subunits: one running at approximately 110 kDa and a second subunit that runs as a slightly diffuse band at approximately 80 kDa [122].

Buffers and Chromatography Media

Buffer A: 0.25 M sucrose, 2 mM EDTA, 20 mM Tris-HCl pH 8.0, 5 mM 2-mercaptoethanol.

Buffer B: 150 mM NaCl, 20 mM Tris-HCl buffer, pH 8.0, 5 mM 2-mercaptoethanol.

Buffer C: 20 mM Tris-HCl buffer, pH 8.0, 5 mM 2-mercaptoethanol.

Buffer D: 1 M NaCl, 20 mM Tris-HCl buffer, pH 8.0, 5 mM 2-mercaptoethanol.

Buffer E: 1 M ammonium sulfate, 5 mM 2-mercaptoethanol.

Buffer F: 0.5 M sucrose, 10 mM imidazol buffer, pH 7.0, 5 mM 2-mercaptoethanol.

Chromatography media are from Pharmacia or Sigma.

Preparation of Rat Liver Microsomes

Rats (male or female, 4–12 weeks old) are starved overnight. They are euthanized and their livers are removed and rinsed in ice-cold buffer A. From this point, all procedures are carried out on ice or in a cold room, except for the elution step from the ConA-Sepharose column. Livers are weighted and minced in a blender with two to four volumes of buffer A containing protease inhibitors (1 mM EDTA, 10 μM leupeptin, 10 μM pepstatin, 10 μM E-64, 10 μM TLCK, 10 μM TPCK, and 100 μM PMSF). The homogenate thus obtained is centrifuged at 10 000 g for 10 min. The supernatant is further centrifuged at 100 000 g for 60 min. The pellet containing the microsomal fraction is resuspended in buffer A and stored at −80 °C.

Extraction of GT and GII from Microsomes

Since both GT and GII are soluble proteins, they can be extracted in the absence of detergents by mechanical disruption of microsomes (10 mg mL^{-1}) in buffer C using sonication, French press, or equivalent method. After mechanical disruption, the soluble fraction is recovered from the supernatant after high-speed centrifugation (60 min at 100 000 g) and precipitated with ammonium sulfate at 50% saturation. The insoluble pellet is resuspended and dialyzed against buffer C. Alternatively, the microsomal fraction can be solubilized with detergents at low concentration. For this, microsomes are resuspended at 10 mg mL^{-1} in buffer C and extracted with 0.1% Triton-X-100 for 30 min on ice. The homogenate is then centrifuged at 100 000 g for 60 min, and the supernatant of the microsomal extraction containing most of the GT and GII activity is saved. Detergent extracts are not fractionated with ammonium sulfate.

DEAE-cellulose

The solubilized fractions obtained from either detergent extraction of microsomes or the ammonium sulfate cut are loaded onto a DEAE-cellulose column equilibrated in buffer C and washed in the same buffer until the Abs280 reaches background. Notice that no detergents are needed from this point on, since both GT and GII are soluble proteins. The column is then eluted with a gradient from 100% buffer C to 50% buffer D over 20 column volumes. The enzymatic activities for GT and GII are measured in the eluate of the DEAE-cellulose column (GT typically elutes before GII). From this point, the fractions containing GT activity are separated from GII activity and are pursued separately.

ConA-Sepharose

The fractions containing GT or GII activity from the DEAE-cellulose eluate are applied separately to ConA-Sepharose columns (5 mL) equilibrated in buffer B supplemented with 1 mM $CaCl_2$, 1 mM $MgCl_2$, and 1 mM $MnCl_2$. After washing in buffer B until the Abs280 reaches background, the column is filled with one volume (approximately 5 mL) of buffer B supplemented with 500 mM α-methyl-mannopyranoside (Sigma) prewarmed at 37 °C. The column is then stopped and kept at 37 °C for 15 min. The elution continues with more prewarmed buffer B supplemented with 500 mM α-methyl-mannopyranoside.

MonoQ

The fractions containing GT or GII activity eluted from their respective ConA-Sepharose columns are diluted fivefold and loaded (separately) onto a MonoQ 5/5 column equilibrated in buffer C. The column is then eluted with a gradient from 100% buffer C to 50% buffer D over 20 column volumes (20 mL), and 1-mL or 0.5-mL fractions are collected.

Gel Filtration

The fractions eluted from the MonoQ step containing GT or GII activity are further purified by gel-filtration chromatography on a Superdex S-200 column (or equivalent), equilibrated, and run in buffer B. At this stage, GII is usually homogenous. If necessary, the MonoQ or gel-filtration steps can be repeated to achieve a homogenous preparation. GII is stored in buffer B at −80 °C.

Phenyl Superose

After the gel-filtration step, GT usually requires further purification using hydrophobic interaction chromatography. The fractions eluted from the gel-filtration column containing GT activity are diluted 10-fold with buffer E, filtered, and loaded onto a Phenyl-Superose column (1 mL) equilibrated in buffer E. The column is eluted at 0.5 mL min^{-1} with a gradient from 100% buffer E to 100% buffer F over 20 mL and then further eluted with another 15 mL of buffer F. GT is strongly retained by the column and typically elutes during the beginning of the wash with buffer F. GT is usually homogenous at this step and can be stored in buffer F at −80 °C. If necessary, the MonoQ or gel-filtration steps can be repeated for further purification, but buffer exchange into buffer F is still recommended for storage of GT.

References

1 Hurtley, S. M. & Helenius, A. (1989). Protein oligomerization in the endoplasmic reticulum. *Annu. Rev. Cell Biol.* **5**, 277–308.

2 Kornfeld, R. & Kornfeld, S. (1985). Assembly of asparagine-linked oligosaccharides. *Annu. Rev. Biochem.* **54**, 631–664.

3 Parodi, A. J. (1993). *N*-glycosylation in trypanosomatid protozoa. *Glycobiology*, **3**, 193–199.

4 Herscovics, A. (2001). Structure and function of Class I α1,2-mannosidases involved in glycoprotein synthesis and endoplasmic reticulum quality control. *Biochimie*, **83**, 757–762.

5 Parodi, A. J. (2000). Protein glucosylation and its role in protein folding. *Annu. Rev. Biochem.* **69**, 69–95.

6 Trombetta. E. S. & Parodi, A. J. (2002). *N*-glycan processing and glycoprotein folding. *Adv. Prot. Chem.* **59**, 303–344.

7 Trombetta, E. S. (2003). The contribution of *N*-glycans and their processing in the endoplasmic reticulum to glycoprotein biosynthesis. *Glycobiology*, **13**, 77R–91R.

8 Trombetta, S., Bosch, M. & Parodi, A. J. (1989). Glucosylation of glycoproteins by mammalian, plant, fungal and trypanosomatid protozoa micro-

somal proteins. *Biochemistry*, **28**, 8108–8116.

9 Zuber, C., Fan, J., Guhl, B., Parodi, A. J., Fessler, J. H., Parker, C. et al. (2001). Immunolocalization of UDP-Glc:glycoprotein glucosyltransferase indicates involvement of pre-Golgi intermediates in protein quality control. *Proc. Natl. Acad. Sci. USA*. **98**, 10771–10715.

10 Conte, I., Labriola, C., Cazzulo, J. J., Docampo, R. & Parodi, A. J. (2003). The interplay between folding facilitating mechanisms in *Trypanosoma cruzi* endoplasmic reticulum. *Mol. Biol. Cell*, **14**, 3529–3540.

11 Pérez, M. & Hirschberg, C. (1986). Topography of glycosylation reactions in the rough endoplasmic reticulum membrane. *J. Biol. Chem.* **261**, 6822–6830.

12 Trombetta, E. S. & Helenius, A. (1999). Glycoprotein reglucosylation and nucleotide sugar utilization in the secretory pathway: identification of a nucleoside diphosphatase in the endoplasmic reticulum. *EMBO J.* **18**, 3282–3292.

13 Failer, B. U., Braun, N. & Zimmermann H. (2002). Cloning, expression, and functional characterization of a Ca^{2+}-dependent endoplasmic reticulum nucleoside diphosphatase. *J. Biol. Chem.* **277**, 36978–36986.

14 Parodi, A. J. (1999). Reglucosylation of glycoproteins and quality control of glycoprotein folding in the endoplasmic reticulum of yeast cells. *Biochim. Biophys. Acta*, **1426**, 287–295.

15 Sousa, M., Ferrero-García, M. & Parodi, A. J. (1992). Recognition of the oligosaccharide and protein moieties of glycoproteins by the UDP-Glc:glycoprotein glucosyltransferase. *Biochemistry*, **31**, 97–105.

16 Grinna, L. S. & Robbins, P. W. (1980). Substrate specificities of rat liver microsomal glucosidases which process glycoproteins. *J. Biol. Chem.* **255**, 2255–2258.

17 Caramelo, J. J., Castro, O., Alonso, L., de Prat-Gay, G. & Parodi, A. J. (2003). UDP-Glc:glycoprotein glucosyltransferase recognizes structured and solvent accessible hydrophobic patches in molten globule-like folding intermediates. *Proc. Natl. Acad. Sci. USA*. **100**, 86–91.

18 Trombetta, E. S. & Helenius, A. (2000). Conformational requirements for glycoprotein reglucosylation in the endoplasmic reticulum. *J. Cell Biol.* **148**, 1123–1129.

19 Fernández, F., D'Alessio, C., Fanchiotti, S. & Parodi, A. J. (1998). A misfolded protein conformation is not a sufficient condition for in vivo glucosylation by the UDP-Glc:glycoprotein glucosyltransferase. *EMBO J.* **17**, 5877–5886.

20 Ritter, C. & Helenius, A. (2000). Recognition of local glycoprotein misfolding by the ER folding sensor UDP-glucose:glycoprotein glucosyltransferase. *Nature Struct. Biol.* **7**, 278–280.

21 Tessier, D. C., Dignard, D., Zapun, A., Radominska-Pandya, A., Parodi, A. J., Begeron, J. J. M. et al. (2000). Cloning and characterization of mammalian UDP-glucose:glycoprotein glucosyltransferase and the development of a specific substrate for this enzyme. *Glycobiology*, **10**, 403–412.

22 Guerin, M. & Parodi, A. J. (2003). The UDP-Glc:glycoprotein glucosyltransferase is organized in at least two tightly bound domains from yeasts to mammals. *J. Biol. Chem.* **278**, 20540–20546.

23 Arnold, S. M. & Kaufman, R. J. (2003). The noncatalytic portion of human UDP-glucose:glycoprotein glucosyltransferase I confers UDP-glucose binding and transferase function to the catalytic domain. *J. Biol. Chem.* **278**, 43320–4328.

24 Helenius, A. (1994). How *N*-linked oligosccharides affect glycoprotein folding in the endoplasmic reticulum. *Mol. Biol. Cell*, **5**, 253–265.

25 Williams, D. B. (1995). Calnexin: a molecular chaperone with a taste for carbohydrate. *Biochem. Cell Biol.* **73**, 123–132.

26 Bergeron, J. J. M., Brenner, M. B., Thomas, D. Y. & Williams, D. B.

(1994). Calnexin: a membrane-bound chaperone of the endoplasmic reticulum. *Trends Biochem. Sci.* **19**, 124–128.

27 Michalak, M., Corbett, E. F., Mesaeli, N., Nakamura, K. & Opas, M. (1999). Calreticulin: one protein, one gene, many functions. *Biochem. J.* **344**, 281–292.

28 Schrag, J. D., Procopio, D. O., Cygler, M., Thomas, D. Y. & Bergeron, J. J. M. (2003). Lectin control of protein folding and sorting in the secretory pathway. *Trends Biochem. Sci.* **28**, 49–57.

29 Oliver, J. D., Llewelyn Roderick, H., Llewelyn, D. H. & High, S. (1999). ERp57 functions as a subunit of specific complexes formed with the ER lectins calreticulin and calnexin. *Mol. Biol. Cell*, **10**, 2573–2582.

30 Molinari, M. & Helenius, A. (1999). Glycoproteins form mixed disulphides with oxidoreductases during folding in living cells. *Nature*, **402**, 90–93.

31 Tatu, U. & Helenius, A. (1997). Interactions between newly synthesized glycoproteins, calnexin and a network of resident chaperones in the endoplasmic reticulum. *J. Cell Biol.* **136**, 555–565.

32 Spiro, R. G., Zhu, Q., Bhoyroo, V. & Söling, H.-D. (1996). Definition of the lectin-like properties of the molecular chaperone, calreticulin, and demonstration of its copurification with endomannosidase from rat liver Golgi. *J. Biol. Chem.* **271**, 11588–11594.

33 Vassilakos, A., Michalak, M., Lehrman, M. A. & Williams, D. B. (1998). Oligosaccharide binding characteristics of the molecular chaperones calnexin and calreticulin. *Biochemistry*, **37**, 3480–3490.

34 Kapoor, M., Srnivas, H., Kandiah, E., Gemma, E., Ellgaard, L., Oscarson, S. et al. (2003). Interactions of substrate with calreticulin, an endoplasmic reticulum chaperone. *J. Biol. Chem.* **278**, 6194–6200.

35 Patil, A. R., Thomas, C. J. & Surolia, A. (2000). Kinetics and the mechanism of interaction of the endoplasmic reticulum chaperone, calreticulin, with monoglucosylated ($Glc_1Man_9GlcNAc_2$) substrate. *J. Biol. Chem.* **275**, 24348–24356.

36 Hammond, C. & Helenius, A. (1997). Quality control in the secretory pathway. *Curr. Opin. Cell Biol.* **7**, 523–529.

37 Trombetta, E. S. & Helenius, A. (1998). Lectins as chaperones in glycoprotein folding. *Curr. Opin. Struct. Biol.* **8**, 587–592.

38 Labriola, C., Cazzulo, J. J. & Parodi, A. J. (1999). *Trypanosoma cruzi* calreticulin is a lectin that binds monoglucosylated oligosaccharides but not protein moieties of glycoproteins. *Mol. Biol. Cell*, **10**, 1381–1394.

39 Hebert, D. N., Zhang, J. X., Chen, W., Foellmer, B. & Helenius, A. (1997). The number and location of glycans on influenza hemagglutinin determine folding and association with calnexin and calreticulin. *J. Cell Biol.* **139**, 613–623.

40 Sadasivan, B., Lehner, P. J., Ortmann, B., Spies, T. & Cresswell, P. (1996). Roles for calreticulin and a novel glycoprotein, tapasin, in the interaction of MHC class I molecules with TAP. *Immunity*, **5**, 103–114.

41 Harris, M. R., Yu, Y. Y. L., Kindle, C. S., Hansen, T. H. & Solheim, J. C. (1998). Calreticulin and calnexin interact with different protein and glycan determinants during the assembly of MHC class I. *J. Immunol.* **160**, 5404–5409.

42 Ou, W.-J., Cameron, P. H., Thomas, D. Y. & Bergeron, J. J. M. (1993). Association of folding intermediates of glycoproteins with calnexin during protein maturation. *Nature*, **364**, 771–776.

43 Degen, E. & Williams, D. B. (1991). Participation of a novel 88-kDa protein in the biogenesis of murine class I histocompatibility molecules. *J. Cell Biol.* **112**, 1099–1115.

44 Cannon, K. & Helenius, A. (1999). Trimming and readdition of glucose to *N*-linked oligosaccharides determines calnexin association of a substrate glycoprotein in living cells. *J. Biol. Chem.* **274**, 7537–7544.

45 Pind, S., Riordan, J. R. & Williams, D. B. (1994). Participation of the endoplasmic reticulum chaperone calnexin (p88, IP90) in the biogenesis of the cystic fibrosis transmembrane conductance regulator. *J. Biol. Chem.* **269**, 12784–12789.

46 Labriola, C., Cazzulo, J. J. & Parodi, A. J. (1995). Retention of glucose residues added by the UDP-Glc:glycoprotein glucosyltransferase delays exit of glycoproteins from the endoplasmic reticulum. *J. Cell Biol.* **130**, 771–779.

47 Hammond, C., Braakman, I. & Helenius, A. (1994). Role of *N*-linked oligosaccharide recognition, glucose trimming, and calnexin in glycoprotein folding and quality control. *Proc. Natl. Acad. Sci. USA.* **91**, 913–917.

48 Schrag, J. D., Bergeron, J. J. M., Li, Y., Bonsova, S., Hahn, M., Thomas, D. Y. et al. (2001). The structure of calnexin, an ER chaperone involved in quality control of protein folding. *Mol Cell*, **8**, 633–644.

49 Zhang, J.-X., Braakman, I., Matlack, K. E. S. & Helenius, A. (1997). Quality control in the secretory pathway: the role of calreticulin, calnexin and BiP in the retention of glycoproteins with C-terminal truncations. *Mol. Biol. Cell*, **8**, 1943–1954.

50 Fernández, F., Jannatipour, M., Hellman, U., Rokeach, L. & Parodi, A. J. (1996). A new stress protein: synthesis of *Schizosaccharomyces pombe* UDP-Glc:glycoprotein glucosyltransferase mRNA is induced under stress conditions but the enzyme is not essential for cell viability. *EMBO J.* **15**, 705–713.

51 Arnold, S. M., Fessler, L. I., Fessler, J. H. & Kaufman, R. J. (2000). Two homologues encoding human UDP-glucose:glycoprotein glucosyltransferase differ in mRNA expression and enzymatic activity. *Biochemistry*, **39**, 2149–2163.

52 Vassilakos, A., Cohen-Doyle, M. F., Peterson, P. A., Jackson, M. R. & Williams, D. B. (1996). The molecular chaperone calnexin facilitates folding and assembly of class I histocompatibility molecules. *EMBO J.* **15**, 1495–1506.

53 Hebert, D. N., Foellmer, B. & Helenius, A. (1996). Calnexin and calreticulin promote folding, delay oligomerization and suppress degradation of influenza hemagglutinin in microsomes. *EMBO J.* **15**, 2961–2968.

54 Bass, J., Chiu, G., Argon, Y. & Steiner, D. F. (1998). Folding of insulin receptor monomers is facilitated by the molecular chaperones calnexin and calreticulin and impaired by rapid dimerization. *J. Cell Biol.* **141**, 637–646.

55 D'Alessio, C., Fernández, F., Trombetta, E. S. & Parodi, A. J. (1999). Genetic evidence for the heterodimeric structure of glucosidase II. The effect of disrupting the subunit-encoding genes on glycoprotein folding. *J. Biol. Chem.* **274**, 25899–25905.

56 Molinari, M. & Helenius, A. (2000). Chaperone selection during glycoprotein translocation into the endoplasmic reticulum. *Science*, **288**, 331–333.

57 Oliver, J. D., van der Wal, F., Bulleid, N. J. & High, S. (1997). Interaction of the thiol-dependent reductase ERp57 with nascent glycoproteins. *Science*, **275**, 86–88.

58 Elliott, J. G., Oliver, J. D. & High, S. (1997). The thiol-dependent reductase ERp57 interacts specifically with *N*-glycosylated integral membrane proteins. *J. Biol. Chem.* **272**, 13849–13855.

59 Farmery, M. R., Allen, S., Allen, A. J. & Bulleid, N. J. (2000). The role of ERp57 in disulfide bond formation during the assembly of major histocompatibility complex class I in a synchronized semipermeabilized cell translation system. *J. Biol. Chem.* **275**, 14933–14938.

60 Zapun, A., Darby, N. J., Tessier, D. C., Michalak, M., Bergeron, J. J. M. & Thomas, D. Y. (1998). Enhanced catalysis of ribonuclease B folding by the interaction of calnexin or calreticu-

lin with ERp57. *J. Biol. Chem.* **273**, 6009–6012.

61 Fanchiotti, S., Fernández, F., D'Alessio, C. & Parodi, A. J. (1998). The UDP-Glc:glycoprotein glucosyltransferase is essential for *Schizosaccharomyces pombe* viability under conditions of extreme endoplasmic reticulum stress. *J. Cell Biol.* **143**, 625–635.

62 Ray, M. K., Young, J., Sundaram, S. & Stanley, P. (1991). A novel glycosylation phenotype expressed by Lec23, a Chinese hamster ovary mutant deficient in α-glucosidase I. *J. Biol. Chem.* **266**, 22818–22825.

63 Reitman, M. L., Trowbridge, L. S. & Kornfeld, S. (1982). A lectin-resistant mouse lymphoma cell line is deficient in glucosidase II, a glycoprotein processing enzyme. *J. Biol. Chem.* **257**, 10357–10363.

64 Balow, J. P., Weissman, J. D. & Kearse, K. P. (1995). Unique expression of major histocompatibility complex class I proteins in the absence of glucose trimming and calnexin association. *J. Biol. Chem.* **270**, 29025–29029.

65 Jakob, C. A., Burda, P., te Heesen, S., Aebi, M. & Roth, J. (1998). Genetic tayloring of *N*-linked oligosaccharides: the role of glucose residues in glycoprotein processing in *Saccharomyces cerevisiae* in vivo. *Glycobiology*, **8**, 155–164.

66 Hammond, C. & Helenius, A. (1994). Quality control in the secretory pathway: retention of a misfolded viral membrane glycoprotein involves cycling between the ER, intermediate compartment, and Golgi apparatus. *J. Cell Biol.* **126**, 41–52.

67 Fischer, P. B., Karlsson, G. B., Butters, T. D., Dwek, R. A. & Platt, F. M. (1996). N-butyldeoxynojirimycin-mediated inhibition of human immunodeficiency virus entry correlates with changes in antibody recognition of the V1–V2 region of gp120. *J. Virol.* **70**, 7143–7152.

68 Petrescu, S. M., Petrescu, A. J., Titu, H. N., Dwek, R. A. & Platt, F. M. (1997). Inhibition of *N*-glycan processing in B16 melanoma cells results in inactivation of tyrosinase but does not prevent its transport to the melanosome. *J. Biol. Chem.* **272**, 15796–15803.

69 Mehta, A., Lu, X., Block, T. M., Blumberg, B. S. & Dwek, R. A. (1997). Hepatitis B virus (HBV) envelope glycoproteins vary drastically in their sensitivity to glycan processing: evidence that alteration of a single *N*-linked glycosylation site can regulate HBV secretion. *Proc. Natl. Acad. Sci. USA* **94**, 1822–1827.

70 Elgaard, L. & Helenius, A. (2003). Quality control in the endoplasmic reticulum. *Nature Rev. Mol. Cell Biol.* **4**, 181–191.

71 Trombetta, E. S. & Parodi, A. J. (2003). Quality control and protein folding in the secretory pathway. *Annu. Rev. Cell Dev. Biol.* **19**, 649–676.

72 Jakob, C. A., Burda, P., Roth, J. & Aebi, M. (1998). Degradation of misfolded endoplasmic reticulum glycoproteins in *Saccharomyces cerevisiae* is determined by a specific oligosaccharide structure. *J. Cell Biol.* **142**, 1223–1233.

73 Marcus, N. Y. & Perlmutter, D. H. (2000). Glucosidase and mannosidase inhibitors mediate increased secretion of mutant α_1-antitrypsin Z. *J. Biol. Chem.* **275**, 1987–1992.

74 De Virgilio, M., Kitzmuller, C., Schwaiger, E., Klein, M., Kreibich, G. & Ivessa, N. E. (1999). Degradation of a short-lived glycoprotein from the lumen of the endoplasmic reticulum: the role of *N*-linked glycans and the unfolded protein response. *Mol. Biol. Cell*, **10**, 4059–4073.

75 Chung, D. H., Ohashi, K., Watanabe, M., Miyasaka, N. & Hirosawa, S. (2000). Mannose trimming targets mutant α_2-plasmin inhibitor for degradation by the proteasome. *J. Biol. Chem.* **275**, 4981–4987.

76 Fagioli, C. & Sitia, R. (2001). Glycoprotein quality control in the endoplasmic reticulum. Mannose trimming by endoplasmic reticulum mannosidase I times the proteasomal degradation of unassembled immunoglobulin subunits. *J. Biol. Chem.* **276**, 12885–12892.

77 Wilson, C. M., Farmery, M. R. & Bulleid, N. J. (2000). Pivotal role of calnexin and mannose trimming in regulating the endoplasmic reticulum-associated degradation of major histocompatibility complex class I heavy chain. *J. Biol. Chem.* **275**, 21224–21232.

78 Nakatsukasa, K., Nishikawa, S., Hosokawa, N., Nagata, K. & Endo, T. (2001). Mnl1p, an α-mannosidase-like protein in yeast *Saccharomyces cerevisiae*, is required for endoplasmic reticulum-associated degradation of glycoproteins. *J. Biol. Chem.* **276**, 8635–8638.

79 Hosokawa, N., Wada, I., Hasegawa, K., Yorihuzi, T., Tremblay, L. O., Herscovics, A. and Nagata, A. (2001). A novel ER α-mannosidase-like protein accelerates ER-associated degradation. *EMBO Rep.* **2**, 415–422.

80 Jakob, C. A., Bodmer, D., Spirig, U., Batig, P., Marcil, A., Dignard, D. et al. (2001). Htm1p, a mannosidae-like protein, is involved in glycoprotein degradation in yeast. *EMBO Rep.* **2**, 423–430.

81 Oda, Y., Hosokawa, N., Wada, I. & Nagata, K. (2003). EDEM as an acceptor of terminally misfolded glycoproteins released from calnexin. *Science*, **299**, 1394–1397.

82 Molinari, M., Calanca, V., Galli, C., Lucca, P. & Paganetti, P. (2003). Role of EDEM in the release of misfolded glycoproteins from the calnexin cycle. *Science*, **299**, 1397–1400.

83 Byrd, J. C., Tarentino, A. L., Maley, F., Atkinson, P. H. & Trimble, R. B. (1982). Glycoprotein synthesis in yeast. Identification of $Man_8GlcNAc_2$ as an essential intermediate in oligosaccharide processing. *J. Biol. Chem.* **257**, 14657–14666.

84 Gonzalez, D. S., Karaveg, K., Vandersall-Nairn, A. S., Lal, A. & Moreman, K. W. (1999). Identification, expression, and characterization of a cDNA encoding human endoplasmic reticulum mannosidase I, the enzyme that catalyzes the first mannose trimming step in mammalian Asn-linked oligosaccharide biosynthesis. *J. Biol. Chem.* **274**, 21375–21386.

85 Tremblay, L. O. & Herscovics, A. (1999). Cloning and expression of a specific human α1,2-mannosidase that trims $Man_9GlcNAc_2$ to $Man_8GlcNAc_2$ isomer B during *N*-glycan biosynthesis. *Glycobiology*, **9**, 1073–1078.

86 Camirand, A., Heysen, A., Grondin, B. & Herscovics, A. (1991). Glycoprotein biosynthesis in *Saccharomyces cerevisiae*. Isolation and characterization of the gene encoding a specific processing alpha-mannosidase. *J. Biol. Chem.* **266**, 15120–15127.

87 Weng, S. & Spiro, R. G. (1993). Demonstration that a kifunensin-resistant α-mannosidase with a unique processing action on *N*-linked oligosaccharides occurs in rat liver endoplasmic reticulum and various cultured cells. *J. Biol. Chem.* **268**, 25656–25663.

88 Zuber, C., Spiro, M. J., Guhl, B., Spiro, R. & Roth, J. (2000). Golgi apparatus immunolocalization of endomannosidase suggests post-endoplasmic reticulum glucose trimming: implications for quality control. *Mol. Biol. Cell*, **11**, 4227–4240.

89 Lal, A., Pang, P., Kalelkar, S., Romero, P. A., Herscovics, A. & Moreman, K. W. (1998). Substrate specificity of recombinant murine Golgi α1,2-mannosidases IA and IB and comparison with endoplasmic reticulum and Golgi processing α1,2-mannosidases. *Glycobiology*, **8**, 981–995.

90 Tremblay, L. O. & Herscovics, A. (2000). Characterization of a cDNA encoding a novel human Golgi α1,2-mannosidase (IC) involved in *N*-glycan biosynthesis. *J. Biol. Chem.* **275**, 31655–31660.

91 Bischoff, J., Liscum, L. & Kornfeld, R. (1986). The use of 1-deoxymannojirimycin to evaluate the role of various α-mannosidases in oligosac-

charide processing in intact cells. *J. Biol. Chem.* **261**, 4766–4774.

92 Herscovics, A., Romero, P. A. & Tremblay, L. O. (2002). The specificity of the yeast and human class I α1,2-mannosidases involved in ER quality control is not as strict as previously reported. *Glycobiology,* **11**, 14G–15G.

93 Tokunaga, F., Brostrom, C., Koide, T. & Arvan, P. (2000). Endoplasmic reticulum (ER)-associated degradation of misfolded *N*-linked glycoproteins is suppressed upon inhibition of ER mannosidase I. *J. Biol. Chem.* **275**, 40757–40764.

94 Ermonval, M., Kitzmuller, C., Mir, A. M., Cacan, R. & Ivessa, N. E. (2001). *N*-glycan structure of a short-lived variant of ribophorin I expressed in the MadIA214 glycosylation-defective cell line reveals the role of a mannosidase that is not ER mannosidase I in the process of glycoprotein degradation. *Glycobiology,* **7**, 565–576.

95 Frenkel, Z., Gregory, W., Kornfeld, S. & Lederkremer, G. (2003). Endoplasmic reticulum-associated degradation of mammalian glycoproteins involves sugar trimming to $Man_{6-5}GlcNAc_2$. *J. Biol. Chem.* **278**, 34119–34124.

96 Hampton, R. Y. (2002). ER-associated degradation in protein quality control and cellular regulation. *Curr. Opin. Cell Biol.* **14**, 476–482.

97 Yoshida, Y., Chiba, T., Tokunaga, F., Kawasaki, H., Iwal, K., Suzuki, T. et al. (2002). E3 ubiquitin ligase that recognizes sugar chains. *Nature,* **418**, 438–442.

98 Hampton, R. Y. (2002). Proteolysis and sterol regulation. *Annu. Rev. Cell Dev. Biol.* **18**, 345–378.

99 Hamman, B. D., Chen, J. C., Johnson, E. E. & Johnson, A. E. (1997). The aqueous pore through the translocon has a diameter of 40–60 A during cotranslational protein translocation at the ER membrane. *Cell,* **89**, 535–544.

100 Menetret, J., Neuhof, A., Morgan, D. G., Plath, K., Rademacher, M., Rapoport, T. O. et al. (2000). The structure of ribosome-channel complexes engaged in protein translocation. *Mol. Cell,* **6**, 1219–1232.

101 Beckmann, R., Spahn, C. M. T., Eswar, N., Helmers, J., Penczek, P. A., Sali, A. et al. (2001). Architecture of the protein conducting channel associated with the translating 80S ribosome. *Cell,* **107**, 361–372.

102 Kowarik, M., Kung, S., Martoglio, B. & Helenius, A. (2002). Protein folding during cotranslational translocation in the endoplasmic reticulum. *Mol. Cell,* **10**, 769–778.

103 Molinari, M., Galli, C., Piccaluga, V., Pieren, M. & Paganetti, P. (2002). Sequential assistance of molecular chaperones and transient formation of covalent complexes during protein degradation from the ER. *J. Cell Biol.* **158**, 247–257.

104 Suzuki, T., Park, H., Hollingworth, N. M., Sternglanz, R. & Lennarz, W. J. (2000). *PNG1*, a yeast gene encoding a highly conserved peptide:*N*-glycanase. *J. Cell Biol.* **149**, 1039–1051.

105 Cacan, R., Dengremont, C., Labiau, O., Kmiécik, D., Mir, A. M. & Verbert, A. (1996). Occurrence of a cytosolic neutral chitobiase activity involved in oligomannoside degradation: a study with Mady-Darby bovine kidney (MDBK) cells. *Biochem. J.* **313**, 597–602.

106 Grard, T., Herman, V., Saint-Pol, A., Kmiécik, D., Labiau, O., Mir, A. M. et al. (1996). Oligomannosides or oligosaccharide-lipids as potential substrates for rat liver cytosolic α–D-mannosidase. *Biochem. J.* **316**, 787–792.

107 Saint-Pol, A., Codogno, P. & Moore, S. H. E. (1999). Cytosol to lysosome transport of free polymannose-type oligosaccharides. Kinetics and specificity studies using rat liver lysosomes. *J. Biol. Chem.* **274**, 13547–13555.

108 Varki, A. (1991). Radioactive tracer techniques in the sequencing of glycoprotein glycans. *FASEB J.* **5**, 226–235.

109 Panneerselvam, K. & Freeze, H. H. (1996). Mannose enters mammalian

synthesis using the fibril-forming collagens. Fibril-forming collagens (also known as fibrillar collagens) are the most abundant of the collagen family, occurring commonly in most animal tissues. For this reason they are probably the best characterized among the collagen protein family. The main features of this group are their large molecule size and large uninterrupted triple helix. The fibrillar collagen molecule itself is a flexible "rod-like" structure that has the ability to associate laterally with similar molecules, giving rise to the well-known fibril structure outside the cell. Inside the cell the non-collagenous extremities known as the propeptide domains maintain the procollagen molecule in a soluble state.

18.1.2
Fibrillar Procollagen

The procollagens are the larger soluble intracellular precursors of the insoluble fibrillar collagen proteins [4, 5]. Members of the fibrillar collagen subgroup show significant homology to each other, and this is also true of their procollagen precursors [6]. This group consists of three major members including types I, II, and III and two smaller members, types V and XI. These collagens form essential constituents of vital structural tissues including bone, skin, blood vessels, and cartilage.

The triple-helical molecules of procollagen are formed from three constituent procollagen chains. The structure of the procollagen molecule is reflected in the amino acid sequence composition of the constituent chains. Complete cDNA and amino acid sequences for many types of procollagen chain are available and may be used in sequence homology analyses to compare and contrast sequences between chains [7–10]. In each procollagen chain, the domains responsible for the formation of the N-propeptide, the N-telopeptide, the triple helix, the C-telopeptide, and the C-propeptide have been identified [11]. The structure of the molecule is illustrated below in Figure 18.1, see p. 649. The characteristic repeating Gly-X-Y motif of the triple-helical domain is wound into the triple helix that is the core part of the trimeric procollagen molecule. While a great deal is known about the structure and properties of the triple helix, the globular propeptides are not so well characterized. The larger of the globular regions is the C-propeptide, which performs specialist functions that are discussed later. The N-propeptide domain is not as well characterized, but some specific roles have been identified. The propeptides are also involved in multiple interactions with various cellular factors that may be involved in the assembly and processing of procollagen molecules. The C-propeptides of the major procollagen types I, II, and III are cleaved by specific C-proteinases immediately upon exiting the cell, and this begins the process of fibrillogenesis.

18.1.3
Expression of Fibrillar Collagens

The complexity and diversity of collagen expression are essential to the developmental pathways that allow a higher organism to develop from its early embryonic stages to the mature adult form. Furthermore, tissue renewal and repair processes that are necessary throughout the life of the organism indicate that some expression pathways operate constitutively while other pathways are induced to repair and replace damaged and worn tissues. Thus, expression of all collagens in terms of both development and maintenance is a tightly controlled operation regulated at both the genetic and biochemical levels. Collagen expression is linked to the expression of many other proteins. Some of these proteins are directly involved in processing procollagen in vivo, such as bone morphogenetic protein (BMP-1) and proteins that interact with immature procollagen such as protein disulfide isomerase (PDI), Hsp47, prolyl hydroxylase, lysyl hydroxylase, and other proteins that are co-expressed along with fibrillar collagens. This ensures that the intracellular procollagens are synthesized, correctly modified, and efficiently transported when required. Some regulatory factors such as TGF-β and IL-6 promote procollagen expression at the gene level. Other factors such as tumor necrosis factor TNF-α and IFN-γ appear to suppress procollagen expression. Thus, there is a whole spectrum of signals that appear to induce or suppress procollagen expression and constitute the physiological feedback mechanisms by which expression is regulated.

18.2 The Procollagen Biosynthetic Process: An Overview

Multi-exon genes such as COL1A1 and COL1A2 that encode the chains that make up type I procollagen reside on different chromosomes. Therefore, the production of fibrillar collagens from collagen genes must necessarily be a highly regulated and coordinated process so that the correct combinations of chains are produced at a given time in order to produce procollagen. Splicing of large mRNA precursors to produce specific mRNAs occurs within the nucleus. The mature mRNAs are transported out into the cytoplasm, where membrane-bound ribosomes can begin the biosynthesis of procollagen.

The biosynthesis of procollagen begins with the translation of specific mRNAs on ribosomal complexes [12–14]. The polypeptide chains are co-translationally inserted into the lumen of the endoplasmic reticulum (ER) [15–17]. An interaction between the translocon Sec61p protein and the lumenal chaperone BiP has been proposed to facilitate the completion of translocation into the ER [18, 19].

Enzymes involved in the biosynthetic process are located within the ER lumenal space [20–26]. Hydroxylation of some proline and lysine residues occurs as the triple-helical domain emerges into the lumen of the ER [27, 28]. Virtually all of the Y position proline residues, as well as some Y position lysine residues, are hydroxylated. This hydroxylation is important to the structural stability of the triple helix that is formed later. Some of the hydroxylysine residues in the triple-helical domains are further modified by *O*-linked glycosylation. This is distinct from the *N*-linked glycosylation that occurs in the C-propeptide domains, the function of which is not clear.

Once translocated, the C-propeptide domain is free to fold, forming its complement of intrachain disulfide bonds [29, 30]. The three polypeptide chains of procollagen must then associate at the C-propeptide domains, bringing the three chains into close proximity and thus allowing the formation of interchain disulfide bonds and nucleation of the triple-helix. The molecule is then able to form a triple-helix in a "zipper-like" fashion [31] starting at the C-terminal end and proceeding towards the N-terminal end of the chains [32]. This folding of the triple helix is rate limited by the conversion of *cis*-hydroxyproline residues to *trans*-hydroxyproline residues [33–36]. Current evidence suggests that only the non-triple-helical chains are substrates for the hydroxylase enzymes and that once the triple-helix is formed no further hydroxylation occurs. The hydroxylase enzymes lose affinity as the triple-helix is formed and this probably serves as the detachment mechanism that allows the triple-helical molecule to escape the clutches of these ER-resident proteins. The putative chaperone Hsp47 is believed to bind to the triple-helix as it is formed. Once helix formation is completed, the N-propeptides of the three constituent chains are able to associate, completing the assembly process. Assembled procollagen molecules are then transported through the secretory pathway via mechanisms that are not yet clear but that could involve vesicular transport or a process of cisternal maturation. A schematic of the biosynthetic process is shown in Figure 18.2.

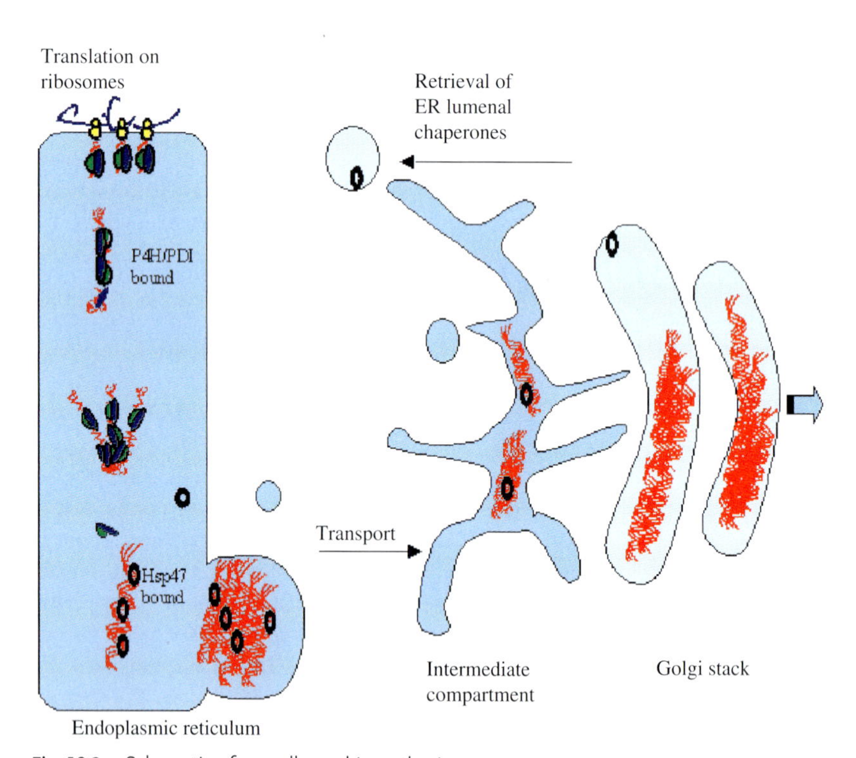

Fig. 18.2. Schematic of procollagen biosynthesis.

18.3
Disulfide Bonding in Procollagen Assembly

Cysteine residues, which form intrachain disulfide linkages, are more highly conserved than other residues in most cysteine-containing proteins [37]. This is unsurprising given that these bonds are extremely important for formation of the native structure of most cysteine-containing proteins. This rule is also applicable to the cysteines in procollagen chains, particularly those in the C-propeptide domain, where a number of intrachain disulfide linkages are thought to occur [6, 38, 39]. The strict conservation of these cysteines between different fibrillar collagen types and between species suggests that the same intrachain links are formed in all the C-propeptide domains and therefore that the tertiary structure of the C-propeptides is similar.

Most of the fibrillar procollagen C-propeptide domains contain eight cysteine residues in conserved positions relative to the C-proteinase cleavage site. These cysteines are numbered sequentially from the N-terminus of the C-propeptide domain. It is notable that both proα2(I) and proα2(V) C-propeptide domains possess only seven of these cysteine residues and that the second cysteine residue position that occurs in the other chains is occupied by a serine residue. Since both proα2(I) and proα2(V) are heterotrimer-forming chains, it was suggested that the lack of this cysteine could account for the inability of these chains to form homotrimers [40]. However, this theory is contradicted by work showing that restoration of a full complement of cysteines in the C-propeptide domain of proα2(I) does not enable them to form homotrimers [41].

Early models suggested that disulfide bonding did not occur while the procollagen chains were nascent and still attached to the polysomes [42]. Disulfide linkages were detected only when chains had been completed and apparently released from the ribosomal complexes. This observation is consistent with the prevailing view that synthesis of the whole C-propeptide domain must be completed before folding of these domains can begin. The last cysteine residue in the procollagen C-propeptide domain is located very close to the C-terminus of the chain; it is probable that this residue is essential to the structure of the C-propeptide and that correct folding cannot occur without it.

It is known that isolated C-propeptide domains can independently associate and form interchain disulfide linkages [30]. Subsequently, it was also shown that a construct with the C-telopeptide and C-propeptide domains from proα1(III) chains linked to a signal peptide folds and assembles into disulfide-linked trimers [43]. Thus, the C-propeptide can be considered as a domain that has a specific role in guiding the formation of the triple-helical procollagen molecule.

Interchain disulfide bonding between C-propeptide domains involves specific sets of cysteine residues on adjacent chains [38, 41, 44]. Early models suggested that only cysteines 5–8 form intrachain links [38] and that cysteines 1–4 were considered to be involved with interchain disulfide linkages [44]. More recent studies analyzing the effect of site-directed mutagenesis of the cysteines in the C-propeptide domain suggest that only cysteine 2 and cysteine 3 are involved in interchain disulfide bond formation [41]. This model indicates that mutations of cysteines 1 and 4 affect the folding of the C-propeptide domain and therefore are probably involved in intrachain linkages. In contrast, the C-propeptide domains in mutants without cys2 or cys3 were folded correctly, suggesting that these residues would normally form interchain disulfide bonds.

18.4
The Influence of Primary Amino Acid Sequence on Intracellular Procollagen Folding

18.4.1
Chain Recognition and Type-specific Assembly

An interesting observation is that procollagen-expressing cells may express multiple types of fibrillar procollagen at a given time. For example, skin fibroblasts may co-express procollagen types I, III, and V [45–50]. Thus, up to six highly homologous chains may be co-expressed and assembled type-specifically into the correct combinations. This discrimination process is illustrated by the fact that proα2(I) chains will associate only with proα1(I) chains to form type I molecules [51] and that no homotrimeric proα2(I) molecules have been reported. This is in contrast to proα1(III) chains, which are solely homotrimer-forming and will not associate with any other type of chain [52, 53]. The formation of heterotypic molecules containing both type V and type XI chains is a known exception to the rule that assembly of fibrillar collagens is type-specific [54]. However, some workers consider types V and XI to be a single collagen type consisting of five polypeptide chains [55–57]. In addition, a highly glycosylated alternatively spliced version of the proα1(II) chain [58, 59] appears to be involved in heterotrimer formation with type XI chains [60, 61]. Despite these complications, it is clear that procollagen chains come together in only certain combinations despite the presence of highly homologous chains being expressed at the same time in the same ER lumenal space.

18.4.2
Assembly of Multi-subunit Proteins

Little is known about how the assembly of multi-subunit proteins, which may involve the interactions of folded and/or partially folded subunits, is achieved. Clearly, a recognition aspect must exist in order for the subunits to recognize each other. Regardless of whether this assembly/recognition is mediated by a catalyst or is an intrinsic property of subunits, the assembly will be unlikely to succeed in the absence of the correct primary information. Complex proteins are often made up of many subunits, which may or may not serve distinct functions. Some subunits can co-assemble with other types of subunits in particular combinations in different circumstances. One example of this versatility is the protein disulfide isomerase (PDI). PDI functions as a catalyst of disulfide bond formation but is also present as a constituent of prolyl-4-hydroxylase (P4H) [62, 63] and the microsomal lipid transfer complex [64–66]. There are important questions regarding how such complex multi-subunit proteins are formed. Is the information for subunit association encoded within the constituent subunits? How are these constituent subunits able to recognize each other and associate together in the correct manner? How is the correct stoichiometric ratio determined? Is the recognition ability encoded within the primary amino acid sequence, or are other proteins in the biosynthetic machinery responsible for "manufacturing" the correct combinations of subunits?

18.4.3 Coordination of Type-specific Procollagen Assembly and Chain Selection

One theory of how chain selection can be achieved is to consider it as a problem of coordinated transcription and translation. These processes must be regulated spatially and temporally to bring about the correct combinations of procollagen chains within the same ER lumenal space [14, 67, 68]. While some studies have suggested that proα1(I) and proα2(I) genes are coordinately transcribed in a ratio close to 2:1 [69, 70], other studies show that the levels of mRNA for these chains deviate appreciably from this ratio [71]. Clearly, some degree of transcriptional coordination is necessary in order to produce heterotrimeric type I procollagen molecules, but this does not necessarily determine the ultimate chain selection or folding events.

Veis and coworkers [72] have shown that the ribosomes reading out mRNA transcripts appear in organized arrays along the ER. This model suggests that monocistronic mRNAs are brought together within the same ER compartment by a complex set of interactions that ensure coordinated synthesis [72]. This implies that some component of the ER membrane participates in the organization of the ribosomes and coordinates the synthesis of nascent chains. Further studies by Kirk [73] and Veis and Kirk [74] support a model of molecular assembly where chain selection and folding are determined by the attachment of the ribosomes to the ER membrane. Hu and coworkers [75] have also proposed that some component of the ER membrane participates in the chain selection process. This proposal is based on the observation that cell-free translations of proα1(I) and proα2(I) are coordinately altered in the presence of added membranes. It has long been postulated that C-propeptide anchoring of the nascent procollagen chains to the ER membrane could be important in chain assembly as this would restrict the freedom of the chains and decrease the concentration of chains required for association [76]. Observations of the behavior of procollagen chains on a monomolecular film suggest that procollagen chains could assemble while associated with the ER membrane [77]. However, this behavior is not a consequence of the interaction of the C-propeptide *N*-linked oligosaccharide chain with the membrane lectin calnexin, as mutagenesis of the site for oligosaccharide attachment does not prevent the resulting non-glycosylated C-propeptide from participating in assembly or secretion of procollagen [78]. It is possible that other membrane-associated proteins could interact with the procollagen C-propeptide and organize trimer assembly at the ER membrane. It is difficult to envisage how a C-propeptide domain can achieve its fully folded state if binding to the ER membrane restricts the C-terminal end. One explanation could be that completed procollagen chains detach from the ER membrane to allow the C-propeptide domain to fold and are subsequently indirectly associated with the membrane via interaction with a membrane-bound chaperone. However, these models fail to explain how ribosomes or other membrane factors distinguish between proα2(I) and proα1(I) chains and bring them together in the correct combinations or why proα2(I) chains cannot form homotrimeric molecules.

18.4.4
Hypervariable Motifs: Components of a Recognition Mechanism That Distinguishes Between Procollagen Chains?

Multi-subunit proteins are crucial to the many complex cellular processes that take place in higher organisms. Clearly, mechanisms that determine the recognition and association of subunits are a vital element in generating multi-subunit proteins. Secondary structure predictions reveal striking similarities between predicted structures for the C-propeptide domains of proα2(I) and proα1(I) chains, despite sequence dissimilarities [38, 79]. Conversely, the same type of prediction suggests a different type of structure for the C-propeptide domain of proα1(II) chains. It has been argued that this could represent the structural basis for chain discrimination during procollagen biosynthesis [38]. Other workers have suggested that the diversity between C-propeptide sequences near the C-proteinase cleavage site might account for the differential chain selection between procollagen chains [80].

More recent studies exploiting advanced in vitro translation technology and genetic engineering have led to a new perspective on how selective association of procollagen chains is achieved. In this system recombinant procollagen chains are expressed in reticulocyte lysate in the presence of semi-permeabilized cells. The utility of this methodology for the analysis of procollagen assembly cannot be understated. Procollagen assembly can be analyzed under controlled experimental conditions while presserving the in vivo environment including the chaperones that are crucial for the assembly process. An experimental strategy based upon the specific exchange of variable sequences between the C-propeptide domains of the homotrimeric proα1(III) chain and the heterotrimer-forming proα2(I) chain has been used to investigate the mechanism of selective chain association [81, 82]. The analysis of procollagen assembly using the chimeric chains expressed in semi-permeabilized cells led to the identification of a specific short, discontinuous sequence of 15 amino acids (GNPELPEDVLDV......SSR) within the proα1(III) C-propeptide, which directs procollagen self-association. This recognition sequence appears to be necessary and sufficient to drive homotrimer formation when placed in the correct context within the proα2(I) C-propeptide. This sequence is presumably responsible for the initial recognition event between chains and is therefore necessary to ensure selective chain association. The variable recognition sequence is interrupted by a central hydrophobic motif, Q(L/M)(T/A)F(L/M)(R/K)L(L/M), that is conserved between different fibrillar procollagen chains. This sequence possibly preserves some element of structural similarity in any recognition site that may be formed at the interface between the interacting C-propeptide domains.

18.4.5
Modeling the C-propeptide

It is known that sequence variations between subunit interfaces determine the specificity of subunit interactions in some proteins. Studies of protein folding and protein-protein interfaces also indicate that similar principles underlie both peptide folding and protein-protein association. In the absence of hard structural data, it is difficult to define the exact interactions that take place during procollagen chain recognition and association. An alternative strategy is to look at computer models based upon information from biophysical studies. Recent low-resolution models based on biophysical studies of the recombinant type III C-propeptide envisage the soluble trimer as "cruciform-shaped" structure with three large lobes and a minor lobe all in the same plane [83]. The larger lobes apparently correspond to each individual propeptide chain, and the minor lobe is proposed to correspond to the junction that links all three C-propeptide domains to the rest of the procollagen molecule. Interestingly, such models place the previously identified recognition sequences at the interface junctions between the individual C-propeptide domains, supporting the idea that such sequences are major determinants of chain selection.

18.4.6
Chain Association

The above models may suggest that recognition domains promote chain selection and association; however, these two events are not necessarily connected. Experimental data indicate that the recognition event is not necessarily followed by an association event. It is possible to envisage a situation where in vivo a certain sequence is responsible for recognition but another neighboring sequence is responsible for the association. Thus, the two events may be uncoupled. There is also sufficient evidence to indicate that trimer formation can be induced in the absence of the recognition sequences. Early sequence comparisons identified a group of aromatic residues within the C-propeptide domain that could be responsible for association among the three chains [84]. Experimental work has shown that the C-propeptide itself can be efficiently replaced by other sequences such as the transmembrane domain of hemagglutinin [43] and the 29-residue foldon domain from T4 bacteriophage fibritin, which are capable of bringing the three chains of procollagen into close proximity [85]. Thus, nonnative combinations of chains can be synthesized in the absence of selective regions. The alpha-helical coiled coil is a very common motif in oligomerizing proteins. These coils are believed to form oligomerization domains in fibrillar and non-fibrillar collagens and also in collagen-like proteins such as C1q. The widespread presence of these domains in collagens and other trimerizing proteins [86] suggests a general role in trimerization and perhaps a specific role in determining the association of procollagen chains in vivo.

18.5 Posttranslational Modifications That Affect Procollagen Folding

18.5.1 Hydroxylation and Triple-helix Stability

Hydroxylation of prolyl and lysyl residues in the triple-helical domain is critical for triple-helix formation and stability. Three types of enzyme are known to participate in hydroxylation of these residues: prolyl 4-hydroxylase (P4H), prolyl 3-hydroxylase (P3H), and lysyl hydroxylase (LH). These hydroxylating enzymes convert the peptidyl-lysine or peptidyl-proline to hydroxylysine or hydroxyproline. Only a small percentage of the peptidyl-prolines in the Y position of the Gly-X-Y triplet are hydroxylated by P3H, while the most important contribution to the stability of the triple-helical procollagen molecule is the hydroxylation of Y-position proline residues in the Gly-X-Y triplet by the P4H enzyme. Formation of the triple helix appears to prevent further hydroxylation [87–89], suggesting that triple-helical procollagen is not a substrate for P4H.

A detailed mechanism for the P4H reaction has been described by a number of workers [90–94]. The hydroxylation reaction requires a number of cofactors, including Fe^{2+}, 2-oxoglutarate, O_2, and ascorbate. The fact that the same cofactors are required for all three types of hydroxylases suggests that the mechanism of action is similar for all three hydroxylating enzymes.

The presence or absence of hydroxylation cofactors can be used experimentally as a tool for the analysis of procollagen production by cells. One example is the chelation of ferrous iron by inhibitors such as $\alpha\alpha'$-dipyridyl. This is a popular method for inhibiting the hydroxylation reaction experimentally. Treatment of procollagen-expressing fibroblasts with this agent leads to retention of unhydroxylated procollagen within the ER lumenal space. This "block" can be reversed by the addition of excess ferrous ions when required. Another method for achieving the same result is to deplete procollagen-producing cells of ascorbate. Ascorbate is an essential cofactor that is required for repeated cycles of enzyme activity. Re-addition of ascorbate to depleted cells restores procollagen secretion. Such strategies are often used in experiments where it is necessary to build up a concentration of unfolded procollagen within the ER and to follow its maturation through the secretory pathway.

18.6 Procollagen Chaperones

18.6.1 Prolyl 4-Hydroxylase

Vertebrate P4H consists of tetramers composed of two α and two β subunits $\alpha2\beta2$ [95–99]. The α-subunit is the catalytic subunit for hydroxylation and is prone to aggregation in the absence of the β-subunit [63, 100]. Thus, one function of the β-subunit in P4H appears to be keeping the α-subunit in the active soluble state [101, 102]. Isoforms of the α-subunit have been identified in human tissue [103], in mouse [104], and in *C. elegans* [105]. The β-subunit is identical to PDI and maintains its PDI activity even as part of the P4H complex [62, 63]. Another function of the β-subunit may be to retain the α-subunit within the ER lumen since the β-subunit has an ER retention motif, whereas the α-subunit does not.

Interestingly, P4H has been found in stable association with procollagen chains possessing a deletion in the triple-helical domain [106], suggesting a chaperone-type role for P4H. Since this result was obtained with antibodies directed against the β-subunit of P4H, this chaperone effect may be mediated by the PDI. However, current evidence suggests that P4H (both α and β subunits) can form stable associations with non-triple helical chains and that it is the major protein that interacts with procollagen chains during early biosynthesis [107]. This binding appears to be conformation-dependent so that P4H is able to distinguish between folded and unfolded procollagen molecules. Taken together with evidence that P4H has reduced affinity for hydroxylated chains [108], it appears that P4H may be a sophisticated hydroxylation- and conformation-sensitive chaperone that is able to mediate the retention of unfolded procollagen chains.

18.6.2 Protein Disulfide Isomerase

PDI is a key cellular folding enzyme that is important for the maturation of several secreted and membrane-associated proteins. This is particularly true for the folding and maturation of procollagen, where PDI is involved in a number of important steps. As well as functioning as part of the P4H complex, this chaperone has multiple other roles in procollagen biosynthesis as well as a general role in biogenesis of other disulfide-bonded proteins. The function of PDI as a component of enzymes such as P4H is independent of its disulfide isomerase activity [109]. PDI has been proposed to act as a general molecular chaperone by binding to unfolded proteins and thereby preventing aggregation [110–112]. This indicates that PDI assists in the refolding of certain denatured proteins in vitro. In other cases PDI appears to have no activity or even anti-chaperone activity with some substrates [113, 114]. PDI also has been shown to interact with newly synthesized proteins [115] and to catalyze the formation of both intrachain and interchain disulfide bonds [116, 117]. Studies using a cross-linking approach in a semi-permeabilized cell system have shown that PDI also appears to have a separate role in the chaperoning of monomeric chains and monomeric C-propeptide domains [118, 119]. The function of this binding appears to be to keep the monomeric chains in solution. These observations support a model where PDI plays a crucial role in binding to the C-propeptide, thereby coordinating heterotrimer assembly. Although the monomeric C-propeptide contains free thiol residues, no mixed disulfides have been detected between PDI and procollagen C-propeptides. Hence, the interaction is probably dependent on hydrophobic interactions.

18.6.3
Hsp47

Prolyl 4-hydroxylase is just one member of a group of chaperones that play a vital role in the biogenesis of procollagen within the cell. These chaperones carry out a number of functions in order to assist with procollagen biosynthesis, including folding, protection, retention, posttranslational modification, and degradation. This group of chaperones interacts with multiple proteins that require assistance with folding and assembly. However, it seems that the variety and complexity of proteins in the ER folding environment requires a second class of more specific protein chaperones that are limited to quality control of particular proteins. Heat shock protein 47 (Hsp47) falls into this second group and is specifically associated with procollagen biosynthesis. This stress-induced chaperone is linked with the production of procollagen in many types of cells. Hsp47 and type IV collagen have been found to decrease during the differentiation of mouse teratocarcinoma cells [120]. Upregulation of Hsp47 also occurs when expression of collagen types I and III is induced during fibrosis [121] and during mechanical and heat stress in embryonic chicken tendon cells [122].

Despite some 20 years of investigation, definitive roles for Hsp47 in procollagen biosynthesis have yet to be elucidated. This is in part due to conflicting evidence regarding the substrate preferences of Hsp47 and also to a lack of understanding of some aspects of procollagen biosynthesis such as how transport from the ER to the Golgi is achieved. The RDEL motif at the carboxy-terminus of Hps47 certainly limits the theater of its operations via the KDEL receptor retention mechanism. Any chaperone-type function must therefore necessarily occur prior to the *cis*-Golgi, where Hsp47 is believed to dissociate from procollagen probably as a result of a pH change that alters its conformation and its affinity for this substrate [123]. The importance of Hsp47 to the biosynthesis of procollagen is illustrated by the dramatic effect of disrupting the Hsp47 genes in a mouse model [124]. This mutation is lethal to the null mouse embryo within 11 days post coitus. The embryo itself is extremely fragile and displays defects in the basal lamina, which compromise basement membrane formation, and apparent defects in associated connective tissues. Most of these effects are likely to be due to the disruption of the extracellular matrix (ECM). Such observations indicate that any secreted collagens from cells lacking Hsp47 may have an abnormal conformation that causes the ECM disruption. The procollagens secreted by these cells were sensitive to proteolytic digestion, indicating that the triple helix may be malformed. Transient transfection of Hsp47 into the knockout cells appears to make the procollagen resistant to proteolysis, demonstrating a possible role in stabilization of the triple helix.

Studies looking at the interaction of Hsp47 with recombinant procollagen chains expressed in semi-permeabilized cells found that only triple-helical procollagen molecules were able to interact with Hsp47 with high affinity [125]. Procollagen chains that were engineered to remain monomeric or were prevented from forming a triple helix by inhibiting hydroxylation did not appear to interact with Hsp47.

This evidence together with evidence from work with collagen peptides [126] shows that Hsp47 has a much higher affinity for the triple-helical form of procollagen.

It has been established that the thermal stability of procollagen within the cell may be higher than that of the isolated protein [127], suggesting that the intracellular environment protects the collagen helix from heat denaturation. It is possible that the Hsp47-procollagen interaction leads to a stabilization of the procollagen triple helix, in particular stabilizing those regions of the helix with lower thermal stability [125]. However, work using triple helices designed to be highly stable without any regions of lower thermal stability shows that Hsp47 is still able to bind to these molecules [128]. Furthermore, this work and other evidence show that a minimum of one Gly-X-Arg triplet in the triple-helical domain is necessary for the binding of Hsp47 to procollagen [129].

An alternative explanation for the purpose of binding of Hsp47 to triple-helical procollagen could be to prevent the lateral association of the chains occurring within the ER. Once the procollagen molecule reaches the Golgi apparatus, it has been suggested to form higher-order aggregates, leading to distension of this organelle [130]. This "aggregate form" could be a necessary intermediate prior to propeptide processing and formation of collagen fibrils. The formation of aggregates within the ER would be an undesirable event, perhaps preventing the transport of procollagen to the Golgi apparatus; hence, the presence of Hsp47 could be required to ensure efficient procollagen transport.

18.6.4
PPI and BiP

Other proteins have been identified that have specific roles during procollagen biosynthesis. One example of this is during the formation of the triple helix. The unfolded procollagen chain contains peptide bonds statistically distributed between *cis* and *trans* configurations. Only the *trans* configuretion is compatible with the structure of the triple helix [131]. Conversion between the *cis* and *trans* configurations is limited by the pyrrolidine ring of the imino acids proline and hydroxyproline, which restricts rotation around the peptide bond. Since both proline and hydroxyproline occur frequently within the triple-helical domain of procollagen chains, the *cis-trans* isomerization reaction imposes a rate-limiting step on the propagation of the triple helix [35, 36]. This rate-limiting step is catalyzed by peptidyl-prolyl *cis-trans* isomerase (PPI), which is identical to the protein known as cyclophilin [132, 133]. This enzyme is inhibited by cyclosporin A (CsA), which has been shown to reduce the rate of triple-helix formation in vitro.

Immunoglobulin-binding protein (BiP), also known as GRP78, is a stress-induced protein implicated in the retention of misfolded proteins [134, 135]. However, it is not clear whether this protein plays a role during the normal biosynthesis of procollagen or whether its role is limited to quality control of misfolded procollagens in vivo. It is known that BiP is constitutively expressed during the production of mutant chains produced in cases of osteogenesis imperfecta [136, 137]. The mutations where BiP is involved affect the C-propeptide and presumably interfere with chain association. The mechanism by which BiP recognizes misfolded C-propeptides is unknown but the binding is clearly one part of the quality-control mechanism that processes incorrectly folded procollagen in vivo.

18.7 Analysis of Procollagen Folding

As stated previously, the early stages of procollagen biosynthesis can be reconstituted in a semi-permeabilized cell system whereby conditions can be manipulated and specific aspects of the procollagen assembly process investigated. This methodology has been successfully employed to look at the assembly process itself as well as the role of chaperones such as Hsp47, PDI, and P4H. The volume of physiologically meaningful data resulting from this approach illustrates the power of this combination of in vivo and in vitro techniques. The technique is also very adaptable and may be used to look at assembly of virtually any protein entering the secretory pathway. Here we illustrate the methodology by describing the techniques as applied to procollagen assembly.

The procedure involves treating cells grown in culture with the detergent digitonin and isolating the cells free from their cytosolic component. The SP cell system allows protein assembly to be studied in an environment that more closely resembles that of the intact cell. As the ER remains morphologically intact, the interactions with endogenous chaperones and the spatial localization of folding are maintained. An in vitro translation system is combined in the presence of the SP cells so that individual components can be manipulated easily, providing a means by which cellular processes can be studied under a variety of conditions. Chemical cross-linking reagents can also be utilized posttranslationally in order to facilitate the study of interaction between proteins within the ER lumen [138, 139].

The basic protocol involves translation of mRNA transcripts encoding particular procollagen chains in a rabbit reticulocyte lysate supplemented with the semi-permeabilized HT1080 cells prepared as outlined below. This cell line is able to carry out the complex co- and posttranslational modifications required for the assembly of procollagen molecules into thermally stable triple helices [140]. The mRNA transcript encoding for the protein of interest is translated in the presence of a radio-labeled amino acid (^{35}S-methionine) such that the protein synthesized can be visualized by autoradiography. As the mRNA can be synthesized in vitro from cloned cDNA, the effect of manipulating the primary amino acid sequence upon folding and assembly can be examined. The procedures for preparing SP cells, transcribing cloned cDNAs, and translating the RNA transcripts are outlined below. In addition we describe some procedures for characterizing procollagen translation products.

18.8 Experimental Part

18.8.1 Materials Required

Transcription in vitro

1. 10 μg linearized plasmid DNA, containing the gene of interest downstream of the appropriate promoter, in RNase-free water
2. 5× Transcription buffer (400 mM HEPES buffer pH7.4, 60 mM $MgCl_2$, 10 mM spermidine)
3. Nucleotide triphosphates (ATP, UTP, CTP, and GTP) (25 mM each) (Roche)
4. 100 mM DTT (Sigma)
5. T3/T7 RNA polymerase (50 U μL^{-1}) (Promega)
6. RNase inhibitor (Promega)

Translation in vitro

1. Flexi rabbit reticulocyte lysate (Promega Corp.)
2. EasyTag ^{35}S-methionine (NEN Dupont)
3. 2.5 M KCl
4. Amino acid mix (minus methionine) (Promega Corp.)

Preparation of semi-permeabilized cells (all reagents are stored at −20 °C.)

1. Phosphate-buffered saline (Biowhittaker)
2. HT1080 cells (75-cm^2 flask of sub-confluent cells)
3. 1× Trypsin-EDTA solution (Biowhittaker)
4. KHM buffer: KOAc 110 mM, MgOAc 2 mM, HEPES 20 mM, pH 7.2
5. HEPES buffer: KOAc 50 mM, HEPES 50 mM, pH 7.2
6. Digitonin (40 mg mL^{-1} in DMSO, stored at −20 °C) (Calbiochem)
7. Soybean trypsin inhibitor (50 mg mL^{-1} in sterile water stored at −20 °C) (Sigma)
8. Trypan blue solution (0.4%)
9. $CaCl_2$ (100 mM, stored at −20 °C)
10. Micrococcal nuclease (1 mg mL^{-1} in sterile water, stored at −20 °C) (Sigma)
11. EGTA (0.4 M, stored at −20 °C).

Protease digestion assay

1. Chymotrypsin (Sigma): make up fresh in SOL buffer
2. Trypsin (Sigma): make up fresh in SOL buffer
3. Pepsin (Sigma): make up in H_2O

4. Triton X-100 (10% in H_2O, stored at 4 °C)
5. SOL buffer: 50 mM Tris-HCl, pH 7.4, 150 mM NaCl, 10 mM EDTA
6. Soybean trypsin inhibitor (50 mg mL^{-1} in sterile water stored at −20 °C) (Sigma)
7. 1 M HCl
8. 1 M Tris
9. 4× SDS-PAGE sample buffer

18.8.2
Experimental Protocols

Method 1: Preparation of SP Cells This procedure is a modification of the protocol used by Plutner and coworkers [141] adapted for the cell-free expression of proteins. Treatment of mammalian cells with the detergent digitonin renders the plasma membrane permeable to the components of the cell-free translation system while retaining the functionally intact ER membrane. This selective permeabilization is a consequence of the cholesterol-binding properties of digitonin. As cholesterol is a minor constituent of the internal membrane system of the cell, the ER and Golgi networks remain intact. Prior to beginning the protocol, refer to note 1.

1. Remove culture medium and rinse HT1080 cells in culture flask with 1 × 10 mL PBS. Drain and add 1 mL of trypsin solution (prewarmed to room temperature) and incubate for 3 min. All cells should now be detached by gently tapping the flask. Add 8 mL of KHM buffer and 20 μL soybean trypsin inhibitor (final concentration 100 μg mL^{-1}) to the tissue culture flask. Transfer cell suspension to a 15-mL Falcon tube on ice.
2. Pellet cells by centrifugation at 1200 rpm for 3 min at 4 °C. Aspirate the supernatant from the cell pellet.
3. Resuspend cells in 6 mL of ice-cold KHM. Add 6 μL digitonin (from 40 mg mL^{-1} stock, i.e., final concentration 40 μg mL^{-1}) and mix immediately by inversion. Incubate on ice for 5 min (see note 2).
4. Adjust the volume to 14 mL with ice-cold KHM and pellet cells by centrifugation as in step 2.
5. Aspirate the supernatant and resuspend cells in 14 mL ice-cold HEPES buffer. Incubate on ice for 10 min and pellet cells by centrifugation as in step 2.
6. Aspirate the supernatant and resuspend cells carefully in 1 mL ice-cold KHM (use a 1-mL Gilson and pipette gently up and down). Place on ice.
7. Transfer a 10-μL aliquot to a separate 1.5-mL microcentrifuge tube and add 10 μL of Trypan Blue.
8. Count cells in a hemocytometer and check for permeabilization.
9. Transfer cells to a 1.5-mL microcentrifuge tube and spin for 30 s at 13 000 g. Aspirate supernatant and resuspend the cells in 100 μL KHM using a pipette.
10. Add 1 μL of 0.1 M $CaCl_2$ and 1 μL of monococcal nuclease and incubate at room temperature for 12 min. This step removes the endogenous mRNA.

11. Add 1 μL of 0.4 M EGTA to chelate the calcium and inactivate the nuclease. Isolate the cells by centrifuging for 30 s in a microcentrifuge and resuspend in 100 μL of KHM.
12. Use approximately 1×10^5 cells per 25-μL translation reaction (approx. 4 μL of the 100 μL obtained).

Method 2: Transcription in Vitro The cDNA encoding the protein of interest must be placed in a suitable expression vector, such as pBluescript SK (Stratagene), upstream of a suitable promoter containing an RNA polymerase-binding site from which transcription is initiated. The cDNA clone must be linearized by an appropriate restriction enzyme to generate a template for mRNA synthesis. This method is a modification of a method described previously [142].

1. Prepare a 100-μL reaction mixture containing 44 μL H_2O, 10 μL linearized DNA (5–10 μg), 20 μL transcription buffer (5×), 10 μL 100 mM DTT, 1 μL RNasin (20 units), and 3 μL of each nucleotide.
2. Add 3 μL of the appropriate RNA polymerase (150 units) and incubate at 37 °C for 2 h (see note 3).
3. The RNA can be extracted with phenol/chloroform 1:1 and then twice with chloroform and precipitate by adding NaOAc, pH 5.2, to a final concentration of 300 mM and three volumes of ethanol. The RNA pellet is resuspended in 100 μL RNase-free H_2O containing 1 mM DTT and 1 μL RNasin.
4. To assess the yield of RNA, a 1-μL aliquot should be removed and analyzed on a 1% agarose gel (see note 4).

Method 3: Translation in Vitro The translation of proteins in vitro can be performed using either wheat germ extracts or rabbit reticulocyte lysates that contain ribosomes, tRNAs, and a creatine phosphate-based energy regeneration system.

1. Prepare a 25-μL reaction mixture containing 17.5 μL Flexi lysate, 0.5 μL amino acid, 0.5 μL KCl, 1.5 μL EasyTag ^{35}S-methionine, 1 μL mRNA, and 4 μL SP cells (see notes 5 and 6). Incubate the translation sample at 30 °C for 60 min and then place on ice.
2. Prepare the translation sample for SDS-PAGE by adding 2 μL of the product to 15 μL SDS-PAGE sample buffer (0.0625 M Tris/HCl pH 6.8, SDS [2% w/v], glycerol [10% v/v], and bromophenol blue) plus 2 μL DTT (1 M) and boiling the sample for 5 min.
3. The samples should be analyzed by SDS-PAGE. After running, the gel should be dried and exposed to autoradiography film (see note 7).

Method 4: Protease Digestion Assay A "protease digestion" assay is used to determine whether the translation products that have been formed within the SP cells contain triple-helical procollagen. A fully folded and correctly aligned triple helix is resistant to proteolysis by a combination of chymotrypsin, trypsin, and pepsin at temperatures below the characteristic melting temperature of a particular triple helix. Misaligned helices or non-helical trimers of procollagen will be digested under the same conditions [143]. The characteristic melting temperature of the helix is a function of the hydroxyproline content and tends to vary between 35 °C and 42 °C for fibrillar collagen molecules.

1. Prepare a 25-μL translation reaction including freshly prepared SP cells as described above. After translation, place the sample on ice and gently disperse the SP cells using a pipette tip.
2. Remove 5 μL of the translation mix as a control to check for translation efficiency.
3. Isolate the SP cells from the remaining translation by centrifugation and wash with KHM twice to remove any remaining translation mixture.
4. Disrupt the cell pellet with a sterile pipette tip in SOL buffer containing 1% v/v Triton X-100. Leave on ice for 30 min to solubilize the SP cells.
5. Centrifuge the solubilized cells at 13 000 rpm for 20 min. Carefully remove the supernatant to a fresh microcentrifuge tube. Remove 5 μL of the supernatant as a non-treated control.
6. Digest the remaining supernatant with chymotrypsin (250 μg mL^{-1}) and trypsin (100 μg mL^{-1}) for 5 min at 30 °C. Add soybean trypsin inhibitor to 500 μg mL^{-1} and then acidify the digest by adding HCl to 100 mM.
7. Treat the sample with Pepsin (100 μg mL^{-1}) for 2 h at 30 °C or overnight at 4 °C.
8. Stop the reaction by neutralization with 100 mM Tris base and boiling 4× SDS-PAGE buffer. Add DTT to 50 mM for separation under reducing conditions.
9. Prepare the untreated samples for electrophoresis by adding SDS-PAGE buffer containing 50 mM DTT.
10. The samples should be separated by SDS-PAGE. After running, the gel should be dried and exposed to autoradiography film (see note 8).

Method 5: Analysis of Disulfide Bond Formation In the case of fibrillar procollagen chains, formation of the correct intrachain disulfide bonds in the carboxy-terminal domain (the C-propeptide) is necessary for the folding of these domains and is a prerequisite for trimer formation. The trimers in turn are stabilized by interchain disulfides. The formation of disulfide bonds during folding can be monitored by trapping folding intermediates using the alkylating reagent *N*-ethyl maleimide (NEM). The formation of intrachain disulfide bonds generally increases the electrophoretic mobility of proteins during SDS-PAGE analysis under non-reducing conditions as compared to reducing conditions. Procollagen molecules stabilized by interchain disulfide bonds have a markedly reduced migration under non-reducing conditions, as the linked chains form a much larger complex.

1. Prepare a 100-μL translation mix and divide this into four aliquots of 25 μL in separate microcentrifuge tubes.

2. At intervals of 15 min, remove one of the tubes, add NEM to a final concentration of 20 mM, and place on ice for the remainder of the time course.
3. Isolate and "wash" the SP cells as described in step 3 of method 4.
4. Solubilize each of the washed cell pellets in 50 μL SDS-PAGE buffer and then transfer 25 μL of each sample into fresh tubes containing 2 μL DTT. Boil the samples for 5 min prior to electrophoresis (see note 9).

Notes:
1. The procedure takes approximately one hour and should be carried out immediately prior to using the SP cells for translation in vitro, as SP cells do not efficiently reconstitute the translocation of proteins after storage. It is also advisable to use a minimum of one 75-cm^2 flask of cells (75–90% confluent) as it is difficult to work with a smaller quantity of cells. The size of the cell pellet will usually decrease during the procedure due to loss of the cell cytosol that is accompanied by a decrease in cell volume.
2. The digitonin concentration has been optimized for permeabilization of HT1080 cells (i.e., the lowest concentration of digitonin that results in 100% permeabilization). If a different cell line is used, the concentration of digitonin required for permeabilization should be assessed by titration. It is not essential to trypan blue-stain each batch of SP cells, although this is recommended if the procedure is not used routinely.
3. The yield of mRNA can be increased by a further addition of RNA polymerase (1 μL) after 1 h.
4. To minimize degradation of the mRNA, the use of sterile pipette tips and microcentrifuge tubes is recommended. If the yield is low or if the RNA is partially degraded, it is possible that apparatus or solutions have been contaminated with RNases.
5. The translation protocol should be optimized for each different mRNA transcript as the optimal salt concentrations (KCl and MgOAc) may vary.
6. Control translations in the absence of SP cells should also be carried out to test the translation efficiency of a new RNA preparation.
7. It may be necessary to denature any secondary structure in the mRNA by heating to 60 °C for 10 min prior to translation. Additional products with molecular weights smaller than the major translation product may be observed due to ribosome binding to "false" start sites downstream of the initiation codon. As natural procollagens are large molecules, a lower percentage gel is appropriate in order to visualize bands representing proteins in the range of 100–350 kDa.
8. Hydroxylation of proline and lysine residues in procollagen results in reduced electrophoretic mobility as compared to proteins predicted to be of the same size. An adaptation of this method can also be used to estimate the melting temperature of a triple-helical procollagen.
9. If reduced and non-reduced samples are to be run on the same gel, they should be separated by a gap of two lanes in order to prevent reduction of the non-reduced samples by DTT that may diffuse across the gel matrix during electrophoresis.

References

1 Kielty, C. M., Hopkinson, I. & Grant, M. E. (1993). The Collagen Family: Structure, Assembly and Organization in the Extracellular Matrix. In *Connective Tissue and its Heritable Disorders*, pp. 103–147. Wiley-Liss, Inc.

2 Kadler, K. (1994). *Protein profile:* Fibril forming collagens (Sheterline, P., ed.), Vol. 1. 2 vols. Academic Press, Oxford.

3 Prockop, D. J. & Kivirikko, K. I. (1995). Collagens – Molecular-Biology, Diseases, and Potentials For Therapy. *Ann Rev Biochem* 64, 403–434.

4 Schofield, J. D. & Prockop, D. J. (1973). Procollagen – a precursor form of collagen. *Clin Orthop* 97, 175–95.

5 Bornstein, P. (1974). The structure and assembly of procollagen – a review. *J Supramol Struct* 2, 108–20.

6 Dion, A. S. & Myers, J. C. (1987). COOH-terminal propeptides of the major human procollagens: structural, functional and genetic comparisons. *J Mol Biol* 193, 127–143.

7 Hofmann, H., Fietzek, P. P. & Kuhn, K. (1980). Comparative analysis of the sequences of the three collagen chains alpha 1(I), alpha 2 and alpha 1(III) Functional and genetic aspects. *J Mol Biol* 141, 293–314.

8 Bernard, M. P., Myers, J. C., Chu, M. L., Ramirez, F., Eikenberry, E. F. & Prockop, D. J. (1983). Structure of a cDNA for the pro alpha 2 chain of human type I procollagen. Comparison with chick cDNA for pro alpha 2(I) identifies structurally conserved features of the protein and the gene. *Biochemistry* 22, 1139–45.

9 Baldwin, C. T., Reginato, A. M., Smith, C., Jimenez, S. A. & Prockop, D. J. (1989). Structure of cDNA clones coding for human type II procollagen. The alpha 1(II) chain is more similar to the alpha 1(I) chain than two other alpha chains of fibrillar collagens. *Biochem J* 262, 521–8.

10 Greenspan, D. S., Cheng, W. & Hoffman, G. G. (1991). The Pro-Alpha-1(V) Collagen Chain – Complete Primary Structure, Distribution of Expression, and Comparison With the Pro-Alpha-1(XI) Collagen Chain. *J Biol Chem* 266, 24727–24733.

11 Kuhn, K. (1984). Structural and Functional Domains of Collagen – a Comparison of the Protein With Its Gene. *Collagen and Related Research* 4, 309–322.

12 Kerwar, S. S., Kohn, L. D., Lapiere, C. M. & Weissbach, H. (1972). In vitro synthesis of procollagen on polysomes. *Proc Natl Acad Sci USA* 69, 2727–31.

13 Kerwar, S. S. (1974). Studies on the nature of procollagen synthesized by chick embryo polysomes. *Arch Biochem Biophys* 163, 609–13.

14 Veis, A. & Brownell, A. G. (1977). Triple-helix formation on ribosome-bound nascent chains of procollagen: deuterium-hydrogen exchange studies. *Proc Natl Acad Sci USA* 74, 902–5.

15 Harwood, R., Grant, M. E. & Jackson, D. S. (1973). The sub-cellular location of inter-chain disulfide bond formation during procollagen biosynthesis by embryonic chick tendon cells. *Biochem Biophys Res Commun* 55, 1188–96.

16 Harwood, R., Bhalla, A. K., Grant, M. E. & Jackson, D. S. (1975). The synthesis and secretion of cartilage procollagen. *Biochem J* 148, 129–38.

17 Harwood, R., Grant, M. E. & Jackson, D. S. (1974). Secretion of procollagen: evidence for the transfer of nascent polypeptides across microsomal membranes of tendon cells. *Biochem Biophys Res Commun* 59, 947–54.

18 Nicchitta, C. V. & G, B. (1993). Lumenal proteins of the mammalian endoplasmic reticulum are required to complete protein translocation. *Cell* 73, 513–21.

19 Sanders, S. L., Whitfield, K. M., Vogel, J. P., Rose, M. D. & Schekman, R. W. (1992). Sec61p and BiP directly facilitate polypeptide translocation into the ER. *Cell* 69, 353–65.

20 Harwood, R., Grant, M. E. &

JACKSON, D. S. (1974). Collagen biosynthesis. Characterization of subcellular fractions from embyonic chick fibroblasts and the intracellular localization of protocollagen prolyl and protocollagen lysyl hydroxylases. *Biochem J* 144, 123–30.

21 HARWOOD, R., GRANT, M. E. & JACKSON, D. S. (1975). Studies on the glycosylation of hydroxylysine residues during collagen biosynthesis and the subcellular localization of collagen galactosyltransferase and collagen glucosyltransferase in tendon and cartilage cells. *Biochem J* 152, 291–302.

22 DIEGELMANN, R. F., BERNSTEIN, L. & PETERKOFSKY, B. (1973). Cell-free collagen synthesis on membrane-bound polysomes of chick embryo connective tissue and the localization of prolyl hydroxylase on the polysome-membrane complex. *J Biol Chem* 248, 6514–21.

23 CUTRONEO, K. R., GUZMAN, N. A. & SHARAWY, M. M. (1974). Evidence for a subcellular vesicular site of collagen prolyl hydroxylation. *J Biol Chem* 249, 5989–94.

24 GUZMAN, N. A., ROJAS, F. J. & CUTRONEO, K. R. (1976). Collagen lysyl hydroxylation occurs within the cisternae of the rough endoplasmic reticulum. *Arch Biochem Biophys* 172, 449–54.

25 FESSLER, J. H. & FESSLER, L. I. (1978). Biosynthesis of procollagen. *Annu Rev Biochem* 47, 129–62.

26 PROCKOP, D., BERG, R., KIVIRIKKO, K. & UITTO, J. (1976). *Biochemistry of Collagen* (RAMACHANDRAN, G. & REDDI, H., Eds.), Plenum Press, New York.

27 UITTO, J. & PROCKOP, D. J. (1974). Hydroxylation of peptide-bound proline and lysine before and after chain completion of the polypeptide chains of procollagen. *Arch Biochem Biophys* 164, 210–7.

28 VUUST, J. & PIEZ, K. A. (1972). A kinetic study of collagen biosynthesis. *J Biol Chem* 247, 856–62.

29 UITTO, V. J., UITTO, J. & PROCKOP, D. J. (1981). Synthesis of Type-I Procollagen – Formation of Interchain Disulfide Bonds Before Complete Hydroxylation of the Protein. *Arch Biochem Biophys* 210, 445–454.

30 DOEGE, K. J. & FESSLER, J. H. (1986). Folding of carboxyl domain and assembly of procollagen I. *J Biol Chem* 261, 8924–35.

31 ENGEL, J. & PROCKOP, D. J. (1991). The Zipper-Like Folding of Collagen Triple Helices and the Effects of Mutations That Disrupt the Zipper. *Ann Rev Biophysics and Biophysical Chem* 20, 137–152.

32 BACHINGER, H. P., FESSLER, L. I., TIMPL, R. & FESSLER, J. H. (1981). Chain assembly intermediate in the biosynthesis of type III procollagen in chick embryo blood vessels. *J Biol Chem* 256, 13193–9.

33 BACHINGER, H. P., BRUCKNER, P., TIMPL, R. & ENGEL, J. (1978). The role of *cis-trans* isomerization of peptide bonds in the coil leads to and comes from triple helix conversion of collagen. *Eur J Biochem* 90, 605–13.

34 BACHINGER, H. P., BRUCKNER, P., TIMPL, R., PROCKOP, D. J. & ENGEL, J. (1980). Folding mechanism of the triple-helix in type III collagen and type III pN-collagen: role of disulfide bridges and peptide bond isomerization. *Eur J Biochem* 106, 619–632.

35 BRUCKNER, P. & EIKENBERRY, E. F. (1984). Formation of the Triple Helix of Type-I Procollagen Incellulo – Temperature-Dependent Kinetics Support a Model Based On *Cis* Reversible *Trans* Isomerization of Peptide-Bonds. *Eur J Biochem* 140, 391–395.

36 BRUCKNER, P., EIKENBERRY, E. F. & PROCKOP, D. J. (1981). Formation of the triple helix of type I procollagen in cellulo. A kinetic model based on *cis-trans* isomerization of peptide bonds. *Eur J Biochem* 118, 607–13.

37 THORNTON, J. M. (1981). Disulfide bridges in globular proteins. *J Mol Biol* 151, 261–87.

38 OLSEN, B. R. (1982). The Carboxyl Propeptides of Procollagens: Structural and Functional Considerations. *New Trends Basement Memb Res*, 225–236.

39 Byers, P. H., Click, E. M., Harper, E. & Bornstein, P. (1975). Interchain disulfide bonds in procollagen are located in a large nontriple-helical COOH-terminal domain. *Proc Natl Acad Sci USA* 72, 3009–13.

40 Koivu, J. (1987). Disulfide bonding as a determinant of the molecular composition of types I, II and III procollagen. *FEBS Lett* 217, 216–20.

41 Lees, J. F. & Bulleid, N. J. (1994). The role of cysteine residues in the folding and association of the COOH-terminal propeptide of types I and III procollagen. *J Biol Chem* 269, 24354–60.

42 Lukens, L. N. (1976). Time of occurrence of disulfide linking between procollagen chains. *J Biol Chem* 251, 3530–8.

43 Bulleid, N. J., Dalley, J. A. & Lees, J. F. (1997). The C-propeptide domain of procollagen can be replaced with a transmembrane domain without affecting trimer formation or collagen triple helix folding during biosynthesis. *EMBO J* 16, 6694–6701.

44 Koivu, J. (1987). Identification of disulfide bonds in carboxy-terminal propeptides of human type I procollagen. *FEBS Lett* 212, 229–32.

45 Lenaers, A. & Lapiere, C. M. (1975). Type III procollagen and collagen in skin. *Biochim Biophys Acta* 400, 121–31.

46 Booth, B. A., Polak, K. L. & Uitto, J. (1980). Collagen biosynthesis by human skin fibroblasts. I. Optimization of the culture conditions for synthesis of type I and type III procollagens. *Biochim Biophys Acta* 607, 145–60.

47 Uitto, J., Booth, B. A. & Polak, K. L. (1980). Collagen biosynthesis by human skin fibroblasts. II. Isolation and further characterization of type I and type III procollagens synthesized in culture. *Biochim Biophys Acta* 624, 545–61.

48 Narayanan, A. S. & Page, R. C. (1983). Biosynthesis and regulation of type V collagen in diploid human fibroblasts. *J Biol Chem* 258, 11694–9.

49 Jimenez, S. A., Williams, C. J., Myers, J. C. & Bashey, R. I. (1986). Increased collagen biosynthesis and increased expression of type I and type III procollagen genes in tight skin (TSK) mouse fibroblasts. *J Biol Chem* 261, 657–62.

50 Kypreos, K. E. & Sonenshein, G. E. (1998). Basic fibroblast growth factor decreases type V/XI collagen expression in cultured bovine aortic smooth muscle cells. *J Cell Biochem* 68, 247–58.

51 Traub, W. & Piez, K. A. (1971). The chemistry and structure of collagen. *Adv Protein Chem* 25, 243–352.

52 Miller, E. J. & Gay, S. (1982). Collagen: an overview. *Methods Enzymol* 82 Pt A, 3–32.

53 Chung, E., Keele, E. M. & Miller, E. J. (1974). Isolation and characterization of the cyanogen bromide peptides from the alpha 1(3) chain of human collagen. *Biochemistry* 13, 3459–64.

54 Kleman, J. P., Hartmann, D. J., Ramirez, F. & Van der Rest, M. (1992). The human Rhabdomyosarcoma cell-line A204 lays down a highly insoluble matrix composed mainly of alpha-1 type XI and alpha-2 type V collagen chains. *Eur J Biochem* 210, 329–335.

55 Mayne, R., Brewton, R. G., Mayne, P. M. & Baker, J. R. (1993). Isolation and characterization of the chains of type-V and type-XI collagen present in bovine vitreous. *J Biol Chem* 268, 9381–9386.

56 Fichard, A., Kleman, J. P. & Ruggiero, F. (1995). Another Look At Collagen-V and Collagen-XI Molecules. *Matrix Biol* 14, 515–531.

57 Olsen, B. R., Winterhalter, K. H. & Gordon, M. K. (1995). FACIT collagens and their biological roles. *Trends Glycosci Glycotech* 7, 115–127.

58 Sandell, L. J., Morris, N., Robbins, J. R. & Goldring, M. B. (1991). Alternatively spliced type II procollagen mRNAs define distinct populations of cells during vertebral development: differential expression of the amino-propeptide. *J Cell Biol* 114, 1307–19.

59 Sandell, L. J., Nalin, A. M. & Reife, R. A. (1994). Alternative splice form of type II procollagen mRNA (IIA) is predominant in skeletal precursors and non-cartilaginous tissues during early mouse development. *Dev Dyn* 199, 129–40.

60 Cremer, M. A., Ye, X. J., Terato, K., Owens, S. W., Seyer, J. M. & Kang, A. H. (1994). Type-XI collagen-induced arthritis in the lewis rat – characterization of cellular and humoral immune-responses to native type-XI, type-V and type-II collagen and constituent alpha-chains. *Journal of Immunology* 153, 824–832.

61 Wu, J. J. & Eyre, D. R. (1995). Structural analysis of cross-linking domains in cartilage type-XI collagen – insights on polymeric assembly. *J Biol Chem* 270, 18865–18870.

62 Koivu, J., Myllyla, R., Helaakoski, T., Pihlajaniemi, T., Tasanen, K. & Kivirikko, K. I. (1987). A single polypeptide acts both as the beta subunit of prolyl 4-hydroxylase and as a protein disulfide-isomerase. *J Biol Chem* 262, 6447–9.

63 Pihlajaniemi, T., Helaakoski, T., Tasanen, K., Myllyla, R., Huhtala, M. L., Koivu, J. & Kivirikko, K. I. (1987). Molecular cloning of the beta-subunit of human prolyl 4-hydroxylase. This subunit and protein disulfide isomerase are products of the same gene. *EMBO J* 6, 643–9.

64 Lamberg, A., Jauhiainen, M., Metso, J., Ehnholm, C., Shoulders, C., Scott, J., Pihlajaniemi, T. & Kivirikko, K. I. (1996). The role of protein disulfide isomerase in the microsomal triacylglycerol transfer protein does not reside in its isomerase activity. *Biochem J* 315, 533–6.

65 Thompson, J., Sack, J., Jamil, H., Wetterau, J., Einspahr, H. & Banaszak, L. (1998). Structure of the microsomal triglyceride transfer protein in complex with protein disulfide isomerase. *Biophys J* 74, A138–A138.

66 Wetterau, J. R., Combs, K. A., Spinner, S. N. & Joiner, B. J. (1990). Protein disulfide isomerase is a component of the microsomal triglyceride transfer protein complex. *J Biol Chem* 265, 9801–7.

67 Brownell, A. G. & Veis, A. (1976). Intracellular location of triple helix formation of collagen. Enzyme probe studies. *J Biol Chem* 251, 7137–43.

68 de Wet, W. J., Chu, M. L. & Prockop, D. J. (1983). The mRNAs for the pro-alpha 1(I) and pro-alpha 2(I) chains of type I procollagen are translated at the same rate in normal human fibroblasts and in fibroblasts from two variants of osteogenesis imperfecta with altered steady state ratios of the two mRNAs. *J Biol Chem* 258, 14385–9.

69 Kosher, R. A., Kulyk, W. M. & Gay, S. W. (1986). Collagen gene expression during limb cartilage differentiation. *J Cell Biol* 102, 1151–6.

70 Vuust, J., Sobel, M. E. & Martin, G. R. (1985). Regulation of type I collagen synthesis. Total pro alpha 1(I) and pro alpha 2(I) mRNAs are maintained in a 2:1 ratio under varying rates of collagen synthesis. *Eur J Biochem* 151, 449–53.

71 Olsen, A. S. & Prockop, D. J. (1989). Transcription of Human Type-I Collagen Genes – Variation in the Relative Rates of Transcription of the Pro-Alpha-1 and Pro-Alpha-2 Genes. *Matrix* 9, 73–81.

72 Veis, A., Leibovich, S. J., Evans, J. & Kirk, T. Z. (1985). Supramolecular assemblies of mRNA direct the coordinated synthesis of type I procollagen chains. *Proc Natl Acad Sci USA* 82, 3693–7.

73 Kirk, T. Z., Evans, J. S. & Veis, A. (1987). Biosynthesis of type I procollagen. Characterization of the distribution of chain sizes and extent of hydroxylation of polysome-associated pro-alpha-chains. *J Biol Chem* 262, 5540–5.

74 Veis, A. & Kirk, T. Z. (1989). The Coordinate Synthesis and Cotranslational Assembly of Type-I Procollagen. *J Biol Chem* 264, 3884–3889.

75 Hu, G., Tylzanowski, P., Inoue, H. & Veis, A. (1995). Relationships

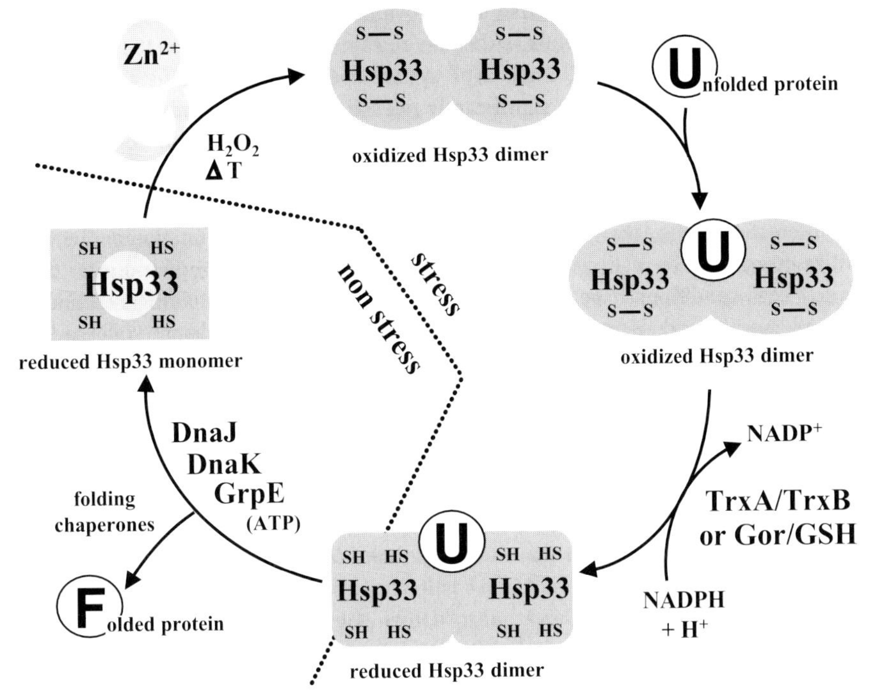

Fig. 19.3. Model for a redox-regulated chaperone network. Active, oxidized Hsp33 dimers act as a chaperone holdases and bind tightly to unfolding protein intermediates to protect them against irreversible aggregation. The thioredoxin or glutathione systems quickly reduce Hsp33–substrate protein complexes. This reduction primes Hsp33 for the interaction with the DnaK/DnaJ/GrpE chaperone system. Upon return to non-stress conditions, the DnaK/DnaJ/GrpE system becomes available again and causes the release of substrate proteins bound to Hsp33. This allows the DnaK/DnaJ/GrpE foldase system, either alone or in concert with the GroEL/GroES system, to refold the substrate proteins. Upon substrate transfer, reduced Hsp33 dimers dissociate into inactive Hsp33 monomers.

seems to be predominantly involved in disulfide bond formation and isomerization of disulfide bonds within secreted proteins. This latter activity is thought to make PDI essential in yeast [52].

To introduce disulfide bonds, PDI needs to be maintained in the oxidized state, which also appears to be the favored redox state of PDI under steady-state conditions in the yeast ER [53]. Re-oxidation of PDI is mediated by an essential protein relay involving the oxidase Ero1p [53], which uses the cofactor FAD to shuttle electrons ultimately onto molecular oxygen [7]. While PDI needs to be oxidized to promote disulfide bond formation, it is required to be in its reduced state to facilitate disulfide isomerization. Even though PDI is considered to be predominantly oxidized in the ER, its transient reduction is thought to be achieved by the presence of sufficient amounts of reduced glutathione (GSH).

PDI has also been discovered in various other cellular locations, such as the cytoplasm, endosomes, and plasma membrane of eukaryotic cells (reviewed in Ref. [54]). Here, the enzyme is less abundant than in the ER and is predominantly in its reduced form. It catalyzes the reduction of protein disulfides and is presumably regenerated by the thioredoxin system using reducing equivalents of NADPH [55].

In addition to being a powerful oxidoreductase, PDI has been shown to support the folding of denatured substrate proteins both in vitro and in vivo. As mentioned before, this polypeptide-binding activity appears to particularly involve the b′ domain. PDI has been found to chaperone the refolding of several proteins such as glyceraldehyde-3-phosphate dehydrogenase (GAPDH) [56], acidic phospholipase A_2, rhodanese [57], and proinsulin [58]. It is now widely accepted that it is this combination of oxidoreductase activity and chaperone activity that makes PDI such a capable and important enzyme [44, 50].

19.4.2 PDI and Redox Regulation

Recently, Rapoport and colleagues reported a novel feature of PDI. The authors found that PDI functions as a redox-dependent chaperone involved in the unfolding of cholera toxin in the ER [20]. Cholera toxin shows a sophisticated mechanism of action, which ultimately leads to a disastrous salt and water secretion from mammalian intestinal epithelial cells due to the uncontrolled opening of chloride channels. The toxin is originally assembled in the periplasm of the bacterium *Vibrio cholerae* and is secreted by the bacterium in an inactive form. In this stage, it consists of one A-subunit surrounded by five B-subunits. The pathogenic component A1 is a fragment of the A-chain. To set this fragment free and unleash its activity, several steps are required. First, the A-chain needs to be cleaved into its A1 and A2 fragments by a protease that is also secreted by *Vibrio cholerae*. Second, the cholera toxin needs to travel backwards through the secretory pathway of its target cell until it reaches the ER. Here, reduction of a crucial intramolecular disulfide bond, which still covalently links the A1 and A2 fragments, is required to release the A1 fragment. The A1 fragment is then unfolded and retro-translocated into the cytoplasm, where it finally takes up its enzymatic activity to ADP-ribosylate the heterotrimeric Gαs protein [59].

Rapoport and coworkers found that PDI plays the key role in the processing of the A1 fragment in the ER [20]. PDI appears to be responsible for the disassembly of the toxin and the unfolding of the A1 fragment, which finally triggers the reduction of the intramolecular disulfide bond and the retrograde translocation of A1 into the cytosol. Interestingly, it appeared that PDI was capable only of binding and subsequently unfolding the A1 fragment when reducing conditions were established and PDI's cysteines were reduced. This unfoldase activity of PDI, however, appeared to be independent of the oxidoreductase activity of PDI but seemed rather to be due to a redox-regulated change in the conformation and chaperone activity of PDI.

In conclusion, the authors proposed a model in which the chaperone activity of

PDI is regulated by the redox state of the protein. According to this model, reduced PDI binds to the A1-chain of cholera toxin and unfolds the polypeptide. Upon oxidation of PDI, conformational changes occur that decrease the substrate-binding affinity in PDI and cause the substrate protein to be released. This model appears to be particularly reasonable for the isomerase activity of PDI, where the chaperone activity of reduced PDI might be required to recognize proteins that are partly unfolded due to incorrect disulfide bonds. Upon binding, PDI would reshuffle these disulfide bonds, become oxidized, and, in turn, release the substrate proteins [20].

The authors also concluded that in analogy to ATP-dependent chaperones like DnaK, which drive protein folding by continuously cycling through high- and low-affinity states based on nucleotide binding, the chaperone activity of PDI could be driven by the continuous cycling through two different redox states [20]. In the case of DnaK, the cycle is further regulated by cofactors such as DnaJ and GrpE. Thus, one could envision respective cofactors for PDI that might affect its redox cycle [20].

In response to this model, Lumb and Bulleid set out to examine the potential redox regulation of PDI's chaperone activity using other known substrate proteins of PDI [60]. They examined the potential redox-dependent interaction of PDI with the C-propeptide of procollagen, a known transient substrate of PDI, as well as with the α-subunit of prolyl 4-hydroxylase (P4H), with which PDI is known to permanently associate to form an active $\alpha_2\beta_2$ tetramer. In both cases, the authors found that the binding of PDI to these proteins was independent of its redox state. The observation that high concentrations of GSSG caused the dissociation of PDI from its substrates was attributed to GSSG acting as a competitor for peptide binding rather than GSSG acting on the potential redox switch of PDI [60]. However, the fact that dissociation of the A1 fragment of cholera toxin from PDI was due to the oxidation-induced change in PDI's chaperone activity became evident when Tsai et al. showed that the natural PDI oxidant Ero1p worked as well as GSSG [61]. This finding supports the model that PDI acts as a redox-regulated chaperone for the A1-fragment of cholera toxin.

So far, cholera toxin is the only substrate in which binding of PDI was found to be redox-regulated. Several reports suggested that PDI's chaperone function is independent from its redox state, especially since redox-inactive variants of PDI still show chaperone activity [58, 62]. Until further investigations are conducted, however, it is impossible to evaluate the general relevance of PDI's redox regulation. Given the complexity of this enzyme, various alternatives seem possible. The future will show the outcome of this lively discussion concerning this attractive model.

19.5
Concluding Remarks and Outlook

This chapter focused on two distinct redox-regulated proteins, eukaryotic PDI and prokaryotic Hsp33. PDI is the first protein that oxidizes/reduces disulfide bonds of substrate proteins and at the same time shows substrate-binding affinities that are regulated by its own redox state. This regulation of substrate-binding activity is thought to support the isomerase and oxidase function of PDI [20].

The redox-regulated chaperone Hsp33 appears to be a chaperone specifically designed to protect cells against severe oxidative stress conditions. Hsp33 is constitutively expressed and present in its inactive monomeric state during non-stress conditions. Once oxidative stress is encountered, however, Hsp33's chaperone activity is quickly turned on via disulfide bond formation. In addition to this posttranslational regulation of its chaperone activity, Hsp33 is also upregulated on transcriptional level. This makes Hsp33 particularly abundant and active under conditions of combined heat and oxidative stress. This dual regulation of Hsp33 reflects its unique, functional location at the interface of heat shock and oxidative stress. We have now found that Hsp33 is the key component of a highly sophisticated chaperone network that serves as a defense system to protect proteins against stress-induced unfolding.

Hsp33's unique regulation and its specific functional mechanism raise the question of why such specific chaperones as Hsp33 are designed for specific stress situations. Are other chaperone systems not capable of performing their regular tasks? Are they simply overwhelmed during severe stress conditions due to the increasing number of unfolded and damaged proteins? Do some of the housekeeping cellular chaperones themselves become victims of severe stress? All efforts to answer these questions seem to be highly valuable for the further understanding of cellular chaperones, cellular stresses, and beyond.

19.6 Appendix – Experimental Protocols

19.6.1 How to Work With Redox-regulated Chaperones in Vitro

19.6.1.1 Preparation of the Reduced Protein Species

One of the most important requirements for working with redox-regulated proteins is the use of homogeneous preparations of either fully reduced or fully oxidized protein. In the case of the reduced protein species, air oxidation needs to be avoided during the purification process and thereafter. Therefore, buffers should always be supplemented with at least 2 mM (fresh) DTT and the protein preparation should be kept on ice. In the case of metal-free proteins, air oxidation can be minimized by the addition of metal chelators such as EDTA ($\sim$1 mM). Often, it is also necessary to freshly reduce the samples before each experiment.

In the case of Hsp33, 100–400 μM of purified Hsp33 is reduced in the presence of 2–8 mM DTT in 40 mM HEPES-KOH (pH 7.5) for 60 min at 37 °C. When necessary, zinc ($ZnCl_2$ in ddH_2O) is added in stoichiometric amounts [21, 35]. It is not recommendable to use excess zinc, because of its nonspecific low-affinity binding (K_D in the micromolar range) to negative charges on protein surfaces, which could affect subsequent assays. In the case of PDI, reduction is achieved by the incubation of PDI in the presence of 1 mM GSH for 30 min at 30 °C [20] or in the presence of 10 mM DTT [60]. After the reduction, excess reductants (and metals) are removed by using a PD10 (Pharmacia) gel filtration column that has been equilibrated in 40 mM HEPES-KOH (pH 7.5). Then, the protein concentration is determined by using the specific extinction coefficient (http://ca.expasy.org/tools/protparam.html) of the respective protein.

To ascertain that the protein has been completely reduced, it is strongly recommended to monitor the redox state of the protein directly. This can be done by determining free thiols using Ellman's reagent [34, 63] or by performing thiol-trapping experiments (see Section 19.6.1.3). In the case of metal-binding proteins, the extent of metal association should also be determined. Zinc coordination, for instance, can be quantified using the PAR/PMPS assay (see Section 19.6.2.1). Other metal associations can be quantified with inductively coupled plasma atomic emission spectroscopy (ICP) analysis. After the complete reduction of the protein has been confirmed, the protein should be used immediately or shock frozen in aliquots and stored at −80 °C.

19.6.1.2 **Preparation of the Oxidized Protein Species**

Suitable agents for direct protein oxidation are oxidized glutathione (GSSG), hydrogen peroxide (H_2O_2), or dipyridyl sulfide (DPS). Oxidation of thiol groups with H_2O_2 is thought to first create sulfenic acid intermediates (R-SOH), which then react with nearby cysteines to form disulfide bonds ([64], reviewed in Ref. [8]). The oxidation process with H_2O_2 can be significantly accelerated in the presence of micromolar concentrations of Fenton reagents such as Fe (II) or Cu (I) [65]. The presence of these transition metals leads to the formation of hydroxyl radicals [65], which then react with thiol groups to create highly reactive thiyl radicals. These then rapidly form disulfide bonds. Because hydroxyl radicals are very reactive, nonspecific amino acid modifications and protein fragmentation can occur. This can be avoided by using the milder oxidant GSSG. After the oxidation process is complete, the oxidant needs to be removed and the extent of oxidation needs to be quantified.

In the case of Hsp33, it is important to start the oxidation with completely reduced, zinc-reconstituted protein, because correct disulfide bond formation requires the presence of the cofactor zinc. Reduced Hsp33 is concentrated using Centricon YM-30 concentrators; then 200 μM Hsp33 is oxidized in the presence of 2 mM H_2O_2 for 240 min at 43 °C. Alternatively, 50 μM of Cu(II) can be added to accelerate the oxidation process (for details see Ref. [35]). Next, H_2O_2 and released zinc are removed by gel filtration columns (PD10, Pharmacia). In the case of PDI, oxidation is usually performed in the presence of GSSG ($\geq$ 1 mM) [20] or in the presence of 1 mM or 5 mM dipyridyl sulfide [60]. To identify the location of the respective disulfide bonds in an oxidized protein species, disulfide mapping should be performed [34, 66].

19.6.1.3 In Vitro Thiol Trapping to Monitor the Redox State of Proteins

To monitor the redox status of proteins, one can make use of the distinct chemical features of reduced thiol groups. Whereas reduced thiols react quickly and quantitatively with so-called "thiol-specific probes", oxidized disulfide-linked thiol groups are not accessible to these reagents.

A commonly used thiol-specific probe, which selectively modifies reduced thiol groups in proteins, is 4-acetamido-4′-maleimidylstilbene-2,2′-disulfonic acid (AMS, Molecular Probes). This reagent is particularly suitable for distinguishing between the reduced and oxidized forms of proteins because the modification of each free thiol group with the AMS moiety adds an additional 500 Da of molecular mass to the respective protein. Thus, AMS-modified proteins migrate slower in SDS-PAGE, depending on the number of accessible, AMS-modified cysteine residues.

In vitro thiol trapping may be used either to confirm complete reduction or oxidation of a protein sample (see above) or to obtain kinetic information about the reduction or oxidation process of a redox-regulated protein. To perform thiol-trapping experiments with Hsp33, a protein solution of 5 μM oxidized or reduced Hsp33 (monomer concentration) is prepared in a suitable buffer such as 40 mM HEPES-KOH (pH 7.5). When the kinetics of the reduction or oxidation reactions are monitored, thiol-disulfide exchange reactions need to be stopped instantaneously. To achieve this, trichloroacetic acid (TCA) to a final concentration of 10% (w/v) is added. The low pH greatly slows all ongoing redox reactions (see review in Chapter 9) and denatures the proteins. This further disrupts any special thiol/disulfide interchange activities of the proteins. The protein samples are then stored on ice for at least 30 minutes and centrifuged (30 min, 16 100 g, 4 °C). The protein pellets are washed in 200 μL ice cold 10% (w/v) TCA followed by a wash in 200 μL 5% (w/v) TCA. The supernatant is quantitatively removed and the protein pellets are resuspended in 20 μL 15 mM AMS in a buffer containing 6 M urea, 200 mM Tris-HCl pH 8.5, 10 mM EDTA, and 0.5% w/v SDS. The trapping reaction is performed at 37 °C for 60 min in the dark under continuous shaking. Subsequently, the samples are analyzed on non-reducing SDS-PAGE.

19.6.2 Thiol Coordinating Zinc Centers as Redox Switches

19.6.2.1 PAR-PMPS Assay to Quantify Zinc

The amount of free zinc in a protein sample can easily be determined spectroscopically using the PAR assay [67]. This assay is based on the complex formation of free zinc with the zinc-chelating dye 4-(2-pyridylazo) resorcinol (PAR). $Zn(PAR)_2$ complexes have an intense red color with an absorption maximum λ_{max} at 500 nm. To account for slight differences in the extinction coefficient of $Zn(PAR)_2$ complexes caused by buffer composition, pH, and ionic strength, the exact ε_{500} should be determined for each buffer system using a zinc standard. PAR has sufficient affinity ($K_a = 10^{12}$ M^{-2} in 40 mM HEPES, pH 7.0) to complex free as well

19.6.3.2 Manipulating and Analyzing Redox Conditions in Vivo

To analyze the redox state of the proteins in vivo, thiol-trapping experiments with AMS can be performed, given that an antibody against the protein of interest is available. To perform thiol trapping in vivo, cells are grown in minimal medium until the desired OD_{600} is reached. Then, the cells are exposed to the respective stress treatment (e.g., 4 mM H_2O_2 to induce oxidative stress, 1 mM diamide to induce disulfide stress [72]). Cells are then precipitated with ice-cold TCA (final concentration of 10% (w/v)) and incubated on ice for at least 30 min. The thiol trapping is then performed as described in Section 19.6.1.3. Western blot analysis is performed to visualize the protein of interest.

To analyze the thiol-disulfide status of redox-regulated proteins in vivo, it is often crucial to establish either reducing or oxidizing conditions in the cell. If the protein of interest is located in the cytosol, it is easy to analyze the protein under the normal reducing conditions, which are maintained by the thioredoxin and glutaredoxin systems. These potent reducing systems, along with other powerful cellular antioxidants (catalase, superoxide dismutase SOD, etc.) that battle oxidative stress, make it difficult, however, to maintain continuous oxidizing conditions in the cell. This is especially obvious for H_2O_2-induced oxidative stress, which leads only to a transient change in the redox environment [18, 39]. To circumvent this problem, mutant *E. coli* strains that lack components of the thioredoxin and/or glutaredoxin system, and are therefore intrinsically oxidatively stressed, can be used [4, 73]. The protein of interest should accumulate in its oxidized state as soon as its physiological redox system is absent from the cells. Detection of the protein in the reduced and oxidized states, depending on the strains and environmental conditions used, not only shows that the protein of interest is redox regulated in vivo but also reveals the physiological redox systems.

Acknowledgements

We thank Drs. James Bardwell, Paul Graf, and Jeannette Winter for critically reading this manuscript. The National Institutes of Health grant GM065318 and a Burroughs Wellcome Fund Career Award to U.J and a Ph.D. scholarship of the Boehringer Ingelheim Fonds to J.H.H supported this work.

References

1 Bukau, B., and Horwich, A. L. (1998). The Hsp70 and Hsp60 chaperone machines. *Cell* **92**, 351–366.
2 Wedemeyer, W. J., Welker, E., Narayan, M., and Scheraga, H. A. (2000). Disulfide bonds and protein folding. *Biochemistry* **39**, 4207–4216.
3 Frand, A. R., Cuozzo, J. W., and Kaiser, C. A. (2000). Pathways for protein disulphide bond formation. *Trends Cell Biol* **10**, 203–210.
4 Prinz, W. A., Aslund, F., Holmgren, A., and Beckwith, J. (1997). The role of the thioredoxin and glutaredoxin pathways in reducing protein disulfide bonds in the

Escherichia coli cytoplasm. *J Biol Chem* **272**, 15661–15667.

5 Collet, J. F., and Bardwell, J. C. (2002). Oxidative protein folding in bacteria. *Mol Microbiol* **44**, 1–8.

6 Noiva, R. (1999). Protein disulfide isomerase: the multifunctional redox chaperone of the endoplasmic reticulum. *Seminars in Cell & Developmental Biology* **10**, 481–493.

7 Tu, B. P., and Weissman, J. S. (2002). The FAD- and O(2)-dependent reaction cycle of Ero1-mediated oxidative protein folding in the endoplasmic reticulum. *Mol Cell* **10**, 983–994.

8 Linke, K., and Jakob, U. (2003). Not every disulfide lasts forever: disulfide bond formation as a redox switch. *Antioxid Redox Signal* **5**, 425–434.

9 Fomenko, D. E., and Gladyshev, V. N. (2003). Identity and functions of CxxC-derived motifs. *Biochemistry* **42**, 11214–11225.

10 Dai, S., Schwendtmayer, C., Schurmann, P., Ramaswamy, S., and Eklund, H. (2000). Redox signaling in chloroplasts: cleavage of disulfides by an iron-sulfur cluster. *Science* **287**, 655–658.

11 Dietz, K. J. (2003). Plant peroxiredoxins. *Annu Rev Plant Biol* **54**, 93–107.

12 Hoyos, B., Imam, A., Korichneva, I., Levi, E., Chua, R., and Hammerling, U. (2002). Activation of c-Raf kinase by ultraviolet light. Regulation by retinoids. *J Biol Chem* **277**, 23949–23957.

13 Knapp, L. T., and Klann, E. (2000). Superoxide-induced stimulation of protein kinase C via thiol modification and modulation of zinc content. *J Biol Chem* **275**, 24136–24145.

14 Leslie, N. R., Bennett, D., Lindsay, Y. E., Stewart, H., Gray, A., and Downes, C. P. (2003). Redox regulation of PI 3-kinase signalling via inactivation of PTEN. *EMBO J* **22**, 5501–5510.

15 Delaunay, A., Isnard, A. D., and Toledano, M. B. (2000). H2O2 sensing through oxidation of the Yap1 transcription factor. *EMBO J* **19**, 5157–5166.

16 Wood, M. J., Andrade, E. C., and Storz, G. (2003). The redox domain of the Yap1p transcription factor contains two disulfide bonds. *Biochemistry* **42**, 11982–11991.

17 Zheng, M., Aslund, F., and Storz, G. (1998). Activation of the OxyR transcription factor by reversible disulfide bond formation. *Science* **279**, 1718–1721.

18 Aslund, F., Zheng, M., Beckwith, J., and Storz, G. (1999). Regulation of the OxyR transcription factor by hydrogen peroxide and the cellular thiol-disulfide status. *Proc Natl Acad Sci USA* **96**, 6161–6165.

19 Jakob, U., Muse, W., Eser, M., and Bardwell, J. C. (1999). Chaperone activity with a redox switch. *Cell* **96**, 341–352.

20 Tsai, B., Rodighiero, C., Lencer, W. I., and Rapoport, T. A. (2001). Protein disulfide isomerase acts as a redox-dependent chaperone to unfold cholera toxin. *Cell* **104**, 937–948.

21 Jakob, U., Eser, M., and Bardwell, J. C. (2000). Redox switch of hsp33 has a novel zinc-binding motif. *J Biol Chem* **275**, 38302–38310.

22 Kang, J. G., Paget, M. S., Seok, Y. J., Hahn, M. Y., Bae, J. B., Hahn, J. S., Kleanthous, C., Buttner, M. J., and Roe, J. H. (1999). RsrA, an anti-sigma factor regulated by redox change. *EMBO J* **18**, 4292–4298.

23 Paget, M. S., Bae, J. B., Hahn, M. Y., Li, W., Kleanthous, C., Roe, J. H., and Buttner, M. J. (2001). Mutational analysis of RsrA, a zinc-binding anti-sigma factor with a thiol-disulphide redox switch. *Mol Microbiol* **39**, 1036–1047.

24 Park, J. S., Wang, M., Park, S. J., and Lee, S. H. (1999). Zinc finger of replication protein A, a non-DNA binding element, regulates its DNA binding activity through redox. *J Biol Chem* **274**, 29075–29080.

25 Wu, X., Bishopric, N. H., Discher, D. J., Murphy, B. J., and Webster, K. A. (1996). Physical and functional sensitivity of zinc finger transcription

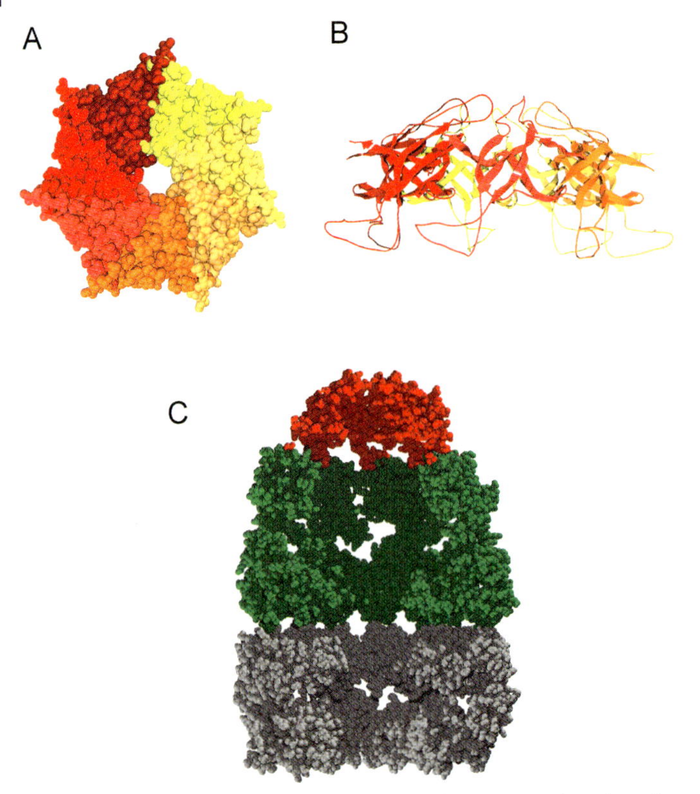

Fig. 20.2. Structures of the co-chaperone GroES (A, B) [17] and of a GroELS bullet complex (C) [21]. (A) GroES, here shown in a top view, consists of seven identical 10-kDa subunits and shares with GroEL the same sevenfold rotational symmetry. (B) Side view of GroES in a ribbon representation. A prominent feature of the GroES structure is the mobile loops, which protrude from the bottom of the heptamer and mediate the interaction with GroEL. (C) Side view of a GroELS complex (cross-section). Binding of GroES (red) to the top ring (green) of the GroEL tetradecamer converts the upper cavity into a large shielded compartment for protein folding. The cavity of the bottom ring (gray) is much smaller in volume.

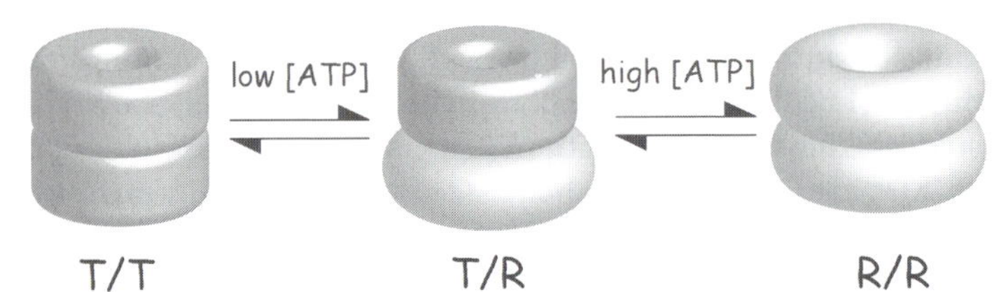

Fig. 20.3. Model for the allosteric transitions within GroEL. Owing to the positive intra-ring cooperativity, all seven subunits of one ring adopt the same conformation. In the absence of ligands (left), both rings are preferentially in the T (tense) state. This conformation displays a high affinity for unfolded polypeptides. Binding of ATP to the bottom ring causes this ring to adopt the R state (middle), in which the affinity for polypeptides is decreased. Because of the negative inter-ring cooperativity, binding of ATP to the second ring is disfavored, and the transition to the R/R state (right) occurs only at high concentrations of nucleotide.

bolic ATP dependence suggested a two-step binding process: ATP initially forms a weak complex with GroEL ($K_D = 4$ mM) before inducing a rapid conformational rearrangement ($t_{1/2} = 5$ ms). In contrast, a GroEL mutant in which a single Trp was inserted in the equatorial domain (F44W) reported three kinetic phases upon rapid mixing with ATP [45], while a similar mutant, Y485W GroEL, reported four distinct kinetic transients [46]. A very rapid (~1 ms) kinetic phase was observed with Y485W GroEL before the main fluorescence quench phase also reported by Y44W GroEL. This phase revealed a bi-sigmoid dependency on ATP concentration reminiscent of that seen in steady-state experiments and presumably represents conformational transitions occurring upon occupying first one heptameric ring with ATP and subsequently the second ring. However, the precise structural changes represented by these kinetic transitions are still under investigation.

In the presence of GroES, ATP binds to GroEL with a greater affinity ($K_{1/2} = 6$ μM) and displays an increased level of cooperativity. Under steady-state conditions, GroES inhibits the rate of ATP hydrolysis to ~35% of that seen with GroEL alone [10, 39, 40, 44]. During the hydrolytic cycle, the association of GroES with GroEL is highly dynamic [47]. GroES associates rapidly with the GroEL-ATP complex (5×10^7 M^{-1} s^{-1}) [44, 48] to form a highly stable complex. Hydrolysis of the *cis* ATP (i.e., that ATP bound to the same ring as GroES) weakens the interaction between GroEL and GroES before ATP binding to the *trans* GroEL ring forces the dissociation of GroES (and any encapsulated polypeptide) from the *cis* GroEL ring [11, 48–50].

20.7 The Reaction Cycle of the GroE Chaperone

The GroEL reaction cycle can be divided into the three distinct stages of polypeptide binding, encapsulation, and ejection [51]. In the first instance, a protein substrate that is unfolded, misfolded, or partially folded is able to interact with the hydrophobic surface on the apical domains of the *trans* ring, which adopts a conformation with a high affinity towards polypeptide substrate similar to the T-state rings found in unliganded GroEL (Figure 20.4). The binding of ATP to this ring induces the rapid dissociation of ligands from the opposite (previously *cis*) ring [48] and permits the binding of GroES to the same ring as the polypeptide. The binding of ATP and GroES results in a large increase in the volume of the central cavity and buries the hydrophobic polypeptide-binding surface within the intersubunit interface (Figure 20.4), thereby displacing the polypeptide substrate into the GroEL cavity where folding is initiated [7, 11]. There is evidence that the polypeptide may undergo some degree of forced mechanical unfolding upon the radial expansion of the protein-binding site [52]. Since the entrance to the cavity is sealed by GroES, the protein substrate may fold without interference by other aggregation-prone polypeptides. However, it is clear that at this stage the polypeptide is still able to interact with the walls of the cavity [53]. ATP hydrolysis in the *cis* ring ($t_{1/2} \sim 10$ s) dictates the lifetime of this enclosed folding compartment by weakening the interaction between GroEL and GroES [11] to prime the chaperone

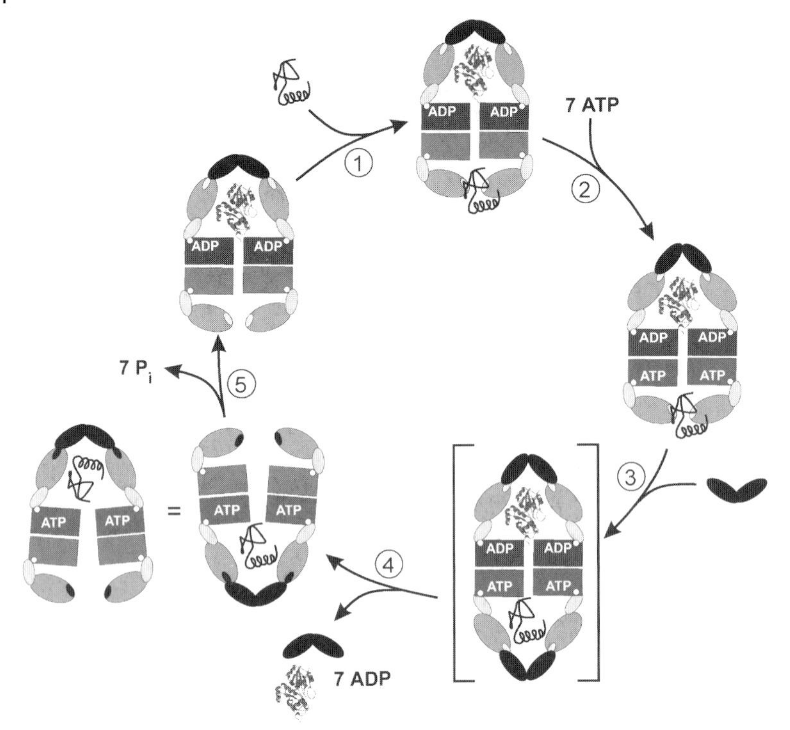

Fig. 20.4. Reaction cycle of the GroE chaperonin from *E. coli*. Although GroEL is composed of two rings, the functional cycle is best described on the level of individual rings, which represent the operational units of GroE. While both rings are active at the same time, they are in different phases of the cycle. Processing of an individual substrate polypeptide requires two revolutions of the GroE cycle, during which the polypeptide remains associated with the same GroEL ring. For graphical reasons, the orientation of the GroE complex is reversed after step 4.

The cycle of GroE-assisted folding can be dissected into three steps: capture, encapsulation/folding, and release. During capture (1), a hydrophobic polypeptide is prevented from aggregation by binding to GroEL. The acceptor ring (bottom ring) is nucleotide-free and therefore has a high affinity for the polypeptide. Binding of ATP (2) and GroES (3) to this ring induces a set of structural changes in GroEL. Most importantly, the affinity for the bound polypeptide is decreased, and it is released into the closed cavity where folding begins. Subsequent hydrolysis of ATP (5) induces a second conformational change in GroEL (top ring), which allows the bottom ring to bind polypeptide and initiate a new cycle. Upon binding of ATP and GroES in the next round, GroES is displaced from the top ring, and the substrate polypeptide is released (4). The formation of the symmetric complex shown in brackets is controversial.

complex for the "ejection" signal from the *trans* ring (Figure 20.4). ATP binding to the *trans* ring is sufficient to eject GroES and the polypeptide from the *cis* ring [11, 47, 49, 50], although this process is accelerated when polypeptide binds to the *trans* ring concomitantly. An intermediate complex with GroES bound to both ends of the GroEL tetradecamer has been observed during assisted protein-folding reactions [54–56]. The role of these so-called "football" intermediates is not clear, as some results suggest that their formation is not required for efficient refolding [57]. One speculative possibility is that they may provide a means for the efficient trapping of a "new" polypeptide substrate to the *trans* ring just prior to ejection of the "old" polypeptide from the *cis* ring (Figure 20.4) [58].

The time in which the ejected polypeptide can fold in this protected environment is limited by the lifetime of the *cis* cavity. In the case of a monomeric protein, such as bovine mitochondrial rhodanese, it may be long enough to reach the fully functional native conformation. However, many substrate polypeptides are active as oligomers and therefore individual subunits must be released from GroEL sufficiently folded such that assembly into the active multimer can take place in bulk solution. The proportion of substrate polypeptide that has folded in the cavity depends upon the intrinsic rate of folding of that particular polypeptide. In most cases, the fraction of molecules that has not yet folded sufficiently to proceed rapidly and efficiently to its native conformation within the lifetime of the cavity will, upon ejection, still be competent to rebind to GroEL and proceed through another reaction cycle, where it will have another chance to refold. In this iterative manner, high yields of fully folded protein substrates can be obtained.

This kind of mechanism exploits several characteristics of the GroE complex:

1. The binding and hydrolysis of ATP serve to modulate the affinity of GroEL for unfolded polypeptide substrate. Systematic studies of the effect of the various GroEL-nucleotide complexes on the refolding rates and yields of a number of protein substrates have been performed [29, 59–62]. Unliganded GroEL has a low affinity for ATP and GroES but a high affinity for unfolded polypeptides, while the GroEL-ATP complex has precisely the reverse properties.
2. The binding of GroES and ATP to a GroEL-polypeptide binary complex ensures efficient displacement of the polypeptide substrate into the central cavity, where it starts to refold. The binding of GroES and ATP switches the chemical nature of the cavity from one lined at the entrance with hydrophobic side chains to one in which this hydrophobic surface is buried. The volume of the cavity is also significantly increased.
3. The positive intra-ring cooperativity with respect to ATP binding may ensure a concerted release of all parts of the polypeptide chain as it becomes displaced into the central cavity. This may be crucial for productive folding in cases where structure formation in one part of the polypeptide cannot proceed before other regions are released from the GroEL apical domains. An optimal level of intra-ring cooperativity appears to have evolved such that the T-to-R transition occurs at a rate that ensures a rapid release of the whole polypeptide chain into the cavity [63].
4. The inherent asymmetry in the GroEL macromolecule enables each heptameric ring to take part in coupled half-reactions. That is, when the central cavity of one ring contains a folding polypeptide encapsulated under GroES (i.e., *cis* ring), the opposite *trans* ring provides the signal for ejection of GroES and polypeptide. The roles of the two rings are then exchanged such that the *trans* ring

now binds GroES and becomes the *cis* ring. This communication between the two rings is essential for efficient protein folding, as indicated by the fact that the single-ring version of GroEL (SR1) cannot rescue a GroEL-depleted strain [64].

20.8 The Effect of GroE on Protein-folding Pathways

GroE is able to exert differing effects on the folding kinetics and efficiency of different protein substrates. In most cases GroE improves the final yield of fully folded protein substrates but does not affect the rate of refolding. There are a number of models have been proposed to account for this phenomenon.

In the first model GroE prevents aggregation by reducing the concentration of aggregation-prone intermediates in bulk solution [65]. This could be done passively by GroEL alone, but at the cost of a significant reduction in the rate of refolding [29, 39]. Dynamically cycling the polypeptide on and off the chaperone using ATP binding and hydrolysis to switch GroEL between conformations with alternately high and low affinity for protein substrate would allow faster rates of refolding to be achieved. An extension of this model is the so-called Anfinsen cage model in which encapsulating the folding protein within the central cavity of GroEL underneath GroES allows the polypeptide an opportunity to fold without the possibility of aggregation. In this instance the GroEL central cavity acts as an "infinite dilution" chamber. Multiple rounds of binding, encapsulation, and release ensure that the polypeptide is bound or encapsulated most of the time, thus preventing aggregation.

An alternative model suggests that the rate-limiting step during the refolding of many larger proteins involves the breaking of incorrect intramolecular interactions that arise during the rapid collapse of the unfolded protein chain. If this step is slow, then the protein chain is likely to aggregate before the isomerization to a productive conformation can occur. The chaperone can therefore improve the efficiency of protein folding by assisting in the unfolding of misfolded conformations [39, 47]. The chaperonin can assist this process in two possible ways. The first way is by annealing the polypeptide to the apical domain binding sites and then mechanically unfolding the polypeptide further upon binding ATP and GroES [52]. The application of this forced unfolding requires interaction with a number of apical domain polypeptide-binding sites on GroEL [66]. Multiple rounds of this forced unfolding as the protein binds and dissociates would ensure a high yield of folded protein. Alternatively, as long as the collapsed protein substrate can explore its available conformational space in rapid equilibrium, effective unfolding can be achieved by a thermodynamic coupling mechanism, since GroEL binds the most unfolded forms tightly [67, 68].

There are some notable instances in which an apparent enhancement in the rate of refolding is observed. Porcine mitochondrial malate dehydrogenase (mMDH) shows a 3.5-fold increase in its apparent rate of refolding with GroEL compared to its spontaneous rate [69, 70], and the folding of bacterial Rubisco is increased fourfold over its spontaneous rate [71–73]. Two hypotheses to explain this have been suggested.

1. Ranson et al. [70] found that the misfolding of mMDH is the result of early steps in aggregation, presumably as small, low-order aggregates form, and that although the equilibrium normally heavily favors the formation of these smaller aggregates, this process is reversible. Remarkably, GroE was able to actively reverse these early aggregation steps even at sub-stoichiometric amounts, suggesting that it acts as a catalyst. Using cycles of binding and release, the chaperonin constantly recycles material from the unproductive aggregation pathway and resupplies the productive folding pathway. As a result the rate of refolding *apparently* increases although the *intrinsic* rate has not changed. A similar model has also been suggested to explain the apparent rate increase observed during the assisted refolding of bacterial Rubisco [72].
2. Alternatively, it has been proposed that confinement of the protein substrate within the narrow GroEL central cavity has the effect of smoothing the energy landscape in order to increase the flux of protein to its native conformation [73]. More recently, molecular simulations of the refolding process have been performed in systems with different accessible volumes, and it was noted that confinement was able to significantly stabilize the protein, leading to an enhancement in the rate of refolding [74].

Recently, the importance of the chemical properties of the central cavity has been strikingly demonstrated in an experiment in which directed evolution was used in an attempt to optimize the central cavity to maximize the efficiency of folding of the green fluorescent protein (GFP). This was achieved by altering the rate of ATP hydrolysis and the inherent allostery of the chaperonin complex and also by shifting the polarity of the central cavity in order to suit the folding pathway of GFP. However this enhanced specialization within GroEL was achieved at the expense of its ability to refold its natural protein substrates, demonstrating that GroE has had to evolve a balance of ATPase rate and central cavity properties in order to perform a very general role as a molecular chaperone [75].

Another important question concerns proteins that are too large to fit into the central cavity underneath GroES. Its volume was calculated to be in the range of 175 000 Å^3, which likely is large enough to accommodate folding intermediates of up to 60 kDa. Can GroEL also assist the folding of larger proteins? Yeast mitochondrial aconitase (82 kDa) was observed to aggregate in chaperonin-deficient mitochondria, indicating an in vivo role for the chaperonin [76]. In vitro mechanistic studies revealed that aconitase does not become encapsulated underneath GroES but interacts only with an open ring. GroES binding to the ring opposite to the protein substrate causes the release of the nonnative substrate into bulk solution, where it can refold. In this way aconitase passes through many cycles of binding and release without ever being encapsulated until it is in a conformation that can form the holo-enzyme [77, 78]. This demonstrates a mechanism that is quite distinct from that found when assisting the folding of proteins less than 60 kDa but that is also efficient at refolding to the native (or near-native) conformation.

20.9 Future Perspectives

Despite the enormous quantity of data now available describing the structure and mechanism of action of the chaperonins, there remain a number of secrets that GroE has yet to give up. For example, GroE is one of the most complex allosteric systems ever studied, and a great deal of work involving a combination of mutagenetic, kinetic, and thermodynamic studies needs to be done before the intricate macromolecular communication can be mapped onto the structure and the role of allostery in the assisted protein-folding reaction is really understood. Additionally, despite the large amount of structural data that has been collected, it is still not certain precisely what conformational ensemble of a polypeptide is recognized by GroEL and whether unfolding upon ATP and GroES binding is a general feature of the chaperonin reaction [33, 52]. Also unknown is the degree to which the nature of the central cavity can play a part in ensuring efficient folding. Does the cavity play an important role in helping the protein reach a folded conformation, or does it act merely as an "infinite dilution cage"? One new approach that may be able to unlock some of these puzzles is the extension of NMR techniques to the study of large macromolecular complexes such as the GroEL-GroES chaperonin [79]. This would permit structural and dynamic information about the chaperonin and any bound substrate. It is clear that an imaginative and multidisciplinary approach will likely be needed to understand the precise molecular details of this extraordinary protein complex.

20.10 Experimental Protocols

Protocol 1: Purification of GroEL and SR1 A large number of plasmids and constructs have been designed for the overexpression of GroEL in *E. coli* [5, 80–82]. They differ mainly in the resistance marker, promoter type, and expression level and in whether or not GroES is co-expressed. It is beyond the scope of this chapter to provide detailed protocols for the fermentation/protein production steps, as they will depend on the individual plasmid. Routinely, *E. coli* strains such as JM109, DH5α, or BL21 derivatives are used for overexpression. After harvesting the cells by centrifugation, the cell pellet may be stored at −20 °C.

The single-ring mutant SR1 was first described by the group of Art Horwich [23]. SR1 cannot support cell viability [83], probably because it is defective in GroES release. Thus, it can be expressed only in *E. coli* strains that also express basal-level wild-type GroEL. Recently, a number of SR1 mutants have been generated that are functional as chaperones because they possess a decreased affinity towards GroES [64]. SR1 can be separated from endogenous GroEL by size-exclusion chromatography, but a small fraction of SR1 heptamers may contain wild-type subunits.

Crude Extract Preparation
1. Thaw cells on ice.
2. Resuspend pellet in 25 mM Tris/HCl pH 7.5, 5 mM DTT, 2 mM EDTA (normally, no protease inhibitors are required).
3. Crack cells with a cell disruptor or a French press.
4. Add 10 mM $MgCl_2$ and 1 U mL^{-1} DNase I; stir at 4 °C for 1 h.
5. Centrifuge (45 000 g, 4 °C, 45 min) and discard pellet.
6. All subsequent steps should be carried out at 4 °C.

Anion-exchange Chromatography
1. Column: Q-Sepharose FF (Amersham Biotech), 150–200 mL of resin, linear flow: 12 cm h^{-1}.
2. Buffer A: 50 mM Tris/HCl pH 7.5, 2 mM DTT, 2 mM EDTA.
3. Buffer B: buffer A + 2 M NaCl.
4. Equilibrate column with at least three column volumes of buffer A.
5. Load crude extract on column.
6. Wash column with two column volumes of 7.5% buffer B.
7. Elute GroEL with a linear gradient of 7.5–30% buffer B in 10 column volumes.
8. Analyze elution profile with SDS-PAGE (12.5% gel), and pool GroEL-containing fractions.
9. Concentrate eluate with ultrafiltration (YM100) to ~30 mL.

Note: If it is additionally intended to purify GroES from a GroELS-overproducing strain, the salt concentration during washing should be lowered to 0% buffer B and the gradient run from 0–30% buffer B. GroES elutes earlier than GroEL, between 100 and 250 mM NaCl. To detect both GroEL and GroES in SDS-PAGE, gradient gels (5–20%) are recommended, but 12.5% gels will work as well.

Size-exclusion Chromatography
1. Column: Sephacryl S-300 HR (Amersham Biosciences), 300 mL of resin, linear flow: 20 cm h^{-1}.
2. Equilibrate with two column volumes of buffer A (see above).
3. Inject ~10 mL of concentrated protein per run, and collect fractions of 3 mL.
4. Analyze elution profile with SDS-PAGE (12.5% gel), and pool GroEL-containing fractions.
5. Concentrate eluate with ultrafiltration (YM100) to ~30 mL.

Note: If it is intended to obtain GroEL without Trp-containing contaminations, it is highly recommended to pool fractions based not only on SDS-PAGE but also on intrinsic fluorescence. The protocol provided below (Protocol 5) is rather time-consuming for screening a large number of samples. A simpler test uses an excita-

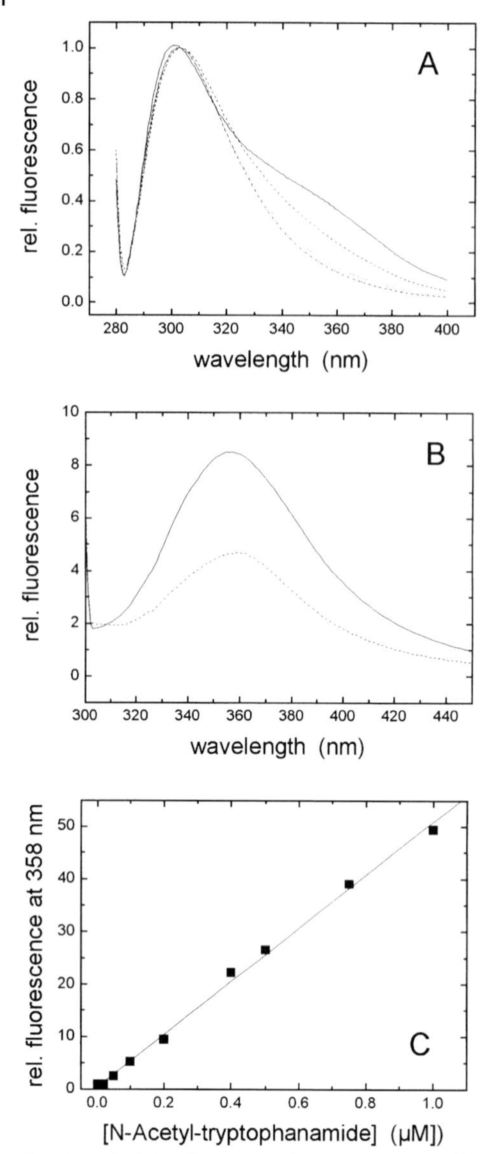

Fig. 20.5. Judging the purity of a GroEL preparation using fluorescence spectroscopy. (A) "Polishing" of GroEL by AffiGel Blue chromatography. Spectra of samples were recorded at 25 °C in 50 mM Tris/HCl pH 7.5, 2 mM DTT, and 2 mM EDTA using wavelengths of $\lambda_{ex} = 276$ nm and $\lambda_{em} = 280$–400 nm. The protein concentration in the samples was ~1 mg mL^{-1}. Spectra were normalized by setting the fluorescence intensity at 304 nm to 1. (–) GroEL pool before AffiGel Blue chromatography, (--) after one passage, (···) after two passages, (-·) and after three passages of AffiGel batch chromatography (see protocol 1). The pronounced shoulder at 350 nm is caused by Trp-containing impurities. (B) The tryptophan contents of a dirty (–) and a pure (--) GroEL sample of equal concentrations (20 μM monomer) were determined by protocol 5. The fluorescence intensity at 358 nm can be used to calculate the tryptophan concentration in the samples using the calibration curve shown in (C).

tion wavelength of 276 nm. Discard any fractions that have a pronounced Trp shoulder in their fluorescence spectrum (Figure 20.5A).

AffiGel Blue Batch Chromatography
This step will remove contaminations – mainly hydrophobic polypeptides – bound to GroEL. Besides the AffiGel Blue resin used in this protocol [84], a number of other dye-based resins have been successfully employed [85, 86]. Also, methods using organic solvents such as methanol or acetone have been developed for "polishing" GroEL [87, 88]. The AffiGel Blue procedure provides a good balance between yield and purity, but some of the alternative procedures may be considered when extremely pure GroEL (as judged by Trp fluorescence) is required.

1. Resin: AffiGel Blue (BioRad), 15 mL of dry resin.
2. Incubate resin in 30 mL 6-M GdmCl in buffer A and tumble gently overnight at 4 °C.
3. Sediment resin by centrifugation (3000 g) and resuspend it in 30 mL of buffer A; repeat five times to remove residual GdmCl.
4. Add GroEL solution to sedimented resin and tumble gently overnight at 4 °C.
5. Sediment resin and transfer the supernatant containing the purified GroEL to a new vial.
6. Record a fluorescence spectrum to determine the amount of Trp contamination; repeat the purification procedure if the purity is not satisfying.
7. Optional: Concentrate the GroEL solution up to 30 mg mL^{-1}.
8. Aliquot and freeze in liquid N_2; store at −80 °C.

Protocol 2: Purification of GroES GroES may be overexpressed alone or along with GroEL [89]. In the latter case, the first purification step should be the anion-exchange chromatography described above, which separates GroEL from GroES. After pooling GroES-containing fractions, the following protocol should be used. Depending on the level of overexpression and on the desired degree of purity, not all of the steps may be necessary.

pH Shift
GroES is quite soluble at low pH and retains its biological activity. Many other proteins, however, will precipitate during this step and can be removed by centrifugation.

1. Lower the pH of the GroES solution by adding one volume of 0.1-M sodium acetate/acetic acid pH 4.6.
2. Incubate for 1 h; spin down precipitated proteins (45 000 g, 45 min, 4 °C), and discard pellet.

Note: The buffering capacity of the GroES solution should be low. Otherwise, the resulting pH will be too high. Perfect for this purpose is 25 mM Tris/HCl buffer at pH 7.5.

Cation-exchange Chromatography

1. Column: SP-Sepharose FF (Amersham Biotech), ~50 mL of resin, linear flow: 12 cm h^{-1}.
2. Buffer A: 50 mM sodium acetate/acetic acid pH 4.6.
3. Buffer B = buffer A + 1 M NaCl.
4. Equilibrate column with at least three column volumes of buffer A.
5. Dilute the supernatant of the pH step 1:3 with H_2O.
6. Load the diluted protein solution on column.
7. Wash column with two column volumes of 0% buffer B.
8. Elute GroES with a linear gradient of 0–30% buffer B in 10 column volumes, fraction size 8 mL.
9. Analyze elution profile with SDS-PAGE (15% gel), and pool GroES-containing fractions.
10. Adjust the pH to 8 with a solution of 1 M Tris.
11. Concentrate eluate with ultrafiltration (YM10) to ~30 mL.

Temperature Shift

This step is quite efficient in removing contaminating proteins, although some GroES may be lost because of co-precipitation.

1. Gently stir the GroES solution in a water bath at 80 °C for 15 min.
2. Spin down precipitated proteins (45 000 g, 45 min, 4 °C) and discard pellet.

Size-exclusion Chromatography

1. Column: Superdex 200 (Amersham Biosciences), 300 mL of resin, linear flow: 20 cm h^{-1}.
2. Buffer C: 50 mM Tris/HCl pH 7.5, 2 mM DTT, 2 mM EDTA.
3. Equilibrate with two column volumes of buffer C.
4. Inject ~10 mL of concentrated protein per run, and collect fractions of 3 mL.
5. Analyze elution profile with SDS-PAGE (15% gel), and pool GroES-containing fractions.
6. Concentrate eluate with ultrafiltration (YM10) to ~30 mL.

Protocol 3: Preparation and Purification of Mixed-ring Complexes Mixed-ring complexes of GroEL have proven to be useful for studying the specific function of the two rings comprising the GroEL particle [11, 50]. In these complexes, one ring consists of wild-type subunits, while the second ring is made up of mutant GroEL subunits that are, e.g., defective in ATP hydrolysis or polypeptide binding.

Incubating a mixture of wild-type $GroEL_{14}$ and mutant $GroEL_{14}$ (G337S/I349E) at 42 °C for 45 min causes the double rings to dissociate and form mixed-ring complexes [83]. Apparently, the interactions between the subunits of the same ring are much stronger than interactions between rings, preventing the formation of complexes that have different types of subunits within the same ring. One important prerequisite is that wild-type and mutant GroEL have to differ sufficiently in their biophysical properties to allow the separation of mixed-ring GroEL from the starting material, i.e., the homotypic tetradecamers.

Protocol 4: Determining the Concentration of GroES and GroEL Both GroEL and GroES have a very low content of aromatic amino acids and do not contain any tryptophans. Thus, they show a relatively weak UV absorbance, and it is extremely important to ensure that their preparations are devoid of any contaminating proteins or peptides. Otherwise, these impurities will contribute over-proportionally to the UV absorbance, and thus the chaperone concentration will be overestimated.

The following protocol is based on the method introduced by Gill and von Hippel [90]. Denaturation of the sample is important because it (1) eliminates any perturbation of the UV absorbance by the protein structure and (2) reduces light scattering by the big GroEL particle.

1. Prepare a solution of 6 M GdmCl, 50 mM Tris/HCl pH 8, and 2 mM DTT (buffer P).
2. Pipet 950 μL of this solution in a 1-cm quartz cuvette.
3. Add 50 μL of protein solution and incubate for >5 min at room temperature.
4. Record a UV spectrum from 240 to 350 nm.
5. As a reference, mix 950 μL buffer P with 50 μL storage buffer and record a spectrum.
6. Subtract the reference from the sample.
7. The absorbance maximum of the resulting spectrum should be around 276 nm.
8. Calculate the protein concentration using the Lambert-Beer law with the following values: GroEL $A_{276\ nm}(0.1\%, 1\ cm) = 0.18$, GroES $A_{276\ nm}(0.1\%, 1\ cm) = 0.14$.

Protocol 5: Judging the Purity of GroEL Preparations by Determining the Tryptophan Content Although a GroEL preparation may look pure on Coomassie or silver-stained SDS gels, it is mandatory to further evaluate the purity using fluorescence spectroscopy [87]. GroEL has a high-affinity binding site for unfolded proteins, which may become contaminated with hydrophobic polypeptides during the purification process. These impurities must be removed since they may interfere with later experiments. The degree of contamination can be judged by measuring the Trp fluorescence of the GroEL sample (Figure 20.5B, C). GroEL itself does not contain Trp, whereas contaminations likely do due to the hydrophobic nature of this amino acid.

1. Prepare a solution of 6 M GdmCl (ultra-pure grade), 50 mM Tris/HCl pH 8, and 2 mM DTT (buffer P); pass it through a 0.2-μm filter and degas it.
2. Prepare five calibration solutions of 0.1, 0.25, 0.5, 0.75, and 1 μM *N*-acetyl-tryptophanamide in buffer P.
3. Prepare a solution of GroEL in buffer P ($\sim$1 mg mL^{-1}).
4. Record fluorescence spectra of buffer P, the protein sample, and the five calibration solutions; set the excitation to 295 nm and the emission to 310–400 nm.
5. Correct both the GroEL spectrum and the calibration spectra for buffer fluorescence; calculate the Trp content of the GroEL preparation using the calibration solutions.

Note: A good preparation of GroEL should have less than one Trp per $GroEL_{14}$. This protocol can also be used to evaluate the quality of your GroES preparation, although this is less crucial because GroES does not contain a binding site for unfolded proteins.

Protocol 6: ATP Hydrolysis by GroEL and Its Regulation by GroES There are several assay systems available to determine enzymatic ATP hydrolysis, all of which have certain merits and disadvantages. In general, continuous assays are to be preferred over discontinuous assays, because pipetting errors are less likely to occur. If a coupled assay is used, additional controls may be required to confirm that an observed effect, e.g., inhibition of ATP turnover by substance X, is really due to an effect of X on your ATPase and not on one of the helper enzymes. Also, an ATP-regenerating system may be of advantage in some cases. It can be added to most of the assay systems except the one based on the detection of α-[^{32}P]-ADP. The following protocol briefly describes a coupled regenerative assay.

Measuring ATP Hydrolysis by GroEL Using a Regenerative Assay
In this assay the ADP produced by the GroEL ATPase is converted back into ATP, as phosphoenol pyruvate (PEP) is converted to pyruvate by pyruvate kinase. The pyruvate is then reduced to lactate by lactate dehydrogenase, with the concomitant oxidation of NADH to NAD^+. The resulting decrease in absorbance at 340 nm provides a convenient optical signal with which to measure ATPase rates.

1. Assay buffer: 50 mM Tris/HCl pH 7.5, 10 mM KCl, and 10 mM $MgCl_2$.
2. Premix: 10 U mL^{-1} L-lactate dehydrogenase (rabbit muscle, Roche), 2 U mL^{-1} pyruvate kinase (rabbit muscle, Roche), 0.2 mM NADH, 2 mM PEP in assay buffer.
3. Equipment: UV-photometer with thermostated cell holder, 1 cm-quartz cuvettes.
4. Incubate 1 μM GroEL (0.8 mg mL^{-1}) in premix for 5 min at 25 °C.
5. Start reaction by adding 1 mM ATP from a 100-mM stock solution adjusted to pH 7–8.
6. Record decrease in absorbance at 340 nm.
7. The rate of turnover can be calculated using $\varepsilon_{340\ nm} = 6200\ M^{-1}\ cm^{-1}$ for NADH.

Owing to the two levels of cooperativity within the GroEL tetradecamer (see Section 20.6), the dependence of the turnover rate constant, k_{app}, on the ATP concentration is complex. At low ATP concentrations, a sigmoidal increase of k_{app} with increasing [ATP] is observed (Figure 20.6). Around 200 μM ATP, k_{app} reaches a maximum of ~4 min^{-1} at 25 °C [10, 40]. A further increase in [ATP] causes a drop in k_{app} by ~50% since the binding of ATP to the second ring inhibits ATP hydrolysis in the first ring [12]. In the case of SR1, this inhibition at high [ATP] is not observed due to the lack of a second ring. An additional complication arises from the fact that ATP hydrolysis by GroEL is highly dependent on K^+ concentration. In the presence of excess GroES, ATP hydrolysis is decreased to 40–50% of

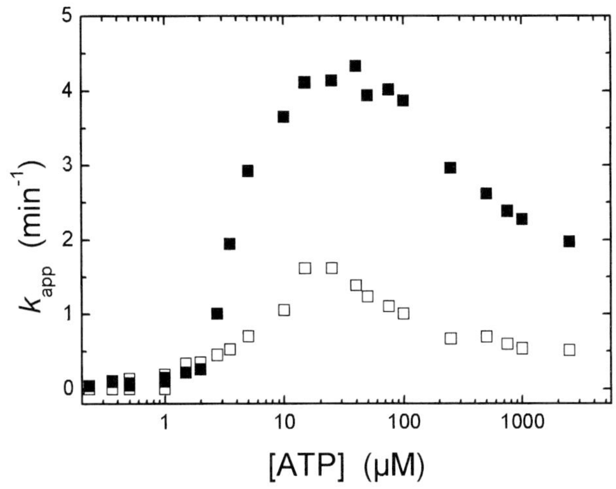

Fig. 20.6. ATP hydrolysis by GroEL and its regulation by GroES. The apparent rate constant of hydrolysis, $k_{app} = d[ATP]/(dt\cdot[GroEL])$ is plotted as a function of ATP concentration. The assay was carried out as described in protocol 6 using a GroEL concentration of 1 μM (monomer). The ATP dependence of k_{app} for GroEL alone (■) shows a bell-shaped behavior and reflects the two levels of cooperativity within the GroEL tetradecamer. When GroES is present at 2 μM (□), the rate of hydrolysis is decreased by ~2.5-fold.

the values measured in the absence of the co-chaperone (Figure 20.6). However, at low concentrations of K^+, inhibition was found to be almost complete [40].

Protocol 7: GroE-mediated Refolding of Malate Dehydrogenase (mMDH) A large number of enzymes have been used as substrates to evaluate the chaperone activity of GroE, including rhodanese (bovine liver), citric synthase (porcine mitochondria), Rubisco (*Rhodospirillum rubrum*), glutamine synthetase (*E. coli*), and DHFR (human). The advantages of malate dehydrogenase (porcine mitochondria) are that it is commercially available and that its assay is easy and cheap (Figure 20.7).

Unfolding of mMDH

Since GdmCl interferes with GroE function [71], urea is used in this protocol to denature mMDH. Depending on your mMDH stock, you may have to concentrate the enzyme prior to unfolding.

1. Remove stabilizing agents such as glycerol or $(NH_4)_2SO_4$ by dialysis or by using a desalting column (buffer: 50 mM Tris/HCl, 2 mM DTT); the concentration of mMDH monomers after this step should be at least 150 μM (5 mg mL^{-1}).
2. To 3 μL of mMDH (150 μM), add 6 μL of 9-M urea, 50 mM Tris/HCl pH 7.5, and 10 mM DTT; allow the protein to unfold for 30 min at 25 °C.

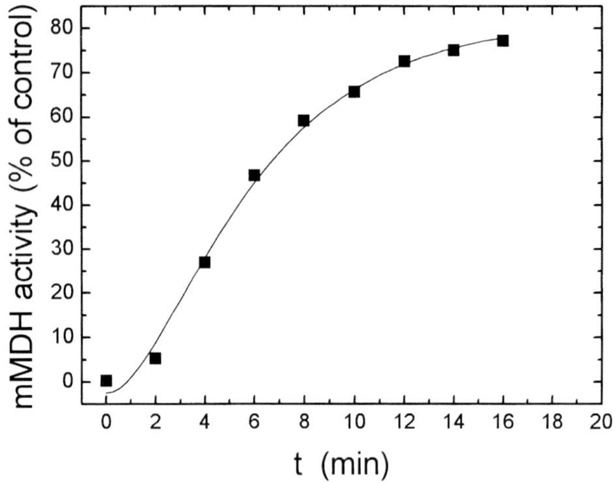

Fig. 20.7. GroE-mediated reactivation of malate dehydrogenase (mMDH). Refolding was carried out as described in protocol 7 using 0.5 μM mMDH, 1 μM $GroEL_{14}$, and 2 μM $GroES_7$ in the refolding assay. At the indicated time points, mMDH activity was monitored at 340 nm using the enzyme assay provided in the protocol. Refolding yields are normalized with respect to the mMDH activity before unfolding.

Refolding of mMDH by the GroE System

The concentration of mMDH in the refolding reaction is 0.5 μM. A twofold excess of GroEL and GroES is used.

1. Prepare a solution of 1 μM $GroEL_{14}$ in 198 μL 50 mM Tris/HCl pH 7.5, 10 mM DTT, 50 mM KCl, 10 mM $MgCl_2$ and incubate at 25 °C for 5 min.
2. Add 2 μL of unfolded mMDH and mix rapidly.
3. After 5 min, start reactivation by adding 2 μM $GroES_7$ and 1 mM ATP.
4. Withdraw aliquots of 20 μL every 2 min and determine the mMDH activity (see below).

To determine the extent of spontaneous refolding, conduct a control experiment without the chaperones. When using SR1 instead of wild-type GroEL, the reactivation protocol has to be modified since SR1 cannot release folded mMDH subunits from its cavity. In this case, incubate the withdrawn aliquots on ice for 5 min. This will release committed mMDH subunits, whereas the non-committed molecules will bind back to SR1. Next, incubate the samples for 15 min at 25 °C to allow dimerization of mMDH and then determine the enzyme activity.

Enzyme Assay for mMDH

mMDH catalyzes the reduction of ketomalonate or oxaloacetate by NADH, which can be followed by the decrease in absorbance at 340 nm using a standard UV spectrophotometer.

1. Assay buffer: 20 mM HEPES/KOH pH 7.3, 50 mM KCl.
2. Substrate: 10 mM ketomalonate; store at −20 °C.
3. Co-enzyme: 10 mM NADH; store at −20 °C.
4. Add 884 μL buffer + 80 μL ketomalonate + 16 μL NADH to a disposable cuvette and incubate at least 5 min at 25 °C.
5. Add 20 μL aliquot from the refolding reaction and mix rapidly.
6. Record $A_{340\ nm}$ for ∼1 min.

Acknowledgments

We thank K. Ruth for practical assistance concerning the described protocols. Support from the Deutsche Forschungsgemeinschaft, SFB594 (S.W.) and the Wellcome Trust (S.G.B.) is gratefully acknowledged.

References

1 Georgopoulos, C., Hendrix, R. W., Casjens, S. R., and Kaiser, A. D. (1973). Host participation in bacteriophage lambda head assembly. *J. Mol. Biol.* **76**, 45–60.

2 Barraclough, R. & Ellis, R. J. (1980). Protein synthesis in chloroplasts. IX. Assembly of newly-synthesized large subunits into ribulose bisphosphate carboxylase in isolated intact pea chloroplasts. *Biochim. Biophys. Acta* **607**, 19–31.

3 Goloubinoff, P., Christeller, J. T., Gatenby, A. A., and Lorimer, G. H. (1989). Reconstitution of active dimeric ribulose bisphosphate carboxylase from an unfolded state depends on two chaperonin proteins and Mg-ATP. *Nature* **342**, 884–889.

4 Cheng, M. Y., Hartl, F. U., Martin, J., Pollock, R. A., Kalousek, F., Neupert, W., Hallberg, R. L., and Horwich, A. L. (1989). Mitochondrial heat-shock protein hsp60 is essential for assembly of proteins imported into yeast mitochondria. *Nature* **337**, 620–625.

5 Braig, K., Otwinowski, Z., Hegde, R., Boisvert, D. C., Joachimiak, A., Horwich, A., and Sigler, P. B. (1994). The crystal structure of the bacterial chaperonin GroEL at 2.8 Å. *Nature* **371**, 578–586.

6 Fenton, W. A., Kashi, Y., Furtak, K., and Horwich, A. L. (1994). Residues in chaperonin GroEL required for polypeptide binding and release. *Nature* **371**, 614–619.

7 Ranson, N. A., Farr, G. W., Roseman, A. M., B., G., Fenton, W. A., Horwich, A. L., and Saibil, H. R. (2001). ATP-bound states of GroEL captured by cryo-electron microscopy. *Cell* **107**, 869–879.

8 Boisvert, D. C., Wang, J., Otwinowski, Z., Horwich, A. L., and Sigler, P. B. (1996). The 2.4 Å crystal structure of the bacterial chaperonin GroEL complexed with ATPγS. *Nat. Struct. Biol.* **3**, 170–177.

9 Roseman, A. M., Chen, S., White, H., Braig, K., and Saibil, H. R. (1996). The chaperonin ATPase cycle: Mechanism of allosteric switching and movements of substrate-binding domains in GroEL. *Cell* **87**, 241–251.

10 Gray, T. E. & Fersht, A. R. (1991). Cooperativity in ATP hydrolysis by GroEL is increased by GroES. *FEBS Lett.* **292**, 254–258.

11 Rye, H. S., Burston, S. G., Fenton, W. A., Beechem, J. M., Xu, Z., Sigler, P. B., and Horwich, A. L.

(1997). Distinct actions of cis and trans ATP within the double ring of the chaperonin GroEL. *Nature* **388**, 792–797.

12 Yifrach, O. & Horovitz, A. (1994). Two Lines of Allosteric Communication in the Oligomeric Chaperonin GroEL are Revealed by the Single Mutation Arg196 → Ala. *J. Mol. Biol.* **243**, 397–401.

13 White, H. E., Chen, S., Roseman, A. M., Yifrach, O., Horovitz, A., and Saibil, H. R. (1997). Structural basis of allosteric changes in the GroEL mutant Arg197 → Ala. *Nat. Struct. Biol.* **4**, 690–693.

14 Yifrach, O. & Horovitz, A. (1998). Mapping the Transition State of the Allosteric Pathway of GroEL by Protein Engineering. *Journal American Chemical Society* **120**, 13262–13263.

15 Ma, J., Sigler, P. B., and Karplus, M. (2000). A dynamic model for the allosteric mechanism of GroEL. *J. Mol. Biol.* **302**, 303–313.

16 Chandrasekhar, G. N., Tilly, K., Woolford, C., Hendrix, R., and Georgopoulos, C. (1986). Purification and properties of the groES morphogenetic protein of Escherichia coli. *J. Biol. Chem.* **261**, 12414–12419.

17 Hunt, J. F., Weaver, A. J., Landry, S. J., Gierasch, L., and Deisenhofer, J. (1996). The crystal structure of the GroES co-chaperonin at 2.8 Å resolution. *Nature* **379**, 37–45.

18 Landry, S. J., Zeilstra-Ryalls, J., Fayet, O., Georgopoulos, C., and Gierasch, L. (1993). Characterization of a functionally important mobile domain of GroES. *Nature* **364**, 255–258.

19 Saibil, H., Dong, Z., Wood, S., and auf der Mauer, A. (1991). Binding of chaperonins. *Nature* **353**, 25–26.

20 Langer, T., Pfeifer, G., Martin, J., Baumeister, W., and Hartl, F. U. (1992). Chaperonin-mediated protein folding: GroES binds to one end of the GroEL cylinder, which accomodates the protein substrate within its central cavity. *EMBO J.* **11**, 4757–4765.

21 Xu, Z., Horwich, A. L., and Sigler, P. B. (1997). The crystal structure of the asymmetric GroEL-GroES-(ADP) ~ 7 chaperonin complex. *Nature* **388**, 741–750.

22 Sakihawa, C., Taguchi, H., Makino, Y., and Yoshida, M. (1999). On the maximum size of proteins to stay and fold in the cavity of GroEL underneath GroES. *J. Biol. Chem.* **274**, 21251–21256.

23 Weissman, J. S., Hohl, C. M., Kovalenko, O., Kashi, Y., Chen, S., Braig, K., Saibil, H. R., Fenton, W. A., and Horwich, A. L. (1995). Mechanism of GroEL action: Productive release of polypeptide from a sequestered position under GroES. *Cell* **83**, 577–587.

24 Chaudhry, C., Farr, G. W., Todd, M. J., Rye, H. S., Brunger, A. T., Adams, P. D., Horwich, A. L., and Sigler, P. B. (2003). Role of the γ-phosphate of ATP in triggering protein folding by GroEL-GroES: function, structure and energetics. *EMBO J.* **22**, 4877–4887.

25 Viitanen, P. V., Gatenby, A. A., and Lorimer, G. H. (1992). Purified chaperonin (GroEL) interacts with the nonnative states of a multitude of *Escherichia coli* proteins. *Protein Sci.* **1**, 363–369.

26 Horwich, A. L., Low, K. B., Fenton, W. A., and Hirshfield, I. N. (1993). Folding In Vivo of Bacterial Cytoplasmic Proteins: Role of GroEL. *Cell* **74**, 909–917.

27 Houry, W. A., Frishman, D., Eckerskorn, C., Lottspeich, F., and Hartl, F. U. (1999). Identification of in vivo substrates of the chaperonin GroEL. *Nature* **402**, 147–154.

28 Lin, Z., Schwarz, F. P., and Eisenstein, E. (1995). The hydrophobic nature of GroEL-substrate binding. *J. Biol. Chem.* **270**, 1011–1014.

29 Badcoe, I. G., Smith, C. J., Wood, S., Halsall, D. J., Holbrook, J. J., Lund, P., and Clarke, A. R. (1991). Binding of a chaperonin to the folding intermediates of lactate dehydrogenase. *Biochemistry* **30**, 9195–9200.

30 Robinson, C. V., Gross, M., Eyles, S. J., Ewbank, J. J., Mayhew, M., Hartl,

F. U., Dobson, C. M., and Radford, S. E. (1994). Conformation of GroEL-bound α-lactalbumin probed by mass spectrometry. *Nature* **372**, 646–651.

31 Zahn, R., Perrett, S., Stenberg, G., and Fersht, A. R. (1996). Catalysis of amide proton exchange by the molecular chaperones GroEL and SecB. *Science* **271**, 642–645.

32 Goldberg, M. S., Zhang, J., Sondek, S., Matthews, C. R., Fox, R. O., and Horwich, A. L. (1997). Native-like structure of a protein-folding intermediate bound to the chaperonin GroEL. *Proc. Natl. Acad. Sci. U.S.A.* **94**, 1080–1085.

33 Chen, J., Walter, S., Horwich, A. L., and Smith, D. L. (2001). Folding of malate dehydrogenase inside the GroEL-GroES cavity. *Nat. Struct. Biol.* 8, 721–728.

34 Landry, S. J. & Gierasch, L. (1991). The chaperonin GroEL binds a polypeptide in an α-helical conformation. *Biochemistry* **30**, 7359–7362.

35 Schmidt, M. & Buchner, J. (1992). Interaction of GroE with an all β-protein. *J. Biol. Chem.* **267**, 16829–16833.

36 Chen, L. & Sigler, P. B. (1999). The crystal structure of a GroEL/peptide complex: plasticity as a basis for substrate diversity. *Cell* **99**, 757–768.

37 Wang, Z., Feng, H., Landry, S. J., Maxwell, J., and Gierasch, L. (1999). Basis of substrate binding by the chaperonin GroEL. *Biochemistry* **38**, 12537–12546.

38 Falke, S., Fisher, M. T., and Gogol, E. P. (2001). Structural changes in GroEL effected by binding of a denatured protein substrate. *J. Mol. Biol.* **308**, 569–577.

39 Jackson, G. S., Staniforth, R. A., Halsall, D. J., Atkinson, T., Holbrook, J. J., Clarke, A. R., and Burston, S. G. (1993). Binding and hydrolysis of nucleotides in the chaperonin catalytic cycle: implications for the mechanism of assisted protein folding. *Biochemistry* **32**, 2554–2563.

40 Todd, M. J., Viitanen, P. V., and Lorimer, G. H. (1993). Hydrolysis of Adenosine 5′-Triphosphate by Escherichia coli GroEL: Effects of GroES and Potassium Ion. *Biochemistry* **32**, 8560–8567.

41 Yifrach, O. & Horovitz, A. (1995). Nested cooperativity in the ATPase activity of the oligomeric chaperonin GroEL. *Biochemistry* **34**, 5303–5308.

42 Monod, J., Wyman, J., and Changeux, J.-P. (1965). On the nature of allosteric transitions: a plausible model. *J. Mol. Biol.* **12**, 88–118.

43 Koshland, D. E., Nemethy, G., and Filmer, D. (1966). Comparison of experimental binding data and theoretical models in proteins containing subunits. *Biochemistry* **5**, 365–385.

44 Burston, S. G., Ranson, N. A., and Clarke, A. R. (1995). The Origins and Consequences of Asymmetry in the Chaperonin Reaction Cycle. *J. Mol. Biol.* **249**, 138–152.

45 Yifrach, O. & Horovitz, A. (1998). Transient Kinetic Analysis of Adenosine 5′-Triphosphate Binding-Induced Conformational Changes in the Allosteric Chaperonin GroEL. *Biochemistry* **37**, 7083–7088.

46 Cliff, M. J., Kad, N. M., Hay, N., Lund, P. A., Webb, M. R., Burston, S. G., and Clarke, A. R. (1999). A Kinetic Analysis of the Nucleotide-induced Allosteric Transitions of GroEL. *J. Mol. Biol.* **293**, 667–684.

47 Todd, M. J., Viitanen, P. V., and Lorimer, G. H. (1994). Dynamics of the chaperonin ATPase cycle: implications for facilitated protein folding. *Science* **265**, 659–666.

48 Rye, H. S., Roseman, A. M., Chen, S., Furtak, K., Fenton, W. A., Saibil, H. R., and Horwich, A. L. (1999). GroEL-GroES Cycling: ATP and Nonnative Polypeptide Direct Alternation of Folding-Active Rings. *Cell* **97**, 325–338.

49 Weissman, J. S., Kashi, Y., Fenton, W. A., and Horwich, A. L. (1994). GroEL-Mediated Protein Folding Proceeds by Multiple Rounds of Binding and Release of Nonnative Forms. *Cell* **78**, 693–702.

50 Burston, S. G., Weissman, J. S., Farr, G. W., Fenton, W. A., and Horwich, A. L. (1996). Release of both native and non-native proteins from a cis-only GroEL ternary complex. *Nature* **383**, 96–98.

51 Ranson, N. A., Burston, S. G., and Clarke, A. R. (1997). Binding, Encapsulation and Ejection: Substrate Dynamics During a Chaperonin-assisted Folding Reaction. *J. Mol. Biol.* **266**, 656–664.

52 Shtilerman, M., Lorimer, G. H., and Englander, S. W. (1999). Chaperonin function: Folding by forced unfolding. *Science* **284**, 822–825.

53 Weissman, J. S., Rye, H. S., Fenton, W. A., Beechem, J. M., and Horwich, A. L. (1996). Characterization of the Active Intermediate of a GroEL-GroES-Mediated Protein Folding Reaction. *Cell* **84**, 481–490.

54 Schmidt, M., Rutkat, K., Rachel, R., Pfeifer, G., Jaenicke, R., Viitanen, P. V., Lorimer, G. H., and Buchner, J. (1994). Symmetric complexes of GroE chaperonins as part of the functional cycle. *Science* **265**, 656–659.

55 Azem, A., Kessel, M., and Goloubinoff, P. (1994). Characterization of a functional GroEL14(GroES7)2 chaperonin hetero-oligomer. *Science* **265**, 653–656.

56 Sparrer, H., Rutkat, K., and Buchner, J. (1997). Catalysis of protein folding by symmetric chaperone complexes. *Proc. Natl. Acad. Sci. U.S.A.* **94**, 1096–1100.

57 Hayer-Hartl, M. K., Ewalt, K. L., and Hartl, F. U. (1999). On the role of symmetrical and asymmetrical chaperonin complexes in assisted protein folding. *Biol. Chem.* **380**, 531–540.

58 Walter, S. (2002). Structure and function of the GroE chaperone. *Cellular and Molecular Life Sciences* **59**, 1589–1597.

59 Fisher, M. T. (1992). Promotion of the in vitro renaturation of dodecameric glutamine synthetase from Escherichia coli in the presence of GroEL (chaperonin-60) and ATP. *Biochemistry* **31**, 3955–3963.

60 Gray, T. E. & Fersht, A. R. (1993). Refolding of barnase in the presence of GroE. *J. Mol. Biol.* **232**, 1197–1207.

61 Staniforth, R. A., Burston, S. G., Atkinson, T., and Clarke, A. R. (1994). Affinity of chaperonin-60 for a protein substrate and its modulation by nucleotides and chaperonin-10. *Biochemical J.* **300**, 651–658.

62 Sparrer, H., Lilie, H., and Buchner, J. (1996). Dynamics of the GroEL-protein complex: effects of nucleotides and folding mutants. *J. Mol. Biol.* **258**, 74–87.

63 Yifrach, O. & Horovitz, A. (1999). Coupling between protein folding and allostery in the GroE chaperonin system. *Proc. Natl. Acad. Sci. U.S.A.* **97**, 1521–1524.

64 Sun, Z., Scott, D. J., and Lund, P. A. (2003). Isolation and characterisation of mutants of GroEL that are fully functional as single rings. *J. Mol. Biol.* **332**, 715–728.

65 Buchner, J., Schmidt, M., Fuchs, M., Jaenicke, R., Rudolph, R., Schmid, F. X., and Kiefhaber, T. (1991). GroE facilitates refolding of citrate synthase by suppressing aggregation. *Biochemistry* **30**, 1586–1591.

66 Farr, G. W., Furtak, K., Rowland, M. B., Ranson, N. A., Saibil, H. R., Kirchhausen, T., and Horwich, A. L. (2000). Multivalent Binding of Nonnative Substrate Proteins by the Chaperonin GroEL. *Cell* **100**, 561–574.

67 Zahn, R., Spitzfaden, C., Ottiger, M., Wüthrich, K., and Plückthun, A. (1994). Destabilization of the complete protein secondary structure on binding to the chaperone GroEL. *Nature* **368**, 261–265.

68 Walter, S., Lorimer, G. H., and Schmid, F. X. (1996). A thermodynamic coupling mechanism for GroEL-mediated unfolding. *Proc. Natl. Acad. Sci. U.S.A.* **93**, 9425–9430.

69 Staniforth, R. A., Cortes, A., Burston, S. G., Atkinson, T., Holbrook, J. J., and Clarke, A. R. (1994). The stability and hydrophobicity of cytosolic and mitochondrial

malate dehydrogenase and their relation to chaperonin-assisted folding. *FEBS Lett.* **344**, 129–135.

70 Ranson, N. A., Dunster, N. J., Burston, S. G., and Clarke, A. R. (1995). Chaperonins can Catalyse the Reversal of Early Aggregation Steps when a Protein Misfolds. *J. Mol. Biol.* **250**, 581–586.

71 Todd, M. J. & Lorimer, G. H. (1995). Stability of the asymmetric *Escherichia coli* chaperonin complex. Guanidine chloride causes rapid dissociation. *J. Biol. Chem.* **270**, 5388–5394.

72 Todd, M. J., Lorimer, G. H., and Thirumalai, D. (1996). Chaperonin-facilitated protein folding: optimization of rate and yield by an iterative annealing mechanism. *Proc. Natl. Acad. Sci. U.S.A.* **93**, 4030–4035.

73 Brinker, A., Pfeifer, G., Kerner, M. J., Naylor, D. J., Hartl, F. U., and Hayer-Hartl, M. K. (2001). Dual function of protein confinement in chaperonin-assisted protein folding. *Cell* **107**, 223–233.

74 Takagi, F., Koga, N., and Takada, S. (2003). How protein thermodynamics and folding mechanisms are altered by the chaperonin cage: molecular simulations. *Proc. Natl. Acad. Sci. U.S.A.* **100**, 11367–11372.

75 Wang, J. D., Herman, C., Tipton, K. A., Gross, C. A., and Weissman, J. S. (2002). Directed evolution of substrate-optimized GroEL/S chaperonins. *Cell* **111**, 1027–1039.

76 Dubaquie, Y., Looser, R., Fünfschilling, U., Jeno, P., and Rospert, S. (1998). Identification of in vivo substrates of the yeast mitochondrial chaperonins reveals overlapping but non-identical requirement for hsp60 and hsp10. *EMBO J.* **15**, 5868–5876.

77 Chaudhuri, T. K., Farr, G. W., Fenton, W. A., Rospert, S., and Horwich, A. L. (2001). GroEL/GroES-Mediated Folding of a Protein Too Large to Be Encapsulated. *Cell* **107**, 235–246.

78 Farr, G. W., Fenton, W. A., Chaudhuri, T. K., Clare, D. K., Saibil, H. R., and Horwich, A. L. (2003). Folding with and without encapsulation by *cis*- and *trans*-only GroEL-GroES complexes. *EMBO J.* **22**, 3220–3230.

79 Fiaux, J., Bertelson, E. B., Horwich, A. L., and Wüthrich, K. (2002). NMR analysis of a 900K GroEL-GroES complex. *Nature* **418**, 207–211.

80 Fayet, O., Louarn, J. M., and Georgopoulos, C. (1986). Suppression of the Escherichia coli dnaA46 mutation by amplification of the groES and groEL genes. *Mol. Gen. Genet.* **202**, 435–445.

81 Jenkins, A. J., March, J. B., Oliver, I. R., and Masters, M. (1986). A DNA fragment containing the groE genes can suppress mutations in the Escherichia coli dnaA gene. *Mol. Gen. Genet.* **202**, 446–454.

82 Ito, K. & Akiyama, Y. (1991). In vivo analysis of integration of membrane proteins in Escherichia coli. *Mol. Microbiol.* **5**, 2243–2253.

83 Horwich, A. L., Burston, S. G., Rye, H. S., Weissman, J. S., and Fenton, W. A. (1998). Construction of single-ring and two-ring hybrid versions of bacterial chaperonin GroEL. *Methods Enzymol.* **290**, 141–146.

84 Fisher, M. T. (1994). The effect of groES on the groEL-dependent assembly of dodecameric glutamine synthetase in the presence of ATP and ADP. *J. Biol. Chem.* **269**, 13629–13636.

85 Blennow, A., Surin, B. P., Ehring, H., McLennan, N. F., and Spangfort, M. D. (1995). Isolation and biochemical characterization of highly purified Escherichia coli molecular chaperone Cpn60 (GroEL) by affinity chromatography and urea-induced monomerization. *Biochim. Biophys. Acta* **1252**, 69–78.

86 Clarke, A. C., Ramanathan, R., and Frieden, C. (1998). Purification of GroEL with Low Fluorescence Background. *Methods Enzymol.* **290**, 100–118.

87 Todd, M. J. & Lorimer, G. H. (1998). Criteria for Assessing the Purity and Quality of GroEL. *Methods Enzymol.* **290**, 135–141.

88 Voziyan, P. A. & Fisher, M. T. (2000). Chaperonin-assisted folding of

glutamine synthetase under non-permissive conditions: off-pathway aggregation propensity does not determine the co-chaperonin requirement. *Protein Sci.* **9**, 2405–2412.

89 Eisenstein, E., Reddy, P., and Fisher, M. T. (1998). Overexpression, Purification, and Properties of GroES from *Escherichia coli*. *Methods Enzymol.* **290**, 119–134.

90 Gill, S. C. & von Hippel, P. H. (1989). Calculation of protein extinction coefficients from amino acid sequence data. *Anal. Biochem.* **182**, 319–326.

21
Structure and Function of the Cytosolic Chaperonin CCT

José M. Valpuesta, José L. Carrascosa, and Keith R. Willison

21.1
Introduction

A decade has passed since the discovery of the eukaryotic cytosolic chaperonin CCT (chaperonin-containing TCP-1; also termed TriC [TCP-1 ring complex] or c-cpn [cytosolic chaperonin]) [1–4]. However, characterization of the chaperonin occurred long after the discovery of one of its components, the t-complex polypeptide-1 (TCP-1), a protein that is abundantly expressed in the germ cells of the mouse testis [5] whose encoding gene was first molecularly cloned from mice [6] and then humans [7]. A history of the genetic analysis of the t-complex and its encoded genes, including TCP-1, was published recently [8]. TCP-1 is highly conserved in lower eukaryotes [9] and was found to be an essential protein involved in microtubule function in yeast [10]. Eventually, primary sequence comparison analyses revealed extensive sequence similarity between the TCP-1 protein and the Group I chaperonins [11, 12] and with thermophilic chaperonins [13], thus setting the scene for the rapid biochemical and functional characterization of the eukaryotic chaperonin. TCP-1, together with seven other homologous proteins, forms part of a chaperonin complex similar to GroEL and the chaperonin from archaea (thermosome or TF55) [1, 2, 4, 14]. All CCT preparations appear as toroidal structures constituted by two superimposed rings [1, 2, 4, 15]. Since then, the eight different subunits that constitute the oligomer have been sequenced from different organisms [16–21], and the biochemical, biophysical, and structural characterization of the cytosolic chaperonin has grown slowly but steadily. However, there is a still a paucity of information concerning the mechanism of action and functional cycle of CCT, especially when compared with the well-defined substrate-folding mechanism of GroEL from *E. coli*, but there is little doubt that CCT is unique when compared with other chaperonins. As will be discussed, its uniqueness lies not only in the fact that it is by far the most complex chaperonin, but also in its folding mechanism and in the range of proteins it assists in folding. The substrate proteins of CCT are critical components in various important cellular processes such as cytoskeletal formation of both microfilaments and microtubules, control of cell cycle progression, signal transduction, gene expression, and proteasome-

Protein Folding Handbook. Part II. Edited by J. Buchner and T. Kiefhaber

ISBN: 3-527-30784-2

associated degradation. This chapter discusses in detail what is known about this exciting molecular chaperone, in particular its structure and composition, functional cycle, and range of substrates and its interaction with other molecular chaperones.

21.2 Structure and Composition of CCT

Chaperonins are oligomeric structures composed of proteins with a ~60-kDa Mw and are divided into two groups according to their origin. Group I chaperonins are present in eubacteria and in eukaryotic organelles of endosymbiotic origin, whereas group II chaperonins are found in archaea and in the eukaryotic cytosol, the location of CCT. All chaperonins share a similar monomeric and oligomeric architecture, although the two groups of chaperonins show important differences. The atomic structure of the monomer is known for some group I and II chaperonins (GroEL from *E. coli* [22, 23] and the thermosome from *Thermoplasma acidophilum* [24], respectively), and they all share a three-domain topology (Figure 21.1, see p. 728): the equatorial domain that holds the nucleotide-binding site and most of the interactions between subunits of the same ring and the opposing one; the apical domain that contains the substrate recognition site; and the intermediate domain that transmits the signal originating in the equatorial domain upon nucleotide binding and hydrolysis to the apical domain, which results in large conformational changes in the latter. The main difference between the two groups of chaperonins resides in an extra helical region (helical protrusion) in group II chaperonins that is located at the tip of the apical domain [24–26] (Figure 21.1D) and that closes the chaperonin cavity during part of its functional cycle. The same role is fulfilled in group I chaperonins by a co-chaperonin called GroES, a small oligomer that is not present in group II chaperonins (Figure 21.1A, B).

The oligomeric structure is similar in both groups of chaperonins – a toroidal structure composed of two rings placed back to back – but major differences reside in their symmetry and composition. Whereas group I chaperonins are homo-oligomeric rings with sevenfold symmetry, the rings in the thermosomes can be composed of one to two different subunits with eightfold symmetry, three different subunits with ninefold symmetry [27], and, in the case of CCT, eight subunits with eightfold symmetry [28] (Figure 21.2B, see p. 728). A further important difference lies in the way the two rings interact (Figure 21.1A, C). Whereas in group II chaperonins each subunit interacts with one from the opposite ring, in group I chaperonins each subunit interacts with two from the opposite ring. This difference in inter-ring connectivity may cause important differences in the signaling between both rings and therefore in the functional cycle of both groups of chaperonins. In fact, inter-ring negative cooperativity in the ATP-induced allosteric transition of CCT is much stronger than that observed for GroEL [29].

No atomic structure has been obtained for the oligomeric CCT, and the structural information available has come mostly from electron microscopy studies. The first structural characterization obtained from image processing of negatively-stained specimens revealed the octameric nature of the CCT structure, a barrel-shaped cylinder with a diameter of ~150 Å and a height of ~160 Å, again different from the more rectangular shape of group I chaperonins [15] (Figure 21.2A–C and Figure 21.3A, see p. 729).

As has been clearly established for many different organisms, CCT is composed of eight different subunits [28]. This has been determined by different techniques such as protein sequencing [14, 16], characterization with different antibodies [30], and mass spectrometry [31]. The mammalian subunits are termed CCTα, CCTβ, CCTγ, CCTδ, CCTε, CCTζ, CCTη, and CCTθ, with CCT1–CCT8 corresponding to their respective counterparts in yeast. The eight CCT polypeptides share ~30% sequence identity in pairwise comparisons within a species, but homologous subunits across different eukaryotic species share a higher degree of similarity, which is indicative of a very early divergence time in the evolution of the CCT gene family (see Section 21.8). The subunits are arranged within each ring in a unique way (Figure 21.2B), and this was determined by characterizing the interaction of CCT microcomplexes of 2 to 3 subunits, which yielded a unique solution out of 5040 possibilities [32]. The ring order was deduced from the observation of the following CCT microcomplexes: (δ–η) (η–α) (α–ε) (ε–ζ) (ζ–β) (β–γ) (θ). Subsequent immunomicroscopy studies with monoclonal antibodies against specific CCT subunits have confirmed this arrangement [33, 34] (Table 21.1) and have shown that only one antibody molecule binds to each of the chaperonin rings [35], reinforcing the idea that a single copy of each subunit is present in each CCT ring and that both rings have the same composition. A ninth CCT subunit, CCTζ-2, has been discovered in vertebrates, which is specific for testis. CCTζ-2 is homologous to CCTζ, and seems to substitute for it in testis CCT, suggesting a specificity for this tissue-specific subunit and, indeed, as will be discussed for the whole chaperonin. An interesting but still obscure aspect regarding the interaction between CCT subunits has to do with the disassembly of the oligomeric structure. In one case, the double-ring has been shown to disassemble into the single-ring species, and this has been hypothesized to serve as a nucleation factor for the generation of new CCT particles [36]. Alternatively, CCT has been described to disassemble into individual subunits, and this has been hypothesized to have a role in the functional cycle of CCT [37].

Tab. 21.1. Summary of monoclonal antibody decoration experiments consistent with the CCT subunit arrangement model of Liou and Willison [32]. Monoclonal antibodies 91A, 23C, 8g, 81a, and PK-5h bind to CCTα subunits (91A and 23C), CCTδ subunits (8G), CCTη subunits (81a), and CCTθ (PK-5h) subunits, respectively.

CCT subunit	*MAb number*	*Substrate bound to CCT-Mab complexes*	*Figure**	*Reference*
CCTα	91A	α-actin	4A,B	33
CCTα	91A	β-actin.sub4	4C,D	33
CCTα	23C	None	1A–C	35
CCTδ	8g	α-actin	4E,F	33
CCTδ	8g	β-actin.sub4	4G,H	33
CCTδ	8g	α/β-tubulin	5A,B	34
CCTδ	8g	β-tubulin-f3	5E,F	34
CCTδ	8g	β-tubulin-f4	5G,H	34
CCTδ	8g	β-actinG150P	3B,C	77
CCTδ	8g	β-tubulin	6B	74
CCTη	81a	α/β-tubulin	Data not shown	34
CCTη	81a	β-tubulin-f5	Data not shown	34
CCTθ	PK-5h	α/β-tubulin	Data not shown	34
CCTθ	PK-5h	β-tubulin-f5	5C,D	34

* Figure number in respective reference.

through heat shock DNA elements and the factors HSF1 and HSF2 [44] as well as through the Ets domain transcription factors under the control of the Ras/MAPK pathway [45]. Further discussion of CCT gene and protein regulation can be found in Ref. [28].

21.4
Functional Cycle of CCT

Cryo-electron microscopy studies have characterized, albeit at low resolution, the two main conformers of CCT. The open conformation, generated in the absence of nucleotide [46] (Figure 21.3A), reveals structural features very similar to those obtained for group I (Figure 21.3C) and other group II chaperonins [47–49], with an open cavity ready to interact with the unfolded substrate. The helical protrusions are not visualized, probably due to the lack of high-resolution information in the reconstructions and to their closeness to the main body of the chaperonin structure in the open form, but they are visible in the closed conformation of CCT, generated in the presence of ATP, because they have undergone large movements (approx. 70° clockwise, viewed from the top) and have locked the chaperonin cavity [46] (Figure 21.3B), as also occurs in other group II chaperonins [47–49]. The structural rearrangements of the apical domains are therefore of a large magnitude [50] and have also been detected by biochemical techniques [51]. These changes are very similar to the ones that in GroEL are induced upon binding of its co-chaperonin GroES and the subsequent closure of the chaperonin cavity (Figure 21.3D). The three-dimensional reconstruction of the CCT closed conformation fits very well with the atomic structure of the thermosome from *Thermoplasma acidophilum* [24]. This thermosome structure has been obtained in the absence of nucleotide, and small-angle neutron-scattering experiments have revealed that the closure of the cavity is induced by the crystallization conditions [52], thus mimicking the conditions generated by the presence of nucleotide. This suggests that the apical domains are very flexible, which is supported by experiments with complexes between CCT and a monoclonal antibody recognizing a C-terminal region of CCTα located in the interior of the cavity which folds actin and tubulin at normal rates in rabbit reticulocyte lysate [35].

Like all chaperonins, CCT is an ATPase that uses ATP binding and hydrolysis to maintain its functional cycle through movements that induce the closure and opening of the ring cavity. These movements are dictated by nested cooperativity that involves intra-ring positive and inter-ring negative cooperativity in ATP binding [53]. Small-angle X-ray scattering (SAXS) experiments suggest that ring closure of CCT requires not ATP binding but ATP hydrolysis to generate a trigonal-bipyramidal transition state of ATP hydrolysis [54]. This study challenges our EM studies, which showed that ATP- or AMPPNP binding closes the lids of CCT [46, 50] and the authors suggested that the AMPPNP we used must have been contaminated by ATP. However, we note that recent studies on *Thermococcus* thermosomes show that ATP- or AMPPNP binding closes the rings even under conditions whereby ATP hydrolysis is completely restricted [56] and that an X-ray structure of the same chaperonin with bound AMPPNP has been determined [57]. Further investigation of the allosteric changes in CCT induced by nucleotide is clearly required [55]. In the case of the cytosolic chaperonin, there are kinetic differences with respect to the better-studied GroEL that suggest that there may be differences in the intrinsic affinities for ATP among the subunits in a ring [29, 55]. Such differences in allosteric behavior would support a genetic model based on the analyses of mutants of the ATP site of several subunits of yeast CCT [58], which have shown that catalytic cooperativity of ATP binding and hydrolysis in CCT takes place in a sequential and hierarchical manner (CCT$\alpha \rightarrow$ CCT$\gamma \rightarrow$ CCT$\beta \rightarrow$ CCTζ; see arrows in Figure 21.4C, E), different from the concerted mechanism described for GroEL [53]. If this turns out to be the case, then closure of the CCT cavity would be sequential and the eight helical protrusions of the apical domains would act like an iris in this process.

21.5 Folding Mechanism of CCT

As in the case of the well-characterized GroEL, the conformational changes undergone by CCT upon ATP binding and hydrolysis are used by the chaperonin to facilitate the folding of substrate proteins. However, there are important differences between the two chaperonins that reside in the mechanism of assisted folding and in

to the bacterial chaperonin, however, in CCT this closure is performed by the movement of the extra helical regions located at the tips of the apical domains. Another important difference resides in the state of the folding substrate after the closure of the ring, which in the case of GroEL is liberated in the cavity, where, free of any unwanted interactions with other proteins, it can fold using the information encoded in its own amino acid sequence. In the case of CCT, electron microscopy, kinetic, and genetic studies have shown that the cytosolic chaperonin follows a mechanism different from the "Anfinsen cage" described for GroEL [72, 73]. Cryo-electron microscopy studies carried out with CCT-actin and CCT-tubulin complexes in the presence of AMP-PNP, a non-hydrolyzable analogue of ATP, have revealed that the movements undergone by the apical domains induced upon nucleotide binding to seal the cavity are coupled to the folding movements of actin and tubulin, which remain bound to the apical domains of the cytosolic chaperonin and adopt a conformation very similar or possibly even identical to their native structure [74] (Figure 21.4D, F). Combining these results with the "sequential" model described above, a "sequential" folding mechanism of actin and tubulin mediated by CCT can be depicted, in which the N-terminal domains of actin and tubulin would be the first ones to undergo the structural rearrangements induced by the sequential movements of the CCT apical domains (see arrows in Figure 21.4C, E). The N-terminal domains of both cytoskeletal proteins bind to CCT with low affinity [34, 67, 71, 75], and as a consequence of the movements of the apical domains, the interaction with them is broken first. On the other hand, the C-terminal regions of actin and tubulin are the ones interacting with CCT with the highest affinity and are also the last ones to respond to the conformational change induced by ATP binding. As a result of these movements, actin and tubulin remain bound to the apical domains of CCT and adopt a conformation very similar or identical to their native structure [74]. In the case of tubulin, the conformational changes undergone by the protein have been found to be coupled to the binding of a GTP molecule to the protein nucleotide-binding site [64, 74, 76], which supports the notion that tubulin is folded on CCT to a state that is competent for GTP binding.

CCT therefore acts as a folding nanomachine that uses the conformational changes undergone by the apical domains upon ATP binding to force the folding of the two cytoskeletal proteins (and probably other substrates) by pushing together the two domains that were previously separated and interacting with two opposed regions of the chaperonin cavity. The movement of the two domains of actin has been corroborated by mutation of conserved residues in the putative hinge region of actin. These experiments have revealed that, upon reduction of the flexibility of this hinge by substituting conserved glycine residues for proline residues, such actin mutants cannot interact with both sides of the CCT cavity and remain bound to CCT [77]. The folding mechanism of CCT seems to be very efficient and sequential and in fact probably does not require multiple rounds of binding and release [76, 78]. This is very different from the more passive release mechanism of chaperonin GroEL (Figure 21.5). This specific mechanism explains why, although GroEL is able to bind actin and tubulin and undergo multiple cycles of release and rebinding, only CCT is able to generate the native conformations of these two cytoskeletal proteins [79].

What is the role of CCT that makes it stringent for the folding of actin and tubulin? Although it is not yet proven, an interesting hypothesis points to the critical step being proper formation of the nucleotide-binding site of both cytoskeletal proteins [34]. CCT may have evolved from a thermosome-like ancestor to help overcome a specific kinetic barrier in the folding of actin and tubulin molecules, which have already reached a certain degree of conformational maturity before interacting with the cytosolic chaperonin [80]. CCT stabilizes an open conformation through the binding of their two topological domains with opposite sides of the chaperonin ring. In the case of actin, heat denaturation experiments have shown that its native conformation depends on the degree to which the nucleotide contributes to the connectivity between the two domains of the protein [81], and most of the bonds that connect the small and large domains occur through the ATP-binding site [82]. The situation is almost the same in tubulins, where GTP provides connectivity between domains [83], and some biochemical results indicate that GTP binding indeed stabilizes the molecule during the CCT-facilitated folding [64].

21.6 Substrates of CCT

Although it was originally believed that members of the actin and tubulin families were the only substrates of CCT, it has now become evident that the cytosolic chaperonin uses its structure to act on a large variety of proteins [84], perhaps as many as 9–15% of newly synthesized proteins [85]. However the folding mechanism described above suggests that the CCT-interacting proteins may have some shared structural or functional patterns and motifs. It is also becoming evident that the conformational changes undergone by CCT during its functional cycle are used not only for the generation of the native structure of certain proteins but also for other functions such as control of quaternary interactions and regulation of and protection from protein degradation (Figure 21.6). Nevertheless, a folding role for CCT has been shown, both in vivo and in vitro, for its major substrates actin and tubulin, since mutations in CCT subunits in yeast cause severe cytoskeletal defects [39, 86–88]. CCT also mediates the folding of other cytoskeletal proteins such as the actin-related proteins (ARPs) [89] and it has been shown to interact with myosin II heavy chain (HMM) [90], cofilin, and actin-depolymerizing factor (ADF1) [91].

Other proteins that are assisted in their folding by CCT are some viral proteins such as the hepatitis B virus capsid [92], the type D retrovirus Gag polyproteins [93], and the Epstein-Barr virus-encoded nuclear protein (EBNA-3) [94]. These are clear examples of opportunistic proteins that have parasitized the functional cycle of CCT to solve their specific folding problems. Other proteins involved in various cellular processes have been shown to be assisted by CCT in their folding, such as luciferase [95], the Gα subunit of the G-transducin complex involved in retinal

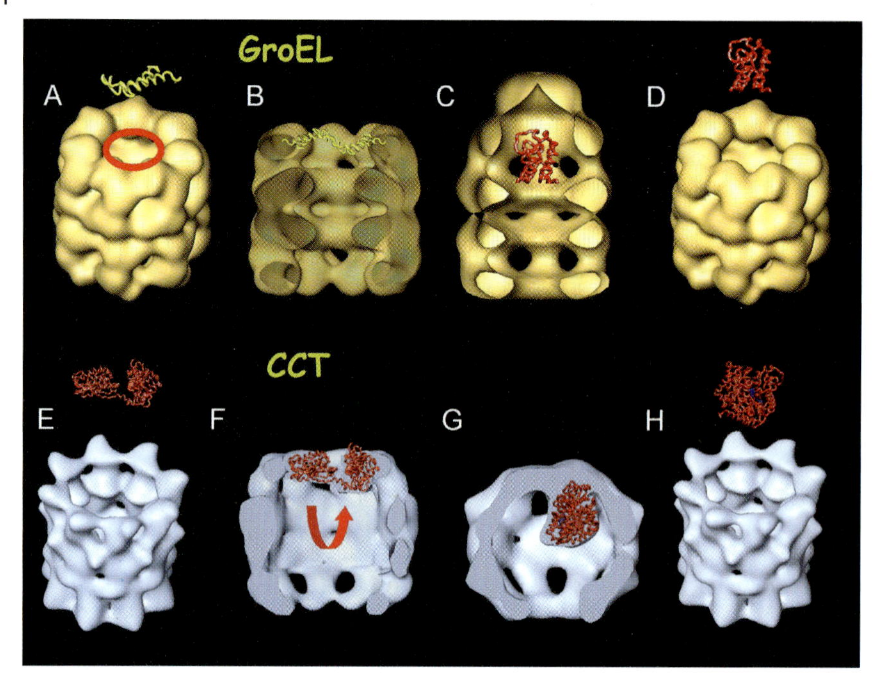

Fig. 21.5. Folding mechanism of GroEL (A–D) and CCT (E–H). In the case of GroEL, the hydrophobic residues located at the entrance of the cavity (circle in (A)) recognize any kind of unfolded polypeptide, provided that hydrophobic residues are exposed at its surface, and bind the polypeptide (B). The large conformational changes generated by ATP binding and the subsequent GroES binding, which occurs in the same region where the polypeptide is interacting with GroEL, liberate the polypeptide into a now much larger and locked cavity (C), where, free of any unwanted interactions, it can fold using the information encoded in the amino acid sequence. Hydrolysis of ATP and conformational changes occurring in the opposing ring liberate the GroES oligomer and the polypeptide, which may have attained its native conformation (D). If this is not the case, the polypeptide will be recognized by a GroEL molecule and would undergo the same again. The mechanism is not very efficient, but it serves a large number of different proteins. In contrast, CCT uses a different mechanism that seems to be very efficient and in which the substrate has already acquired a certain degree of conformation (E) before the interaction occurs between specific regions of the unfolded polypeptide and specific CCT subunits (F). The conformational changes undergone by the CCT apical domains upon ATP binding close the cavity and force the folding of the polypeptide (G). Subsequent hydrolysis and nucleotide release opens the chaperonin cavity and releases the folded protein (H).

phototransduction [76], and cyclin E, a protein implicated in the control of the cell cycle [96]. In the latter case, it has been proposed that interaction with CCT is necessary to generate a conformation that is apt to form a stable and functional complex with its partner protein, Cdk2. This also seems to be the situation with the interaction of CCT with other proteins involved in activation or repression of gene expression, such as the histone deacetylases Hos2 [97] and HDAC3 [98].

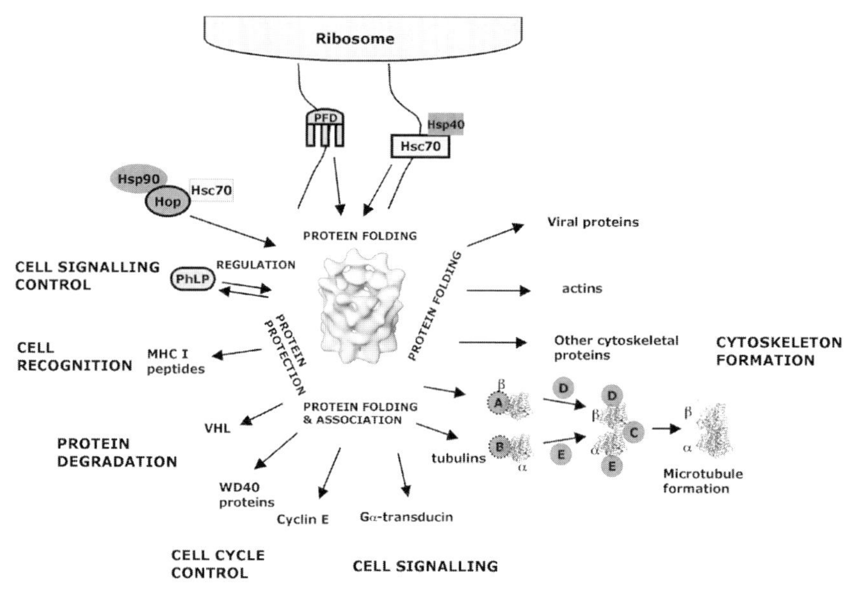

Fig. 21.6. Function of CCT and its interaction with its substrates and co-chaperones. The figure tries to convey the modes of interaction of CCT with different proteins to promote their folding, their quaternary association, their protection, or their presence in the cytosol. CCT seems to be at the core, through these interactions, of various important cellular processes.

The interaction of HDAC3 with CCT seems to be stringent for generating a stable and functional complex with a partner protein, the nuclear receptor co-repressor SMRT. Another example of this kind of interaction is the case of the von Hippel-Lindau tumor suppressor protein (VHL), a protein involved in several aspects of cell proliferation and tumor formation. The most well known role of VHL is as the substrate-recognition component of the VCB-Cul2 E3 ubiquitin protein–ligase complex, involved, like other E3 ubiquitin–ligase complexes, in protein degradation processes [99]. VHL forms a stable complex with the elongin BC complex, formed by the elongins B and C [100], but for this to occur, VHL must first interact with CCT, which helps this protein to adopt a conformation that can be recognized by the elongin BC complex [101]. This conformation is not stable, and in the absence of the elongin BC complex, the protein rebinds to the cytosolic chaperonin [101]. It is clear that in this case (and probably in the case of HDAC3), the conformational changes undergone by CCT during its functional cycle do not generate a native conformation but rather a specific transient one that is necessary for the interaction between and subsequent stabilization by other proteins. The fact that the VHL–elongin BC complex can be assembled in *E. coli* through co-expression of the three components and then crystallized supports the assembly model [100]. As often reiterated [102], the fact that native actins and tubulins cannot be obtained by expression in bacterial systems can be explained by their stringent requirement for CCT.

Other major CCT substrates are a large set of proteins containing WD40 repeats, a degenerate motif comprising 44–60 amino acid residues that typically contains a GH dipeptide at its N-terminal domain and the WD dipeptide at the C-terminus [103]. These WD40 repeats fold into four β-stranded domains, and together multiple repeats fold up into ring structures. The interaction between CCT and these proteins has been found by proteomic techniques using tagged open-reading frames to pull down multi-protein complexes [104], which has resulted in the identification of a group of at least 21 proteins that interact with the CCT oligomer (17% of the known WD40 proteins in yeast [84]). These proteins are implicated in various and important cellular processes, especially in cell cycle control and protein degradation. Most WD40 proteins possess in their polypeptide chains seven such WD40 repeats [103], and crystal structures of three such seven-bladed propellers have been determined so far. Curiously, a large number of the WD40 CCT-interacting proteins characterized so far have molecular masses larger than 60 kDa, the estimated mass that could be encapsulated into the CCT cavity, and therefore it is unlikely that these proteins are folded entirely inside the CCT cavity. It is therefore feasible with respect to these large proteins that the cytosolic chaperonin has other roles besides folding assistance and that it could be involved in regulating the proteins' activity by controlling their liberation or association with other proteins. Some recent studies show that some of these proteins indeed interact with CCT and that this interaction is necessary for their correct function. This is the case of the proteins Cdc20 and Cdh1, which are both implicated in the control of cell cycle through activation of the anaphase-promoting complex [105, 106], or Cdc55, a regulatory subunit of a PP2A protein phosphatase that acts in the cell cycle checkpoint inhibiting exit from mitosis in response to spindle or kinetochore damage [106, 107]. Other WD40 proteins shown in these studies to interact with CCT are TLE2 [106], Step4, Pex7, Prp46, and Sec7 [107], confirming that CCT interacts with several members of this structurally defined class of proteins. The interaction between the WD40 proteins and CCT is of a specific nature and, in the case of Cdc20, occurs through a region encompassed by specific WD40 repeats [105]. It is intriguing that certain WD40 proteins do not interact with CCT [106, 107], and this points towards a selective and specific CCT-binding mechanism as determined for actins and tubulins.

A role for CCT different from folding or regulation of protein association seems to be the reason behind the interaction between CCT and PhLP (phosducin-like protein). PhLP is involved, like its homologue phosducin, in regulation of cell signaling through binding to the $G\beta\gamma$-protein complex and blocking its interaction with the $G\alpha$-subunit [108]. PhLP has been shown to interact with CCT in its native form, and its overexpression in cells inhibits the actin-folding activity of the cytosolic chaperonin [109]. It is therefore likely that this interaction between PhLP and CCT may be a way to control CCT activity or, alternatively, to regulate the availability of PhLP to modulate G-protein signaling.

Finally, another role for CCT has been described recently, namely, protection against degradation of proteolytic intermediates of the MHC class I in the processing pathway until their delivery in the cell surface [110]. This is an exciting discovery that further extends the already complex cellular functions of the cytosolic chaperonin.

21.7 Co-chaperones of CCT

It is increasingly evident that a host of molecular chaperones work in conjunction in protein-folding pathways. Numerous studies have revealed the intricacy in the function of chaperones such as Hsp90, Hsp70, Hsp60, Hsp40, and Hsp10 [111], and CCT is no exception to this rule, since CCT's interaction with several co-chaperones has been firmly established. These chaperones interact with the unfolded polypeptide either before or after the association with CCT. The best-characterized member of the former case, with the regard to its interaction with CCT, is the co-chaperone prefoldin [112, 113] (PFD; also termed GIM), a hetero-hexameric protein that exists in archaeabacterial and eukaryotic organisms and that in the latter has been shown to be involved in the folding of actins and tubulins [107, 112, 113] to the extent that its presence increases by at least fivefold the throughput of actin folded by CCT in vivo [78]. In this reaction, PFD transfers the substrate to CCT through a mechanism that involves a physical interaction between PFD and CCT [112]. This interaction, which has been observed by cryo-electron microscopy studies (Figure 21.7B), seems to occur through specific PFD and CCT subunits [114]. The three-dimensional reconstruction of the complex formed by eukaryotic PFD and unfolded actin also suggests that PFD delivers the unfolded actin to CCT in a quasi-folded conformation, ready to undergo the last step in its folding process in the chaperonin cavity [114] (Figure 21.7C).

CCT has also been shown to interact with members of the Hsp70/Hsc70 family (hereafter referred to as Hsc70); either directly or through association with the Hsp70/Hsp90 organizing protein (Hop, also termed p60), which links the cytosolic chaperonin with two of the major chaperone systems, the Hsp90 and Hsp70/Hsp40 systems (Figure 21.6). The interaction between CCT and Hsc70 has been known for a long time [1]. Proteomic analysis of CCT substrates has revealed that Hsc70 interacts with the cytosolic chaperonin [38], and several substrates are known to be transported by Hsc70 to CCT, including, among others, luciferase [95, 115], the tumor suppressor protein VHL [116], and several WD40 proteins [107]. Hop is a member of a structurally unrelated group of proteins including Bag-1, Hap46, and Hip, which associate with Hsc70 chaperones and regulate their activity [115, 117]. However, unlike the rest of the family members, Hop also binds to the chaperone Hsp90 [118, 119], and the complex formed by the three chaperones is involved in protein folding, especially in processes that have to do with the formation of mature progesterone and glucocorticoid receptors [118, 120]. Hop also interacts with CCT, and the interaction occurs through its C-terminal region [115], a domain that is different from the Hsc70-binding domain (N-terminus) and the Hsp90-binding domain (central part of the protein sequence). The interaction between CCT and Hop is dependent on the presence of nucleotide, and bind-

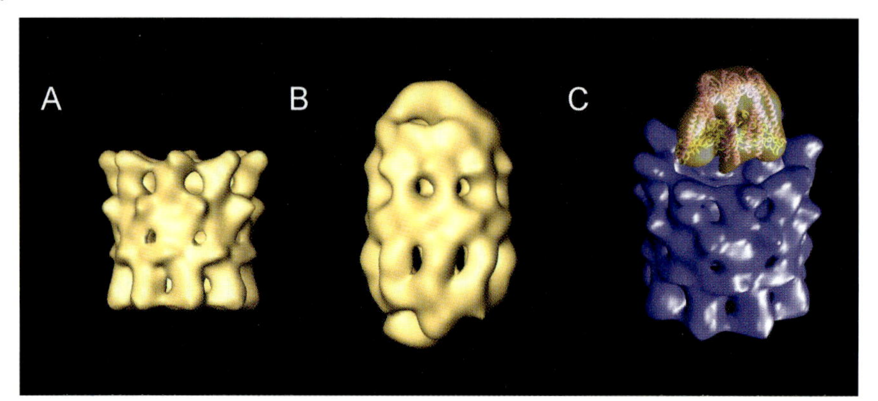

Fig. 21.7. Interaction between CCT and its co-chaperone PFD. (A, B) Three-dimensional reconstructions, obtained by cryo-electron microscopy and image processing, of the PFD-free CCT (A) and of the CCT-PFD symmetric complex (B) as described by Martín-Benito et al. [114]. (C) Model of the interaction between CCT and the complex formed by eukaryotic PFD and unfolded actin, obtained by electron microscopy and image processing. The atomic model docked into this three-dimensional reconstruction corresponds to the atomic structure of the homologous archaeal PFD from *Methanobacterium thermoautotrophicum* [154] (pdb code 1fxk) and the atomic model of CCT-bound actin, as suggested in Llorca et al. [34] and also depicted in Figure 21.4C. The model tries to describe the interaction between the complex formed by the co-chaperone PFD and its cargo, the unfolded actin, with the chaperonin CCT.

ing of this chaperone to CCT increases the chaperonin ADP/ATP exchange activity but decreases its substrate-binding affinity, thus suggesting for Hop a regulatory role of the cytosolic chaperonin [115].

While actin folded by CCT is readily incorporated into filaments, tubulin is folded by the cytosolic chaperonin to a unstable conformation that requires the assistance of several cofactors (A, B, C, D, and E) that facilitate the formation of the assembly-competent α/β tubulin dimers, necessary for the formation of the microtubules [121, 122] (Figure 21.6). According to the gathered biochemical information, after the sequential interaction of α-tubulin with cofactors B and E, and β-tubulin with cofactors A and D, a complex is formed that contains cofactors C, D, E, and α- and β-tubulin and that generates the formation and release of a stable α,β-tubulin heterodimer after the induced GTP hydrolysis of β-tubulin [102, 121, 123]. There is a paucity of structural information regarding these cofactors; only the atomic structure of cofactor A has been elucidated [124, 125], which reveals a rod-like structure, slightly convex in shape, formed by three α-helices. The central part of the structure contains the β-tubulin-binding domain, composed of a mixture of hydrophilic and hydrophobic residues [125]. Curiously, the structure of cofactor A strikingly resembles that of the BAG domain, and it has been hypothesized that both proteins could belong to the same family of chaperone cofactors [125]. Little is known about the interaction between CCT and cofactors A and B, and it is possible that the two cofactors only bind to the tubulin monomer after the chaperonin folding cycle. Although there is biochemical evidence for the interaction between cofactor A and CCT [126], some authors deny that such interaction occurs [127].

21.8 Evolution of CCT

It is clear that gene duplication is the main force behind the evolution of group II chaperonins [16, 128, 129]. In archaeal species complex patterns of gene duplication, gene loss, and gene conversion have allowed the appearance of one, two, or three homologous subunits in different thermosomes. Gene duplication has been hypothesized to occur as the result of the appearance of mutations in the intra-ring subunit domain of one subunit followed by compensatory changes in an evolved duplicate subunit, and this could have produced a tendency towards hetero-oligomerism, even in the absence of specialized roles for the duplicate subunits. This latter view is reinforced by the facts that duplicate chaperonin genes have occasionally been lost in evolution [130] and that gene conversions can homogenize the region encoding the thermosome putative substrate-binding domains [129].

It is clear that CCT has evolved from archaebacterial precursors because of the close sequence homology, but, in contrast to thermosomes, the eukaryotic chaperonin is the result of a series of duplications that occurred probably only once, at the very beginning of the evolution of the eukaryotic genome, approximately two billion years ago [16, 17, 131]. This is supported among other things by the fact that all eight CCT subunits are present not only in all the protists but also in amitochondriate protist lineages, which are believed to have diverged early in eukarya evolution [128]. Phylogenetic analysis of different eukaryotic organisms (animals, fungi, plants, and amitochondriate protists) reveals that the different CCT subunits group together with high statistical significance, which reinforces the notion that gene duplications took place in the ancestors of these organisms [131]. A recent comparison of synonymous versus non-synonymous nucleotide differences in CCT genes shows strong positive selection occurring after each duplication event to provide sub-functionalization of the new subunits [132].

What is the order of the duplication events? Although there is not yet a clear view of the original process that produced the eight subunits, due to the very long time since it happened, phylogenetic analyses reveal that there is more recent common ancestry between some CCT subunits, i.e., the CCTδ and CCTε groups and the CCTα, CCTη, and CCTβ groups [131, 132]. However, it is clear in the case of the two different vertebrate CCTζ subunits (1 and 2) that posses a high percentage of identity [19] (81%) that they are more related to each other than to CCTζ subunits from other organisms. The gene duplication that gave rise to these two genes occurred recently, around 200 million years ago, compared to the origin of CCT, which occurred two billion years ago [16]. As discussed above, CCTζ-2 is expressed only in testis, and it has been suggested that it originated to assist in the folding of specific testis proteins [19]. This agrees with the hypothesis that CCT hetero-oligomerization occurred at the same time as the appearance of the cytoskeleton [17, 80, 133, 134]. The existence of a developed cytoskeleton in eukaryotes has given rise to functions that are unique to them, including, among others, chromosome segregation, locomotion, and phagocytosis, the latter undoubtedly involved in the endosymbiotic processes that generated and conserved mitochondria and chloroplasts in the eukaryotic cell. The evolution of CCT towards hetero-oligomerization and specialization has been paralleled by its co-chaperone PFD, which has evolved from a simple oligomer in archaea (composed of two differrent and homologous proteins, α and β) to a more complex assembly in eukarya (six different proteins, two α-like and four β-like). This co-evolution has likely served to generate a more efficient folding machinery for actins and tubulins through a mechanism involving specific interactions between defined regions of the cytoskeletal proteins with particular CCT and PFD subunits [17, 33, 34, 135]. In the case of CCT, several results point towards confirmation of this hypothesis, and the immunomicroscopy experiments with CCT-actin and CCT-tubulin complexes cited above reveal that CCT interacts with the two cytoskeletal proteins through specific subunits [33, 34]. The interaction between CCT and the two proteins occurs in the upper part of the apical domain, just below the helical protrusion in the case of actin and encompassing part of this protrusion in the case of tubulin [34], and it is in this region where sequence- and structure-based comparison analyses have revealed a region of highly conserved, subunit-specific residues, most of them charged [26, 131]. These findings strengthen the notion of a specialized function of the CCT subunits. In the case of actin and tubulin, sequence-comparison and structural analyses have shown that the domains of both cytoskeletal proteins putatively involved in CCT binding are absent or greatly modified in their corresponding prokaryote homologues FtsA/MreB and FtsZ, and most of these domains are also implicated in the formation of actin and tubulin polymers [65]. This further supports the coevolution of CCT with its two main substrates in the maintenance of the cytoskeletal structures that are the essence of all eukaryotic organisms.

What about the evolution of CCT interaction by the many CCT-binding proteins other than actins and tubulins? In the case of the large group from the family of WD40 proteins [84, 104–107], these proteins are found almost exclusively in eukaryotes and are involved generally in processes that are fundamental and unique to these organisms [103]. This CCT-binding subfamily of WD40 proteins probably also appeared at the beginning of the eukaryote evolution, similar to actins, tubulins, and CCT. Since WD40 repeats are not found in Archaea, it is likely that this fold entered the eukaryotic lineage by lateral transfer from a eubacterium, as is proposed for the actin/hexokinase/Hsp70 fold [28, 80]. It is likely that CCT, by virtue of its assistance in the folding of all these important proteins, sits at the heart of many processes that are at the core of eukaryotic cell function (Figure 21.6). Subsequent to the establishment of the CCT system at the very beginning of eukaryotic evolution, interactions between CCT and other proteins may have occurred as a result of selection of new functions and also from opportunistic processes such as the assistance of CCT in the folding of viral proteins.

21.9
Concluding Remarks

All the functional and structural results obtained so far point to CCT being a unique chaperonin, different from all the others in its functional mechanism and in the cellular roles in which it is involved. Besides its involvement in the folding of actin and tubulin, and therefore in the formation of the cytoskeleton, CCT is implicated in the folding or quaternary association of a host of proteins that are essential in a multitude of cellular processes such as cell cycle control, signal transduction, transcription, protection from degradation, etc. Many exciting issues remain to be uncovered, in particular, a more-detailed characterization of the functional cycle of CCT and its interaction with the new substrates that are now continuously being discovered.

21.10
Experimental Protocols

21.10.1
Purification

In the early phase of CCT discovery and characterization, various laboratories established protocols for purifying CCT from mammalian testis extracts [1, 2, 4]. CCT holo-complex is present at around 3×10^5 copies per cell in mouse testis germ cell preparations [1]. CCT has also been purified from rabbit reticulocyte lysate [2, 14, 136].

Bovine testis CCT is purified over Q-Sepharose HR (Pharmacia) followed by sucrose gradient size fractionation and affinity chromatography on an agarose-bound ATP column; a combination of the Gao et al. [1] and Frydman et al. [4] procedures led to the protocol of Melki and Cowan [137], and a detailed description is to be found in their paper. This protocol allows processing of up to 0.5 kg tissue. These bovine CCT preparations are active in actin and tubulin folding assays [1, 137]. Recently, bovine CCT was used to study steady-state ATP binding and hydrolysis [29], and this modified protocol involves sucrose gradient size fractionation, MonoQ HR 5/5 chromatography, HiTrap heparin chromatography, and final desalting using PD-10 Sephadex. This protocol was modified yet again, by replacing the MonoQ step with C-8 ATP-agarose to produce large amounts of CCT, for time-resolved fluorescence emission analysis of ATP-induced allosteric transitions, from 150-g aliquots of bovine testis and is described in useful detail in Kafri and Horovitz [55].

Mouse testis CCT has been used extensively in our laboratory for biochemical and structural studies. We always use sucrose gradient size fractionation as a first step, since this gives an immediate 900-fold enrichment [1] because of the large size of CCT (nearly 900 kDa). Presently, our general preferred protocol for purification of CCT from testis cells, tissue culture cell lines, or rabbit reticulocyte lysate consists of sucrose density gradient centrifugation, anion-exchange chromatography (elution 0.1–0.3 M NaCl), heparin chromatography (elution 0.5–0.7 M NaCl), and buffer exchange by ultrafiltration [138]. The final material is 95% pure and particulate and is non-aggregated under the electron microscope [46, 50].

Antibody reagents recognizing a CCT subunit such as TCP-1 [1] (MAb 91A) or a pan-CCT reagent [30] (UM1) provide the most straightforward method to follow CCT during purification and have been used by many laboratories. However, once CCT is reasonably enriched it can be easily followed by its signature pattern of ~60-kDa polypeptides on SDS gels [1, 136].

We refer readers to protocols for purification of rabbit reticulocyte lysate CCT from Norcum [136] and Cowan [139]. The Norcum protocol was used by Szpikowska et al. [51] to examine MgATP-induced conformational changes in rabbit CCT.

21.10.2
ATP Hydrolysis Measurements

In early studies various steady-state ATP hydrolysis assays, measuring phosphate release from [γ-^{32}P]ATP, were established for CCT (reviewed in Ref. [17]). These gave values of between 14.3 and 19.3 nmol mg^{-1} min^{-1} at 37 °C (14–19 molecules of ATP hydrolyzed per CCT complex per minute). Recent studies from Horovitz and colleagues [29] measured initial rates of ATP hydrolysis by bovine testis CCT as a function of ATP concentration using a [γ-^{32}P]ATP and a ^{32}P-phosphate release quantitation assay first established for GroEL [140]. The value of k_{cat} for ATP hydrolysis by one ring and by both rings of CCT was 0.0119 sec^{-1} and 0.0183 sec^{-1}, respectively [29]. Kafri and Horovitz [55] have recently described stopped-flow analysis of ATP binding and hydrolysis by CCT using excitation at 280 nm and recording tryptophan emission beyond 320 nm using a cutoff filter.

21.10.3
CCT Substrate-binding and Folding Assays

CCT-mediated actin and tubulin folding assays have been established, in the main, by Cowan's laboratory [2, 137] and have been summarized by him in experimental detail [139]. They are analogous to the classic folding assays performed with GroEL and involve dilution of chemically denatured substrate proteins into buffer containing chaperonin. The appearance of native, folded proteins is monitored through native gel electrophoresis.

Several laboratories, including our own, have made use of rabbit reticulocyte lysate–based in vitro transcription/translation systems to monitor actin and tubulin folding [68, 71].

21.10.4
Electron Microscopy and Image Processing

Electron microscopy is widely used as a method for retrieving structural information of biological samples up to near-atomic resolution, and it is ideally suited for the analysis of macromolecular complexes. To achieve maximal resolution, samples have to be imaged without any contrasting agent; the current method of attaining good structural preservation while allowing the operation under the conditions of electron microscopes is cryo-electron microscopy. In this method, samples are rapidly frozen at liquid nitrogen temperature with the use of cryogenic reagents and then transferred inside the microscope, where they can be imaged using limited electron dosage to prevent specimen destruction. Images obtained with this method preserve structural information but are extremely noisy. This, together with the fact that transmission electron images are two-dimensional projections of the objects under study (Figure 21.8A), implies that the data obtained by electron microscopy have to be treated in a mathematical way to retrieve the three-dimensional information.

CCT is a multimeric assembly that exhibits a certain degree of symmetry. Nevertheless, most of the procedures applied for extracting structural information from electron micrographs of CCT are essentially independent from those symmetry

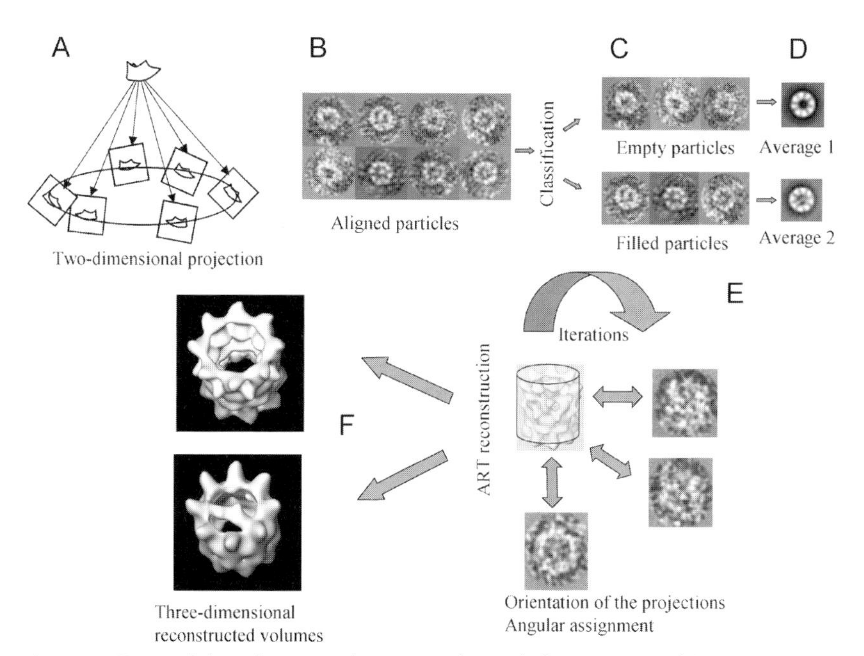

Fig. 21.8. Two- and three-dimensional image processing of single particles. (A) Generation of two-dimensional projections from a three-dimensional object following a conical geometry. (B) Gallery of images obtained after boxing and alignment of particles selected from electron micrographs. (C) Homogeneous subsets of views corresponding to different particle families obtained after classification of two mixed populations. (D) Average images obtained after averaging of the two different populations. (E) Three-dimensional reconstruction protocol including iterative angular assignment for the orientation of two-dimensional projections, volume reconstruction, and re-projection of the three-dimensional volume in projections using the determined orientations. (F) Three-dimensional volume representation including filtering at the calculated resolution.

constraints and hence are considered as "single-particle" processing procedures [141]. The first step in image processing is the digitization of the micrographs. Micrographs are converted in arrays of squared pixels with a size around 2–4 Å. The actual images corresponding to the chaperonin views are then selected in the computer and boxed (Figure 21.8B). Due to the fact that molecular aggregates are randomly distributed in solution, particles under study offer a complex set of projection views. The images selected are centered by subjecting them to translational and rotational alignment. This step is important and care has to be taken not to bias the alignment by using patterns; therefore, the use of free-pattern alignment algorithms is instrumental for this purpose [142, 143].

Once the images are aligned, a most critical analysis is the determination of variations among the different images. These differences can arise from biochemical differences, partial destruction of the particles under study, or, most relevant, different views of the object due to the out-of-plane orientation variations of the particles. Usually, there are a number of preferential views that produce different families of projections that must be sorted out [144]. The separation of the different populations can be carried out by a number of classification procedures (supervised, non-supervised, self-organizing, etc.). In our studies with CCT, we have used self-organizing maps based on neural networks [145] that provide a reduction in the number of vector images and a significant increment in the signal-to-noise ratio. This procedure facilitates the use of other subsequent classification methods, if required. Once the images are divided in homogeneous sets of views (Figure 21.8C), they can be averaged to render statistically significant data representing each two-dimensional projection of the objects under study (Figure 21.8D). Interpretation of the averaged views is fully dependent on the estimation of the resolution attained. This critical value depends upon the sample properties, the quality of the imaging method, and the number of averaged views, among other factors, and it is estimated using different criteria. We have used the spectral signal-to-noise ratio (SSNR), as proposed by Unser et al. [146]. Once the resolution is determined, the averages are filtered up to that value to remove those structural features inconsistent with the resolution level.

Reconstruction of the three-dimensional volume of the object is based on the use of the different two-dimensional projections obtained in the microscope. In principle, the random orientation of the particles in the thin ice layer used for cryo-electron microscopy is sufficient for rendering enough different views of the object so as to reconstruct the original volume. In other cases, different strategies for collecting data must be devised. One of the most popular is the random conical tilting method based on the collection of pairs of tilted and untilted micrographs [147]. Most of the reconstruction procedures are based on the projection theorem formulated by De Rosier and Klug [148]. The first problem in reconstruction is the definition of the orientation of the projections with respect to the three-dimensional original object (Figure 21.8E). We have used the angular refinement algorithms provided by the suite of programs called SPIDER [149], which determine the characteristic orientation parameters for each projection. Once these parameters are assigned, there are different approaches to reconstruct the three-dimensional volume [141, 150]. We have used an algebraic reconstruction approximation [151] that renders a three-dimensional map that can be used as a reference to re-project two-dimensional views using the orientation data obtained during the first angular refinement. These projections can then be compared to the experimental ones, and, carrying out several iterations, a refinement of the structure can be approximated. Once the iterations converge to a stable solution, this volume is taken as final representation of the object (Figure 21.8F). In cases where the particles exhibit a certain degree of symmetry, it can be used in either the initial reference model production or the final stages of the processing (representation of the final volume).

The interpretation of the reconstructed volume is very dependent on the threshold chosen for representation, which must account for the expected mass of the object, as well as on the resolution attained. As in the case of the two-dimensional averaged images, the determination of the resolution is still an unsolved issue. Different methods have been proposed [152] and modified along the years. We actually use the Fourier shell correlation coefficient calculated from two independent reconstructions using the Bsoft program package [153]. Final volumes for interpretation must be filtered up to that resolution to remove those features corresponding to frequencies above the resolution threshold.

References

1 Lewis, V. A., Hynes, G. M., Zheng, D., Saibil, H. & Willison, K. R. (1992). T-complex polypeptide-1 is a subunit of a heteromeric particle in the eukaryotic cytosol. *Nature* **358**, 249–252.

2 Gao, Y., Thomas, J. O., Chow, R. L., Lee, G. H. & Cowan, N. J. (1992). A cytoplasmic chaperonin that catalyzes beta-actin folding. *Cell* **69**, 1043–1050.

3 Yaffe, M. B., Farr, G. W., Miklos, D., Horwich, A. L., Sternlicht, M. L. & Sternlicht, J. (1992). TCP1 complex is a molecular chaperone in tubulin biogenesis. *Nature* **358**, 245–248.

4 Frydman, J., Nimmesgern, E., Erdjument-Bromage, H., Wall, J. S., Tempst, P. & Hartl, F. U. (1992). Function in protein folding of TRiC, a cytosolic ring complex containing TCP-1 and structurally related subunits. *EMBO J.* **11**, 4767–4778.

5 Silver, L. M., Artz, K. & Bennett, D. (1979). A major testicular cell protein specified by a mouse T/t complex gene. *Cell* **17**, 275–284.

6 Willison, K. R., Dudley, K. & Potter, J. (1986). Molecular cloning and sequence analysis of a haploid expressed gene encoding *t*-complex polypeptide 1. *Cell* **44**, 727–738.

7 Willison, K. R., Kelly, A., Dudley, K., Goodfellow, P., Spurr, N., Groves, V., Gorman, P., Sheer, D. & Trowsdale, J. (1987). The human homologue of the mouse *t*-complex-gene *TCP1* is located on chromosome 6 but not near the HLA region. *EMBO J.* **6**, 1867–1874.

8 Willison, K. R. & Lyon, M. F. (2000). A UK-centric history of studies on the mouse *t*-complex. *Int. J. Dev. Biol.* **44**, 57–63.

9 Ursic, D. & Ganetzky, B. (1988). A

tional regulation of the cytosolic chaperonin theta subunit gene, Cctq, by Ets domain transcription factors Elk Sap-1a, and Net in the absence of serum response factor. *J. Biol. Chem.* **278**, 30642–30651.

46 Llorca, O., Smyth, M. G., Carrascosa, J. L., Willison, K. R., Radermacher, M., Steinbacher, S., Valpuesta, J. M. (1999). 3D reconstruction of the ATP-bound form of CCT reveals the asymmetric folding conformation of a type II chaperonin. *Nat. Struct. Biol.* **6**, 639–642.

47 Nitsch, M., Walz, J., Typke, D., Klumpp, M., Essen, L. O. & Baumeister, W. (1998). Group II chaperonin in an open conformation examined by electron tomography. *Nat. Struct. Biol.* **5**, 855–857.

48 Schoehn, G., Quaite-Randall, E., Jimenez, J. L., Joachimiak, A. & Saibil, H. R. (2000). Three conformations of an archaeal chaperonin, TF55 from *Sulfolobus shibatae*. *J. Mol. Biol.* **296**, 813–819.

49 Schoehn, G., Hayes, M., Cliff, M., Clarke, A. R. & Saibil, H. R. (2000). Domain rotations between open, closed and bullet-shaped forms of the Thermosome, an archael chaperonin. *J. Mol. Biol.* **301**, 323–332.

50 Llorca, O., Smyth, M. G., Marco, S., Carrascosa, J. L., Willison, K. R. & Valpuesta, J. M. (1998). ATP binding induces large conformational changes in the apical and equatorial domains of the eukaryotic cytoplasmic chaperonin (CCT). *J. Biol. Chem.* **273**, 10091–10094.

51 Szpikowska, B. K., Swiderek, K. M., Sherman, M. A. & Mas, M. T. (1998). MgATP binding to the nucleotide-binding domains of the eukaryotic cytoplasmic chaperonin induces conformational changes in the putative substrate-binding domains. *Prot. Sci.* **7**, 1524–1530.

52 Gutsche, I., Holzinger, J., Rauh, N., Baumeister, W. & May, R. P. (2001). ATP-induced structural change of the thermosome is temperature-dependent. *J. Struct. Biol.* **135**, 139–146.

53 Horovitz, A., Fridmann, Y., Kafri, G. & Yifrach, O. (2001). Allostery in chaperonins. *J. Struct. Biol.* **135**, 104–114.

54 Meyer, A. S., Gillespie, J. R., Walther, D., Millet, I. S., Doniach, S. & Frydman, J. (2003) Closing the folding chamber of the eukaryotic chaperonin requires the transition state of ATP hydrolysis. *Cell* **113**, 369–381.

55 Kafri, G. & Horovitz, A. (2003). Transient kinetic analysis of ATP-induced allosteric transitions in the eukaryotic chaperonin containing TCP-1. *J. Mol. Biol.* **326**, 981–988.

56 Iizuka, R., Yoshida, T., Shomura, Y., Miki, K., Maruyama, T., Odaka, M. & Yohda, M. (2003) ATP binding is critical for the conformational change from an open to closed state in archaeal group II chaperonin. *J. Biol. Chem.* **278**, 44959–44965.

57 Shomura, Y., Yoshida, T., Iizuka, R., Maruyama, M., Yohda, M. & Miki, K. (2004) Crystal structures of the group II chaperonin from *Thermococcus* strain KS-1: steric hindrance by the substituted amino acid, and inter-subunit rearrangement between two crystal forms. *J. Mol. Biol.* **335**, 1265–1278.

58 Lin, P. & Sherman, F. (1997). The unique hetero-oligomeric nature of the subunits in the catalytic cooperativity of the yeast Cct chaperonin complex. *Proc. Natl. Acad. Sci. USA* **94**, 10780–10785.

59 Gómez-Puertas, P., Martín-Benito, J., Carrascosa, J. L., Willison, K. R. & Valpuesta, J. M. (2004). The substrate recognition mechanisms in chaperonins. *J. Mol. Recog.* **17**, 85–94.

60 Fenton, W. A., Kashi, Y., Furtak, K. & Horwich, A. L. (1994). Residues in chaperonin GroEL required for polypeptide binding and release. *Nature.* **371**, 614–619.

61 Buckle, A. M., Zahn, R., Fersht, A. R. (1997). A structural model for GroEL-polypeptide recognition. *Proc. Natl. Acad. Sci. USA.* **94**, 3571–3575.

62 Chen, L. & Sigler, P. B. (1999). The crystal structure of a GroEL/peptide complex: plasticity as a basis for substrate recognition. *Cell.* **99**, 757–768.

63 FARR, G. W., FURTAK, K., ROWLAND, M. B., RANSON, N. A., SAIBIL, H. R., KIRCHHAUSEN, T. & HORWICH, A. L. (2000). Multivalent binding of nonnative substrate proteins by the chaperonin GroEL. *Cell.* **100**, 561–573.

64 TIAN, G., VAINBERG, I. E., TAP, W. D., LEWIS, S. A. & COWAN, N. J. (1995). Quasi-native chaperonin-bound intermediates in facilitated protein folding. *J. Biol. Chem.* **270**, 23910–23913.

65 LLORCA, O., MARTIN-BENITO, J., GOMEZ-PUERTAS, P., RITCO-VONSOVICI, M., WILLISON, K. R., CARRASCOSA, J. L. & VALPUESTA, J. M. (2001). Analysis of the interaction between the eukaryotic chaperonin CCT and its substrates actin and tubulin. *J Struct Biol.* **135**, 205–218.

66 ROMMELAERE, H., DE NEVE, M., MELKI, R., VANDEKERCKHOVE, J. & AMPE, C. (1999). The cytosolic class II chaperonin CCT recognizes delineated hydrophobic sequences in its target proteins, *Biochemistry* **38**, 3246–3257.

67 HYNES, G. M. & WILLISON, K. R. (2000). Individual subunits of the chaperonin containing TCP-1 (CCT) mediate interactions with binding sites located on subdomains of β-actin, *J. Biol. Chem.* **275**, 18985–18994.

68 MCCORMACK, E., ROHMAN, M. & WILLISON, K. R. (2001). Mutational screen identifies critical amino acid residues of β-actin mediating interaction between folding intermediates and cytosolic chaperonin TCP-1. *J. Struct. Biol.* **135**, 185–197.

69 DOBRZYNSKI, J. K., STERNLICHT, M. L., FARR, G. W. & STERNLICHT, H. (1996). Newly-synthesized β-tubulin demonstrates domain-specific interactions with the cytosolic chaperonin. *Biochemistry.* **35**, 15870–15882.

70 DOBRZYNSKI, J. K., STERNLICHT, M. L., PENG, I., FARR, G. W. & STERNLICHT, H. (2000). Evidence that β-tubulin induces a conformation change in the cytosolic chaperonin which stabilizes binding: implications for the mechanism of action. *Biochemistry.* **39**, 3988–4003.

71 RITCO-VONSOVICI, M. & WILLISON, K. R. (2000). Defining the eukaryotic chaperonin-binding sites in human tubulins. *J. Mol. Biol.* **304**, 81–98.

72 ELLIS, R. J. (1994). Opening and closing the Anfinsen cage. *Curr. Biol.* **4**, 633–635.

73 WEISSMANN, J. S., RYE, H. S., FENTON, W. A., BEECHEM, J. M. & HORWICH, A. L. (1996). Characterization of the active intermediate of a GroEL-GroES mediated protein folding reaction, *Cell* **84**, 481–490.

74 LLORCA, O., MARTIN-BENITO, J., GRANTHAM, J., RITCO-VONSOVICI, M., WILLISON, K. R., CARRASCOSA, J. L. & VALPUESTA, J. M. (2001). The 'sequential allosteric ring' mechanism in the eukaryotic chaperonin-assisted folding of actin and tubulin. *EMBO J.* **20**, 4065–4075.

75 MCCALLUM, C. D., DO, H., JOHNSON, A. E. & FRYDMAN, J. (2000). The interaction of the chaperonin tailless complex polypeptide 1 (TCP1) ring complex (TRiC) with ribosome-bound nascent chains examined using photo-cross-linking. *J. Cell Biol.* **149**, 591–601.

76 FARR, G. W., SCHARL, E. C., SCHUMACHER, R. J., SONDEK, S., & HORWICH, A. L. (1997). Chaperonin-mediated folding in the eukaryotic cytosol proceeds through rounds of release of native and nonnative forms. *Cell.* **89**, 927–937.

77 MCCORMACK, E., LLORCA, O., CARRASCOSA, J. L., VALPUESTA, J. M. & WILLISON, K. R. (2001). Point mutations in a hinge linking the small and large domains of β-actin result in trapped folding intermediates bound to cytosolic chaperonin containing TCP-1. *J. Struct. Biol.* **135**, 198–204.

78 SIEGERS, K., WALDMANN, T., LEROUX, M. R., GREIN, K., SHEVCHENKO, A., SCHIEBEL, E. & HARTL, F. U. (1999). Compartmentation of protein folding in vivo: sequestration of non-native polypeptide by the chaperonin-GimC system. *EMBO J.* **18**, 75–84.

79 TIAN, G., VAINBERG, I. E., TAP, W. D.,

Lewis, S. A., & Cowan, N. J. (1995). Specificity in chaperonin-mediated protein folding. *Nature* **375**, 250–253.

80 Willison, K. R. (1999). Composition and function of the eukaryotic cytosolic chaperonin-containing TCP1 in *Molecular Chaperones and Folding Catalysts* (Ed. Bernd Bukau) (Harwood Academic Publishers, Amsterdam), pp. 555–571.

81 Schüler, H., Lindberg, U., Schutt, C. E. & Karlsson, R. (2000). Thermal unfolding of G-actin monitored with the Dnase I-inhibition assay. *Eur. J. Biochem.*, **267**, 476–486.

82 Kabsch, W., Mannherz, H. G., Suck, D., Pai, E. F. & Holmes, K. C. (1990). Atomic structure of the actin: DNase I complex. *Nature*, **347**, 37–44.

83 Nogales, E., Wolf, S. G. & Downing, K. H. (1998). Structure of the $\alpha\beta$tubulin dimer by electron crystallography. *Nature*, **391**, 199–203.

84 Valpuesta, J. M., Martín-Benito, J., Gómez-Puertas, P., Carrascosa, J. L. & Willison, K. R. (2002). Structure and function of a protein folding machine: the eukaryotic cytosolic chaperonin CCT. *FEBS letters* **529**, 11–16.

85 Thulasariman, V., Yang, C. F. & Frydman, J. (1999). In vivo newly translated polypeptides are sequestered in a protective folding environment. *EMBO J.* **18**, 85–95.

86 Chen, X., Sullivan, D. S. & Huffaker, T. C. (1994). Two yeast genes with similarities to TCP-1 are required for microtubule and actin function in vivo. *Proc. Natl. Acad. Sci. USA* **91**, 9111–9115.

87 Vinh, D. B. & Drubin, D. G. (1994). A yeast TCP-1-like protein is required for actin function in vivo. *Proc. Natl. Acad. Sci. USA.* **91**, 9116–9120.

88 Miklos, D., Caplan, S., Martens, D., Hynes, G., Pitluk, Z., Barrell, B., Horwich, A. L. & Willison, K. R. (1994). Primary structure and function of a second essential member of heterooligomeric TCP1 chaperonin complex of yeast. *Proc. Natl. Acad. Sci. USA* **91**, 2743–2747.

89 Melki, R., Vainberg, I. E., Chow, R. L. & Cowan, N. J. (1993). Chaperonin-mediated folding of vertebrate actin-related protein and gamma-tubulin. *J. Cell Biol.* **122**, 1301–1310.

90 Srikakulam, R. & Winkelmann, D. A. (1999). Myosin II folding is mediated by a molecular chaperone. *J. Biol. Chem.* **274**, 27265–27273.

91 Melki, R., Batelier, G., Soulie, S. & Williams, R. C. Jr. (1997). Cytoplasmic chaperonin containing TCP-1: Structural and functional characterization. *Biochemistry* **36**, 5817–5826.

92 Lingappa, J. R., Martin, R. L., Wong, M. L., Ganem, D., Welch, W. J. & Lingappa, V. R. (1994). A eukaryotic cytosolic chaperonin is associated with a high molecular weight intermediate in the assembly of hepatitis B virus capsid, a multimeric particle. *J. Cell. Biol.* **125**, 99–111.

93 Hong, S., Choi, G., Park, S., Chung, A. S., Hunter, E. & Rhee, S. S. (2001). Type D retrovirus Gag polyprotein interacts with the cytosolic TriC. *J. Virol.* **75**, 2526–2534.

94 Kashuba, E., Pokrovskaja, K., Klein, G. & Szekely, L. (1999). Epstein-Barr virus-encoded nuclear protein EBNA-3 interacts with the ε-subunit of the T-complex protein 1 chaperonin complex. *J. Human. Virol.* **2**, 33–37.

95 Frydman, J., Nimmesgem, E., Ohtsuka, K. & Hartl, F. U. (1994). Folding of nascent polypeptide chains in a high molecular mass assembly with molecular chaperones. *Nature.* **370**, 111–117.

96 Won, K. A., Schumaker, R. J., Farr, G. W., Horwich, A. L. & Reed, S. I. (1998). Maturation of human cyclin E requires the function of eukaryotic chaperonin CCT. *Mol. Cell. Biol.* **18**, 7584–7589.

97 Pijnappel, W. W. M. P., Schaft, D., Rogue, V. A., Shevchenko, A., Tekotte, H., Wilm, M., Rigaut, G., Séraphin, B., Aasland, R. & Stewart, A. F. (2001). The *S. cerevisiae* SET3 complex includes two histone deacetylases, Hos2 and Hst1, and is a meiotic-specific repressor of the sporulation gene program. *Genes & Dev.* **15**, 2991–3004.

98 Guenther, M. G., Yu, J., Kao, G. D., Yen, T. J. & Lazar, M. A. (2002). Assembly of the SMRT-histone deacetylase 3 repression complex requires the TCP-1 ring complex. *Genes & Dev.* **16**, 3130–3135.

99 Kaelin, W. G. (1999). Cancer: many vessels, faulty genes. *Nature* **399**, 203–204.

100 Stebbins, C. E., Kaelin, W. G. & Pavletich, N. P. (1999). Structure of the VHL-elongingC-elonginB complex: implications for VHL tumor suppressor function. *Science* **284**, 455–461.

101 Feldman, D. E., Thulasariman, V., Ferreyra, R. G. & Frydman, J. (1999). Formation of the VHL-elongin BC tumor suppressor complex is mediated by the chaperonin TriC. *Mol. Cell.* **4**, 1051–1061.

102 Cowan, N. J. & Lewis, S. A. (2002) Type II chaperonins, prefoldin and the tubulin-specific chaperones. *Advances in Prot. Chem.* **59**, 73–104.

103 Smith, T. F., Gaitaztes, C., Saxena, K. & Neer, E. J. (1999). The WD40 repeat: a common architecture for diverse functions. *TIBS* **24**, 181–185.

104 Ho, Y. et al. (2002). Systematic identification of protein complexes in *Saccharomyces cerevisiae* by mass spectrometry. *Nature* **415**, 180–183.

105 Camasses, A., Bodganova, A., Shevchenko, A. & Zachariae, W. (2003). The CCT chaperonin promotes activation of the anaphase-promoting complex through the generation of functional Cdc20. *Mol. Cell.* **12**, 87–100.

106 Passmore, L. A., McCormack, E. A., Au, S. W. N., Paul, A., Willison, K. R., Harper, J. W. & Barford, D. (2003). Doc1 mediates the activity of the anaphase-promoting complex by contributing to substrate recognition. *EMBO J.* **22**, 786–796.

107 Siegers, K., Bölter, B., Schwarz, J. P., Böttcher, U. M. K., Guha, S. & Hartl, F. U. (2003). TRiC/CCT cooperates with different upstream chaperones in the folding of distinct protein classes. *EMBO J.* **22**, 5230–5240.

108 Gaudet, R., Bohm, A. & Sigler, P. B. (1996). Crystal Structure at 2.4 Å Resolution of the Complex of Transducin $\beta\gamma$ and Its Regulator, Phosducin. *Cell* **87**, 577–588.

109 McLaughlin, J. N., Thulin, C. D., Hart, S. D., Resing, K. A., Ahn, N. G. & Willardson, B. M. (2002). Regulatory interaction of phosducin-like protein with the cytosolic chaperonin complex. *Proc. Natl. Acad. Sci. USA* **99**, 7962–7967.

110 Kunisawa, J. & Shastri, N. (2003). The group II chaperonin TriC protects proteolytic intermediates from degradation in the MHC class I antigen processing pathway. *Mol. Cell* **12**, 565–576.

111 Mogk, A., Bukau, B. & Deuerling, E. (2001). Cellular functions of cytosolic *E. coli* chaperones. In *Molecular Chaperones in the cell* (Ed. Peter Lund) (Oxford University Press, Oxford), pp. 3–34.

112 Geissler, S., Siegers, K. & Schiebel, E. (1998). A novel protein complex promoting formation of functional alpha- and gamma-tubulin. *EMBO J.* **17**, 952–966.

113 Vainberg, I. E., Lewis, S. A., Rommelaere, H., Ampe, C., Vandekerckhove, J., Klein, H. L. & Cowan, N. J. (1998). Prefoldin, a chaperone that delivers unfolded proteins to cytosolic chaperonin. *Cell* **93**, 863–873.

114 Martín-Benito, J., Boskovic, J., Gómez-Puertas, P., Carrascosa, J. L., Simons, C., Lewis, S. A., Bartolini, F., Cowan, N. C. & Valpuesta, J. M. (2002). Structure of eukaryotic prefoldin and of its complexes with unfolded actin and the cytosolic chaperonin CCT. *EMBO J.* **21**, 6377–6386.

115 Gebauer, M., Melki, R. & Gehring, U. (1998). The chaperone cofactor Hop/p60 interacts with the cytosolic chaperonin-containing TCP-1 and affects its nucleotide exchange and protein folding activities. *J. Biol. Chem.* **273**, 29475–29480.

116 Melville, M. W., McClellan, A. J., Meyer, A. S., Darveau, A. & Frydman, J. (2003). The Hsp70 and TriC/CCT

chaperone systems cooperate in vivo to assemble the von Hippel-Lindau tumor suppressor complex. *Mol. Cell. Biol.* **23**, 3141–3151.

117 Song, J. & Morimoto, R. I. (2001) Hsp70 chaperones networks: the role of regulatory co-chaperones in coordinating stress responses with cell growth and death. In *Molecular Chaperones in the cell* (Ed. Peter Lund) (Oxford University Press, Oxford), pp. 142–163.

118 Chen, S., Prapapanich, V., Rimerman, R. A., Honoré, B. & Smith, D. F. (1996). Interactions of p60, a mediator of progesterone receptor assembly, with heat-shock proteins hsp90 and hsp70. *Mol. Endocrinol.* **10**, 682–693.

119 Johnson, B. D., Schumacher, R. J., Ross, E. D. & Toft, D. O. (1998). Hop modulates Hsp70/Hsp90 interactions in protein folding. *J. Biol. Chem.* **273**, 3679–3686.

120 Dittmar, K. D. & Pratt, W. B. (1997). Folding of the glucocorticoid receptor by the reconstituted Hsp90-based chaperone machinery. The initial Hsp90.p60. hsp70-dependent step is sufficient for creating the steroid binding conformation. *J. Biol. Chem.* **272**, 13047–13054.

121 Tian, G., Huang, Y., Rommelaere, H., Vandekerckhove, J., Ampe, C. & Cowan, N. J. (1996). Pathway leading to correctly-folded β-tubulin. *Cell* **86**, 287–296.

122 Tian, G., Lewis, S. A., Feierbach, B., Stearns, T., Rommelaere, H., Vandekerckhove, J., Ampe, C. & Cowan, N. J. (1997). Tubulin subunits exists in an activated conformational state generated and maintained by protein cofactors. *J. Cell. Biol.* **138**, 821–832.

123 López-Fanarraga, M., Avila, J., Guasch, A., Coll, M. & Zabala, J. C. (2001) Postchaperonin tubulin folding cofactors and their role in microtubule dynamics. *J. Struct. Biol.* **135**, 219–229.

124 Steinbacher, S. (1999). Crystal structure of the postchaperonin β-tubulin binding cofactor Rbl2p. *Nature Struct. Biol.* **6**, 1029–1032.

125 Guasch, A., Aloria, K., Pérez, R., Avila, J., Zabala, J. C. & Coll, M. (2002). Three dimensional structure of human tubulin chaperone cofactor A. *J. Mol. Biol.* **318**, 1139–1149.

126 Gao, Y., Melki, R., Walden, P. D., Lewis, S. A., Ampe, C., Rommelaere., Vandekerckhove, J. & Cowan, N. J. (1994). A novel co-chaperonin that modulates the ATPase activity of cytosplasmic chaperonin. *J. Cell. Biol.* **125**, 989–996.

127 Melki, R., Rommelaere, H., Leguy, R., Vandekerckhove, J. & Ampe, C. (1996). Cofactor A is a molecular chaperone required for β-tubulin folding: functional and structural characterization. *Biochem.* **32**, 10422–10435.

128 Archibald, J. M., Logsdon, J. M. Jr. & Ford Doolittle, W. (2000). Origin and evolution of eukarotic chaperonins: phylogenetic evidence for ancient duplications in CCT genes. *Mol. Biol. Evol.* **17**, 1456–1466.

129 Archibald, J. M. & Roger, A. J. (2002). Gene duplication and gene conversion shape the evolution of archaeal chaperonins. *J. Mol. Biol.* **316**, 1041–1050.

130 Archibald, J. M., Logsdon, J. M. Jr. & Ford Doolittle, W. (1999). Recurrent paralogy in the evolution of archaeal chaperonin. *Curr. Biol.* **9**, 1053–1056.

131 Archibald, J. M., Blouin, C. & Ford Doolittle, W. (2001). Gene duplication and the evolution of group II chaperonins: implications for structure and function. *J. Struct. Biol.* **135**, 157–169.

132 Fares, M. A. & Wolfe, K. H. (2003) Positive selection and subfunctionalization of duplicated CCT chaperonin subunits. *Mol. Biol. Evol.* **20**, 1588–1597.

133 Willison, K. R. and Horwich, A. L. (1996) Structure and function of chaperonins. In *The Chaperonins* (Ed. R. J. Ellis) 107–135 (Academic Press).

134 Leroux, M. R. & Hartl, F. U. (2000). Versatility of the cytosolic chaperonin TriC/CCT. *Curr. Biol.* **10**, R260–R264.

135 Rommelaere, H., De Neve, M.,

Neirynck, K., Peelaers, D., Waterschoot, D., Goethals, M., Fraeyman, N., Vandekerckhove, J. & Ampe, C. (2001). Prefoldin recognition motifs in the nonhomologous proteins of the actin and tubulin families. *J Biol Chem.* **276**, 41023–41028.

136 Norcum, M. T. (1996). Novel isolation method and structural stability of a eukaryotic chaperonin: the TCP-1 ring complex from rabbit reticulocyte. *Protein Sci*, **5**, 1366–1375.

137 Melki, R. & Cowan, N. J. (1994). Facilitated folding of actins and tubulins occurs via a nucleotide-dependent interaction between cytoplasmic chaperonin and distinctive folding intermediates. *Mol. Cell. Biol.* **14**, 2895–2904.

138 Rohman, M. (1999). Biochemical characterization of chaperonin containing TCP-1 (CCT) PhD Thesis, University of London, England.

139 Cowan, N. J. (1998) Mammalian cytosolic chaperonin. *Methods Enzymol.* **290**, 230–241.

140 Horovitz, A., Bochkareva, E. S., Kovalenko, O., and Girshovich, A. S. (1993). Mutation Ala2 → Ser destabilizes intersubunit interactions in the molecular chaperone GroEL. *J. Mol. Biol.* **231**, 58–64.

141 Frank, J. (1996). Three-dimensional Reconstruction. *Three-dimensional Electron Microscopy of Macromolecular Assemblies.* Academic Press, San Diego, pp. 182–246.

142 Penczek, P., Radermacher, M. & Frank, J. (1992). Three-dimensional reconstruction of single particles embedded in ice. *Ultramicroscopy* **40**, 33–53.

143 Marco, S., Chagoyen, M., de la Fraga, L. G., Carazo, J. M. & Carrascosa, J. L. (1996). A variant of the "random approximation" of the reference-free alignment algorithm. *Ultramicroscopy* **66**, 5–10.

144 van Heel, M. & Frank, J. (1981). Use of multivariate statistics in analysing the images of biological macromolecules. *Ultramicroscopy* **6**, 187–194.

145 Marabini, R. & Carazo, J. M. (1994). Pattern recognition and classification of images of biological macromolecules using artificial neural networks. *Biophys. J.* **66**, 1804–1814.

146 Unser, M., Trus, B. L. & Steven, A. C. (1987). A new resolution criterion based on spectral signal-to-noise ratios. *Ultramicroscopy* **23**, 39–51.

147 Radermacher, M. (1988). Three-dimensional reconstruction of single particles from random and non-random tilt series. *J. Electron Microsc. Tech.* **9**, 359–394.

148 De Rosier, D. & Klug, A. (1968). Reconstruction of three-dimensional structures from electron micrographs. *Nature.* **217**, 130–134.

149 Frank, J., Radermacher, M., Penczek, P., Zhu, J., Li, Y., Ladjadj, M. & Leith, A. (1996). SPIDER and WEB: processing and visualization of images in 3D electron microscopy and related fields. *J. Struct. Biol.* **116**, 190–199.

150 van Heel, M., Gowen, B., Matadeen, R., Orlova, E. V., Finn, R., Pape, T., Cohen, D., Stark, H., Schmidt, R., Schatz, M. & Patwardhan, A. (2000). Single-particle electron cryo-microscopy: towards atomic resolution. *Quart. Rev. Biophys.* **33**, 307–369.

151 Marabini, R., Herman, G. T. & Carazo, J. M. (1998). 3D reconstruction in electron microscopy using ART with smooth spherically symmetric volume elements (blobs). *Ultramicroscopy* **72**, 53–65.

152 Saxton, W. O. & Baumeister, W. (1982). The correlation averaging of a regularly arranged bacterial cell envelope protein. *J. Microsc.* **127**, 127–138.

153 Heymann, J. B. (2001). Bsoft: image and molecular processing in electron microscopy. *J. Struct. Biol.* **133**, 156–169.

154 Siegert, R., Leroux, M. R., Scheufler, C., Hartl, F. U. & Moarefi, I. (2000). Structure of the molecular chaperone prefoldin: unique interaction of multiple coiled coil tentacles with unfolded proteins. *Cell* **103**, 621–632.

22
Structure and Function of GimC/Prefoldin

Katja Siegers, Andreas Bracher, and F. Ulrich Hartl

22.1
Introduction

The hetero-oligomeric protein complex GimC/prefoldin, present in eukarya and archaea, is a critical co-factor of TRiC/CCT-assisted folding of actin and tubulins [1, 2]. The identification of homologues in mammals, plants, nematodes, and archaea and complementation studies in yeast revealed that GimC/prefoldin is a highly conserved component of the protein-folding machinery. While GimC/prefoldin in archaea is formed by only two types of distinct subunits, it is comprised of six distinct subunits in both yeast and mammalian cells. Unlike the GroES-like co-chaperones of group I chaperonins in bacteria, GimC/prefoldin plays a more active role in protein folding by interacting with unfolded proteins and stabilizing them against aggregation for subsequent folding by the eukaryotic chaperonin. GimC has been shown to assist in the transfer of nascent polypeptide chains to TRiC/CCT in vitro and in vivo and to prevent the premature release of nonnative chaperonin substrates into the cytosol [1, 3, 4]. GimC/prefoldin appears to have a special role in assisting TRiC in the folding of its major substrates, newly synthesized actin, and tubulins, and this function cannot be performed by other nascent chain-binding chaperones, such as Hsp70 [5]. A direct interaction of GimC/prefoldin with TRiC [1, 3, 6] may facilitate the delivery of actin and tubulins in a defined orientation relative to the subunit topology of the chaperonin ring [7]. It has been shown that eukaryotic and archaeal GimC/prefoldin has the general properties of a molecular chaperone, and, similar to the Hsp70 system, archaeal GimC/prefoldin complexes have been reported to prevent or retard the aggregation of a number of nonnative proteins and to stabilize them for subsequent folding [8, 9]. A combined deletion of genes encoding the nascent chain-binding chaperones GimC/prefoldin and Ssb-type Hsp70 in yeast results in a pronounced synthetic growth defect [5], resembling the drastic effect described for a combined trigger factor and DnaK deletion in the *E. coli* system [10, 11]. This points to the existence of a certain degree of functional redundancy among nascent chain-binding chaperones, in addition to the specialized function of eukaryotic GimC/prefoldin in TRiC/CCT-assisted folding of actin and tubulins. The simpler and more abundant archaeal version of

Protein Folding Handbook. Part II. Edited by J. Buchner and T. Kiefhaber

ISBN: 3-527-30784-2

GimC/prefoldin may have a more general function in stabilizing nascent polypeptides since archaea do not contain an extensive actin and tubulin cytoskeleton, and some archaea lack Hsp70 altogether [12].

22.2 Evolutionary Distribution of GimC/Prefoldin

Eukaryotic GimC/prefoldin complexes like bovine prefoldin and yeast GimC consist of six distinct protein subunits termed Pfd1–Pfd6 and Gim1–Gim6, respectively [1–3]. For each component, a closely related homologue has been found in the nematode *Caenorhabditis elegans*, suggesting that they are likely to be conserved in eukaryotes. Gim1 (Pfd6) and Gim5 (Pfd5) homologues have been identified in nearly all known archaeal genomes [1, 2].

A comprehensive phylogenetic analysis of Gim/prefoldin proteins from eukaryotes and archaea revealed that all sequences of Gim/prefoldin form two separate classes of proteins, represented by Gim2/5 (Pfd3/5) (denoted as the α class) and Gim1/3/4/6 (Pfd6/4/2/1) (β class). Unlike eukaryotes, however, archaeal genomes contain only one gene from each class. An exemplary sequence alignment of *Methanobacterium thermoautotrophicum* GimC/prefoldin subunits with one representative yeast homologue each is shown in Figure 22.1.

22.3 Structure of the Archaeal GimC/Prefoldin

In the crystal structure of the GimC/prefoldin from the archaeon *Methanobacterium thermoautotrophicum*, two α and four β subunits form a hexameric complex resembling the shape of a jellyfish (Figure 22.2) [13]. Its main body consists of two eight-stranded β barrels from which six long tentacle-like two-helical bundles protrude. As was correctly predicted [8], all subunits assume self-symmetric secondary structures with α-helical regions containing heptad repeats typical for coiled-coil structures at the chain termini and either two or four β-strands in the center. This arrangement folds into a hairpin-like structure with a 60–70 Å long two-helix bundle at its ends and either one or two β-hairpins at the tip, respectively. The core of the complex is composed of two α subunits, which contribute one β-hairpin to each β-barrel. The β-hairpins of two α chains complete each β-barrel.

The α-helical coiled coils forming the six long tentacles border a large central cavity (Figure 22.2). In the crystal structure, access to the cavity is not blocked by any structured elements. The individual coiled-coil rods expose mostly charged and polar side chains to the solvent, and there are virtually no interactions between the coiled coils from different subunits. In contrast to this, the distal regions of the coiled coils are partially unwound and therefore expose hydrophobic patches (Figure 22.2). These regions have been shown to be required for binding of nonnative proteins.

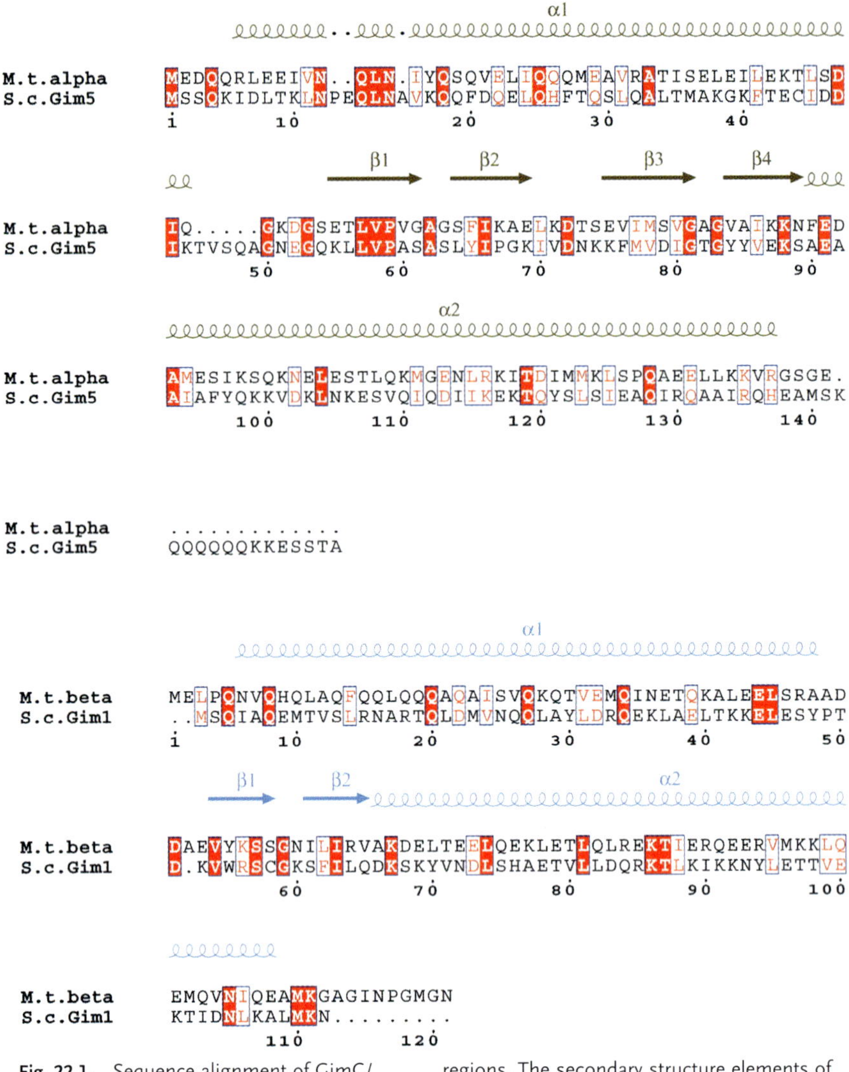

Fig. 22.1. Sequence alignment of GimC/prefoldin subunits from *M. thermoautotrophicum* and *S. cerevisiae*. *M. thermoautotrophicum* α and β subunits were aligned with yeast Gim5 and Gim1, respectively. Sequence conservation is indicated by red background for identical residues and blue boxes around homologous regions. The secondary structure elements of *M. thermoautotrophicum* GimC/prefoldin are shown on top of the alignment. Coils and arrows stand for α-helices and β-strands, respectively. The figure was created with ESPript [23].

The structure does not provide any evidence for the presence of a nucleotide-binding site, which is in accordance with the apparent lack of an ATP-regulated protein function for both archaeal and eukaryotic GimC/prefoldin [1, 8].

Electron microscopic studies indicated recently that eukaryotic GimC/prefoldin possesses an architecture similar to that of archaeal GimC, with six arms pro-

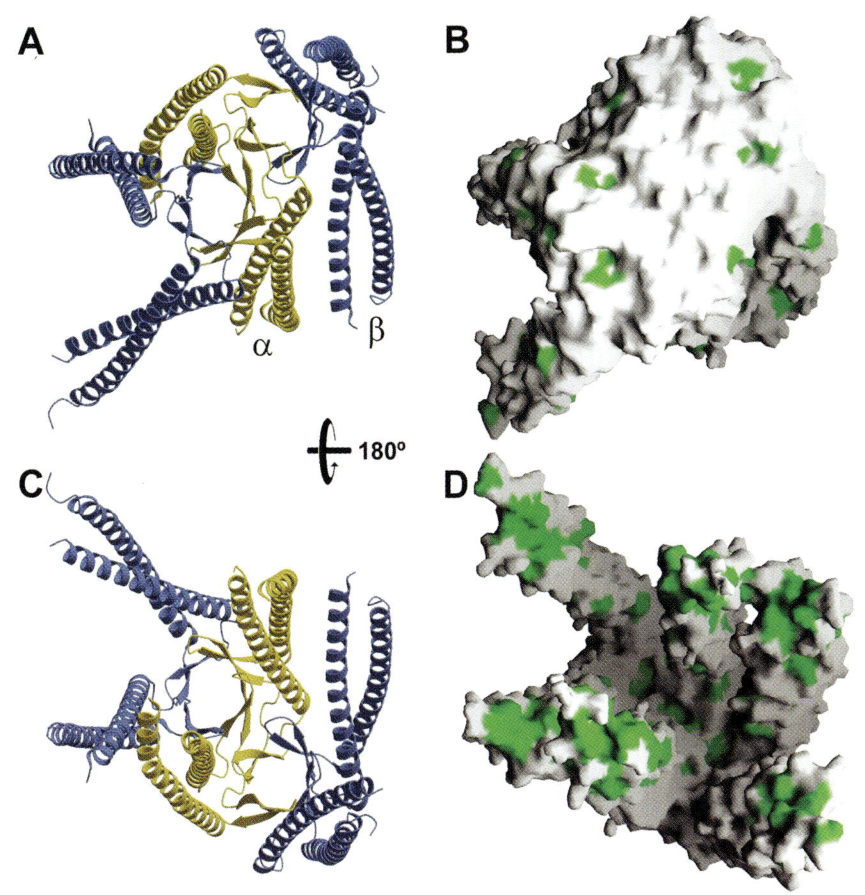

Fig. 22.2. Crystal structure of GimC/prefoldin from *M. thermoautotrophicum* [13]. In panels A and C, α and β subunits are depicted as yellow and blue ribbons, respectively. A molecular surface of GimC/prefoldin in the same orientation is shown on the right (panels B and D). Hydrophobic residues are highlighted in green. The bottom row shows the same representations of the GimC/prefoldin complex after 180° rotation. Ribbon diagrams were made using the programs Molscript and Raster-3D [24, 25]. The surface representations were generated with Grasp [26].

truding from the base of the oligomer [6]. Three-dimensional reconstruction of the same oligomer complexed with an unfolded actin molecule suggests that the prefoldin-bound target protein has a defined conformation that seems to interact with the tips of the chaperone's tentacles.

22.4
Complexity of the Eukaryotic/Archaeal GimC/Prefoldin

Biochemical and yeast complementation studies revealed that archaeal and eukaryotic GimC/prefoldin are likely to form a very similar hexameric arrangement [8].

The general architecture of archaeal and eukaryotic GimC/prefoldin is likely to be conserved, based on the finding that the archaeal α and β subunits are the founding members of the two classes of Gim subunits: the α class comprises subunits Gim2 (Pfd3) and Gim5 (Pfd5), and the β class comprises subunits Gim1 (Pfd6), Gim3 (Pfd4), Gim4 (Pfd2), and Gim6 (Pfd1) of GimC/prefoldin. Accordingly, the eukaryotic complex would consist of a heterodimer of Gim2 (Pfd3) and Gim5 (Pfd5), forming a core with which the Gim1 (Pfd6), Gim3 (Pfd4), Gim4 (Pfd2), and Gim6 (Pfd1) subunits associate.

That the archaeal and eukaryotic Gim/prefoldin complexes are structurally related is supported by the finding that both Gimα and Gimβ from *Methanobacterium thermoautotrophicum* can be co-immunoprecipitated with yeast Gim2 when expressed in strains lacking *GIM5* and *GIM1*, respectively, and that the archaeal Gimβ is able to partially complement defects in two yeast β class subunits, namely, Gim1 and Gim4 [8].

As reported [2], yeast Δ*gim1* and yeast Δ*gim5* strains are sensitive to the microtubule-depolymerizing drug benomyl. This phenotype in the Δ*gim1* yeast can be fully complemented by transforming the strain with plasmids expressing yeast or mouse *GIM1 (PFD6)*. Expression of archaeal *GIM*β (the *GIM1* homologue) in this strain was also shown to partially rescue the benomyl sensitivity of the yeast mutant (Figure 22.3) [8]. This effect is specific, as no visible changes in growth behavior can be observed upon expression of archaeal *GIM*α (the *GIM5* homologue) in the Δ*gim1* strain, and co-expression of both archaeal genes in the same strain is no more effective than expression of *GIM*β alone. Remarkably, archaeal *GIM*β was also found to be able to partially complement a yeast strain deleted for *GIM4*, encoding another member of the β class of GimC subunits, despite the low overall sequence homology between the two proteins.

Co-immunoprecipitation experiments with antibodies against yeast Gim2, an α

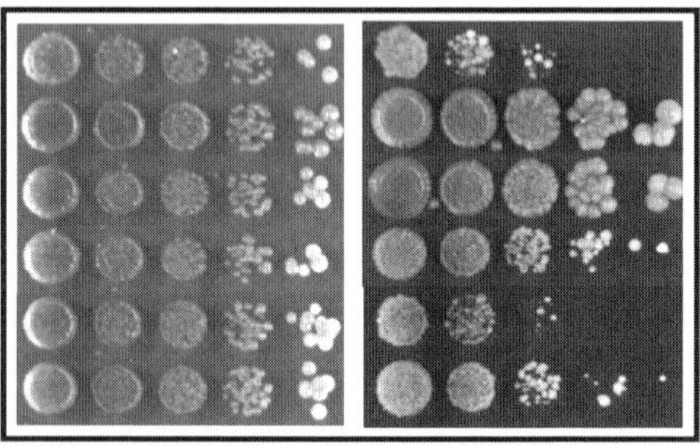

Fig. 22.3. Complementation analyses of *S. cerevisiae* Δ*gim1* [8]. Δ*gim1* mutants are highly sensitive to a 2.5-µg mL^{-1} concentration of the microtubule depolymerizing drug benomyl. This phenotype is rescued entirely by expression of yeast (Sc) and murine (Mm) *GIM1* and is partially complemented by expression of archaeal *MtGIM*β.

class subunit, revealed that the positive effect of archaeal *GIMβ* expression in the yeast Δ*gim1* strain can be directly attributed to the incorporation of Gimβ into the Gim1-deficient yeast GimC complex. In contrast, expression of Gimα was found to have no detectable rescuing effect in the Δ*gim5* strain, even though the archaeal protein can be co-immunoprecipitated with yeast Gim2 in Gim5-deficient yeast cells.

Together, these data confirmed that the archaeal Gimα and Gimβ are structural homologues of the yeast Gim5 and Gim1/Gim4 proteins, respectively, and that the function of archaeal Gimβ is similar to that of the yeast subunits Gim1 and Gim4. The reason for the inability of archaeal Gimα to partially complement the Δ*gim5* strain may be due to the fact that although Gimα can associate with Gim2 (of the α class), other yeast subunits of the β class may not assemble properly onto the archaeal Gimα protein due to possible structural differences in the yeast and archaeal subunits. It was also observed that co-expressing both archaeal *GIM* genes has no effect on the benomyl sensitivity of a yeast strain entirely lacking endogenous GimC. Since the archaeal protein complex appears to be functional at the growth temperature of yeast, these data suggest that eukaryotic GimC/prefoldin may have evolved specialized functions that cannot be performed by its archaeal counterpart.

22.5 Functional Cooperation of GimC/Prefoldin With the Eukaryotic Chaperonin TRiC/CCT

GimC/prefoldin has been reported to interact with TRiC in vivo as well as in vitro [1, 3]. Three-dimensional reconstruction from electron microscopic images of that complex suggests a symmetric arrangement with one GimC/prefoldin hexamer covering the axial pore of each ring in the chaperonin complex [6]. Consistent with the participation of GimC/prefoldin and TRiC in shared protein-folding pathways, deletion of *GIM* genes in yeast results in phenotypes related to defects in the actin and tubulin cytoskeleton. These phenotypes are essentially identical to those observed for various TRiC subunit mutants [1–3, 14]. In yeast strains deleted of one or more *GIM* genes, the folding of actin by TRiC is delayed approximately five-fold relative to wild-type cells (Figure 22.4). This reduced rate of folding correlates with a decreased rate of actin release from TRiC and an increased release of non-native forms of actin into the cytosol [3]. Interestingly, GimC/prefoldin can bind denatured forms of actin and tubulin, suggesting that it plays a more active role during protein folding than does GroES, the co-chaperone of the group I chaperonin, which does not interact with nonnative polypeptides. However, the association of GimC/prefoldin with TRiC appears to be more transient than the interaction of GroES with GroEL, possibly reflecting the existence of dynamic ternary complexes during folding processes assisted by the eukaryotic chaperonin system. Both GimC and TRiC may bind to their substrates during translation [4, 15], and a role for GimC/prefoldin in targeting polypeptides to TRiC has been suggested [1, 4]. On the other hand, functional GimC is not essential for viability in budding yeast [2, 3], although the efficiency of the interaction of the chaperonin with its most abun-

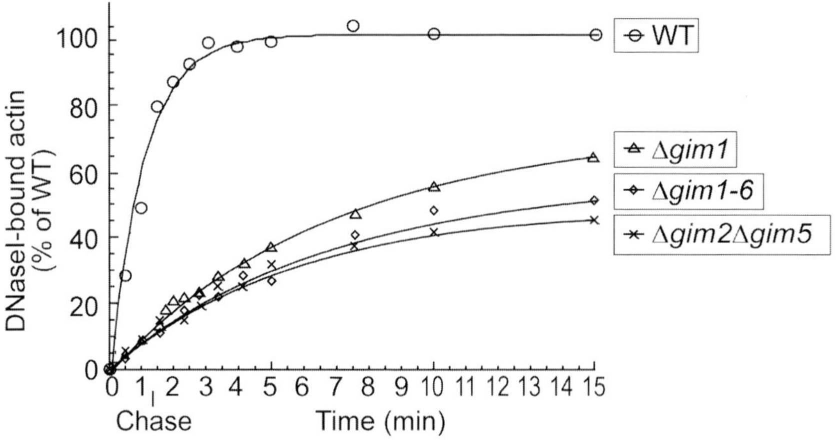

Fig. 22.4. Yields and rate of actin folding are reduced in GimC-deficient cells. Precipitation of newly synthesized, labeled actin from wild-type cells and various *gim*-mutants with DNase I-Sepharose. Native, DNaseI-bound actin was precipitated at the indicated time points and quantified.

dant substrate proteins, actin and tubulin, has been shown to be clearly reduced in the absence of functional GimC/prefoldin [5].

Removal of the hydrophobic patches at the distal parts of the coiled coils in archaeal GimC/prefoldin interferes with model substrate binding [13]. Studies on complexes reconstituted from partially truncated subunits indicate that the distal regions of the coiled coils in both classes of GimC/prefoldin subunits (α and β) are required for full chaperone activity. Substrate binding to GimC/prefoldin appears to require both classes of subunits at the same time, indicating multivalent substrate binding. The contribution of the four β subunits present in the complex was found to be greater than the contribution of the α dimer.

The eukaryotic chaperonin TRiC is thought to mediate protein folding by eventually forming an enclosed folding chamber around the nonnative substrate [16, 17], similar to the mechanism of GroEL/GroES. However, according to the current model, opening and closing of the TRiC cage is mediated by the ATP-dependent movement of helical extensions of the TRiC subunits, not by the binding and release of a separate GroES cofactor [16–19]. It was found that the capacity of TRiC to fold actin rapidly and efficiently in a sequestered environment depends on the activity of GimC/prefoldin [3, 5]. Thus, GimC/prefoldin may represent a general co-chaperone of group II chaperonins, necessary for efficient chaperonin-assisted folding. In contrast to GroES, GimC function is essential for cell growth only at low temperatures or when the function of TRiC is also compromised [2, 3].

Eukaryotic GimC/prefoldin binds to nascent chains of actin and tubulin in a cell-free translation extract and is thought to target these substrate proteins to TRiC [4]. Sucrose gradient fractionation of cell lysates revealed the association of substantial amounts of GimC and TRiC with translating ribosomes [5], extending the findings from in vitro translation extracts [4, 15] to the in vivo situation. The association of the chaperones with translating ribosomes is abolished by treatment of the lysates with RNase A, which causes dissociation of polysome complexes and ribosome release of nascent chains.

The major known nascent chain-binding chaperone in the yeast cytosol is the ribosome-associated Ssb-type Hsp70, which interacts more or less unspecifically with newly synthesized polypeptide chains on translating ribosomes. However, the association of GimC and TRiC with translating ribosomes was found to be undiminished in a *SSB1/2* deletion strain [5], excluding the possibility that the Ssb-type Hsp70, Ssb1/2p, recruits GimC/TRiC to nascent chains.

Immunoprecipitation of actin and tubulin from ATP-depleted lysates of wild-type yeast cells results in the co-precipitation of Ssb1/2p, GimC, and TRiC [5]. It can be calculated from the amount of co-precipitated chaperonin that actin and α-tubulin each occupy approximately 15–20% of total TRiC, suggesting that about 50–60% of the chaperonin capacity is devoted to the folding of actin and α- and β-tubulin. The amount of TRiC and GimC associated with actin or tubulin was found to be independent of the presence of Ssb1/2p, consistent with the results of the polysome gradient analysis. In contrast, deletion of *GIM* genes results in a ~60% reduction in the amount of TRiC-bound actin or tubulin, supporting the proposed role of GimC in actin/tubulin delivery to the chaperonin [4]. Thus, although both GimC/prefoldin and the nascent chain-binding yeast Hsp70, Ssb1/2p, interact with newly synthesized actin and tubulin, only GimC/prefoldin is required for the efficient recruitment of TRiC to these substrates.

Analysis of actin folding in wild-type and mutant yeast cells revealed that in vivo efficient actin folding on TRiC is critically dependent on the hetero-oligomeric co-chaperone GimC (Figure 22.4) [3, 5]. By interacting with folding intermediates and with TRiC, GimC was shown to accelerate actin folding at least fivefold and to prevent the premature release of nonnative folding intermediates from the chaperonin. In GimC-deficient yeast cells, the kinetics of actin folding and transit through TRiC is drastically slowed, suggesting that under these conditions folding involves multiple cycles of chaperonin action. This chaperonin cycling is inefficient and is accompanied by the loss of about 50% of actin chains into the cytosol in a nonnative state, where they fail to fold. Based on its ability to bind directly to nonnative substrate polypeptides and to TRiC, GimC may retain nonnative substrate proteins on TRiC and promote the formation of folding intermediates, resulting in acceleration of folding.

Removal of the nascent chain-binding Hsp70 Ssb in a yeast GimC/prefoldin mutant background does not further reduce the efficiency of actin folding or aggravate the impairment of the actin cytoskeleton. Likewise, loss of Ssb function does not increase the sensitivity of yeast GimC/prefoldin mutants towards the microtubule-destabilizing drug benomyl, suggesting that tubulin folding is also independent of the Ssb chaperones. Furthermore, overexpression of other cytosolic chaperones, such as yeast *SSB*, *SSA*, or *YDJ1*, does not alleviate the phenotypes observed for GimC/prefoldin mutants, demonstrating that the efficient folding of actin and tubulin requires the specific cooperation of TRiC with GimC/prefoldin [5].

However, it has been shown that TRiC also cooperates with other cytosolic chaperones, such as Hsp70, in the folding or assembly of other substrate proteins. In mammalian cells the constitutively expressed Hsc70 was found to cooperate with TRiC in the folding or assembly of the von Hippel-Lindau tumor-suppressor complex [20, 21]. Furthermore, a number of newly identified TRiC substrates, which belong to the family of WD-40 repeat proteins [5, 22], were found to require a specific posttranslational interaction with the Ssb-type Hsp70 of yeast. In the case of the WD-40 protein Cdc55p, it was observed that GimC/prefoldin can at least partially compensate for the loss of the Ssb-type Hsp70 in a deletion strain. In contrast to this, the Ssb chaperones were found to be unable to substitute for GimC in actin or tubulin folding, indicating the requirement of a specialized chaperone function for these highly abundant substrate proteins.

The ability of yeast GimC/prefoldin to at least partially replace Ssb1/2p in the folding of Cdc55p, and presumably other TRiC-dependent WD-40 substrates, demonstrates that although eukaryotic GimC/prefoldin has a specialized function in actin/tubulin folding, it can also fulfill a general chaperone function. GimC/prefoldin is present in all eukaryotic and archaeal genomes sequenced so far, but in contrast to eukaryotic cells, archaea do not contain an extensive actin and tubulin cytoskeleton and in certain cases lack the abundant nascent chain-binding Hsp70 system altogether [12]. Therefore, the simpler and more abundant archaeal version of GimC/prefoldin may have a more general function in stabilizing nascent polypeptides in these organisms.

22.6 Experimental Protocols

22.6.1 Actin-folding Kinetics

Measuring actin-folding kinetics in yeast cells involves the preparation of yeast spheroplasts and precipitation of folded actin from radio-labeled yeast lysates with DNase I beads.

Preparation of Spheroplasts Yeast cells are grown in 100-mL cultures to mid-log phase ($OD_{600} \sim 0.5$); harvested by centrifugation (3000 g, 5 min); resuspended in 5 mL of synthetic complete medium (0.67% yeast nitrogen base, 2% glucose), minus methionine and cysteine (SC-M-C), containing 1.2 M sorbitol and 30 mM DTT, pH 7.5; and incubated for 10 min at room temperature. Subsequently, cells are harvested as above and resuspended in the same medium without DTT containing 0.5 mg mL^{-1} Zymolyase 100T (ICN Biochemicals) and incubated for 30–60 min at 30 °C until conversion of the cells to spheroplasts is above 90%. The spheroplasts are harvested by centrifugation (1000 g, 15 min); washed twice in SC-M-C, 1.2 M sorbitol, pH 5.5; resuspended in 2–5 mL of the same buffer; and incubated at 30 °C for radio-labeling.

Radiolabeling of Yeast Cells and Actin Folding Pulse-chase radio-labeling of spheroplasts is performed with 100 μCi mL^{-1} [^{35}S]-methionine/cysteine ProMix (Amersham) followed by a chase with cycloheximide (0.36 mM). At various time points, aliquots of 250 μL of spheroplasts (100–200 μg of protein) are withdrawn, diluted 1:1 in cold 2× lysis buffer (2× PBS [20 mM Na_2HPO_4, 3.6 mM KH_2PO_4, 5.4 mM KCl, 280 mM NaCl] pH 7.4, 10 mM EDTA, 1% TWEEN-20, 2× complete protease inhibitors [Roche]), mixed, and immediately frozen in liquid nitrogen. At the end of the chase, all reactions are thawed on ice and cell extracts are cleared by centrifugation (20 000 g, 4 °C, 15 min).

To precipitate folded actin from the supernatant, 30 μL of DNase I-beads (1:1 suspension in 1× PBS, pH 7.4) (obtained by cross-linking purified DNase I [Sigma] to CNBr-activated Sepharose 4B-CL [Amersham] as described by the manufacturer) is added to the cleared lysates followed by incubation at 4 °C for 1 h. For the precipitation DNase I is in excess over the actin present. Beads are washed extensively once with 500 μL buffer W1 (1% Triton-X-100, 0.5% DOC, 150 mM NaCl, 50 mM Tris/HCl pH8, 5 mM EDTA) and twice with 500 μL buffer W2 (1% Triton-X-100, 500 mM NaCl, 50 mM Tris/HCl pH 8.0, 5 mM EDTA), followed by PBS, prior to SDS-PAGE and phosphorimager analysis.

22.6.2 Prevention of Aggregation (Light-scattering) Assay

Rabbit muscle actin (200 μM) (Sigma), lysozyme (160 μM) (Sigma), or bovine rhodanese (100 μM) (highly purified; Sigma), is prepared in denaturing buffer (6 M guanidine hydrochloride, 20 mM Tris pH 8.0, 100 mM NaCl, 1 mM $MgCl_2$, and 50 mM DTT in the case of lysozyme). The unfolded proteins are diluted 100-fold in buffer B (20 mM Tris pH 8.0, 100 mM NaCl, 1 mM $MgCl_2$) alone, as a reference for maximum aggregation or containing various amounts of GimC/prefoldin, or in IgG as negative control. Aggregation is measured at 25 °C over a period of 10 min by following the increase in light scattering at 320 nm. Alternatively, the effect of GimC/prefoldin on thermally induced aggregation of a 1 μM solution of native substrate proteins in buffer B can be monitored at 320 nm in a thermostated cuvette preheated to 43 °C.

22.6.3 Actin-binding Assay

To determine chaperone activity of purified chaperones, the separation of TRiC/CCT or GimC/prefoldin-bound radio-labeled actin is accomplished by native polyacrylamide electrophoresis.

Radio-labeled denatured actin is obtained by expressing cDNAs of yeast or mouse *β*-actin to high levels using the *E. coli* strain BL21 (DE3) pLysS in the presence of 1 mCi mL^{-1} [^{35}S]-methionine/cysteine, rifampicin (0.2 mg mL^{-1}), and IPTG (0.5 mM). The labeled actin is solubilized from isolated inclusion bodies using 8 M urea, 20 mM Tris-HCl pH 7.5, and 10 mM DTT.

GimC/prefoldin-actin complexes are formed by diluting denatured radio-labeled actin 100-fold to a final concentration of ~0.13 μM into 50 mM Tris-HCl pH 7.5, 50 mM NaCl, and 2 mM EDTA containing GimC/prefoldin (0.15 μM), followed by incubation for 30 min at 30 °C. Samples are separated on a non-denaturing 4.5% polyacrylamide, 80 mM MOPS-KOH pH 7.0 gel containing 1 mM Mg-ATP at 4 °C for 2.5 h.

Acknowledgements

We thank Dr. S. Grallath for critically reading the manuscript. Work in the authors' laboratory is supported by the Deutsche Forschungsgemeinschaft.

References

1 Vainberg, I. E., Lewis, S. A., Rommelaere, H., Ampe, C., Vandekerckhove, J., Klein, H. L., Cowan, N. J. (1998). Prefoldin, a chaperone that delivers unfolded proteins to cytosolic chaperonin. *Cell*, **93**, 863–873.

2 Geissler, S., Siegers, K., Schiebel, E. (1998). A novel protein complex promoting formation of functional α- and γ-tubulin. *EMBO J.*, **17**, 952–966.

3 Siegers, K., Waldmann, T., Leroux, M. R., Grein, K., Shevchenko, A., Schiebel, E., Hartl, F. U. (1999). Compartmentation of protein folding in vivo: sequestration of non-native polypeptide by the chaperonin-GimC system. *EMBO J.*, **18**, 75–84.

4 Hansen, W. J., Cowan, N. J., Welch, W. J. (1999). Prefoldin-nascent chain complexes in the folding of cyto-skeletal proteins. *J. Cell Biol.*, **145**, 265–277.

5 Siegers, K., Boelter, B., Schwarz, J. P., Boettcher, U. M. K., Guha, S., Hartl, F. U. (2003). TRiC/CCT cooperates with different upstream chaperones in the folding of distinct protein classes. *EMBO J.*, **22**, 5230–5240.

6 Martin-Benito, J., Boskovic, J., Gomez-Puertas, P., Carrascosa, J. L., Simons, C. T., Lewis, S. A., Bartolini, F., Cowan, N. J., Valpuesta, J. M. (2002). Structure of eukaryotic prefoldin and of its complexes with unfolded actin and the cytosolic chaperonin CCT. *EMBO J.*, **21**, 6377–6386.

7 Llorca, O., McCormack, E. A., Hynes, G., Grantham, J., Cordell, J., Carrascosa, J. L., Willison, K. R., Fernandez, J. J., Valpuesta, J. M. (1999). Eukaryotic type II chaperonin CCT interacts with actin through specific subunits. *Nature*, **402**, 693–696.

8 Leroux, M., Fändrich, M., Klunker, D., Siegers, K., Lupas, A. N., Brown, J. R., Schiebel, E., Dobson, C., Hartl, F. U. (1999). MtGimC, a novel archaeal chaperone related to the eukaryotic chaperonin cofactor GimC/prefoldin. *EMBO J.*, **18**, 6730–6743.

9 Okochi, M., Yoshida, T., Maruyama, T., Kawarabayasi, Y., Kikuchi, H., Yohda, M. (2002). Pyrococcus prefoldin stabilizes protein-folding intermediates and transfers them to chaperonins for correct folding. *Biochem. Biophys. Res. Comm.* **4**, 769–774.

10 Deuerling, E., Schulze-Specking, A., Tomoyasu, T., Mogk, A., Bukau, B. (1999). Trigger factor and DnaK cooperate in folding of newly synthesised proteins. *Nature*, **400**, 693–696.

11 Teter, S. A., Houry, W. A., Ang, D.,

Tradler, T., Rockabrand, D., Fischer, G., Blum, P., Georgopoulos, C., Hartl, F. U. (1999). Polypeptide flux through bacterial Hsp70: DnaK cooperates with trigger factor in chaperoning nascent chains. *Cell*, **97**, 755–765.

12 Ruepp, A., Rockel, B., Gutsche, I., Baumeister, W., Lupas, A. N. (2001). The chaperones of the archaeon *Thermoplasma acidophilum*. *J. Struct. Biol.*, **135**, 126–138.

13 Siegert, R., Leroux, M. R., Scheufler, C., Hartl, F. U., Moarefi, I. (2000). Structure of the molecular chaperone prefoldin: Unique interaction of multiple coiled coil tentacles with unfolded proteins. *Cell*, **103**, 621–632.

14 Stoldt, V., Rademacher, F., Kehren, V., Ernst, J. F., Pearce, D. A., Sherman, F. (1996) Review: the Cct eukaryotic chaperonin subunits of *Saccharomyces cerevisiae* and other yeasts. *Yeast*, **12**, 523–529.

15 McCallum, C. D., Do, H., Johnson, A. E., Frydman, J. (2000). The interaction of the chaperonin tailless complex polypeptide 1 (*TCP1*) ring complex (TRiC) with ribosome-bound nascent chains examined using photo-cross-linking. *J. Cell Biol.*, **149**, 591–601.

16 Klumpp, M., Baumeister, W., Essen, L.-O. (1997). Structure of the substrate binding domain of the thermosome, an Archaeal Group II chaperonin. *Cell*, **91**, 263–270.

17 Ditzel, L., Lowe, J., Stock, D., Stetter, K. O., Huber, H., Huber, R., Steinbacher, S. (1998). Crystal structure of the thermosome, the archaeal chaperonin and homolog of CCT. *Cell*, **93**, 125–138.

18 Llorca, O., Smyth, M. G., Marco, S., Carrascosa, J. L., Willison, K. R., Valpuesta, J. M. (1998). ATP binding induces large conformational changes in the apical and equatorial domains of the eukaryotic chaperonin containing TCP-1 complex. *J. Biol. Chem.*, **273**, 10091–10094.

19 Meyer, A. S., Gillespie, J. R., Walther, D., Millet, I. S., Doniach, S., Frydman, J. (2003). Closing the folding chamber of the eukaryotic chaperonin requires the transition state of ATP hydrolysis. *Cell*, **113**, 369–381.

20 Feldman, D. E., Thulasiraman, V., Ferreyra, R. G., Frydman, J. (1999). Formation of the VHL-elongin BC tumor suppressor complex is mediated by the chaperonin TRiC. *Molecular Cell*, **4**, 1051–1061.

21 Melville, M. W., McClellan, A. J., Meyer, A. S., Darveau, A., Frydman, J. (2003). The Hsp70 and TRiC/CCT chaperone systems cooperate in vivo to assemble the von Hippel-Lindau tumor suppressor complex. *Mol. Cell. Biol.*, **23**, 3141–3151.

22 Camasses, A., Bogdanova, A., Shevchenko, A., Zachariae, W. (2003). The CCT chaperonin promotes activation of the anaphase-promoting complex through the generation of functional Cdc20. *Molecular Cell*, **12**, 87–100.

23 Gouet, P., Courcelle, E., Stuart, D. I., Metoz, F. (1999). ESPript: multiple sequence alignments in PostScript. *Bioinformatics*, **15**, 305–308.

24 Kraulis, P. (1991). MOLSCRIPT: a program to produce both detailed and schematic plots of protein structures. *J. Appl. Cryst.*, **24**, 946–950.

25 Merritt, E. A., Bacon, D. J. (1997). Raster3D photorealistic graphics. *Methods Enzymol.*, **277**, 505–524.

26 Nicholls, A., Sharp, K. A., Honig, B. (1991). Protein folding and association: insights from the interfacial and thermodynamic properties of hydrocarbons. *Proteins*, **11**, 281–296.

23
Hsp90: From Dispensable Heat Shock Protein to Global Player

Klaus Richter, Birgit Meinlschmidt, and Johannes Buchner

23.1
Introduction

Hsp90 is a cytosolic molecular chaperone that has been found in context with many signal transduction pathways in higher eukaryotes [1]. Its participation in these processes is still enigmatic but is thought to involve specific conformational changes in its substrate proteins, many of which are protein kinases and transcription factors. Because these conformational changes are required to confer activity to the substrates, Hsp90 has become a global player in the signal transduction network of eukaryotic cells. In contrast, the prokaryotic homologue of Hsp90 (HtpG) and the compartmentalized versions of Hsp90 from mitochondria, chloroplasts, and the endoplasmatic reticulum are thought to be predominantly folding helpers for the proteins that reside in or pass through these cellular structures. The evolution of Hsp90 function may be the result of the ever more complex set of cytosolic partner proteins that are known to associate in a substrate-specific way with Hsp90. In this chapter specific focus is put on the ATPase cycle of Hsp90, the interplay of Hsp90 with partner proteins or substrates, and the specific ways to investigate these complex macromolecular assemblies.

23.2
The Hsp90 Family in Vivo

23.2.1
Evolutionary Relationships within the Hsp90 Gene Family

About 130 genes have been unambiguously assigned to the Hsp90 gene family so far. Most of them account for cytosolic variants of Hsp90. Usually, one cytosolic Hsp90 gene is found in prokaryotes, but gene duplications have arisen in eukaryotes that have led to two Hsp90 genes in yeast [2], two Hsp90 genes in mammals [3], and four Hsp90 genes in *Arabidopsis thaliana* [4]. The functional differences between these proteins, and thus the reasons for duplications, are unknown, espe-

Protein Folding Handbook. Part II. Edited by J. Buchner and T. Kiefhaber

ISBN: 3-527-30784-2

cially as the degree of homology between them is very high (as much as 98% in yeast). However, differences in the regulation exist. Sequence alignments show that all the Hsp90 genes share a common organization consisting of conserved domains connected by flexible linkers [5, 6] (Figure 23.1B). Analysis of the individual domains has seen considerable progress in the past few years concerning both structure and function. The N-terminal domain is the nucleotide-binding site of the protein, whereas the C-terminal domain hosts the dimerization site of Hsp90. The middle domain seems to be involved in substrate binding and ATP hydrolysis (Figure 23.1B). Based on the many sequences available for Hsp90 and its homologous proteins, it is possible to trace the evolution of this protein family throughout the organismal kingdom.

The phylogenetic tree of the primary sequences of Hsp90 family members reveals interesting aspects of the evolutionary relationship (Figure 23.1A). First, eukaryotic (including chloroplast and ER members) differ from prokaryotic (including mitochondrial) Hsp90 proteins by a massive extension of the linker region between the N-terminal and the middle domain. This linker consists predominantly (about 90%) of charged amino acids and starts around amino acid 210. It has a maximal length of 92 amino acids in human Hsp90 genes, of about 40 amino acids in yeast Hsp90, and of seven amino acids in HtpG from *E. coli*. Its function is unknown and its presence is not required for yeast cell growth [7]. Second, cytosolic eukaryotic Hsp90 proteins differ from other eukaryotic Hsp90s by a C-terminal extension. This region, which includes the MEEVD motif at the C-terminal end of the protein, serves as the binding site for cytosolic partner proteins containing TPR motifs [8, 9]. This region is also not required for yeast cell growth [7]. Third, Grp94, the ER-resident species, shows an N-terminal extension of about 60 amino acids following the N-terminal signal sequence for ER import. The function and origin of this region are unknown, as it lacks any homology to known protein domains. Interestingly, the evolutionary tree indicates that the Hsp90s in chloroplasts [10] and ER are descendants of cytosolic variants and thus have evolved more recently, while the mitochondrial Hsp90 gene is of ancient, bacterial origin [6]. One feature that shows this origin is the presence of the linker between the N-terminal and the middle domain in the chloroplast and ER-resident Hsp90 isoforms, which is missing in the mitochondrial Hsp90 homologue [11]. The ER-resident Grp94 is found only in multicellular organisms, suggesting either a more recent development or a secondary loss of this gene by single-celled eukaryotes. Duplications of Hsp90 appear to have happened on multiple occasions during evolution, indicating a strong correlation of organismal evolution and Hsp90 evolution. Interestingly, in archaea Hsp90 genes have not been detected yet.

23.2.2
In Vivo Functions of Hsp90

Investigating the importance of a gene for an organism means in particular investigating phenotypes upon deletion of this gene. Knocking out HtpG, the bacterial Hsp90 homologue, produced a modest heat-sensitive phenotype in *E. coli* [12] but

A

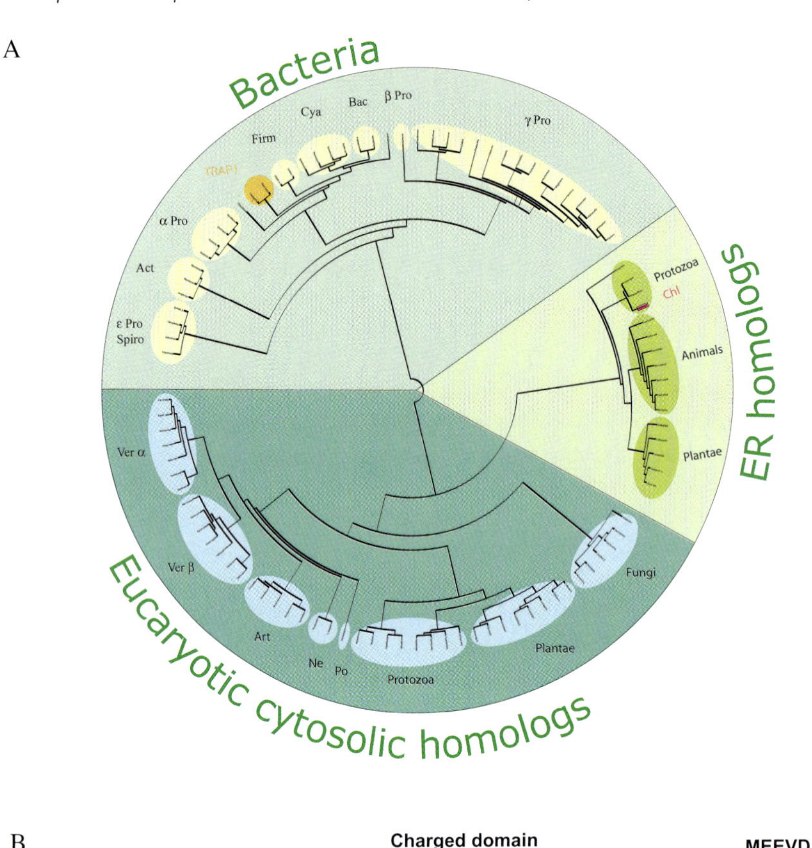
Bacteria
Cya
Bac
β Pro
γ Pro
Firm
TRAP1
α Pro
Act
ε Pro
Spiro
Protozoa
Chl
Animals
Plantae
ER homologs
Ver α
Ver β
Art
Ne
Po
Protozoa
Plantae
Fungi
Eucaryotic cytosolic homologs

B

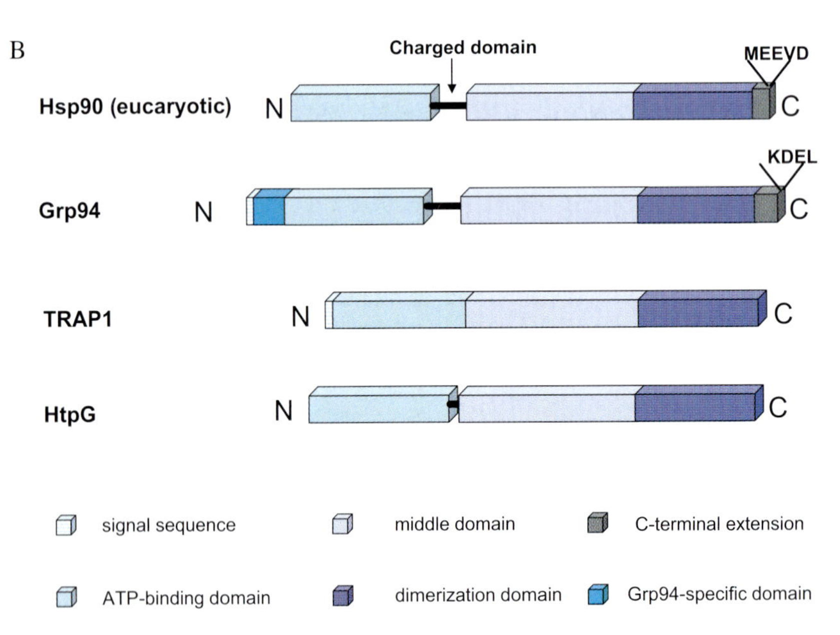
Charged domain
MEEVD
Hsp90 (eucaryotic)
N
C
KDEL
Grp94
N
C
TRAP1
N
C
HtpG
N
C
signal sequence
middle domain
C-terminal extension
ATP-binding domain
dimerization domain
Grp94-specific domain

a lack of thermoprotective functions in cyanobacteria [13]. As no substrate of HtpG is known to date, the role of HtpG in bacteria is largely enigmatic. A slight evolutionary advantage could be detected that leads to an outgrowth of HtpG-deficient bacteria by those that carry a functional HtpG gene [12]. In lower eukaryotes, like yeast, Hsp90 is essential, as knocking out both cytosolic Hsp90 genes results in the loss of viability [2]. The reasons for this are unknown, but it has been observed that the cell cycle of yeast is arrested at both the G_1/S and the G_2/M phases upon deletion of Hsp90s [14], which highlights the involvement of Hsp90 in cell cycle control [15, 16]. This view is further strengthened by the identification of Cdk4 as a substrate of Hsp90 in yeast [17]. Several other kinases, such as dsRNA-dependent protein kinase [18] and the cell cycle kinase Wee1 [19], were found to require functional Hsp90 in yeast. Hsp90 in higher eukaryotes was found to be involved in many different pathways, as best exemplified by the study of Rutherford and Lindquist [20]. Here, genotype-specific phenotypes could be generated by the reduction of Hsp90 levels in *Drosophila*. Hsp90 appeared to buffer the accumulation of mutations in target genes and thus allowed for the invisible accumulation of these events. Partial reduction of Hsp90 function resulted in the manifestation of the previously silent phenotypes and, consequently, led to a diverse set of developmentally defective flies [20]. Comparable effects were later found in *Arabidopsis*, suggesting a similar scenario in animals and plants [21].

In mammals, knockout studies indicated that at least Hsp90β is essential and that its knockout leads to defects in placental development [22]. Hsp90α is also required for embryonic development, as inhibition of the Hsp90 protein results in defects in muscle cell development in zebra fish [23]. For Grp94, on the other hand, studies with B-cell lines indicate an involvement in innate immunity [24]. Several other reports on Grp94 function suggest interactions of the major ER chaperones BiP and Grp94 in processing, folding, and quality control during protein production in the endoplasmic reticulum [25, 26]. In addition, there are reports of an active role of Grp94 in the loading of peptides onto MHCI complexes during the immune response [27].

Much less is known about the in vivo function of the recently discovered Hsp90 homologues of mitochondria and chloroplasts. Also, a shortened Hsp90 homologue, Hsp90N, lacking the N-terminal ATP-binding site, has recently been identified in human cell lines [28]. The function of this protein is still unknown.

Fig. 23.1. Overview of the Hsp90 family. (A) Evolutionary tree of the Hsp90 family. The evolutionary tree of the Hsp90 family has been calculated based on 120 Hsp90 sequences that are publicly accessible. The tree has been constructed as outlined in the Appendix. In principle, it is in agreement with previously constructed trees [5, 6]. ER: endoplasmic reticulum; Verα: vertebrata Hsp90α; Verβ: vertebrata Hsp90β; Art: Arthropoda; Ne: Nematoda; Po: Polifera; Chl: Chloroplast; Spiro: Spirochaetes; Pro: Proteobacteria; Firm: Firmicutes; Cya: Cyanobacteria; Bac: Bacteroides. (B) Schematic representation of the domain organization of the Hsp90 family. Functional equivalent domains are labeled with the same colors.

23.2.3
Regulation of Hsp90 Expression and Posttranscriptional Activation

To understand the function of proteins, it is helpful to investigate the specific requirements for their expression. Thus, the strongly enhanced expression of many Hsp90 isoforms as a response to heat stress hints at an involvement in the protective system of the cell. Conditions that lead to the expression of the Hsp90 gene in prokaryotes include exposure to ethanol, toxic substances, and radicicol as well as other harmful conditions [29, 30]. The regulation of HtpG expression is understood in detail only for heat shock. Here, the heat shock factor sigma 32 determines transcriptional control [31]. Upon heat-induced activation of sigma 32, the heat shock element upstream of the *htpG* gene is used as a docking site of the transcription factor, leading to strongly induced expression of the target protein [32]. To ensure coupling of transcriptional control of chaperones to the presence of unfolded proteins, sigma 32 is negatively regulated by DnaK/Hsp70 and other chaperones. This might add an additional level of control to the regulation, as it guarantees that expression is induced only if a lack of free chaperones is observed. Interestingly, the association of HtpG with sigma 32 is also described [33].

Very similar regulation patterns have been observed for the Hsp90 genes of eukaryotes. Regulation here is performed by the heat shock factor (HSF) [34]. This protein, while not sharing homology with sigma 32, behaves in a comparable way. Here, heat induces trimerization of the protein [35], leading to an active transcription factor, which binds to heat shock elements (HSEs) found in the promoters of all major heat shock proteins, sometimes in multiple redundant sequences [36–38].

If more than one cytosolic Hsp90 isoform exists, the expression is differentially regulated. In yeast, only Hsp82 is stress-regulated, while the 98% identical Hsc82 is constitutively expressed; its levels increase only modestly following stress [2]. Similar results were obtained for mammalian Hsp90. Here, Hsp90β is expressed constitutively and is only slightly upregulated following heat shock and interleukin-dependent signal transduction [39, 40]. The modest upregulation upon heat shock as well as the constitutive expression require a HSE sequence, which is found within the first intron of the gene [40]. Hsp90α instead is strongly heat-inducible, as it contains several HSEs in its 5′ upstream promoter sequence [41]. In addition, it is also under control of interleukin-induced signaling pathways [41, 42]. Other, less well-understood mechanisms – involving the signaling proteins PKCε [43] or STAT1 [44] – are involved in the regulation of Hsp90 expression and appear to couple the expression to an accurate control machinery. Interestingly, recent reports suggest that several chromosomal copies exist for each of the Hsp90 genes in humans [44].

The only other Hsp90 homologue for which the regulation of expression is understood in some detail is Grp94. Grp94 is controlled, among other pathways, by the unfolded protein response (UPR), which connects the appearance of unfolded proteins in the ER with the expression of the major chaperones in this compartment. The pathway starts with the recognition of unfolded proteins by the ER-resident transmembrane kinase Ire1 and finally results in the activation of the transcription factor Xbp1 [45]. Here as well, molecular chaperones participate in the activation process as negative regulators of Ire1 [46, 47]. Similar pathways exist in lower eukaryotes as well [48].

Thus, throughout the whole animal kingdom, Hsp90 appears as a protein that, although highly expressed even under non-stress conditions, can be strongly induced following either heat stress or chemical stress. Especially the observed coupling of overexpression to the available chaperone activity, as is observed throughout the activation process of the required transcription factors, strongly hints to the origin of Hsp90 as a folding helper and participant in the homeostasis of protein stability in the cell. In yeast, for example, Hsp90 represents 1–2% of the cytosolic protein under permissive conditions and significantly more under stress conditions. Interestingly, only 1/20th of the amount of Hsp90 present under permissive conditions is required to ensure yeast cell growth, while significantly more Hsp90 is required under stress conditions [2].

23.2.4
Chemical Inhibition of Hsp90

In addition to knockout strategies, natural Hsp90 inhibitors were used to imitate the loss of Hsp90 functions in cell culture and some eukaryotic species [49]. Those mostly used are radicicol and geldanamycin (Figure 23.2). Radicicol (formerly monorden) is a product of the fungus *Neocosmospora tenuicristata* [50]. Geldanamycin, an ansamycin, is derived from *Streptomyces hygroscopicus* var. *geldanus* [51].

Several substances that are similar to geldanamycin were also found in streptomyces strains, including macbecin [52, 53] and herbimycin A/B [54, 55]. All these inhibitors appear similar in their efficiency to inhibit Hsp90 in vitro, and strong similarities exist in their in vivo effects [56, 57]. Geldanamycin originally was thought to be an inhibitor of tyrosine kinases, but this effect later was proved to be indirect via the inhibition of Hsp90 [58]. It was observed that incubation with Hsp90 inhibitors results in the rapid degradation of many substrate proteins [59, 60]. The potent inhibition of Hsp90 function is the result of a strong affinity for the N-terminal ATP-binding domain ($K_D = 19$ nM for radicicol, $K_D = 1.2$ μM for geldanamycin), which makes them competitive inhibitors of ATP. ATP binds at least 300-fold weaker to Hsp90 [61, 62]. Thus, the binding of these substances to Hsp90 inhibits the ATPase-dependent functions of Hsp90.

The strong effect of these substances on the tumorigenic growth of cell lines is probably the result of disruption of Hsp90-substrate complexes that are important for cell division. As many of these substrates are oncogenic or associated with pathways leading to cell growth and cell division, the antitumor potential of these substances has beem employed in clinical studies [63]. Especially one derivative of geldanamycin, 17-allylamino, 17-demethoxygeldanamycin (17AAG), shows promising results in the treatment of melanoma patients [64–66]. In particular, the ability of this inhibitor to simultaneously block multiple pathways required for tumor growth by disrupting the Hsp90 chaperone system holds great promise for cancer

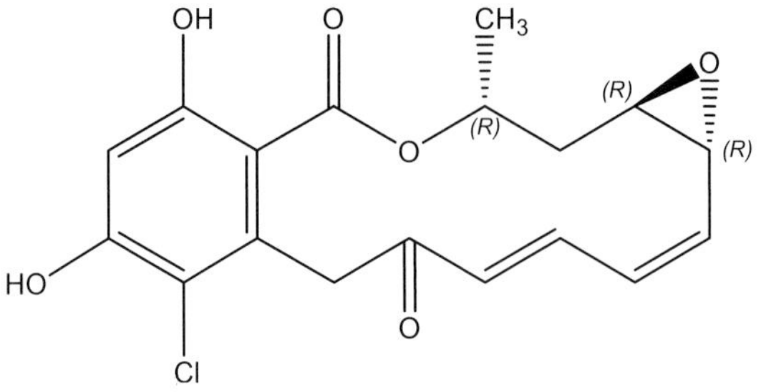

Radicicol

R1 $CH{=}CHCH_2NH$

R2 CH_3O

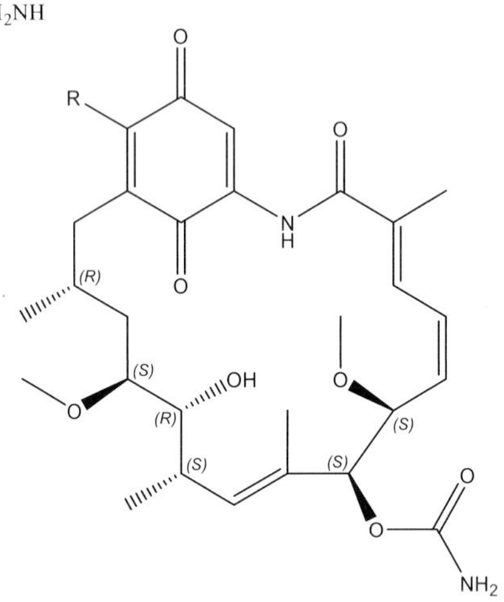

Geldanamycin

Fig. 23.2. Chemical structure of geldanamycin and radicicol. R2: Geldanamycin; R1: 17-allylamino, 17-demethoxygeldanamycin.

therapy [49]. Further, these substances appear to accumulate in cancer cells to concentrations far higher than in normal cells [67]. Considerable interest has arisen in identifying further artificial Hsp90 inhibitors, leading to structure-based strategies in the development of new compounds [68]. First progress in the form of potential lead structures based on purine-scaffolds has already been reported [69, 70].

23.2.5
Identification of Natural Hsp90 Substrates

The phenotypes observed after Hsp90 deletion in eukaryotes are likely the result of diminished activation of the Hsp90 substrate proteins. Many substrates were found in specific, stable complexes with Hsp90 by co-purification or co-precipitation. This implies that, in sharp contrast to Hsp70, the interaction with this chaperone is long-lived. These substrates are derived from several different and seemingly unrelated protein families [1, 71]. Among them are protein kinases of the tyrosine and threonine/serine kinase family, such as Src, Raf, Mek, Cdk4. As these proteins are at the heart of signaling decisions in eukaryotes, Hsp90 is involved in many important cellular processes. Some other substrates can be grouped together into the large class of transcription factors, although no sequence homology exists between these proteins. Hsp90 targets are, e.g., p53 [72], steroid hormone receptors (SHR), and heat shock factor [73]. Several other unrelated proteins have also been found in complex with Hsp90. The known Hsp90 substrates are listed in Table 23.1.

Strikingly, regarding the strong response of Hsp90 expression to heat shock, none of the substrates identified to date gives any hint towards the chaperoning activity of Hsp90 at elevated temperatures. All of the known substrates require Hsp90 function even at permissive temperatures. Therefore, it may well be that the substrate range changes significantly at elevated temperatures.

23.3
In Vitro Investigation of the Chaperone Hsp90

23.3.1
Hsp90: A Special Kind of ATPase

The crystal structure of the N-terminal domain of Hsp90 was solved in 1997 [61, 74] (Figure 23.3A). These and subsequent studies revealed a new type of binding site for ATP and for the inhibitor geldanamycin [61, 75]. The only related fold known at that time was that of GyraseB [76].

As is evident from the structures with ADP and ATP, the nucleotide is bound in a cleft formed by α-helices on top of an eight-stranded β-sheet [75]. Four α-helices participate directly in the binding of the nucleotide, while three others surround them. Binding induces a special kinked conformation of the nucleotide. The adenine and ribose moieties are hidden inside the nucleotide-binding cleft, while the phosphate groups are oriented towards the surface of this domain. The γ-phosphate is highly unordered in the structure of the ATP-protein complex, leading to the assumption that it might be completely solvent-exposed [75]. This orientation of the nucleotide within the N-terminal domain is shared by other members of this ATPase family. The family of related proteins has grown and now includes, among others, the proteins GyraseB, MutL, and Pts and, with some restrictions, the histidine kinases EnvZ and CheA. Collectively they are called GHKL-ATPases [77]. These proteins share very similar 3D structures of their ATP-binding sites, although sequence homology is rather low. With the exception of the histidine kinases, the ATP-binding domains are located in the N-terminal part of the respective protein. They show a dimeric architecture, for which a C-terminal dimerization site is responsible. The surface exposure of the γ-phosphate [75] allows complex

Tab. 23.1. Hsp90 substrate proteins.[1]

	Protein	***References***
Transcription	Androgen receptor	212, 213
	Aryl hydrocarbon (Ah) receptor	214, 215
	CAR, transducer protein	216
	Ecdysone receptor	217
	Estrogen receptor	212, 218
	Glucocorticoid receptor	219
	Heme activator protein (Hap1)	220
	HSF-1	221
	Hypoxia-inducible factor-1α	222
	Mineralocorticoid receptor	223
	MTG8 myeloid leukemia protein	224
	p53	72
	PPARα (PPARβ); peroxisome proliferator-activated receptor	225, 226
	Progesterone receptor	227, 228
	Retinoid receptor	229
	Sim	230
	Stat3; Signal transducer and activator of transcription	231
	SV40 large T antigen	232
	Tumor promotor-specific binding protein	233
	v-erbA	234
	water mold Achlya steroid (antheridiol) receptor	235
Polymerases	Telomerase	236
	Hepatitis B virus reverse transcriptase	237
	DNA-polymerase α	191
Kinases	3-Phosphoinositide-dependent kinase-1; PDK1	238
	Akt	239
	Aurora B	240
	Bcr-Abl	59
	Calmodulin-regulated eEF-2 kinase	241
	Casein kinase II	103
	Cdc2	242
	Cdk4	17, 162
	Cdk6	243
	Cdk9	244
	Chk1	245
	c-Mos	246
	Epidermal growth factor receptor	247
	ErbB2	248
	Flt3	249
	Focal adhesion kinase	250
	GRK2	251
	Hck	252
	Heme-regulated eIF-2α kinase	253, 254
	IκB kinases $\alpha, \beta, \gamma, \varepsilon$	255
	Insulin receptor	256
	Insulin-like growth factor receptor	257

Tab. 23.1. *(continued)*

	Protein	***References***
	Ire1	258
	Kinase suppressor of ras (KSR)	165
	Lkb1	259, 260
	Lymphoid cell kinase $p56^{lck}$	261
	MAK-related kinase	262
	Male germ cell-associated kinase MAK	262
	MEK (MAP kinase kinase)	247
	MEKK1 and MEKK3	263
	Mik1	264
	Mitogen-activated protein kinase MOK	262
	MRK	240
	Nucleophosmin-Anaplastic Lymphoma Kinase	265
	Perk	258
	Phosphatidylinositol 4-kinase	266
	Pim-1	267
	PKR	18
	Platelet-derived growth factor receptor	268
	Polo mitotic kinase	269
	Raf family kinases: v-Raf, c-Raf, B-Raf, Gag-Mil, Ste11	7, 270, 271, 272, 273
	Receptor-interacting protein (RIP)	255
	Sevenless PTK	274
	Swe1	264
	Translation initiation factor kinase Gcn2	275
	Tropomyosin related kinase B (trkB)	276
	v-fes	277
	v-fps	277
	v-fgr, c-fgr	261, 278
	v-Src, c-Src	163, 279, 280
	v-yes	277
	VEGFR2	250
	Wee1	19
Others	Actin	281
	Apaf-1	282
	Apoprotein B	283
	Atrial natriuretic peptide receptor	284
	Bid	285
	calmodulin	286
	Calponin	287
	Centrin/centrosome	288
	Cna2 (catalytic subunit of calcineurin)	289
	CFTR	290
	Ctf13/Skp1 component of CBF3	291
	Endothelial NOS	292
	Ether-a-gogo-related cardiac potassium channel	293
	Fanconi anemia group C protein (FACC protein)	294
	G protein $\beta\gamma$	295

strand-swapping mechanism, apparently a loop consisting of two α-helices has to be moved within the N-terminal domain and claps over the ATP-binding site. This "ATP lid" then forms part of the receptor region for the N-terminal strand of the other domain. It directly interacts with the intruding strand of the other N-terminal domain and is therefore considered to be involved in the hydrolysis mechanism. In addition, complex formation between the γ-phosphate of AMP-PNP and the middle domain is evident in both crystal structures. In both cases, a specific lysine residue appears to serve the function of the γ-phosphate receptor. Concluding from these structures, the corresponding residue is expected to be in the region of amino acids 320–400 of Hsp90. Based on the crystal structure of the middle domain of Hsp90, residue Glu381 was suggested as the acceptor site of the γ-phosphate [80]. However, sequence alignments suggest other possible residues in that region. Further studies on the interaction between the N-terminal and the middle domain are needed for a conclusive understanding of participating segments and the order of conformational changes within this molecular machine.

23.3.2
The ATPase Cycle of Hsp90

Because the crystal structures present a rather static picture of one particular step during the ATPase reaction, much effort has been put into the dissection of the steps that lead to ATP hydrolysis by Hsp90. Yeast Hsp90 hydrolyzes ATP with a turnover of about one per minute [81, 82]. Early studies attempted to define the minimum requirements of Hsp90 to perform ATP hydrolysis. These studies, involving different fragments starting from the N-terminal ATP-binding site, showed the critical requirement of regions outside the ATP-binding pocket for ATP hydrolysis [83, 84]. Only 2% of the wild-type ATPase activity was obtained for the isolated N-terminal domain. A slight increase in ATPase activity (about 12% of wt activity) was observed if the middle domain was present in addition to the N-terminal domain [83, 84]. Higher concentrations of this fragment or its covalent dimerization by a C-terminal cysteine bridge lead to an increase in the ATPase activity [85, 86]. The latter approach led to wild-type ATPase activity, proving the necessity of dimer formation for hydrolysis [86]. Experiments involving heterodimeric variants of Hsp90 helped to identify the regions within Hsp90 that mediate the contacts in the dimer that are important for hydrolysis. These studies proved the transient N-terminal dimerization within the Hsp90 dimer during the ATPase reaction [85] that had been proposed by earlier studies [83, 87]. Further deletion studies imply that the critical region for formation of N-terminal dimers during the ATPase reaction of Hsp90 involves the first 24 amino acids of Hsp90 [88], matching the orientation observed in the crystal structure of the homologous ATPases.

Based on these studies, a cycle of coordinated conformational changes has been proposed for the ATP hydrolysis reaction of Hsp90 (Figure 23.5). This cycle starts with the binding of two ATP molecules to the N-terminal domains of Hsp90. Kinetic studies showed that ATP binding is a fast process, especially compared to the turnover of Hsp90 [84]. A series of conformational changes results in the trapping of the ATP molecule as observed by Weikl et al. [84] for yeast Hsp90. Early investigations already had pointed to ATP-induced conformational changes within Hsp90 [89]. Several studies suggest that N-terminal dimerization and subsequent activation of the ATPase activity are critical steps during the ATPase reaction [83, 85, 87]. The activation is thought to be – in analogy to other GHKL-ATPases and based on biochemical data – the result of a strand-swapping reaction involving the very N-terminal amino acids [88]. Evidence is accumulating that the rate-limiting step of the hydrolysis reaction is the conformational change leading to the N-terminally dimerized state [88]. While it has not yet been possible to exactly map the order of conformational changes, mutagenesis studies allow defining of regions that help to facilitate this reaction [88]. A rational model for the N-terminal dimerization reaction can be envisioned. First, the swapping strand has to be released from its own domain, by breaking interactions of the α-helix and the accompanying ATP lid. Based on dynamics studies using NMR spectroscopy, it is assumed that ATP binding might facilitate the N-terminal dimerization reaction. The closure of the ATP lid after ATP binding might set this strand free [90]. One mutant – called Δ8-Hsp90 and which lacks the first eight amino acids – was found to form the N-terminal dimers much more efficiently and is thus especially useful for the investigation of the ATPase cycle. The rate-limiting step is shifted here from formation of the N-terminal dimer to the opening of the N-terminal dimer after hydrolysis [88]. These steps might facilitate the interaction with the middle domain to form the fully functional ATPase-active state. Unfortunately, it is not yet possible to determine whether the interaction with the middle domain precedes or follows the N-terminal dimerization reaction.

This cycle of conformational changes is based mostly on studies with yeast Hsp90, but, judging from the high degree of homology, it is assumed to be identical for the other Hsp90-like proteins. The ATPase activities determined for HtpG from *E. coli* (0.4 min^{-1}), yeast Hsp90 (0.5 min^{-1}), chicken TRAP1 (0.4 min^{-1}), and chicken Grp94 (0.4 min^{-1}) [11, 81, 91, 92] are quite similar, while human Hsp90 (0.04 min^{-1}) is significantly decelerated [91].

Interestingly, a second nucleotide-binding site has been suggested recently in the C-terminal domain of Hsp90 [93, 94]. The involvement of this site in the ATPase reaction and the specificity of this site have not yet been established. At present it is also possible that this site is the second part of the ATPase-active site that is thought to form contacts with the γ-phosphate.

23.3.3
Interaction of Hsp90 with Model Substrate Proteins

Compared to the ATPase reaction, the main functions of Hsp90 – the binding and processing of target proteins – are not understood sufficiently. Several studies attempted to identify the substrate-binding site using nonnative proteins such as thermally destabilized citrate synthase [82, 95–98], luciferase [97, 99], rhodanese [100], chemically destabilized insulin [82, 101], and thermally inactivated casein

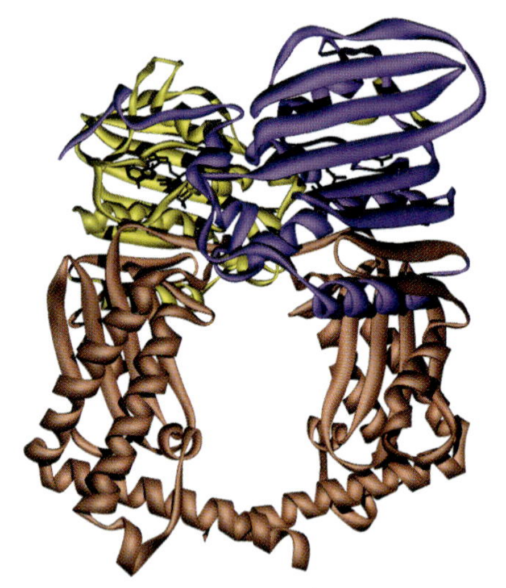

Fig. 23.4. Crystal structure of a dimeric GyraseB fragment. The crystal structure of GyraseB in the AMP-PNP complexed N-terminal dimerized conformation was solved by Brino et al. [78]. The coloring of the structure highlights the mechanistic features of the N-terminal dimerization reaction. The swapped strand is colored purple, the N-terminal ATP-binding site is yellow, and the middle domain, including parts of the dimerization site, is red. The PDB accession number for this structure is 1EI1.

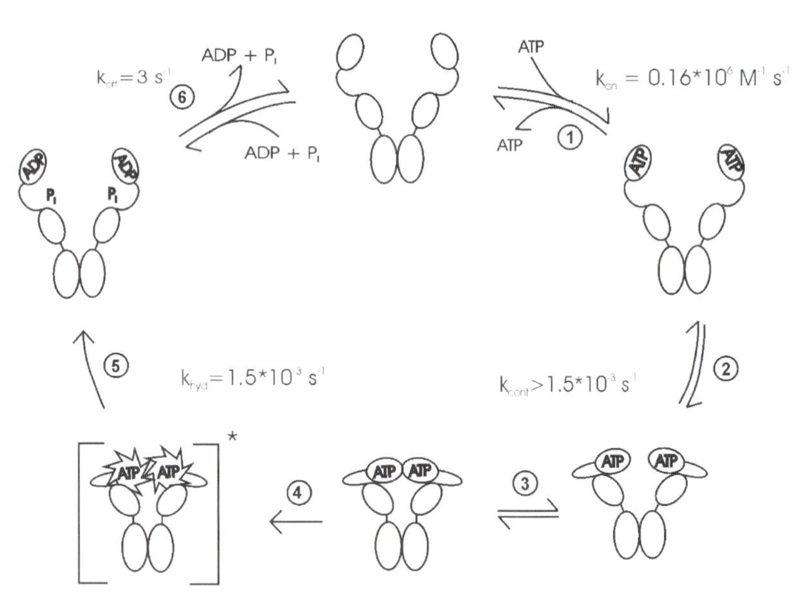

Fig. 23.5. ATPase cycle of Hsp90. The ATPase cycle of Hsp90 combines biochemical and structural evidence. The numbers are based on the work of Weikl et al. [84] and represent kinetic studies of yeast Hsp90 performed at 25 °C. ATP hydrolysis is supposed to happen within the N-terminal dimerized state as indicated by the asterisk. There is no evidence so far for the order of the conformational changes.

kinase [102–104]. These studies led to the identification of chaperone-active sites in all major domains of Hsp90 and thus show the difficulty in understanding Hsp90 interaction with substrate proteins. Many of these assays are based on the aggregation of proteins and thus represent rather complicated systems. This excludes complete understanding of the molecular interactions that influence these processes and thus does not allow obtaining of quantitative data. Based on these studies, the chaperone activity of Hsp90 originally was proposed to reside in the N-terminal domain and in a construct comprising the middle and the C-terminal domain [82, 100]. An N-terminal peptide-binding site has been described based on interactions with reduced insulin B-chain and several peptides including the viral VSV8 peptide [82]. The C-terminal construct instead was found to be more promiscuous with respect to interaction partners. Additionally, the interaction of Hsp90 with substrates in the C-terminal region was found to be independent of ATP binding or hydrolysis. One study addressed the specific state of the substrate during the interaction process. Here, using citrate synthase as a model target protein, it was observed that the Hsp90-interacting species is native-like and thus does not represent a fully unfolded or aggregated protein [95]. This is in agreement with in vivo studies, which suggest that Hsp90 is not involved in de novo protein folding [105]. Additionally, using the Hsp90 homologue from the ER, Grp94, attempts to identify substrate-binding sites were made [106–108]. This protein is thought to bind peptides tightly, which allows their co-purification [106, 109]. The interaction of Grp94 with peptides is still not understood completely, although recent data highlight the possibility of an N-terminal peptide-binding site that is controlled by nucleotide binding and radicicol [106, 109].

23.3.4 Investigating Hsp90 Substrate Interactions Using Native Substrates

Although more than 100 authentic Hsp90 substrates have been identified to date, the large-scale purification of these proteins presents great challenges. Therefore, unlike for GroE, it has not yet been possible to reconstitute a nucleotide-dependent processing of substrate proteins in the case of Hsp90. So far, in particular the protein kinases Lck and eIF2α kinase [110], the transcription factor MyoD [111, 112], the reverse transcriptase of the hepatitis B virus [113, 114], and the ligand-binding domain (LBD) of the steroid hormone receptor [91, 115, 116] have been used in interaction studies. The first three proteins were investigated in combination with truncated Hsp90 mutants. The resulting complex interaction patterns did not allow an unambiguous identification of a binding site. Studies by McLaughlin et al. regarding the effect of the LBD on the ATPase of Hsp90 showed a coupling of the ATPase cycle to substrate binding. Binding to Hsp90 appeared only weak; nevertheless, the LBD was found to stimulate the ATPase activity of human Hsp90 by a factor of 200 [91]. The mechanistic aspects of this stimulation remain unknown; however, studies like these will ultimately lead to a deeper understanding of the reactions involved in the processing of protein substrates.

It can be anticipated that the conformational changes that were observed as necessary for the hydrolysis of ATP might be translated into the change of substrate conformation. This would require the movement of two substrate-binding sites against each other in an ATP-dependent reaction. The unambiguous identification of these sites remains to be achieved to further understand the chaperone function of Hsp90.

23.4
Partner Proteins: Does Complexity Lead to Specificity?

Several partner proteins of Hsp90 were consistently identified in Hsp90-substrate complexes of eukaryotes, and most of them were later found to bind directly to Hsp90. The investigation of these complexes, in the absence of substrates, led to important information about the assembly of the Hsp90 machinery. All partner proteins described to date are found only in eukaryotic cells and thus might present the evolutionary adaptation of the Hsp90 system to an increasing number of substrates. For some of the partner proteins, pronounced substrate specificity has been observed.

23.4.1
Hop, p23, and PPIases: The Chaperone Cycle of Hsp90

The first partner proteins were detected in complexes with SHRs. These macromolecular assemblies were described in 1966 as 9S receptor complexes [117], implying that several proteins in addition to the SHR were assembled into these structures. Subsequently, Hsp90 and some of its partner proteins were identified as the key components [118]. Attempts to reconstitute these complexes led to the observation that the capacity of SHRs to bind hormones is strictly connected to the assembly into Hsp90-containing protein structures. It additionally became evident that the assembly process appears to be a chronological progression through several distinct complexes [119, 120]. Dependent on the involvement of different partner proteins, these complexes were termed "early complex," "intermediate complex," and "mature complex." The partner proteins were identified as Hop [121], p23 [122], Cyp40 [123], FKBP51, and FKBP52 [124, 125]. Yeast proteins with considerable homology were identified for Hop (Sti1), Cyp40 (Cpr6, Cpr7), and p23 (Sba1) [126]. Detailed studies by Smith and coworkers led to a chaperone cycle (Figure 23.6) that describes the chronological interaction of these partner proteins with Hsp90-SHR complexes [120]. Proteins of the "early complex" were identified to be Hsp70 [127] and Hsp40 [128, 129]. In the "intermediate complex," the proteins Hsp70, Hop, and Hsp90 were found, while proteins of the "late complex" are Hsp90, p23, and FKBP51, FKBP52, or Cyp40 [130]. Similar heterocomplexes can be found in yeast and mammals [126]. The progression time through the cycle has been estimated to be about 5 min.

Recent biochemical experiments have shown that the ATP cycle is the potential driving force behind this sequence of interactions. Based on studies using the yeast system, a thermodynamically valid cycle for the exchange of partner proteins was proposed [131]. The first biochemical evidence for the chaperone cycle in yeast came from investigations using the yeast proteins homologous to Hop and Cyp40 [132]. It was shown that the protein Sti1 is a high-affinity inhibitor of Hsp90 ATPase and that Cpr6 could replace Sti1 from Hsp90. The binding site for both proteins was identified to be in the C-terminal region [132]. Further information about the inhibitory mechanism of Sti1 was obtained by the identification of an additional weak binding site in the N-terminal domain of Hsp90 [9, 133]. The inhibitory mechanism was found to be noncompetitive concerning ATP. Interestingly, Sti1 inhibits the N-terminal dimerization reaction required for efficient ATP hydrolysis [133]. However, ATP can still bind to this inhibited conformation. This static complex can be resolved efficiently by the combined action of ATP, Sba1, and Cpr6 [130, 131]. Sba1 was found to bind only to an N-terminal dimerized conformation of Hsp90 that can be accumulated by using the non-hydrolyzable ATP analogue AMP-PNP [83, 134, 135] or by Hsp90 mutants in which the rate-limiting step is shifted from N-terminal dimerization to ATP hydrolysis [131]. The binding constant of Sba1 to this conformation is in the nanomolar range. Binding results in a decrease of the turnover by about 60% [131, 136].

Thus, it is possible to describe the chaperone cycle as an exchange of partner proteins that is driven by the conformational changes of the Hsp90 molecular chaperone. As these rearrangements are the result of the hydrolysis reaction, the exchange of partner proteins itself is driven by ATP turnover. It can be envisioned that substrate turnover, at least in the presence of these partner proteins, is achieved by conformational changes that are driven by ATP hydrolysis (Figure 23.7).

Of particular interest is how the progression of the substrate through the individual complexes affects the conformation of the substrate. For SHRs, the substrate enters the chaperone cycle probably by interaction with Hsp40 proteins [137]. These proteins recognize hydrophobic surfaces on nonnative proteins with relatively low specificity. The interaction of Hsp40 proteins with Hsp70 allows a transfer of the substrate to Hsp70. Hsp70, itself an ATPase, interacts with Hop. For the yeast system, it was shown that Sti1 activates the ATPase activity of yeast Hsp70 tremendously and thus might facilitate the processing of the substrate in this early complex [138]. Substrate transfer to Hsp90 occurs in the intermediate complex and might be accomplished by ATP hydrolysis of the Hsp70 component [139]. Whether the substrate is already active in this complex is not clear. The exchange of cofactors leading to the mature complex clearly results in an active SHR [71]. Hormone binding to this complex is allowed and results in the release of SHR from the chaperoning machinery [71]. Progression through the chaperone cycle may occur even in the absence of hormones, and it might be speculated that ATP is turned over in the p23-containing mature complex. This is followed by release of p23 and the release of SHR. If the SHR is hormone-free, it will again be a client for the reloading machinery of Hsp40/Hsp70. Thus, the low p23-inhibited ATPase activity of Hsp90 in the mature complex would function as a timer, controlling the active state of the SHR.

In addition to the partner proteins mentioned in the context of the SHRs, many

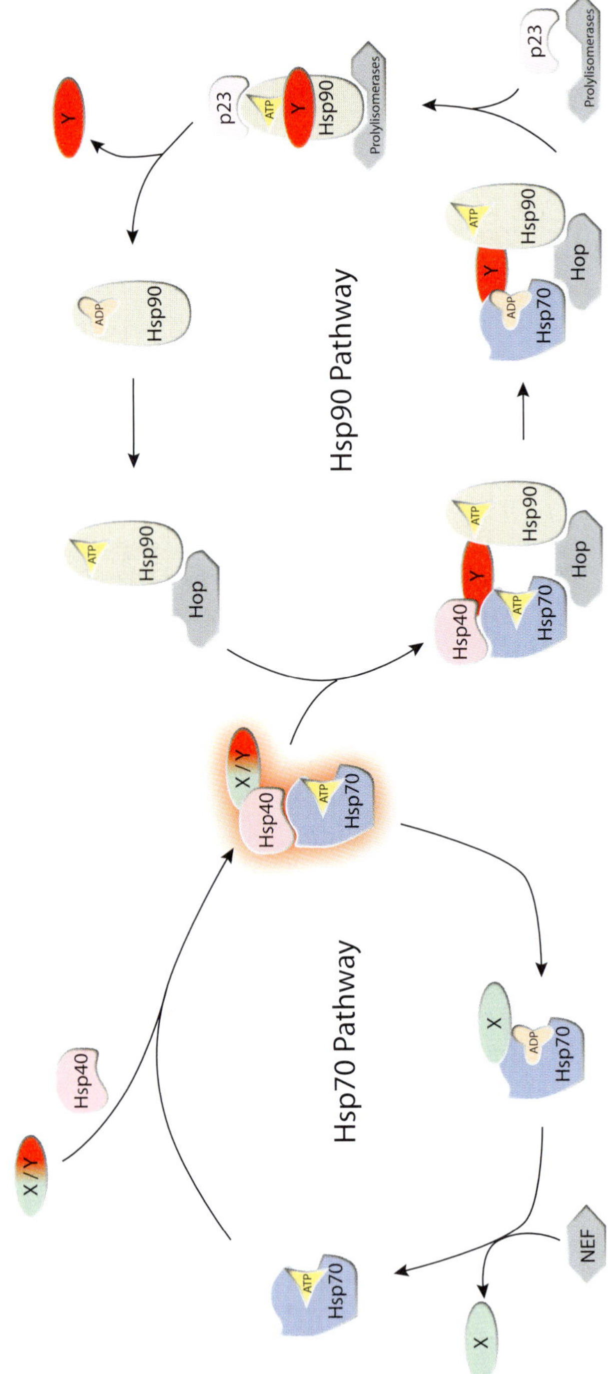

Fig. 23.6. The chaperone cycle of Hsp70/Hsp90. This cycle was established by Smith et al. [120] for the substrate protein glucocorticoid receptor. The substrate is transferred from Hsp70 to Hsp90 and further processed there to reach its activated state. This cycle may look different for other substrates as, especially in the case of kinases, different partner proteins can be found in Hsp90-containing chaperone complexes (see also Wegele et al. [210]).

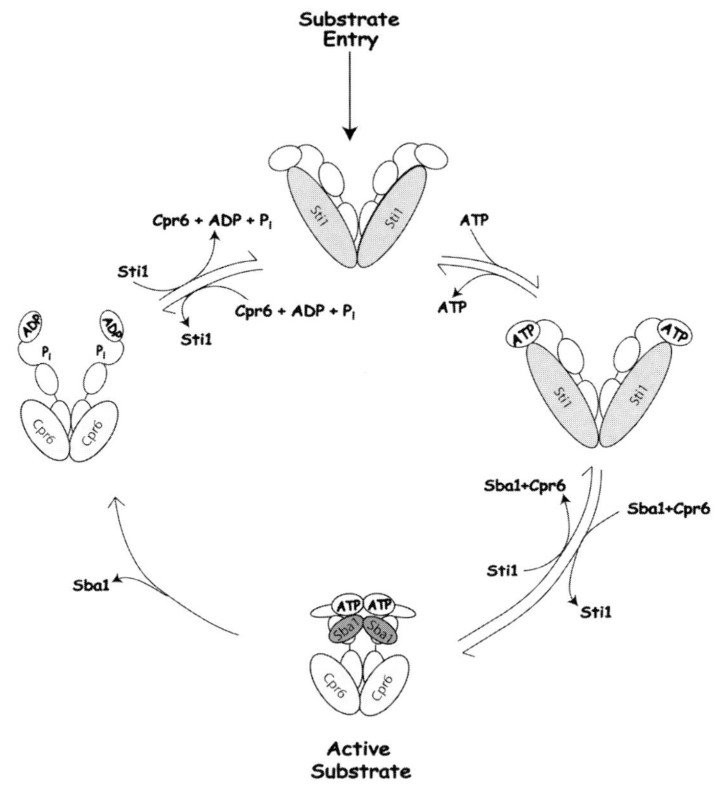

Fig. 23.7. ATPase and co-chaperone binding. Based on the conformational changes during the ATPase cycle, it is possible to integrate the co-chaperones into this cycle. Some of the co-chaperones, such as p23 and Sti1, clearly have different affinities for specific conformations of Hsp90. Therefore, the exchange of co-chaperones as observed in the chaperone cycle may be the result of ATP-induced conformational changes of Hsp90.

more are known in the yeast and mammalian systems. The individual functions of these proteins in the context of the Hsp90 chaperone activities are, with some exceptions, unknown. A striking feature is the presence of TPR domains in a number of partner proteins (Figure 23.8).

23.4.2
Hop/Sti1: Interactions Mediated by TPR Domains

The protein Hop has been identified as outlined above by the specific interaction with Hsp90-containing SHR complexes. Further experiments resulted in the observation that this protein greatly enhances the ability to assemble the SHR heterocomplexes from purified proteins [140]. This probably is due to its ability to work as an adaptor protein between Hsp90 and Hsp70. Three TPR domains have been identified in the protein Hop. Each of these domains is composed of three tandem

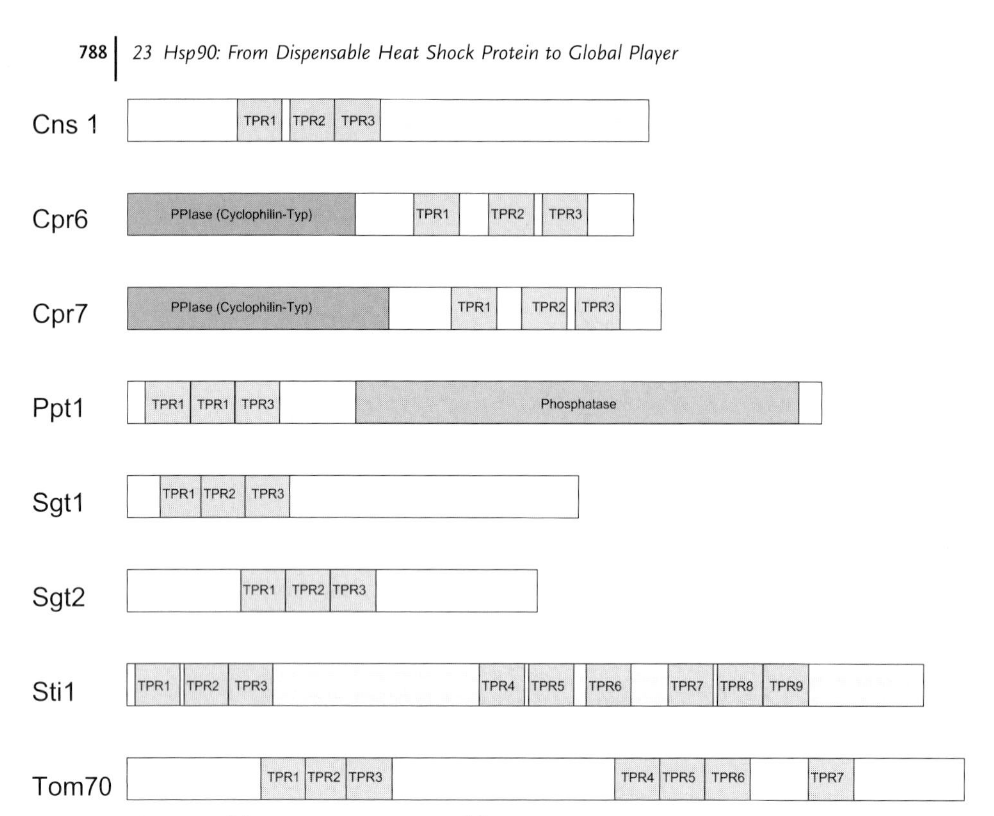

Fig. 23.8. Schematic representation of the TPR-containing Hsp90 partner proteins in yeast and the localization of the TPR domains within the protein.

TPR motifs. A TPR motif is formed structurally by two antiparallel α-helices connected by a turn (Figure 23.9). Therefore, a full TPR motif is built from six (in some cases seven) α-helices that form a binding site for a peptide [8, 141].

In the case of Hop, the specific interactors of the TPR motifs are known only partially. TPRI (the most N-terminal) binds Hsp70, and TPRIIa (the middle domain) binds Hsp90; the function of TPRIIb is not yet fully understood [8, 142, 143]. The same arrangement can be found in the yeast homologue of Hop, Sti1. In both cases, the main interacting peptide is the very C-terminal extension of the chaperones, which is formed by the amino acid sequence MEEVD. As described before, Hop/Sti1 function not only includes binding to Hsp90 and Hsp70. Sti1 is known to inhibit the ATPase activity of yeast Hsp90, and this inhibitory interaction has been found to be the result of two binding sites [132, 133]. Therefore, a more complex interaction pattern is found in this case. The same is true for the interaction of Sti1 with the yeast Hsp70 protein Ssa1. Sti1 strongly activates the ATPase activity of this protein by accelerating the hydrolysis reaction, indicating a more complex interaction as well [138]. None of these effects have been observed for

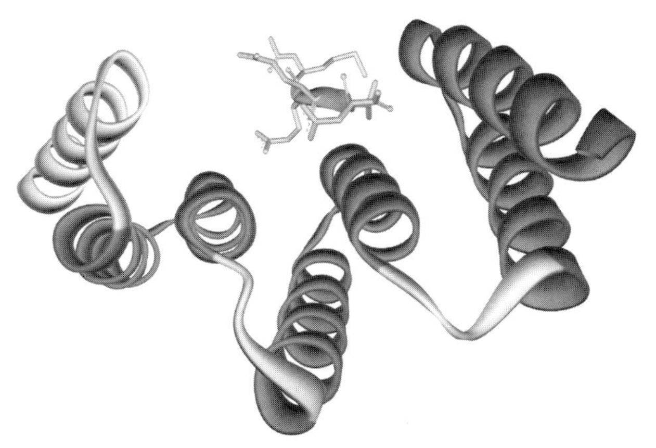

Fig. 23.9. TPR binding to the C-terminal peptide of Hsp90. The TPR domain found in many of the Hsp90 co-chaperones binds with considerable affinity to the C-terminal peptide of Hsp90. The crystal structure of the middle TPR domain of Hop in complex with the pentapeptide MEEVD was solved by Scheufler et al. [8]. Its PDB accession number is 1ELR.

the human Hsp90 system, which indicates that the regulation of the Hsp90 system has been subject to evolutionary changes.

23.4.3
p23/Sba1: Nucleotide-specific Interaction with Hsp90

The protein p23 was also identified in the context of the chaperone cycle of Hsp90 [122]. Its presence leads to a much higher yield of active SHRs [144]. The participation of p23 in the heterocomplexes remarkably stabilizes these assemblies and thus keeps SHRs in the active conformation much longer [134, 144]. This is achieved by specifically binding to a nucleotide-induced conformation of Hsp90 [130, 135]. This conformation has been shown to be N-terminal dimerized [87]. Truncation of the dimerization site resulted in a complete loss of p23-binding activity [9].

p23 is a small, acidic protein. It is composed of an antiparallel β-sandwich built from 10 strands (Figure 23.10). This domain has been shown to be responsible for the interaction with Hsp90 [145, 146]. The C-terminal of this domain is a flexible extension of about 70 amino acids. This is the chaperone-active site of p23, but it does not influence Hsp90 binding [145, 146]. Interestingly, the N-terminal structured part of p23 shares strong similarity with the structure of the small heat shock protein Hsp16 from *Methanococcus jannaschii*. The significance of this observation is unknown, but, based on the chaperone activity of p23, it may well be that they share a common origin.

Fig. 23.10. Structure of p23. The crystal structure of p23 was solved by Weaver et al. [146]. Its PDB accession number is 1EJF.

23.4.4
Large PPlases: Conferring Specificity to Substrate Localization?

Analysis of SHR-containing heterocomplexes resulted in the identification of the proteins FKBP51, FKBP52, and Cyp40. These proteins have a domain structure that is composed of a PPIase domain of either the FKBP type (FKBP51, FKBP52) or the cyclophilin type (Cyp40). This peptidyl-prolyl *cis/trans* isomerase activity [147, 148] is known to accelerate the *cis/trans* isomerization of peptide bonds prior to proline residues. Due to the specific conformation of proline, these bonds have a remarkably slow rotation rate and thus need assistance to flip from the *cis* to the *trans* position and vice versa. The PPIase activity of the large PPIases is quite low compared to related enzymes. In addition to the PPIase domain, they contain a TPR domain that mediates the interaction with Hsp90 (Figure 23.11).

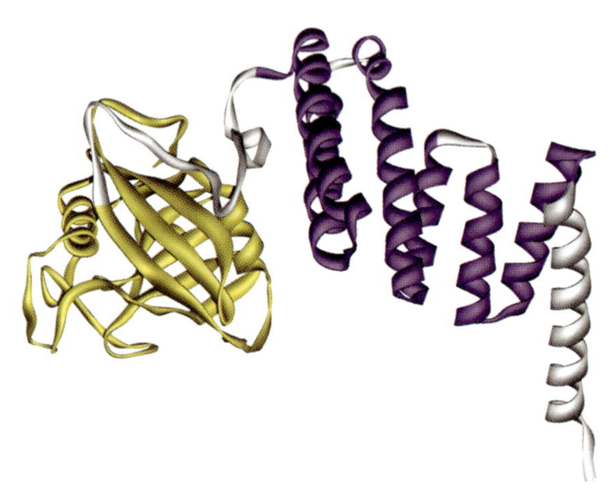

Fig. 23.11. Structure of FKBP51. The crystal structure of FKBP51 was solved by Taylor et al. [211]. Its PDB accession number is 1IHG. Highlighted in different colors is the domain organization of this protein. The N-terminal PPlase domain (yellow) is followed by a TPR-containing domain (purple) that serves as interaction site with Hsp90.

All these PPIases were found to bind to the C-terminal end of Hsp90 and the accompanying sequences [9, 133]. Hsp90 binding by the TPR-containing large PPIases is governed not only by the TPR domain but also by other C-terminal sequences outside this binding site [149, 150]. In addition, the appearance of the specific PPIase in substrate complexes seems to be determined by the substrate. Especially the duplication of these proteins in higher eukaryotes and the appearance of two classes of PPIases (the FKBPs and the cyclophilins) as TPR-containing Hsp90 co-chaperones raise important questions about the specific function of these proteins for specific substrates. Interestingly, the large PPIases come in different numbers in different organisms – two cyclophilins in yeast, one FKBP in *C. elegans*, and two FKBPs plus one cyclophilin in mammals. This may represent a further indication for the evolutionary adaptation of the Hsp90 system towards specific chaperone requirements. Analysis of the SHR-containing complexes showed the presence of all three PPIases in these complexes, although probably not simultaneously [151].

Differences regarding specific functions of the individual large PPIases have been reported. In yeast, Cpr6 is nonessential, whereas deletion of Cpr7 results in a slow-growth phenotype. The reasons for this are unknown [152]. In mammals, special focus has been set on differences between FKBP51 and FKBP52. Both proteins show chaperoning activity [153, 154] and both bind to the C-terminus of Hsp90 with comparable affinity [148]. One specific difference might be the appearance of a nuclear localization signal in the primary sequence of FBKP52. Indeed, FKBP52 appears to be localized to the nucleus, and specific interaction with the transcription factor and Hsp90 substrate HSF1 has been reported [155]. Another interesting feature is the complex formation between FKBP52 and dynein [156, 157]. This interaction may be required for the transport of Hsp90 substrates to the nucleus [158]. This has been demonstrated recently for the transport of transcription factor and the Hsp90 substrate p53 from the cytosol to the nucleus, which occurs in a complex containing Hsp90 and FKBP52 [159]. For SHRs, which also require a shuttling system from the cytosol to the nucleus, a similar dependency on FKBP52 has been observed [160]. Therefore, evidence is accumulating that the differential interaction of Hsp90 substrate complexes with PPIases also determines the cellular localization of the active substrate. Thus, they add specificity to the Hsp90 system.

23.4.5
Pp5: Facilitating Dephosphorylation

Pp5/PPT is a phosphatase that associates with Hsp90 based on TPR regions and their interaction with the C-terminus of Hsp90 [132]. The specific function of Pp5 in the context of Hsp90 substrate complexes is unknown, although several reports indicate that Pp5 as well might be involved in the activation process of SHRs [161]. Binding, at least in vitro, appears to occur in competition with FKBPs and other TPR-containing proteins [132]. The structure of the TPR-domain of Pp5 was solved before any other TPR structure was known [141].

The involvement of phosphorylation on the functionality of the Hsp90 system and on the activity of many Hsp90 substrate proteins has not been investigated in detail. Thus, it is unknown what specific function the phosphatase might serve. Interestingly, many of the known substrates of Hsp90, such as the heat shock factor and several kinases, are clearly regulated by phosphorylation and dephosphorylation, and thus there are ample possibilities to include phosphatases into the regulatory functions of Hsp90-containing protein complexes.

23.4.6
Cdc37: Building Complexes with Kinases

Of particular interest in the context of protein kinase substrates is the co-chaperone Cdc37. Cdc37 was identified as a part of Hsp90 complexes with Src kinase in eukaryotes [162, 163]. In these complexes, Cdc37 is thought to bind to Hsp90 with its C-terminal part and to the kinase with its N-terminal part [164]. This interaction seems to confer specificity to the otherwise promiscuous Hsp90-substrate interaction. Cdc37-containing complexes were sometimes found to be multi-protein complexes in the megadalton range, involving several other proteins [165]. In addition, the possibility exists that Cdc37 is also able to work independently in the activation process of kinase targets [166].

Besides the many kinases that are known to associate with Hsp90 and Cdc37, it is still not clear exactly what the function of these two proteins in the context of the activation of kinases is. Src kinase appears to be shuttled by the system from the nascent state to the cell membrane, where the kinase becomes activated upon myristoylation and insertion of the N-terminus into the plasma membrane [163, 167, 168]. Disruption of the Hsp90 system leaves these substrates inactive and prone to degradation. The same has been observed for most other tyrosine kinases of the Src family, many serine/threonine kinases, and several kinases involved in cell cycle control such as Cdk4 [169]. Experiments using heterologously expressed Src kinase in yeast showed that yeast Cdc37 can substitute for human Cdc37 in the activation of the kinase. This is remarkable, as significant homology between yeast and human Cdc37 is restricted to 25 amino acids in the N-terminal part of the protein [170].

The interaction of Cdc37 with Hsp90 results in low-affinity inhibition of the ATPase activity [171]. Recently, the crystal structure of human Cdc37 with the N-terminal domain of yeast Hsp90 was solved [172]. The observed interaction between the N-terminal domain of Hsp90 and the co-crystallized Cdc37 fragment is remarkable (Figure 23.12, see p. 794). Cdc37 binds as a dimeric protein to the two N-terminal domains of Hsp90. This orientation prevents the N-terminal dimerization reaction required for ATP hydrolysis by Hsp90 and explains the inhibitory effect of Cdc37 on the ATPase activity of Hsp90 [172]. Its direct interaction site is the ATP lid, which has been implicated in the ATP hydrolysis reaction. Thus, as with p23 and Hop/Sti1, a mechanism has been found to couple the binding of a co-chaperone to a distinct step in the ATPase cycle of Hsp90. According to genetic experiments, Cdc37 shares overlapping functions with Sti1 in vivo and thus appears to function in the loading of substrates onto Hsp90 complexes [173].

This simplistic view of Cdc37 as a kinase-specific cofactor of Hsp90 is challenged by recent discoveries [174]. In addition to functions in the activation of Hsp90 substrates other than kinases, Cdc37 seems to have Hsp90-independent functions. Furthermore, it has been found to form complexes with the Hsp90 co-chaperone Sti1, even in the absence of Hsp90 [175]. In addition, Cdc37 is known to be a molecular chaperone on its own [176].

23.4.7
Tom70: Chaperoning Mitochondrial Import

Tom70 is an integral part of the mitochondrial import machinery. It serves as a receptor of preproteins in the outer mitochondrial membrane. The substrates to be imported into mitochondria are transferred from Tom70 to the mitochondrial import channel. Tom70 was found to associate with Hsp90 via its TPR domain [177]. The association allows delivering certain preproteins from the cytosol onto the Tom70 receptor as a first step of mitochondrial import. This interaction could be shown to be required for the transport of several targets into the mitochondrial matrix [177]. Further experiments showed that the ATPase activity of Hsp90 is involved. Interestingly, both Hsp70 and Hsp90 are able to interact with the single TPR domain of Tom70 in yeast and mammals [177]. The individual affinities of these chaperones for Tom70 vary in different organisms. In humans, Hsp90 appears to be the preferred cytosolic chaperone to deliver preproteins, whereas in yeast most preproteins are delivered by cytosolic Hsp70 [177]. Thus, this example indicates that besides the well-documented involvement of Hsp90 in the activation of signal transducers, some general folding functions remain to be discovered. The specific function and localization of Hsp90 in complex with its substrate protein may be governed to a large degree by its association with a specific co-chaperone [178].

23.4.8
CHIP and Sgt1: Multiple Connections to Protein Degradation

Hsp90 has long been speculated to be involved in protein degradation in addition to its role in protein folding [60]. In particular, the fast degradation observed for several substrates after Hsp90 inhibition hints towards this possibility [59, 60]. The recent identification of the co-chaperone CHIP in mammalian cells has intensified these considerations, as this protein apparently directs substrates towards degradation by the proteasome [179]. CHIP is composed of two domains: a TPR domain mediating the interaction with Hsp70 and Hsp90 and an ubiquitin-ligase domain [180]. The latter domain is involved in the attachment of ubiquitin chains to degradation-prone proteins [180]. Therefore, it can be envisioned that via this

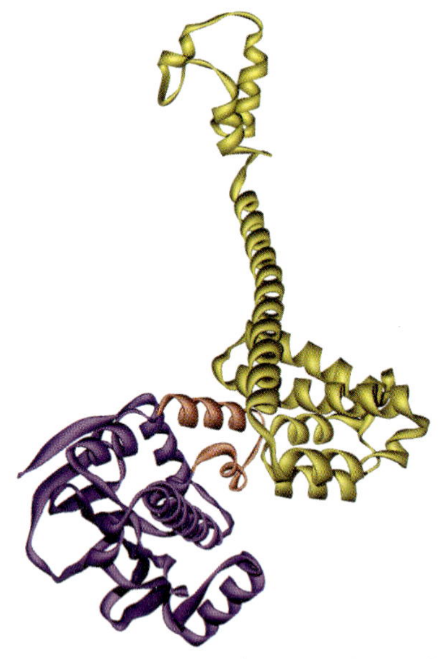

Fig. 23.12. Interaction between Cdc37 and Hsp90. The crystal structure of the complex between the N-terminal domain of Hsp90 and a fragment of the co-chaperone Cdc37 was reported by Roe et al. [172]. In this structure the C-terminal part of Cdc37 (yellow) interacts with the N-terminal domain of Hsp90 (purple). The primary interaction site is the ATP lid (red). The PDB accession number of this structure is 1US7.

partner protein, Hsp90 substrates are marked for degradation. How the decision between folding and degradation is made is the subject of intense research.

Sgt1, another Hsp90 co-chaperone that was identified recently, is also involved in protein degradation. Sgt1 is part of the SCF (Skp1-Cul1-F-box) ubiquitin-ligase complex [181] and was found to bind to Hsp90 [182]. Disturbance of its interaction with Hsp90 results in defects in genomic defense mechanisms of plants [183, 184]. The interaction of Sgt1 with Hsp90 is not fully understood. Recent results indicate that the domain of Sgt1, which binds to Hsp90, is similar to p23. Binding seems to be slightly ATP-dependent [185, 186]. Interestingly, the TPR domain also present in Sgt1 appears not to be involved in Hsp90 binding. This new co-chaperone seems to localize primarily to the kinetochore, and therefore a nuclear function of Hsp90 has to be envisioned in the context of this chaperone [181].

23.4.9
Aha1 and Hch1: Just Stimulating the ATPase?

In the late 1990s, two temperature-sensitive Hsp90 variants were used in yeast to identify cellular suppressors of the *ts*-associated defects [187]. Overexpression of three proteins was found to decrease the *ts* effects. These proteins were identified as Cns1, Ssf1, and Hch1 [187]. Subsequent biochemical investigations led to the determination that Hch1 is a stimulator of the ATPase activity of Hsp90 [136]. Although binding was found to be weak (in the high micromolar range), the effect on the turnover number of Hsp90 is strong, leading to a 20-fold increase in ATP hydrolysis. Using bioinformatics approaches, a homologue of Hch1 (Aha1) was identified and shown to bind with greater affinity to Hsp90 [136]. It also stimulates the ATPase activity of Hsp90. Cellular roles for these proteins are unknown to date, although yeast experiments using heterologously expressed v-Src as an Hsp90 substrate indicate that the interaction between Hsp90 and Aha1 is important for the maturation of Hsp90 substrates [136]. While Hch1 is a fairly small protein composed of 140 amino acids, Aha1 consists of two domains. The N-terminal domain of Aha1 is homologous to Hch1, and the C-terminal domain is of unknown origin. The isolated N-terminal domain of Aha1 is sufficient to bind to and stimulate the ATPase activity of Hsp90. While Aha1 is present in all eukaryotes, Hch1 has so far been found only in yeast.

The crystal structure of the complex between Hsp90 and Aha1 suggests a mechanism for the stimulation of the ATPase activity of Hsp90 by Aha1 [188] (Figure 23.13). Aha1 binds to the middle domain of Hsp90 [80, 189]. This leads to the reorientation of a loop involved in ATP hydrolysis in other GHKL-ATPases as well. The loop is thought to interact with the γ-phosphate of the bound nucleotide, and this interaction appears to be facilitated by Aha1 [188].

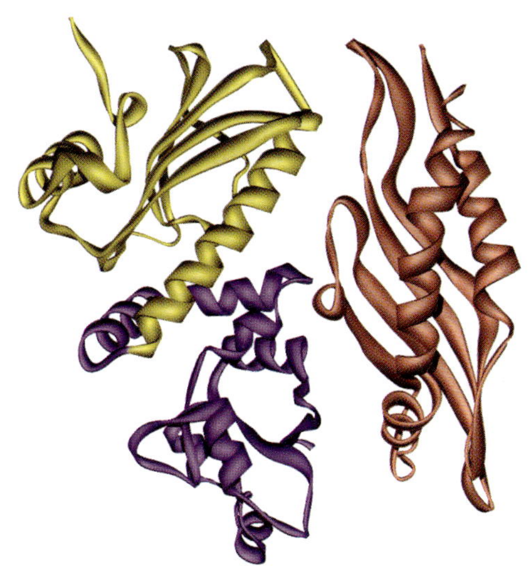

Fig. 23.13. Interaction of Aha1 with Hsp90. The crystal structure of this complex was solved by Meyer et al. [188]. Aha1 (red) interacts with the middle domain of Hsp90. The interaction sites within the middle domain reside in the ATPase-interacting part (yellow) as well as in the part supposed to interact with substrates (purple). The PDB accession number of this structure is 1USU.

23.4.10
Cns1, Sgt2, and Xap2: Is a TPR Enough to Become an Hsp90 Partner?

Several other proteins from *S. cerevisiae* contain TPR domains with homology to those of Cpr6, Hop, or Ppt1. In particular, for the TPR proteins Cns1, Sgt2, and mammalian Xap2, an interaction with either Hsp90 or Hsp70 or both has been described. Cns1, which was identified as a suppressor of temperature-sensitive Hsp90 phenotypes [187], has been detected as an Hsp90 co-chaperone in different contexts [190]. The interaction has been confirmed in *Drosophila* [191]. Recent biochemical investigations showed that Cns1, like the co-chaperone CHIP, is able to interact with both Hsp90 and Hsp70 with its single TPR domain [192]. In addition, it was found to strongly stimulate the ATPase activity of yeast Hsp70 [192]. In contrast to Sti1, the binding of Cns1 to Hsp90 does not influence its ATPase activity. Interestingly, Cns1 is one of the few essential co-chaperones in yeast [190, 193]. In this context, a close genetic relationship between Cns1 and Cp7 was reported [193] and a direct interaction between the two proteins was observed [194].

Xap2 contains a TPR domain and interacts with Hsp90 during the chaperoning of the substrate reverse transcriptase from hepatitis B virus [195, 196] and the dioxin receptor, a cytosolic receptor with some functional similarity to the steroid receptors [197]. In both cases, the function of Xap2 is unknown and the participation of further proteins in this process is very likely.

Using bioinformatics approaches, several other proteins with TPR domains have been identified. The characterization of their function in the Hsp90 chaperone cycle is just beginning. These include Sgt2, Cdc23, Cdc26, Tom34, Tpr2, and others [141, 198, 199]. Sequence homology studies so far have failed to predict a clear differentiation between Hsp90- and Hsp70-binding motifs within the TPR domains. Ultimately, an investigation of binding affinities between Hsp90/Hsp70 and these co-chaperones will have to be performed to fully establish the Hsp90 network within the cell.

23.5
Outlook

The evolution of the Hsp90 system from an obviously single, nonessential protein to an essential network of specific substrates, co-chaperones, and cellular functions highlights the importance of this protein system for the eukaryotic cells. In recent years, great progress has been achieved regarding the identification of new participants in the system, the biochemistry of the ATPase reaction, the interaction with partner proteins, and the identification of new substrates. Nevertheless, despite over 2000 publications on Hsp90, a clear understanding of its cellular function is still lacking.

Many of the studies performed with Hsp90 to date address the interaction of partner proteins in the absence of substrates. The functional aspects of the co-chaperone–Hsp90 interactions are beginning to emerge for the best-described partner proteins. For the others, information is scarce. It thus is of utmost importance to integrate the partner proteins into the ATPase cycle and to determine which partner proteins associate to heterogenic complexes and which bind mutually exclusively to Hsp90. These investigations in combination with in vitro studies on substrates will ultimately result in the elucidation of critical functions of the Hsp90 chaperone machine.

23.6 Appendix – Experimental Protocols

For the investigation of the molecular chaperone Hsp90, a set of techniques, ranging from in vivo techniques to biophysics, has been established to address the specific functions of this protein.

23.6.1 Calculation of Phylogenetic Trees Based on Protein Sequences

Phylogenetic trees for the evolutionary investigation of the Hsp90 protein family can be calculated using a combination of freely available software packages. Sequences related to Hsp90 are obtained by BLAST searches (http://www.ncbi.nlm.nih.gov/BLAST/) using either full-length Hsp90 or partial sequences thereof as input. Preliminary sequence alignment is done using the ClustalW alignment algorithm (http://www.ebi.ac.uk/clustalw/). This alignment is then refined manually to include all the necessary structural and functional information that can be derived from the literature, such as mutational studies and structural and biochemical information.

The BioEdit software (http://www.mbio.ncsu.edu/BioEdit/bioedit.html) allows manual adjustment of a large number of sequences. Subsequently, the adjusted alignment is incorporated into the Phylip format.

Phylip (http://evolution.genetics.washington.edu/phylip.html) allows calculating of evolutionary trees based on several available algorithms. These include parsimony and bootstrap calculation of distances. Rooted or un-rooted trees can be calculated, depending on the selection of specific out-groups. The resulting tree can be imported into Adobe Illustrator or any other postscript-compatible program for further graphical rearrangements.

23.6.2 Investigating the in Vivo Effect of Hsp90 Mutations in *S. cerevisiae*

The yeast homologues of Hsp90, i.e., Hsp82 and Hsc82, are essential proteins, and the knockout of both of these proteins is lethal [2]. Therefore, the investigation of specific point mutations of Hsp90 fragments with respect to the essential function of Hsp90 is possible. For this experiment, a yeast strain lacking both genomic Hsp90 versions is used. As a knockout of both genomic Hsp90 copies is lethal, viability of the yeast strain before transformation with the plasmid-encoded Hsp90 variant has to be ensured. Therefore, wt Hsp82 is supplied on a plasmid containing the URA-selection marker. In the experiment, the plasmid-encoded wt Hsp90 is exchanged for a plasmid-encoded mutant Hsp90. Only in the case that the mutant Hsp90 is fully functional is growth of the yeast strain ensured.

The fragment or *hsp90* gene carrying the mutation is subcloned into a yeast shuttle vector, e.g., the vector p2HGal or any other yeast expression vector. This vector uses the gene HIS3 as a selection marker. The Hsp82 variant of interest is expressed under the control of the galactose promoter. Other combinations of selection marker (TRP, LEU, or ADE) and promoter (constitutive GPD promoter or Cu-inducible promoter) are available as well and might be equally useful. To start expression of the Hsp90 variant, cells are plated on medium with galactose as the only carbon source. Residual glucose will lead to the repression of the galactose promoter. Lack of galactose also leads to repressed expression. Because the reduction of intracellular glucose levels will require time after shifting to galactose-containing minimal medium, it is advised that yeasts be shifted to raffinose medium for several hours to allow complete depletion of intracellular glucose before they finally are transferred to the galactose medium. This will allow complete depletion of the intracellular glucose levels and thus will lead to the immediate expression of the galactose-controlled gene after exposure to galactose. Generally, it is good practice to prove the expression of the Hsp90 variant by Western blotting to avoid being misled by negative results due to low expression levels.

After transformation with the p2HGal-encoded Hsp90 variant, yeasts are grown on medium lacking HIS and URA and containing galactose as the carbon source. This selection ensures the maintenance of both plasmids. It is possible to counter-select against URA-containing plasmids by using 1% 5-fluororotic acid (5-FOA) in the medium [200]. This substance is converted into a toxic compound by the URA gene used as selection marker on the wt Hsp90 plasmid. Therefore, a strong selection pressure is applied for the loss of this plasmid, due to the accumulation of toxic compounds in the presence of the URA gene. With the loss of this plasmid, the wt Hsp90 gene is lost. As a consequence, cell proliferation is possible only if the supplied Hsp90 variant on the p2HGal plasmid is functional.

Additionally, in some cases, a dominant negative effect of truncated versions of Hsp90 has been observed in yeast, leading to slower growth, if these variants were overexpressed [201]. To investigate this, the same plasmid as for the complementation experiment can be used. Overexpression is started by using galactose-containing medium, and the growth of the yeast strain containing the truncated version is compared to a strain containing the wt version or an empty plasmid.

23.6.3
Well-characterized Hsp90 Mutants

A number of mutations in Hsp90 have been described that are particularly useful for the investigation of functional aspects. The specific amino acid numbers refer to yeast Hsp82. Conserved residues can be found in homologous Hsp90 proteins.

The first set of mutations affects the ATPase activity of Hsp90. This activity is essential in vivo, and amino acid residues involved in this reaction have been identified [81, 202]. Amino acid Asp79 participates in the binding of the nucleotide. It makes contacts at the base of the binding pocket [75]. Its mutation to Asn (D79N) leaves the structure of the mutant fully intact but strongly reduces the ability of the Hsp90 mutant to bind ATP [81, 202]. Thus, it is possible to use this mutant to investigate the importance of ATP binding for a specific reaction under investigation. Another particularly useful amino acid is Glu33. This residue participates in the hydrolysis reaction leading to ADP. The mutation of this residue to alanine (E33A) leads to an Hsp90 variant that binds ATP with mostly unchanged affinity but is unable to hydrolyze it [81, 202]. The conformational changes involved in ATP hydrolysis are blocked at the point of hydrolysis. Used in combination, the two mutants allow differentiation between the effects of nucleotide binding and hydrolysis.

A second set of mutations involves the regions implicated in the N-terminal dimerization reaction. As such mutants, deletions of the first 8, 16, and 24 amino acids were characterized [88]. These mutations can be used to investigate the N-terminal dimerization and the effects of this reaction on the binding of partner proteins. For several of the partner proteins, especially those that influence ATPase activity, N-terminal dimerization would affect their binding. The investigation of these differences relies on the known behavior of these mutants [133]. Full-length Hsp90 does not form N-terminal dimers stably unless the nucleotide AMP-PNP is used. Δ8-Hsp90 forms stable N-terminal dimers in the presence of ATP and AMP-PNP [88]. The rate-limiting step of hydrolysis is therefore shifted from the formation of N-terminal dimers to the dissociation of the N-terminal domains, allowing investigation of binding to this conformation specifically. The deletion of the first 24 amino acids results in the elimination of the ability to form N-terminal dimers.

Another set of mutations is particularly useful for in vivo studies. Several mutations were identified that show temperature-sensitive effects [105, 187]. These mutations can be used to investigate essential cellular functions of Hsp90, as the yeast strains are viable at permissive temperature and inhibit growth at elevated temperatures. One such mutation is G170D. This mutant is viable only below 34 °C [105]. Other temperature-sensitive mutants are T101I, E381K, and G313S [105]. These mutations also have been used to perform suppressor screens in order to identify new components of the Hsp90 system.

For the interaction of Hsp90 with the large number of TPR proteins, a mutant that lacks the C-terminal extension MEEVD has been used extensively. Exchanging this protein for the wt Hsp90 in yeast leads to no detectable phenotype [7]. Especially for the investigation of in vitro interactions, this mutant has proven to be a great control, as the deletion of these residues leads to complete inhibition of the binding of several partner proteins [9, 133].

Last but not least, it is worth mentioning that the addition of a His_6 tag at the N-terminus of Hsp90 has been used extensively to facilitate purification. This modification of the protein apparently does not influence the activity of Hsp90 in vivo

[126]. It allows co-purification of Hsp90 partner proteins. In vitro, the addition of the His_6 tag from the pET28 vector results in an ATPase activity about twice as high as the wt Hsp90 (unpublished data). This might be the result of changes in the N-terminal dimerization properties that are controlled to a large extent by the very N-terminal amino acids (see Section 23.3.1).

23.6.4
Investigating Activation of Heterologously Expressed Src Kinase in *S. cerevisiae*

It is widely accepted that the function of Hsp90 is the activation of a large number of substrate proteins. Yeast lacks many known Hsp90 target proteins found in higher eukaryotes. Therefore, well-known mammalian Hsp90 targets have been expressed in the cytosol of yeast. One of the best studied mammalian Hsp90 substrates is the tyrosine kinase c-Src and its oncogenic variant v-Src [163]. These proteins can be expressed in yeast from a plasmid [168]. Both proteins require Hsp90 to gain activity. Interestingly, the function of mammalian Hsp90 can be substituted by the endogenous yeast Hsp90 system [167, 168]. While overexpression of c-Src does not lead to a phenotype, v-Src overexpression is lethal, as the amount of uncontrolled tyrosine phosphorylation is obviously damaging in yeast [168]. This assay allows investigation of the functionality of non-lethal Hsp90 mutations and the influence of partner proteins on Src maturation in yeast. In addition, if non-functional Hsp90 mutants are concerned, sometimes the overexpression of these in a wt yeast background results in damaging effects on the activation of the substrate protein studied, for example, as partner proteins are depleted from the functional Hsp90 system of the cell [201]. Thus, this assay nicely complements the growth assays described before.

To determine the activity of the Src protein, global cellular tyrosine phosphorylation is measured by Western blotting [167, 168]. To ensure comparability between different yeast strains, equal amounts of cells are loaded onto the SDS-PAGE. Detection is performed using a commercially available anti-phosphotyrosine antibody.

23.6.5
Investigation of Heterologously Expressed Glucocorticoid Receptor in *S. cerevisiae*

Another known target of Hsp90 in mammals are SHRs, such as the glucocorticoid receptor (GR) or the progesterone receptor. These proteins require the function of Hsp90 and several of its partner proteins for their hormone binding activity. Like Src kinase, GR can be expressed in yeast [105, 160, 203]. Its activity can be detected using a GR-inducible reporter gene construct (β-galactosidase), which is on a second plasmid that has been co-transformed into the yeast strain. The activity of GR as a transcription factor after external addition of hormone is measured by the activity of GR-induced β-galactosidase in a reporter gene assay. This assay allows addressing of questions similar to those of the Src kinase assay. Due to the fact that Hsp90 substrates do not share obvious functional or structural homology, performing of both assays is advised, as mutations or truncations in Hsp90 may affect the substrates in different ways.

To perform this assay, the GR is expressed from a yeast plasmid. The activity of the GR is switched on by the addition of deoxycorticosterone (DOC), which acts as a GR agonist. This substance represents a long-lived, membrane-permeable analogue of the glucocorticoid hormone. To allow quantitation, it has to be ensured that equal amounts of protein extract are used, and great care has to be employed to use standardized conditions for the preparation of the samples. The activity of the reporter gene is detected using X-gal as substrate in an OMPG test. The production of a blue color is quantified at 495 nm in a spectrometer. Investigation of the β-galactosidase activity at different optical densities (OD_{600}) greatly increases the resolving power of this assay. The result indicates whether the Hsp90 system in the specific yeast strain is sufficient to ensure activation of the GR receptor. Using this assay, it has been shown that some Hsp90 fragments, when overexpressed, do influence the activity of the GR in a negative way [201], matching the dominant negative effects also observed in some growth assays.

23.6.6
Investigation of Chaperone Activity

Most of the investigations regarding the chaperoning activity of Hsp90 have relied on traditional chaperone assays. The specific protocols to perform chaperone activity measurements are described elsewhere in this book (see Chapter 5). Several of these assays give positive results for Hsp90 as well. In particular, Hsp90 markedly slows down the thermal inactivation and aggregation of citrate synthase [98]. Also, the aggregation of insulin upon reduction of its disulfide bonds is inhibited by Hsp90 [82]. Assays employed to investigate the potential of Hsp90 to suppress aggregation of proteins also include the proteins rhodanese and luciferase [97, 100]. The quantitation of aggregation relies on either UV absorbance measurements at 360–600 nm in a spectrometer or the measurement of light scattering. To detect light scattering, a conventional fluorescence instrument can be used. The excitation and the emission wavelength are set to the same value (between 360 and 600 nm). Generally, lower protein concentrations can be used in light-scattering setups, as the sensitivity is much higher. For citrate synthase aggregation, 2 μM of citrate synthase gives sufficient signal in the UV/VIS measurement, while only 150 nM citrate synthase has to be used in the light-scattering measurement. Due to the very complex events leading to aggregation, it is impossible to perform a quantitative analysis of the aggregation curve.

The thermal inactivation of citrate synthase is a useful assay that allows a more quantitative description of the chaperoning activity of Hsp90. The assay can be analyzed using a single exponential function to determine the rate constant for the inactivation reaction [95]. This rate constant varies depending on the buffer conditions. In 40 mM HEPES/KOH, pH 7.5, its value is usually 0.4 min^{-1} at 43 °C. In the presence of Hsp90, this can be slowed down to 0.04 min^{-1}, indicating that

where RT represents the retention time, K_D the dissociation constant, $RT_{Monomer}$ and RT_{Dimer} the retention times of the monomer and the dimer, and c_t the protein concentration in the injection solution.

23.6.10
Investigation of Binding Events Using Changes of the Intrinsic Fluorescence

The intrinsic fluorescence of the protein can be a very useful parameter where the binding of substrates to proteins is concerned. A variety of different ligands such as ADP, ATP, geldanamycin, and radicicol bind to the N-terminal domain of Hsp90. Some of these ligands, such as radicicol, induce a huge change in intrinsic fluorescence of yeast Hsp90 upon binding, especially if the isolated N-terminal domain is investigated [88]. The advantage of this method is the low amount of protein required for the investigation (about 1 μM). In addition, here the binding event is determined in an equilibrium setup of unmodified binding partners, as no labeling or coupling is required.

However, in the case of radicicol, binding constants cannot be derived from this titration, as binding is much too strong ($K_D = 19$ nM), with a dissociation constant far below the required protein concentration. As binding is stoichiometric under these conditions, this assay can be used to estimate the number of active sites that are present in the protein preparation. Therefore, the binding curve has to be evaluated according to the following functions:

$$F = F_{ML} + (F_M - F_{ML}) \cdot \frac{[L]}{K_D + [L]} \quad (3)$$

$$[L] = \frac{-(K_D + [M]_t - [L]_t) + \sqrt{(K_D + [M]_t - [L]_t)^2 + 4 \cdot [L]_t \cdot K_D}}{2} \quad (4)$$

where F represents the fluorescence, F_{ML} the fluorescence of the liganded macromolecule, F_M the fluorescence of the unliganded macromolecule, K_D the dissociation constant, [L] the concentration of the ligand, and [M] the concentration of the macromolecule.

Other ligands do not induce a strong change in the intrinsic fluorescence of yeast Hsp90, although binding of ADP to the N-terminal domain can be observed, especially if higher salt concentrations are used [88]. For the titration with nucleotides, it is very important to avoid inner-filter effects, as these compounds show strong absorbance, even at wavelengths of 280 nm. Therefore, use of the highest possible excitation wavelength (295 nm) is advised. Otherwise, the inner-filter effect has to be taken into account, which is possible, based on the knowledge of the extinction coefficient of the nucleotide at the excitation wavelength. The resulting titration curve can be analyzed using the standard Longmuir adsorption isotherm, as binding is rather weak, with a K_D value far above the protein concentration used. The equation used for this calculation is Eq. (3).

23.6.11
Investigation of Binding Events Using Isothermal Titration Calorimetry

To analyze protein-ligand interactions in the Hsp90 system, isothermal titration calorimetry (ITC) has proved to be extremely useful in the determination of both stoichiometries and thermodynamic parameters. In this setup, the heat change upon binding of the ligand or partner protein is detected by the ITC instrument for several injections of ligand to the protein solution (usually around 35 injections are automatically performed). The successive saturation of the protein leads to a binding curve that can be analyzed by using the appropriate equation, which in most cases will be the binding of one ligand to one binding site. All parameters can be derived from one ITC experiment with great confidence, if sufficient protein ligand (usually 15 μM in the cell and 150 μM in the injection device) is used to generate a sufficiently high signal. Great care should be taken to match the buffers of the two reaction solutions in order to avoid signals arising from dilution effects of buffer components.

Experiments have been performed for the binding of radicicol, geldanamycin, ADP, and AMP-PNP to the N-terminal domain of Hsp90 [62, 81, 101]. ITC has been used to determine the stoichiometries and the affinities of the binding sites. Especially in the case of small ligands, ITC may in addition provide very useful information regarding the nature of the binding event. The binding enthalpy can be correlated with the specific interactions within the binding pocket and thus yield some structural information. Therefore, it is useful to compare either the interactions of compounds with limited changes in their functional groups with wt Hsp90 or different point mutations of Hsp90 with one ligand to gain understanding about the interactions within the binding pocket that lead to the specificity and affinity of the interaction.

Regarding the binding of partner proteins, ITC was successfully used for the systems Sti1-Hsp90, Cpr6-Hsp90, Aha1-Hsp90, Cdc37-Hsp90, and Sba1-Hsp90-AMP-PNP [80, 131, 132, 172]. The stoichiometry of these complexes is 1:1 and binding constants have been determined to be between 100 nM and 10 μM. The only disadvantage of these measurements is the high protein concentrations required, which usually range from 10 μM to 100 μM in the 1.5-mL cell and far higher in the injection device. Therefore, methods that rely on smaller amounts of proteins and produce results with similar quality are interesting.

23.6.12
Investigation of Protein-Protein Interactions Using Cross-linking

Several of the interactions within the Hsp90 system can be observed by chemical cross-linking. Glutardialdehyde and the cleavable cross-linker DTSSP have been used as cross-linkers [145]. In both cases, it is possible to visualize the cross-linking success on SDS-PAGE. For better resolution, it may be advisable to use Western blotting subsequently to identify the composition of specific bands after cross-linking.

One example in which cross-linking in particular is useful is the interaction between Hsp90 and Sba1/p23. Here, extensive investigations revealed a nucleotide-dependent association of the proteins [207]. Cross-linking was used to monitor optimum binding conditions. For both cross-linkers, the buffer must not contain free amine groups, as these would react with the cross-linker. Therefore, this experiment usually is performed in HEPES buffer (e.g., 40 mM HEPES, pH 7.5, 150 mM KCl, 5 mM $MgCl_2$). The overall volume of the reaction can be as low as 20 μL. Next, 1.6 μL of 2.5% glutardialdehyde is added to the proteins (usually 5 μg of each) and cross-linking is performed for 5 min at room temperature. Then, 5 μL of 1 M Tris, pH 8.5, is added to stop the reaction. Separation of the cross-linking species is achieved on SDS-PAGE. Depending on the size of the protein components and the cross-linking products, gradient gels, ranging from 4% to 12% acrylamide, may be used. However, limitations exist for this technique, as cross-linking artificially selects for species that have lysine residues in the right positioning. Another limitation that has been observed frequently is artificial positive results if high concentrations of cross-linker and/or proteins are used or long reaction times are employed.

23.6.13 Investigation of Protein-Protein Interactions Using Surface Plasmon Resonance Spectroscopy

Surface plasmon resonance spectroscopy (SPR) has been used recently to investigate Hsp90 partner protein interactions in detail. The application of this relatively new technique requires only small amounts of protein, and the experiments are fast and reproducibility is high. The detected signal directly monitors a binding/dissociation event. The high sensitivity, on the other hand, imposes some difficulties as well, as artificial signals arise sometimes. This makes it necessary to carefully establish binding assays and to perform the appropriate controls.

SPR theory has been reviewed extensively by others [208, 209]. In short, a light beam is sent onto a gold slide, which is on top of a dextran matrix to which the protein is coupled. The light intruding into the gold slide is reflected without changing the angle. When the angle of the intruding light is varied, under one specific angle, a resonance phenomenon is observed that leads to the production of an evanescent wave within the gold slide and to the excitation of the plasmons within this slide. Under this angle, a reflection is no longer observed, as the light energy dissipates into the chip surface. This angle, on the other hand, is very sensitive to the weight of the gold slide and the refractive index of the solution within a short distance of the gold slide. As this refractive index changes upon binding of protein to the chip surface, a change in this angle is observed. This change is directly related to the "resonance units" representing the output signal. Thus, any binding/dissociation event can be quantified. Therefore, it is possible to couple one protein covalently to the sensor surface (most often via lysine residues) and apply another protein onto the chip surface to investigate binding of a potential interaction partner. The flow-through system allows defined association and dissociation kinetics to be obtained. The combination of equilibrium approaches, where the signal height is plotted against the concentration of the applied protein, with kinetic approaches, where the observed kinetics is analyzed, offers unique approaches to study interactions.

Although theoretically very straightforward, the technique imposes some challenges. First, the protein has to be coupled to the sample cell. Several chips differing in their surfaces and coupling methods are available. As such, streptavidin-coated chips can be used to couple biotinylated macromolecules, Ni-NTA chips can be used to couple His_6-tagged proteins, and CM5 chips are available to couple covalently primary amines. While coupling in the first two cases requires only applying the protein onto the matrix, in the last case, the COOH groups of the CM5 matrix have to be activated first. This is done by EDC/NHS solution, which is part of the BiaCore coupling kit. The activated groups readily react with primary amines, e.g., lysine residues. To perform this reaction with satisfactory efficiency, the affinity of the protein towards the acidic matrix should be increased by lowering the pH of the coupling buffer below the isoelectric point of the protein. This sometimes imposes some uncertainty regarding the native state of the coupled protein. In the case of yeast Hsp90, 40 mM potassium acetate, pH 4.5, has been used successfully for coupling. Hsp90 coupled in this way has good affinity towards Sti1 and Cpr6 and several other TPR-containing proteins [147, 192]. Specific binding curves are obtained using different concentrations of partner proteins, until at sufficiently high concentrations no further increase in resonance units is observed due to complete saturation of the Hsp90 bound to the chip surface. This saturation behavior is a first indication of a specific binding event. As running buffer 40 mM HEPES, pH 7.5, 150 mM KCl, 1 mM EDTA, and 0.002% Tween20 is recommended, although other buffer composition can also be used. In some cases, it may be advantageous to couple the partner protein instead of Hsp90. One such example is Sba1. The binding of Sba1 requires specific conformations of Hsp90 that are induced by nucleotides, and it is apparently not possible to couple Hsp90 in a conformation to allow Sba1 binding. In addition, the coupling of Sba1 allows the use of different nucleotides to influence the Hsp90 conformations accordingly and investigation of the influence of nucleotides on the interaction between Sba1 and Hsp90.

Additional types of assays are competitive experiments. The coupled protein can be used as competitor. This allows investigation of an equilibrium between a soluble interaction partner and the chip-bound interaction partner. By further increasing the concentration of the soluble interaction partner, it should be possible to completely suppress the binding to the chip surface, as all the protein is already complexed in the injection solution [133, 147]. This is possible for all the interactions described, and generally this is a very powerful method to prove the specificity of the interaction. In addition, this method allows investigation of the binding affinities by an independent titration. Here, the direct and the indirect titration should lead to similar K_D values. It is not necessary that the competitor be the chip-bound protein itself. Another binding partner can be added, and if binding between the chip-bound protein (e.g., Sba1) and the third protein (e.g., Sti1) is competitive, Sti1 will reduce the amount of Hsp90 binding to the coupled Sba1. If the third protein binds by forming a ternary complex (e.g., Cpr6), the binding signal will further increase above the value recorded for Hsp90 alone [131].

Acknowledgements

The authors would like to acknowledge Didier Picard for critically reading the manuscript. We thank Stefan Walter for help with the equations and Benjamin Bösl for drawing of chemical structures. In addition, we would like to thank Lin Mueller and Harald Wegele for providing graphic material. This work was supported by grants of the DFG and the Fonds der chemischen Industrie.

References

1 Picard, D. (2002). Heat-shock protein 90, a chaperone for folding and regulation. *Cell Mol. Life Sci.*, **59**, 1640–1648.
2 Borkovich, K. A., Farrelly, F. W., Finkelstein, D. B., Taulien, J., and Lindquist, S. (1989). hsp82 is an essential protein that is required in higher concentrations for growth of cells at higher temperatures. *Mol. Cell Biol.*, **9**, 3919–3930.
3 Hickey, E., Brandon, S. E., Smale, G., Lloyd, D., and Weber, L. A. (1989). Sequence and regulation of a gene encoding a human 89-kilodalton heat shock protein. *Mol. Cell Biol.*, **9**, 2615–2626.
4 Krishna, P. and Gloor, G. (2001). The Hsp90 family of proteins in Arabidopsis thaliana. *Cell Stress. Chaperones.*, **6**, 238–246.
5 Gupta, R. S. (1995). Phylogenetic analysis of the 90 kD heat shock family of protein sequences and an examination of the relationship among animals, plants, and fungi species. *Mol. Biol. Evol.*, **12**, 1063–1073.
6 Emelyanov, V. V. (2002). Phylogenetic relationships of organellar Hsp90 homologs reveal fundamental differences to organellar Hsp70 and Hsp60 evolution. *Gene*, **299**, 125–133.
7 Louvion, J. F., Warth, R., and Picard, D. (1996). Two eukaryote-specific regions of Hsp82 are dispensable for its viability and signal transduction functions in yeast. *Proc. Natl. Acad. Sci. U.S.A*, **93**, 13937–13942.
8 Scheufler, C., Brinker, A., Bourenkov, G., Pegoraro, S., Moroder, L., Bartunik, H., Hartl, F. U., and Moarefi, I. (2000). Structure of TPR domain-peptide complexes: critical elements in the assembly of the Hsp70-Hsp90 multichaperone machine. *Cell*, **101**, 199–210.
9 Chen, S., Sullivan, W. P., Toft, D. O., and Smith, D. F. (1998). Differential interactions of p23 and the TPR-containing proteins Hop, Cyp40, FKBP52 and FKBP51 with Hsp90 mutants. *Cell Stress. Chaperones.*, **3**, 118–129.
10 Schmitz, G., Schmidt, M., and Feierabend, J. (1996). Characterization of a plastid-specific HSP90 homologue: identification of a cDNA sequence, phylogenetic descendence and analysis of its mRNA and protein expression. *Plant Mol. Biol.*, **30**, 479–492.
11 Felts, S. J., Owen, B. A., Nguyen, P., Trepel, J., Donner, D. B., and Toft, D. O. (2000). The hsp90-related protein TRAP1 is a mitochondrial

protein with distinct functional properties. *J. Biol. Chem.*, **275**, 3305–3312.

12 Bardwell, J. C. and Craig, E. A. (1988). Ancient heat shock gene is dispensable. *J. Bacteriol.*, **170**, 2977–2983.

13 Tanaka, N. and Nakamoto, H. (1999). HtpG is essential for the thermal stress management in cyanobacteria. *FEBS Lett.*, **458**, 117–123.

14 Morano, K. A., Santoro, N., Koch, K. A., and Thiele, D. J. (1999). A *trans*-activation domain in yeast heat shock transcription factor is essential for cell cycle progression during stress. *Mol. Cell Biol.*, **19**, 402–411.

15 Endo, N., Tashiro, E., Umezawa, K., Kawada, M., Uehara, Y., Doki, Y., Weinstein, I. B., and Imoto, M. (2000). Herbimycin A induces G1 arrest through accumulation of p27(Kip1) in cyclin D1-overexpressing fibroblasts. *Biochem. Biophys. Res. Commun.*, **267**, 54–58.

16 Shiotsu, Y., Neckers, L. M., Wortman, I., An, W. G., Schulte, T. W., Soga, S., Murakata, C., Tamaoki, T., and Akinaga, S. (2000). Novel oxime derivatives of radicicol induce erythroid differentiation associated with preferential G(1) phase accumulation against chronic myelogenous leukemia cells through destabilization of Bcr-Abl with Hsp90 complex. *Blood*, **96**, 2284–2291.

17 Dai, K., Kobayashi, R., and Beach, D. (1996). Physical interaction of mammalian CDC37 with CDK4. *J. Biol. Chem.*, **271**, 22030–22034.

18 Donze, O., Abbas-Terki, T., and Picard, D. (2001). The Hsp90 chaperone complex is both a facilitator and a repressor of the dsRNA-dependent kinase PKR. *EMBO J.*, **20**, 3771–3780.

19 Aligue, R., Akhavan-Niak, H., and Russell, P. (1994). A role for Hsp90 in cell cycle control: Wee1 tyrosine kinase activity requires interaction with Hsp90. *EMBO J.*, **13**, 6099–6106.

20 Rutherford, S. L. and Lindquist, S. (1998). Hsp90 as a capacitor for morphological evolution. *Nature*, **396**, 336–342.

21 Queitsch, C., Sangster, T. A., and Lindquist, S. (2002). Hsp90 as a capacitor of phenotypic variation. *Nature*, **417**, 618–624.

22 Voss, A. K., Thomas, T., and Gruss, P. (2000). Mice lacking HSP90beta fail to develop a placental labyrinth. *Development*, **127**, 1–11.

23 Lele, Z., Hartson, S. D., Martin, C. C., Whitesell, L., Matts, R. L., and Krone, P. H. (1999). Disruption of zebrafish somite development by pharmacologic inhibition of Hsp90. *Dev. Biol.*, **210**, 56–70.

24 Randow, F. and Seed, B. (2001). Endoplasmic reticulum chaperone gp96 is required for innate immunity but not cell viability. *Nat. Cell Biol.*, **3**, 891–896.

25 Gass, J. N., Gifford, N. M., and Brewer, J. W. (2002). Activation of an Unfolded Protein Response during Differentiation of Antibody-secreting B Cells. *J. Biol. Chem.*, **277**, 49047–49054.

26 Meunier, L., Usherwood, Y. K., Chung, K. T., and Hendershot, L. M. (2002). A subset of chaperones and folding enzymes form multiprotein complexes in endoplasmic reticulum to bind nascent proteins. *Mol. Biol. Cell*, **13**, 4456–4469.

27 Binder, R. J. and Srivastava, P. K. (2004). Essential role of CD91 in representation of gp96-chaperoned peptides. *Proc. Natl. Acad. Sci. U.S.A.*

28 Grammatikakis, N., Vultur, A., Ramana, C. V., Siganou, A., Schweinfest, C. W., Watson, D. K., and Raptis, L. (2002). The role of Hsp90N, a new member of the Hsp90 family, in signal transduction and neoplastic transformation. *J. Biol. Chem.*, **277**, 8312–8320.

29 Heitzer, A., Mason, C. A., Snozzi, M., and Hamer, G. (1990). Some effects of growth conditions on steady state and heat shock induced htpG gene expression in continuous cultures of Escherichia coli. *Arch. Microbiol.*, **155**, 7–12.

30 Aranda, A., Querol, A., and del

Olmo, M. (2002). Correlation between acetaldehyde and ethanol resistance and expression of HSP genes in yeast strains isolated during the biological aging of sherry wines. *Arch. Microbiol.*, **177**, 304–312.

31 Zhou, Y. N., Kusukawa, N., Erickson, J. W., Gross, C. A., and Yura, T. (1988). Isolation and characterization of Escherichia coli mutants that lack the heat shock sigma factor sigma 32. *J. Bacteriol.*, **170**, 3640–3649.

32 Cowing, D. W., Bardwell, J. C., Craig, E. A., Woolford, C., Hendrix, R. W., and Gross, C. A. (1985). Consensus sequence for Escherichia coli heat shock gene promoters. *Proc. Natl. Acad. Sci. U.S.A*, **82**, 2679–2683.

33 Nadeau, K., Das, A., and Walsh, C. T. (1993). Hsp90 chaperonins possess ATPase activity and bind heat shock transcription factors and peptidyl prolyl isomerases. *J. Biol. Chem.*, **268**, 1479–1487.

34 Zarzov, P., Boucherie, H., and Mann, C. (1997). A yeast heat shock transcription factor (Hsf1) mutant is defective in both Hsc82/Hsp82 synthesis and spindle pole body duplication. *J. Cell Sci.*, **110 (Pt 16)**, 1879–1891.

35 Ahn, S. G., Liu, P. C., Klyachko, K., Morimoto, R. I., and Thiele, D. J. (2001). The loop domain of heat shock transcription factor 1 dictates DNA-binding specificity and responses to heat stress. *Genes Dev.*, **15**, 2134–2145.

36 Gross, D. S., Adams, C. C., English, K. E., Collins, K. W., and Lee, S. (1990). Promoter function and in situ protein/DNA interactions upstream of the yeast HSP90 heat shock genes. *Antonie Van Leeuwenhoek*, **58**, 175–186.

37 Gross, D. S., English, K. E., Collins, K. W., and Lee, S. W. (1990). Genomic footprinting of the yeast HSP82 promoter reveals marked distortion of the DNA helix and constitutive occupancy of heat shock and TATA elements. *J. Mol. Biol.*, **216**, 611–631.

38 Erkine, A. M., Adams, C. C., Gao, M., and Gross, D. S. (1995). Multiple protein-DNA interactions over the yeast HSC82 heat shock gene promoter. *Nucleic Acids Res.*, **23**, 1822–1829.

39 Ripley, B. J., Stephanou, A., Isenberg, D. A., and Latchman, D. S. (1999). Interleukin-10 activates heat-shock protein 90beta gene expression. *Immunology*, **97**, 226–231.

40 Shen, Y., Liu, J., Wang, X., Cheng, X., Wang, Y., and Wu, N. (1997). Essential role of the first intron in the transcription of hsp90beta gene. *FEBS Lett.*, **413**, 92–98.

41 Zhang, S. L., Yu, J., Cheng, X. K., Ding, L., Heng, F. Y., Wu, N. H., and Shen, Y. F. (1999). Regulation of human hsp90alpha gene expression. *FEBS Lett.*, **444**, 130–135.

42 Dugyala, R. R., Claggett, T. W., Kimmel, G. L., and Kimmel, C. A. (2002). HSP90alpha, HSP90beta, and p53 expression following in vitro hyperthermia exposure in gestation day 10 rat embryos. *Toxicol. Sci.*, **69**, 183–190.

43 Wu, J. M., Xiao, L., Cheng, X. K., Cui, L. X., Wu, N. H., and Shen, Y. F. (2003). PKC epsilon is a unique regulator for hsp90 beta gene in heat shock response. *J. Biol. Chem.*, **278**, 51143–51149.

44 Sreedhar, A. S., Kalmar, E., Csermely, P., and Shen, Y. F. (2004). Hsp90 isoforms: functions, expression and clinical importance. *FEBS Lett.*, **562**, 11–15.

45 Rutkowski, D. T. and Kaufman, R. J. (2004). A trip to the ER: coping with stress. *Trends Cell Biol.*, **14**, 20–28.

46 Okamura, K., Kimata, Y., Higashio, H., Tsuru, A., and Kohno, K. (2000). Dissociation of Kar2p/BiP from an ER sensory molecule, Ire1p, triggers the unfolded protein response in yeast. *Biochem. Biophys. Res. Commun.*, **279**, 445–450.

47 Kimata, Y., Kimata, Y. I., Shimizu, Y., Abe, H., Farcasanu, I. C., Takeuchi, M., Rose, M. D., and Kohno, K. (2003). Genetic evidence for a role of BiP/Kar2 that regulates Ire1 in response to accumulation of

unfolded proteins. *Mol. Biol. Cell*, **14**, 2559–2569.

48 Patil, C. and Walter, P. (2001). Intracellular signaling from the endoplasmic reticulum to the nucleus: the unfolded protein response in yeast and mammals. *Curr. Opin. Cell Biol.*, **13**, 349–355.

49 Neckers, L. and Ivy, S. P. (2003). Heat shock protein 90. *Curr. Opin. Oncol.*, **15**, 419–424.

50 Horakova, K. and Betina, V. (1977). Cytotoxic activity of macrocyclic metabolites from fungi. *Neoplasma*, **24**, 21–27.

51 Johnson, R. D., Haber, A., and Rinehart, K. L., Jr. (1974). Geldanamycin biosynthesis and carbon magnetic resonance. *J. Am. Chem. Soc.*, **96**, 3316–3317.

52 Muroi, M., Izawa, M., Kosai, Y., and Asai, M. (1980). Macbecins I and II, new antitumor antibiotics. II. Isolation and characterization. *J. Antibiot. (Tokyo)*, **33**, 205–212.

53 Tanida, S., Hasegawa, T., and Higashide, E. (1980). Macbecins I and II, new antitumor antibiotics. I. Producing organism, fermentation and antimicrobial activities. *J. Antibiot. (Tokyo)*, **33**, 199–204.

54 Omura, S., Miyano, K., Nakagawa, A., Sano, H., Komiyama, K., Umezawa, I., Shibata, K., and Satsumabayashi, S. (1984). Chemical modification and antitumor activity of herbimycin A. 8,9-Epoxide, 7,9-cyclic carbamate, and 17 or 19-amino derivatives. *J. Antibiot. (Tokyo)*, **37**, 1264–1267.

55 Omura, S., Iwai, Y., Takahashi, Y., Sadakane, N., Nakagawa, A., Oiwa, H., Hasegawa, Y., and Ikai, T. (1979). Herbimycin, a new antibiotic produced by a strain of Streptomyces. *J. Antibiot. (Tokyo)*, **32**, 255–261.

56 Uehara, Y., Hori, M., Takeuchi, T., and Umezawa, H. (1986). Phenotypic change from transformed to normal induced by benzoquinonoid ansamycins accompanies inactivation of p60src in rat kidney cells infected with Rous sarcoma virus. *Mol. Cell Biol.*, **6**, 2198–2206.

57 Uehara, Y. (2003). Natural product origins of Hsp90 inhibitors. *Curr. Cancer Drug Targets.*, **3**, 325–330.

58 Whitesell, L. and Cook, P. (1996). Stable and specific binding of heat shock protein 90 by geldanamycin disrupts glucocorticoid receptor function in intact cells. *Mol. Endocrinol.*, **10**, 705–712.

59 An, W. G., Schulte, T. W., and Neckers, L. M. (2000). The heat shock protein 90 antagonist geldanamycin alters chaperone association with p210bcr-abl and v-src proteins before their degradation by the proteasome. *Cell Growth Differ.*, **11**, 355–360.

60 Schneider, C., Sepp-Lorenzino, L., Nimmesgern, E., Ouerfelli, O., Danishefsky, S., Rosen, N., and Hartl, F. U. (1996). Pharmacologic shifting of a balance between protein refolding and degradation mediated by Hsp90. *Proc. Natl. Acad. Sci. U.S.A*, **93**, 14536–14541.

61 Stebbins, C. E., Russo, A. A., Schneider, C., Rosen, N., Hartl, F. U., and Pavletich, N. P. (1997). Crystal structure of an Hsp90-geldanamycin complex: targeting of a protein chaperone by an antitumor agent. *Cell*, **89**, 239–250.

62 Roe, S. M., Prodromou, C., O'Brien, R., Ladbury, J. E., Piper, P. W., and Pearl, L. H. (1999). Structural basis for inhibition of the Hsp90 molecular chaperone by the antitumor antibiotics radicicol and geldanamycin. *J. Med. Chem.*, **42**, 260–266.

63 Neckers, L. (2002). Hsp90 inhibitors as novel cancer chemotherapeutic agents. *Trends Mol. Med.*, **8**, S55–S61.

64 Workman, P. (2004). Combinatorial attack on multistep oncogenesis by inhibiting the Hsp90 molecular chaperone. *Cancer Lett.*, **206**, 149–157.

65 Kelland, L. R., Sharp, S. Y., Rogers, P. M., Myers, T. G., and Workman, P. (1999). DT-Diaphorase expression and tumor cell sensitivity to 17-allylamino, 17-demethoxygeldanamycin, an inhibitor of heat shock protein 90. *J. Natl. Cancer Inst.*, **91**, 1940–1949.

66 Sausville, E. A., Tomaszewski, J. E.,

and Ivy, P. (2003). Clinical development of 17-allylamino, 17-demethoxygeldanamycin. *Curr. Cancer Drug Targets.*, **3**, 377–383.

67 Kamal, A., Thao, L., Sensintaffar, J., Zhang, L., Boehm, M. F., Fritz, L. C., and Burrows, F. J. (2003). A high-affinity conformation of Hsp90 confers tumour selectivity on Hsp90 inhibitors. *Nature*, **425**, 407–410.

68 Rowlands, M. G., Newbatt, Y. M., Prodromou, C., Pearl, L. H., Workman, P., and Aherne, W. (2004). High-throughput screening assay for inhibitors of heat-shock protein 90 ATPase activity. *Anal. Biochem.*, **327**, 176–183.

69 Chiosis, G., Timaul, M. N., Lucas, B., Munster, P. N., Zheng, F. F., Sepp-Lorenzino, L., and Rosen, N. (2001). A small molecule designed to bind to the adenine nucleotide pocket of Hsp90 causes Her2 degradation and the growth arrest and differentiation of breast cancer cells. *Chem. Biol.*, **8**, 289–299.

70 Chiosis, G., Lucas, B., Huezo, H., Solit, D., Basso, A., and Rosen, N. (2003). Development of purine-scaffold small molecule inhibitors of Hsp90. *Curr. Cancer Drug Targets.*, **3**, 371–376.

71 Pratt, W. B. and Toft, D. O. (2003). Regulation of signaling protein function and trafficking by the hsp90/hsp70-based chaperone machinery. *Exp. Biol. Med. (Maywood.)*, **228**, 111–133.

72 Blagosklonny, M. V., Toretsky, J., Bohen, S., and Neckers, L. (1996). Mutant conformation of p53 translated in vitro or in vivo requires functional HSP90. *Proc. Natl. Acad. Sci. U.S.A*, **93**, 8379–8383.

73 Zou, J., Guo, Y., Guettouche, T., Smith, D. F., and Voellmy, R. (1998). Repression of heat shock transcription factor HSF1 activation by HSP90 (HSP90 complex) that forms a stress-sensitive complex with HSF1. *Cell*, **94**, 471–480.

74 Prodromou, C., Roe, S. M., Piper, P. W., and Pearl, L. H. (1997). A molecular clamp in the crystal structure of the N-terminal domain of the yeast Hsp90 chaperone. *Nat. Struct. Biol.*, **4**, 477–482.

75 Prodromou, C., Roe, S. M., O'Brien, R., Ladbury, J. E., Piper, P. W., and Pearl, L. H. (1997). Identification and structural characterization of the ATP/ADP-binding site in the Hsp90 molecular chaperone. *Cell*, **90**, 65–75.

76 Wigley, D. B., Davies, G. J., Dodson, E. J., Maxwell, A., and Dodson, G. (1991). Crystal structure of an N-terminal fragment of the DNA gyrase B protein. *Nature*, **351**, 624–629.

77 Dutta, R. and Inouye, M. (2000). GHKL, an emergent ATPase/kinase superfamily. *Trends Biochem. Sci.*, **25**, 24–28.

78 Brino, L., Urzhumtsev, A., Mousli, M., Bronner, C., Mitschler, A., Oudet, P., and Moras, D. (2000). Dimerization of Escherichia coli DNA-gyrase B provides a structural mechanism for activating the ATPase catalytic center. *J. Biol. Chem.*, **275**, 9468–9475.

79 Ban, C., Junop, M., and Yang, W. (1999). Transformation of MutL by ATP binding and hydrolysis: a switch in DNA mismatch repair. *Cell*, **97**, 85–97.

80 Meyer, P., Prodromou, C., Hu, B., Vaughan, C., Roe, S. M., Panaretou, B., Piper, P. W., and Pearl, L. H. (2003). Structural and functional analysis of the middle segment of hsp90: implications for ATP hydrolysis and client protein and cochaperone interactions. *Mol. Cell*, **11**, 647–658.

81 Panaretou, B., Prodromou, C., Roe, S. M., O'Brien, R., Ladbury, J. E., Piper, P. W., and Pearl, L. H. (1998). ATP binding and hydrolysis are essential to the function of the Hsp90 molecular chaperone in vivo. *EMBO J.*, **17**, 4829–4836.

82 Scheibel, T., Weikl, T., and Buchner, J. (1998). Two chaperone sites in Hsp90 differing in substrate specificity and ATP dependence. *Proc. Natl. Acad. Sci. U.S.A*, **95**, 1495–1499.

83 Prodromou, C., Panaretou, B., Chohan, S., Siligardi, G., O'Brien, R., Ladbury, J. E., Roe, S. M., Piper,

P. W., and PEARL, L. H. (2000). The ATPase cycle of Hsp90 drives a molecular 'clamp' via transient dimerization of the N-terminal domains. *EMBO J.*, **19**, 4383–4392.

84 WEIKL, T., MUSCHLER, P., RICHTER, K., VEIT, T., REINSTEIN, J., and BUCHNER, J. (2000). C-terminal regions of Hsp90 are important for trapping the nucleotide during the ATPase cycle. *J. Mol. Biol.*, **303**, 583–592.

85 RICHTER, K., MUSCHLER, P., HAINZL, O., and BUCHNER, J. (2001). Coordinated ATP hydrolysis by the Hsp90 dimer. *J. Biol. Chem.*, **276**, 33689–33696.

86 WEGELE, H., MUSCHLER, P., BUNCK, M., REINSTEIN, J., and BUCHNER, J. (2003). Dissection of the contribution of individual domains to the ATPase mechanism of Hsp90. *J. Biol. Chem.*, **278**, 39303–39310.

87 CHADLI, A., BOUHOUCHE, I., SULLIVAN, W., STENSGARD, B., MCMAHON, N., CATELLI, M. G., and TOFT, D. O. (2000). Dimerization and N-terminal domain proximity underlie the function of the molecular chaperone heat shock protein 90. *Proc. Natl. Acad. Sci. U.S.A*, **97**, 12524–12529.

88 RICHTER, K., REINSTEIN, J., and BUCHNER, J. (2002). N-terminal residues regulate the catalytic efficiency of the Hsp90 ATPase cycle. *J. Biol. Chem.*, **277**, 44905–44910.

89 GRENERT, J. P., SULLIVAN, W. P., FADDEN, P., HAYSTEAD, T. A., CLARK, J., MIMNAUGH, E., KRUTZSCH, H., OCHEL, H. J., SCHULTE, T. W., SAUSVILLE, E., NECKERS, L. M., and TOFT, D. O. (1997). The amino-terminal domain of heat shock protein 90 (hsp90) that binds geldanamycin is an ATP/ADP switch domain that regulates hsp90 conformation. *J. Biol. Chem.*, **272**, 23843–23850.

90 DEHNER, A., FURRER, J., RICHTER, K., SCHUSTER, I., BUCHNER, J., and KESSLER, H. (2003). NMR Chemical Shift Perturbation Study of the N-Terminal Domain of Hsp90 upon Binding of ADP, AMP-PNP, Geldanamycin, and Radicicol. *Chembiochem.*, **4**, 870–877.

91 MCLAUGHLIN, S. H., SMITH, H. W., and JACKSON, S. E. (2002). Stimulation of the weak ATPase activity of human hsp90 by a client protein. *J. Mol. Biol.*, **315**, 787–798.

92 OWEN, B. A., SULLIVAN, W. P., FELTS, S. J., and TOFT, D. O. (2002). Regulation of heat shock protein 90 ATPase activity by sequences in the carboxyl terminus. *J. Biol. Chem.*, **277**, 7086–7091.

93 SOTI, C., RACZ, A., and CSERMELY, P. (2002). A Nucleotide-dependent molecular switch controls ATP binding at the C-terminal domain of Hsp90. N-terminal nucleotide binding unmasks a C-terminal binding pocket. *J. Biol. Chem.*, **277**, 7066–7075.

94 GARNIER, C., LAFITTE, D., TSVETKOV, P. O., BARBIER, P., LECLERC-DEVIN, J., MILLOT, J. M., BRIAND, C., MAKAROV, A. A., CATELLI, M. G., and PEYROT, V. (2002). Binding of ATP to heat shock protein 90: evidence for an ATP-binding site in the C-terminal domain. *J. Biol. Chem.*, **277**, 12208–12214.

95 JAKOB, U., LILIE, H., MEYER, I., and BUCHNER, J. (1995). Transient interaction of Hsp90 with early unfolding intermediates of citrate synthase. Implications for heat shock in vivo. *J. Biol. Chem.*, **270**, 7288–7294.

96 NEMOTO, T. K., ONO, T., and TANAKA, K. (2001). Substrate-binding characteristics of proteins in the 90 kDa heat shock protein family. *Biochem. J.*, **354**, 663–670.

97 JOHNSON, B. D., CHADLI, A., FELTS, S. J., BOUHOUCHE, I., CATELLI, M. G., and TOFT, D. O. (2000). Hsp90 chaperone activity requires the full-length protein and interaction among its multiple domains. *J. Biol. Chem.*, **275**, 32499–32507.

98 WIECH, H., BUCHNER, J., ZIMMERMANN, R., and JAKOB, U. (1992). Hsp90 chaperones protein folding in vitro. *Nature*, **358**, 169–170.

99 THULASIRAMAN, V. and MATTS, R. L. (1996). Effect of geldanamycin on the kinetics of chaperone-mediated renaturation of firefly luciferase in

rabbit reticulocyte lysate. *Biochemistry*, **35**, 13443–13450.

100 Young, J. C., Schneider, C., and Hartl, F. U. (1997). In vitro evidence that hsp90 contains two independent chaperone sites. *FEBS Lett.*, **418**, 139–143.

101 Scheibel, T., Siegmund, H. I., Jaenicke, R., Ganz, P., Lilie, H., and Buchner, J. (1999). The charged region of Hsp90 modulates the function of the N-terminal domain. *Proc. Natl. Acad. Sci. U.S.A*, **96**, 1297–1302.

102 Csermely, P., Miyata, Y., Soti, C., and Yahara, I. (1997). Binding affinity of proteins to hsp90 correlates with both hydrophobicity and positive charges. A surface plasmon resonance study. *Life Sci.*, **61**, 411–418.

103 Miyata, Y. and Yahara, I. (1992). The 90-kDa heat shock protein, HSP90, binds and protects casein kinase II from self-aggregation and enhances its kinase activity. *J. Biol. Chem.*, **267**, 7042–7047.

104 Miyata, Y. and Yahara, I. (1995). Interaction between casein kinase II and the 90-kDa stress protein, HSP90. *Biochemistry*, **34**, 8123–8129.

105 Nathan, D. F., Vos, M. H., and Lindquist, S. (1997). In vivo functions of the Saccharomyces cerevisiae Hsp90 chaperone. *Proc. Natl. Acad. Sci. U.S.A*, **94**, 12949–12956.

106 Vogen, S., Gidalevitz, T., Biswas, C., Simen, B. B., Stein, E., Gulmen, F., and Argon, Y. (2002). Radicicol-sensitive peptide binding to the N-terminal portion of GRP94. *J. Biol. Chem.*, **277**, 40742–40750.

107 Wearsch, P. A. and Nicchitta, C. V. (1997). Interaction of endoplasmic reticulum chaperone GRP94 with peptide substrates is adenine nucleotide-independent. *J. Biol. Chem.*, **272**, 5152–5156.

108 Wearsch, P. A., Voglino, L., and Nicchitta, C. V. (1998). Structural transitions accompanying the activation of peptide binding to the endoplasmic reticulum Hsp90 chaperone GRP94. *Biochemistry*, **37**, 5709–5719.

109 Gidalevitz, T., Biswas, C., Ding, H., Schneidman-Duhovny, D., Wolfson, H. J., Stevens, F., Radford, S., and Argon, Y. (2004). Identification of the N-terminal Peptide Binding Site of Glucose-regulated Protein 94. *J. Biol. Chem.*, **279**, 16543–16552.

110 Scroggins, B. T., Prince, T., Shao, J., Uma, S., Huang, W., Guo, Y., Yun, B. G., Hedman, K., Matts, R. L., and Hartson, S. D. (2003). High affinity binding of Hsp90 is triggered by multiple discrete segments of its kinase clients. *Biochemistry*, **42**, 12550–12561.

111 Shue, G. and Kohtz, D. S. (1994). Structural and functional aspects of basic helix-loop-helix protein folding by heat-shock protein 90. *J. Biol. Chem.*, **269**, 2707–2711.

112 Shaknovich, R., Shue, G., and Kohtz, D. S. (1992). Conformational activation of a basic helix-loop-helix protein (MyoD1) by the C-terminal region of murine HSP90 (HSP84). *Mol. Cell Biol.*, **12**, 5059–5068.

113 Hu, J., Toft, D., Anselmo, D., and Wang, X. (2002). In vitro reconstitution of functional hepadnavirus reverse transcriptase with cellular chaperone proteins. *J. Virol.*, **76**, 269–279.

114 Beck, J. and Nassal, M. (2003). Efficient Hsp90-independent in vitro activation by Hsc70 and Hsp40 of duck hepatits B virus reverse transcriptase, an assumed Hsp90 client protein. *J. Biol. Chem.*

115 Young, J. C. and Hartl, F. U. (2000). Polypeptide release by Hsp90 involves ATP hydrolysis and is enhanced by the co-chaperone p23. *EMBO J.*, **19**, 5930–5940.

116 Scherrer, L. C., Dalman, F. C., Massa, E., Meshinchi, S., and Pratt, W. B. (1990). Structural and functional reconstitution of the glucocorticoid receptor-hsp90 complex. *J. Biol. Chem.*, **265**, 21397–21400.

117 Toft, D. and Gorski, J. (1966). A receptor molecule for estrogens: isolation from the rat uterus and

preliminary characterization. *Proc. Natl. Acad. Sci. U.S.A*, **55**, 1574–1581.

118 Sanchez, E. R., Meshinchi, S., Schlesinger, M. J., and Pratt, W. B. (1987). Demonstration that the 90-kilodalton heat shock protein is bound to the glucocorticoid receptor in its 9S nondeoxynucleic acid binding form. *Mol. Endocrinol.*, **1**, 908–912.

119 Smith, D. F., Schowalter, D. B., Kost, S. L., and Toft, D. O. (1990). Reconstitution of progesterone receptor with heat shock proteins. *Mol. Endocrinol.*, **4**, 1704–1711.

120 Smith, D. F. (1993). Dynamics of heat shock protein 90-progesterone receptor binding and the disactivation loop model for steroid receptor complexes. *Mol. Endocrinol.*, **7**, 1418–1429.

121 Smith, D. F., Sullivan, W. P., Marion, T. N., Zaitsu, K., Madden, B., McCormick, D. J., and Toft, D. O. (1993). Identification of a 60-kilodalton stress-related protein, p60, which interacts with hsp90 and hsp70. *Mol. Cell Biol.*, **13**, 869–876.

122 Johnson, J. L., Beito, T. G., Krco, C. J., and Toft, D. O. (1994). Characterization of a novel 23-kilodalton protein of unactive progesterone receptor complexes. *Mol. Cell Biol.*, **14**, 1956–1963.

123 Owens-Grillo, J. K., Hoffmann, K., Hutchison, K. A., Yem, A. W., Deibel, M. R., Jr., Handschumacher, R. E., and Pratt, W. B. (1995). The cyclosporin A-binding immunophilin CyP-40 and the FK506-binding immunophilin hsp56 bind to a common site on hsp90 and exist in independent cytosolic heterocomplexes with the untransformed glucocorticoid receptor. *J. Biol. Chem.*, **270**, 20479–20484.

124 Renoir, J. M., Pahl, A., Keller, U., and Baulieu, E. E. (1993). Immunological identification of a 50 kDa Mr FK506-binding immunophilin as a component of the non-DNA binding, hsp90 and hsp70 containing, heterooligomeric form of the chick oviduct progesterone receptor. *C. R. Acad. Sci. III*, **316**, 1410–1416.

125 Smith, D. F., Baggenstoss, B. A., Marion, T. N., and Rimerman, R. A. (1993). Two FKBP-related proteins are associated with progesterone receptor complexes. *J. Biol. Chem.*, **268**, 18365–18371.

126 Chang, H. C. and Lindquist, S. (1994). Conservation of Hsp90 macromolecular complexes in Saccharomyces cerevisiae. *J. Biol. Chem.*, **269**, 24983–24988.

127 Perdew, G. H. and Whitelaw, M. L. (1991). Evidence that the 90-kDa heat shock protein (HSP90) exists in cytosol in heteromeric complexes containing HSP70 and three other proteins with Mr of 63,000, 56,000, and 50,000. *J. Biol. Chem.*, **266**, 6708–6713.

128 Johnson, J. L. and Craig, E. A. (2000). A role for the Hsp40 Ydj1 in repression of basal steroid receptor activity in yeast. *Mol. Cell Biol.*, **20**, 3027–3036.

129 Kimura, Y., Yahara, I., and Lindquist, S. (1995). Role of the protein chaperone YDJ1 in establishing Hsp90-mediated signal transduction pathways. *Science*, **268**, 1362–1365.

130 Johnson, J., Corbisier, R., Stensgard, B., and Toft, D. (1996). The involvement of p23, hsp90, and immunophilins in the assembly of progesterone receptor complexes. *J. Steroid Biochem. Mol. Biol.*, **56**, 31–37.

131 Richter, K., Walter, S., and Buchner, J. (2004). Sba1 connects the ATPase reaction of Hsp90 to the progression of the chaperone cycle. *manuscript submitted.*

132 Prodromou, C., Siligardi, G., O'Brien, R., Woolfson, D. N., Regan, L., Panaretou, B., Ladbury, J. E., Piper, P. W., and Pearl, L. H. (1999). Regulation of Hsp90 ATPase activity by tetratricopeptide repeat (TPR)-domain co-chaperones. *EMBO J.*, **18**, 754–762.

133 Richter, K., Muschler, P., Hainzl, O., Reinstein, J., and Buchner, J. (2003). Sti1 is a non-competitive inhibitor of the Hsp90 ATPase. Binding prevents the N-terminal

dimerization reaction during the atpase cycle. *J. Biol. Chem.*, **278**, 10328–10333.

134 Johnson, J. L. and Toft, D. O. (1995). Binding of p23 and hsp90 during assembly with the progesterone receptor. *Mol. Endocrinol.*, **9**, 670–678.

135 Sullivan, W. P., Owen, B. A., and Toft, D. O. (2002). The influence of ATP and p23 on the conformation of hsp90. *J. Biol. Chem.*

136 Panaretou, B., Siligardi, G., Meyer, P., Maloney, A., Sullivan, J. K., Singh, S., Millson, S. H., Clarke, P. A., Naaby-Hansen, S., Stein, R., Cramer, R., Mollapour, M., Workman, P., Piper, P. W., Pearl, L. H., and Prodromou, C. (2002). Activation of the ATPase activity of hsp90 by the stress-regulated cochaperone aha1. *Mol. Cell*, **10**, 1307–1318.

137 Hernandez, M. P., Chadli, A., and Toft, D. O. (2002). HSP40 binding is the first step in the HSP90 chaperoning pathway for the progesterone receptor. *J. Biol. Chem.*, **277**, 11873–11881.

138 Wegele, H., Haslbeck, M., Reinstein, J., and Buchner, J. (2003). Sti1 is a novel activator of the Ssa proteins. *J. Biol. Chem*, **278**, 25970–25976.

139 Morishima, Y., Kanelakis, K. C., Murphy, P. J., Shewach, D. S., and Pratt, W. B. (2001). Evidence for iterative ratcheting of receptor-bound hsp70 between its ATP and ADP conformations during assembly of glucocorticoid receptor.hsp90 heterocomplexes. *Biochemistry*, **40**, 1109–1116.

140 Morishima, Y., Kanelakis, K. C., Silverstein, A. M., Dittmar, K. D., Estrada, L., and Pratt, W. B. (2000). The Hsp organizer protein hop enhances the rate of but is not essential for glucocorticoid receptor folding by the multiprotein Hsp90-based chaperone system. *J. Biol. Chem.*, **275**, 6894–6900.

141 Das, A. K., Cohen, P. W., and Barford, D. (1998). The structure of the tetratricopeptide repeats of protein phosphatase 5: implications for TPR-mediated protein-protein interactions. *EMBO J.*, **17**, 1192–1199.

142 Brinker, A., Scheufler, C., Von Der, M. F., Fleckenstein, B., Herrmann, C., Jung, G., Moarefi, I., and Hartl, F. U. (2002). Ligand discrimination by TPR domains. Relevance and selectivity of EEVD-recognition in Hsp70 × Hop × Hsp90 complexes. *J. Biol. Chem.*, **277**, 19265–19275.

143 Odunuga, O. O., Hornby, J. A., Bies, C., Zimmermann, R., Pugh, D. J., and Blatch, G. L. (2003). Tetratricopeptide repeat motif-mediated Hsc70-mSTI1 interaction. Molecular characterization of the critical contacts for successful binding and specificity. *J. Biol. Chem.*, **278**, 6896–6904.

144 Johnson, J. L. and Toft, D. O. (1994). A novel chaperone complex for steroid receptors involving heat shock proteins, immunophilins, and p23. *J. Biol. Chem.*, **269**, 24989–24993.

145 Weikl, T., Abelmann, K., and Buchner, J. (1999). An unstructured C-terminal region of the Hsp90 co-chaperone p23 is important for its chaperone function. *J. Mol. Biol.*, **293**, 685–691.

146 Weaver, A. J., Sullivan, W. P., Felts, S. J., Owen, B. A., and Toft, D. O. (2000). Crystal structure and activity of human p23, a heat shock protein 90 co-chaperone. *J. Biol. Chem.*, **275**, 23045–23052.

147 Mayr, C., Richter, K., Lilie, H., and Buchner, J. (2000). Cpr6 and Cpr7, two closely related Hsp90-associated immunophilins from Saccharomyces cerevisiae, differ in their functional properties. *J. Biol. Chem.*, **275**, 34140–34146.

148 Pirkl, F. and Buchner, J. (2001). Functional analysis of the Hsp90-associated human peptidyl prolyl *cis/trans* isomerases FKBP51, FKBP52 and Cyp40. *J. Mol. Biol.*, **308**, 795–806.

149 Carrigan, P. E., Nelson, G. M., Roberts, P. J., Stoffer, J., Riggs, D. L., and Smith, D. F. (2004). Multiple Domains of the Co-chaperone Hop

Are Important for Hsp70 Binding. *J. Biol. Chem.*, **279**, 16185–16193.

150 Cheung-Flynn, J., Roberts, P. J., Riggs, D. L., and Smith, D. F. (2003). C-terminal sequences outside the tetratricopeptide repeat domain of FKBP51 and FKBP52 cause differential binding to Hsp90. *J. Biol. Chem.*, **278**, 17388–17394.

151 Nair, S. C., Rimerman, R. A., Toran, E. J., Chen, S., Prapapanich, V., Butts, R. N., and Smith, D. F. (1997). Molecular cloning of human FKBP51 and comparisons of immunophilin interactions with Hsp90 and progesterone receptor. *Mol. Cell Biol.*, **17**, 594–603.

152 Duina, A. A., Marsh, J. A., Kurtz, R. B., Chang, H. C., Lindquist, S., and Gaber, R. F. (1998). The peptidyl-prolyl isomerase domain of the CyP-40 cyclophilin homolog Cpr7 is not required to support growth or glucocorticoid receptor activity in Saccharomyces cerevisiae. *J. Biol. Chem.*, **273**, 10819–10822.

153 Bose, S., Weikl, T., Bugl, H., and Buchner, J. (1996). Chaperone function of Hsp90-associated proteins. *Science*, **274**, 1715–1717.

154 Pirkl, F., Fischer, E., Modrow, S., and Buchner, J. (2001). Localization of the chaperone domain of FKBP52. *J. Biol. Chem.*, **276**, 37034–37041.

155 Bharadwaj, S., Ali, A., and Ovsenek, N. (1999). Multiple components of the HSP90 chaperone complex function in regulation of heat shock factor 1 In vivo. *Mol. Cell Biol.*, **19**, 8033–8041.

156 Silverstein, A. M., Galigniana, M. D., Kanelakis, K. C., Radanyi, C., Renoir, J. M., and Pratt, W. B. (1999). Different regions of the immunophilin FKBP52 determine its association with the glucocorticoid receptor, hsp90, and cytoplasmic dynein. *J. Biol. Chem.*, **274**, 36980–36986.

157 Galigniana, M. D., Radanyi, C., Renoir, J. M., Housley, P. R., and Pratt, W. B. (2001). Evidence that the peptidylprolyl isomerase domain of the hsp90-binding immunophilin FKBP52 is involved in both dynein interaction and glucocorticoid receptor movement to the nucleus. *J. Biol. Chem.*, **276**, 14884–14889.

158 Davies, T. H., Ning, Y. M., and Sanchez, E. R. (2002). A new first step in activation of steroid receptors: hormone-induced switching of FKBP51 and FKBP52 immunophilins. *J. Biol. Chem.*, **277**, 4597–4600.

159 Galigniana, M. D., Harrell, J. M., O'Hagen, H. M., Ljungman, M., and Pratt, W. B. (2004). Hsp90-binding immunophilins link p53 to dynein during p53 transport to the nucleus. *J. Biol. Chem.*

160 Riggs, D. L., Roberts, P. J., Chirillo, S. C., Cheung-Flynn, J., Prapapanich, V., Ratajczak, T., Gaber, R., Picard, D., and Smith, D. F. (2003). The Hsp90-binding peptidylprolyl isomerase FKBP52 potentiates glucocorticoid signaling in vivo. *EMBO J.*, **22**, 1158–1167.

161 Silverstein, A. M., Galigniana, M. D., Chen, M. S., Owens-Grillo, J. K., Chinkers, M., and Pratt, W. B. (1997). Protein phosphatase 5 is a major component of glucocorticoid receptor.hsp90 complexes with properties of an FK506-binding immunophilin. *J. Biol. Chem.*, **272**, 16224–16230.

162 Stepanova, L., Leng, X., Parker, S. B., and Harper, J. W. (1996). Mammalian p50Cdc37 is a protein kinase-targeting subunit of Hsp90 that binds and stabilizes Cdk4. *Genes Dev.*, **10**, 1491–1502.

163 Brugge, J. S. (1986). Interaction of the Rous sarcoma virus protein pp60src with the cellular proteins pp50 and pp90. *Curr. Top. Microbiol. Immunol.*, **123**, 1–22.

164 Grammatikakis, N., Lin, J. H., Grammatikakis, A., Tsichlis, P. N., and Cochran, B. H. (1999). p50(cdc37) acting in concert with Hsp90 is required for Raf-1 function. *Mol. Cell Biol.*, **19**, 1661–1672.

165 Stewart, S., Sundaram, M., Zhang, Y., Lee, J., Han, M., and Guan, K. L. (1999). Kinase suppressor of Ras forms a multiprotein signaling complex and modulates MEK

Natsoulis, G., and Fink, G. R. (1987). 5-Fluoroorotic acid as a selective agent in yeast molecular genetics. *Methods Enzymol.*, **154**, 164–175.

201 Scheibel, T., Weikl, T., Rimerman, R., Smith, D., Lindquist, S., and Buchner, J. (1999). Contribution of N- and C-terminal domains to the function of Hsp90 in Saccharomyces cerevisiae. *Mol. Microbiol.*, **34**, 701–713.

202 Obermann, W. M., Sondermann, H., Russo, A. A., Pavletich, N. P., and Hartl, F. U. (1998). In vivo function of Hsp90 is dependent on ATP binding and ATP hydrolysis. *J. Cell Biol.*, **143**, 901–910.

203 Picard, D., Khursheed, B., Garabedian, M. J., Fortin, M. G., Lindquist, S., and Yamamoto, K. R. (1990). Reduced levels of hsp90 compromise steroid receptor action in vivo. *Nature*, **348**, 166–168.

204 Ali, J. A., Jackson, A. P., Howells, A. J., and Maxwell, A. (1993). The 43-kilodalton N-terminal fragment of the DNA gyrase B protein hydrolyzes ATP and binds coumarin drugs. *Biochemistry*, **32**, 2717–2724.

205 Lanzetta, P. A., Alvarez, L. J., Reinach, P. S., and Candia, O. A. (1979). An improved assay for nanomole amounts of inorganic phosphate. *Anal. Biochem.*, **100**, 95–97.

206 Kornberg, A., Scott, J. F., and Bertsch, L. L. (1978). ATP utilization by rep protein in the catalytic separation of DNA strands at a replicating fork. *J. Biol. Chem.*, **253**, 3298–3304.

207 Freeman, B. C., Felts, S. J., Toft, D. O., and Yamamoto, K. R. (2000). The p23 molecular chaperones act at a late step in intracellular receptor action to differentially affect ligand efficacies. *Genes Dev.*, **14**, 422–434.

208 Malmqvist, M. (1993). Surface plasmon resonance for detection and measurement of antibody-antigen affinity and kinetics. *Curr. Opin. Immunol.*, **5**, 282–286.

209 Szabo, A., Stolz, L., and Granzow, R. (1995). Surface plasmon resonance and its use in biomolecular interaction analysis (BIA). *Curr. Opin. Struct. Biol.*, **5**, 699–705.

210 Wegele, H., Muller, L., and Buchner, J. (2004). Hsp70 and Hsp90-a relay team for protein folding. *Rev. Physiol Biochem. Pharmacol.*

211 Taylor, P., Dornan, J., Carrello, A., Minchin, R. F., Ratajczak, T., and Walkinshaw, M. D. (2001). Two structures of cyclophilin 40: folding and fidelity in the TPR domains. *Structure. (Camb.)*, **9**, 431–438.

212 Joab, I., Radanyi, C., Renoir, M., Buchou, T., Catelli, M. G., Binart, N., Mester, J., and Baulieu, E. E. (1984). Common non-hormone binding component in non-transformed chick oviduct receptors of four steroid hormones. *Nature*, **308**, 850–853.

213 Veldscholte, J., Berrevoets, C. A., Brinkmann, A. O., Grootegoed, J. A., and Mulder, E. (1992). Anti-androgens and the mutated androgen receptor of LNCaP cells: differential effects on binding affinity, heat-shock protein interaction, and transcription activation. *Biochemistry*, **31**, 2393–2399.

214 Denis, M., Cuthill, S., Wikstrom, A. C., Poellinger, L., and Gustafsson, J. A. (1988). Association of the dioxin receptor with the Mr 90,000 heat shock protein: a structural kinship with the glucocorticoid receptor. *Biochem. Biophys. Res. Commun.*, **155**, 801–807.

215 Perdew, G. H. (1988). Association of the Ah receptor with the 90-kDa heat shock protein. *J. Biol. Chem.*, **263**, 13802–13805.

216 Yoshinari, K., Kobayashi, K., Moore, R., Kawamoto, T., and Negishi, M. (2003). Identification of the nuclear receptor CAR:HSP90 complex in mouse liver and recruitment of protein phosphatase 2A in response to phenobarbital. *FEBS Lett.*, **548**, 17–20.

217 Arbeitman, M. N. and Hogness, D. S. (2000). Molecular chaperones activate the Drosophila ecdysone receptor, an RXR heterodimer. *Cell*, **101**, 67–77.

218 Sabbah, M., Redeuilh, G., and Baulieu, E. E. (1989). Subunit composition of the estrogen receptor. Involvement of the hormone-binding domain in the dimeric state. *J. Biol. Chem.*, **264**, 2397–2400.

219 Ben Or, S. (1989). Evidence that 5 S intermediate state in glucocorticoid receptor transformation contains hsp90 in addition to the steroid-binding protein. *J. Steroid Biochem.*, **33**, 899–906.

220 Zhang, L., Hach, A., and Wang, C. (1998). Molecular mechanism governing heme signaling in yeast: a higher-order complex mediates heme regulation of the transcriptional activator HAP1. *Mol. Cell Biol.*, **18**, 3819–3828.

221 Ali, A., Bharadwaj, S., O'Carroll, R., and Ovsenek, N. (1998). HSP90 interacts with and regulates the activity of heat shock factor 1 in Xenopus oocytes. *Mol. Cell Biol.*, **18**, 4949–4960.

222 Minet, E., Mottet, D., Michel, G., Roland, I., Raes, M., Remacle, J., and Michiels, C. (1999). Hypoxia-induced activation of HIF-1: role of HIF-1alpha-Hsp90 interaction. *FEBS Lett.*, **460**, 251–256.

223 Rafestin-Oblin, M. E., Couette, B., Radanyi, C., Lombes, M., and Baulieu, E. E. (1989). Mineralocorticosteroid receptor of the chick intestine. Oligomeric structure and transformation. *J. Biol. Chem.*, **264**, 9304–9309.

224 Komori, A., Sueoka, E., Fujiki, H., Ishii, M., and Kozu, T. (1999). Association of MTG8 (ETO/CDR), a leukemia-related protein, with serine/threonine protein kinases and heat shock protein HSP90 in human hematopoietic cell lines. *Jpn. J. Cancer Res.*, **90**, 60–68.

225 Sumanasekera, W. K., Tien, E. S., Davis, J. W., Turpey, R., Perdew, G. H., and Vanden Heuvel, J. P. (2003). Heat shock protein-90 (Hsp90) acts as a repressor of peroxisome proliferator-activated receptor-alpha (PPARalpha) and PPARbeta activity. *Biochemistry*, **42**, 10726–10735.

226 Sumanasekera, W. K., Tien, E. S., Turpey, R., Vanden Heuvel, J. P., and Perdew, G. H. (2003). Evidence that peroxisome proliferator-activated receptor alpha is complexed with the 90-kDa heat shock protein and the hepatitis virus B X-associated protein 2. *J. Biol. Chem.*, **278**, 4467–4473.

227 Catelli, M. G., Binart, N., Jung-Testas, I., Renoir, J. M., Baulieu, E. E., Feramisco, J. R., and Welch, W. J. (1985). The common 90-kd protein component of non-transformed '8S' steroid receptors is a heat-shock protein. *EMBO J.*, **4**, 3131–3135.

228 Schuh, S., Yonemoto, W., Brugge, J., Bauer, V. J., Riehl, R. M., Sullivan, W. P., and Toft, D. O. (1985). A 90,000-dalton binding protein common to both steroid receptors and the Rous sarcoma virus transforming protein, pp60v-src. *J. Biol. Chem.*, **260**, 14292–14296.

229 Holley, S. J. and Yamamoto, K. R. (1995). A role for Hsp90 in retinoid receptor signal transduction. *Mol. Biol. Cell*, **6**, 1833–1842.

230 McGuire, J., Coumailleau, P., Whitelaw, M. L., Gustafsson, J. A., and Poellinger, L. (1995). The basic helix-loop-helix/PAS factor Sim is associated with hsp90. Implications for regulation by interaction with partner factors. *J. Biol. Chem.*, **270**, 31353–31357.

231 Shah, M., Patel, K., Fried, V. A., and Sehgal, P. B. (2002). Interactions of STAT3 with caveolin-1 and heat shock protein 90 in plasma membrane raft and cytosolic complexes: preservation of cytokine signaling during fever. *J. Biol. Chem.*

232 Miyata, Y. and Yahara, I. (2000). p53-independent association between SV40 large T antigen and the major cytosolic heat shock protein, HSP90. *Oncogene*, **19**, 1477–1484.

233 Hashimoto, Y. and Shudo, K. (1991). Cytosolic-nuclear tumor promoter-specific binding protein: association with the 90 kDa heat shock protein and translocation into nuclei by

treatment with 12-O-tetradecanoyl-phorbol 13-acetate. *Jpn. J. Cancer Res.*, **82**, 665–675.

234 Privalsky, M. L. (1991). A subpopulation of the v-erb A oncogene protein, a derivative of a thyroid hormone receptor, associates with heat shock protein 90. *J. Biol. Chem.*, **266**, 1456–1462.

235 Brunt, S. A., Perdew, G. H., Toft, D. O., and Silver, J. C. (1998). Hsp90-containing multiprotein complexes in the eukaryotic microbe Achlya. *Cell Stress. Chaperones.*, **3**, 44–56.

236 Holt, S. E., Aisner, D. L., Baur, J., Tesmer, V. M., Dy, M., Ouellette, M., Trager, J. B., Morin, G. B., Toft, D. O., Shay, J. W., Wright, W. E., and White, M. A. (1999). Functional requirement of p23 and Hsp90 in telomerase complexes. *Genes Dev.*, **13**, 817–826.

237 Hu, J. and Seeger, C. (1996). Hsp90 is required for the activity of a hepatitis B virus reverse transcriptase. *Proc. Natl. Acad. Sci. U.S.A*, **93**, 1060–1064.

238 Fujita, N., Sato, S., Ishida, A., and Tsuruo, T. (2002). Involvement of Hsp90 in signaling and stability of 3-phosphoinositide-dependent kinase-1. *J. Biol. Chem.*, **277**, 10346–10353.

239 Sato, S., Fujita, N., and Tsuruo, T. (2000). Modulation of Akt kinase activity by binding to Hsp90. *Proc. Natl. Acad. Sci. U.S.A*, **97**, 10832–10837.

240 Miyata, Y. and Nishida, E. (2004). CK2 controls multiple protein kinases by phosphorylating a kinase-targeting molecular chaperone, Cdc37. *Mol. Cell Biol.*, **24**, 4065–4074.

241 Palmquist, K., Riis, B., Nilsson, A., and Nygard, O. (1994). Interaction of the calcium and calmodulin regulated eEF-2 kinase with heat shock protein 90. *FEBS Lett.*, **349**, 239–242.

242 Munoz, M. J. and Jimenez, J. (1999). Genetic interactions between Hsp90 and the Cdc2 mitotic machinery in the fission yeast Schizosaccharomyces pombe. *Mol. Gen. Genet.*, **261**, 242–250.

243 Mahony, D., Parry, D. A., and Lees, E. (1998). Active cdk6 complexes are predominantly nuclear and represent only a minority of the cdk6 in T cells. *Oncogene*, **16**, 603–611.

244 O'Keeffe, B., Fong, Y., Chen, D., Zhou, S., and Zhou, Q. (2000). Requirement for a kinase-specific chaperone pathway in the production of a Cdk9/cyclin T1 heterodimer responsible for P-TEFb-mediated tat stimulation of HIV-1 transcription. *J. Biol. Chem.*, **275**, 279–287.

245 Arlander, S. J., Eapen, A. K., Vroman, B. T., McDonald, R. J., Toft, D. O., and Karnitz, L. M. (2003). Hsp90 inhibition depletes Chk1 and sensitizes tumor cells to replication stress. *J. Biol. Chem.*, **278**, 52572–52577.

246 Fisher, D. L., Mandart, E., and Doree, M. (2000). Hsp90 is required for c-Mos activation and biphasic MAP kinase activation in Xenopus oocytes. *EMBO J.*, **19**, 1516–1524.

247 Stancato, L. F., Silverstein, A. M., Owens-Grillo, J. K., Chow, Y. H., Jove, R., and Pratt, W. B. (1997). The hsp90-binding antibiotic geldanamycin decreases Raf levels and epidermal growth factor signaling without disrupting formation of signaling complexes or reducing the specific enzymatic activity of Raf kinase. *J. Biol. Chem.*, **272**, 4013–4020.

248 Xu, W., Mimnaugh, E., Rosser, M. F., Nicchitta, C., Marcu, M., Yarden, Y., and Neckers, L. (2001). Sensitivity of mature Erbb2 to geldanamycin is conferred by its kinase domain and is mediated by the chaperone protein Hsp90. *J. Biol. Chem.*, **276**, 3702–3708.

249 Minami, Y., Kiyoi, H., Yamamoto, Y., Yamamoto, K., Ueda, R., Saito, H., and Naoe, T. (2002). Selective apoptosis of tandemly duplicated FLT3-transformed leukemia cells by Hsp90 inhibitors. *Leukemia*, **16**, 1535–1540.

250 Masson-Gadais, B., Houle, F., Laferriere, J., and Huot, J. (2003). Integrin alphavbeta3, requirement for VEGFR2-mediated activation of SAPK2/p38 and for Hsp90-dependent

phosphorylation of focal adhesion kinase in endothelial cells activated by VEGF. *Cell Stress. Chaperones.*, **8**, 37–52.

251 Luo, J. and Benovic, J. L. (2003). G protein-coupled receptor kinase interaction with Hsp90 mediates kinase maturation. *J. Biol. Chem.*, **278**, 50908–50914.

252 Scholz, G., Hartson, S. D., Cartledge, K., Hall, N., Shao, J., Dunn, A. R., and Matts, R. L. (2000). p50(Cdc37) can buffer the temperature-sensitive properties of a mutant of Hck. *Mol. Cell Biol.*, **20**, 6984–6995.

253 Matts, R. L. and Hurst, R. (1989). Evidence for the association of the heme-regulated eIF-2 alpha kinase with the 90-kDa heat shock protein in rabbit reticulocyte lysate in situ. *J. Biol. Chem.*, **264**, 15542–15547.

254 Rose, D. W., Wettenhall, R. E., Kudlicki, W., Kramer, G., and Hardesty, B. (1987). The 90-kilodalton peptide of the heme-regulated eIF-2 alpha kinase has sequence similarity with the 90-kilodalton heat shock protein. *Biochemistry*, **26**, 6583–6587.

255 Lewis, J., Devin, A., Miller, A., Lin, Y., Rodriguez, Y., Neckers, L., and Liu, Z. G. (2000). Disruption of hsp90 function results in degradation of the death domain kinase, receptor-interacting protein (RIP), and blockage of tumor necrosis factor-induced nuclear factor-kappaB activation. *J. Biol. Chem.*, **275**, 10519–10526.

256 Takata, Y., Imamura, T., Iwata, M., Usui, I., Haruta, T., Nandachi, N., Ishiki, M., Sasaoka, T., and Kobayashi, M. (1997). Functional importance of heat shock protein 90 associated with insulin receptor on insulin-stimulated mitogenesis. *Biochem. Biophys. Res. Commun.*, **237**, 345–347.

257 Jerome, V., Leger, J., Devin, J., Baulieu, E. E., and Catelli, M. G. (1991). Growth factors acting via tyrosine kinase receptors induce HSP90 alpha gene expression. *Growth Factors*, **4**, 317–327.

258 Marcu, M. G., Doyle, M., Bertolotti, A., Ron, D., Hendershot, L., and Neckers, L. (2002). Heat shock protein 90 modulates the unfolded protein response by stabilizing IRE1alpha. *Mol. Cell Biol.*, **22**, 8506–8513.

259 Boudeau, J., Deak, M., Lawlor, M. A., Morrice, N. A., and Alessi, D. R. (2003). Heat-shock protein 90 and Cdc37 interact with LKB1 and regulate its stability. *Biochem. J.*, **370**, 849–857.

260 Nony, P., Gaude, H., Rossel, M., Fournier, L., Rouault, J. P., and Billaud, M. (2003). Stability of the Peutz-Jeghers syndrome kinase LKB1 requires its binding to the molecular chaperones Hsp90/Cdc37. *Oncogene*, **22**, 9165–9175.

261 Hartson, S. D. and Matts, R. L. (1994). Association of Hsp90 with cellular Src-family kinases in a cell-free system correlates with altered kinase structure and function. *Biochemistry*, **33**, 8912–8920.

262 Miyata, Y., Ikawa, Y., Shibuya, M., and Nishida, E. (2001). Specific association of a set of molecular chaperones including HSP90 and Cdc37 with MOK, a member of the mitogen-activated protein kinase superfamily. *J. Biol. Chem.*, **276**, 21841–21848.

263 Cissel, D. S. and Beaven, M. A. (2000). Disruption of Raf-1/heat shock protein 90 complex and Raf signaling by dexamethasone in mast cells. *J. Biol. Chem.*, **275**, 7066–7070.

264 Goes, F. S. and Martin, J. (2001). Hsp90 chaperone complexes are required for the activity and stability of yeast protein kinases Mik1, Wee1 and Swe1. *Eur. J. Biochem*, **268**, 2281–2289.

265 Bonvini, P., Gastaldi, T., Falini, B., and Rosolen, A. (2002). Nucleophosmin-anaplastic lymphoma kinase (NPM-ALK), a novel Hsp90-client tyrosine kinase: down-regulation of NPM-ALK expression and tyrosine phosphorylation in ALK(+) CD30(+) lymphoma cells by the Hsp90 antagonist 17-allylamino, 17-demethoxygeldanamycin. *Cancer Res.*, **62**, 1559–1566.

266 Flanagan, C. A. and Thorner, J. (1992). Purification and characterization of a soluble phosphatidylinositol 4-kinase from the yeast Saccharomyces cerevisiae. *J. Biol. Chem.*, **267**, 24117–24125.

267 Mizuno, K., Shirogane, T., Shinohara, A., Iwamatsu, A., Hibi, M., and Hirano, T. (2001). Regulation of Pim-1 by Hsp90. *Biochem. Biophys. Res. Commun.*, **281**, 663–669.

268 Sakagami, M., Morrison, P., and Welch, W. J. (1999). Benzoquinoid ansamycins (herbimycin A and geldanamycin) interfere with the maturation of growth factor receptor tyrosine kinases. *Cell Stress. Chaperones.*, **4**, 19–28.

269 de Carcer, G., do Carmo, A. M., Lallena, M. J., Glover, D. M., and Gonzalez, C. (2001). Requirement of Hsp90 for centrosomal function reflects its regulation of Polo kinase stability. *EMBO J.*, **20**, 2878–2884.

270 Jaiswal, R. K., Weissinger, E., Kolch, W., and Landreth, G. E. (1996). Nerve growth factor-mediated activation of the mitogen-activated protein (MAP) kinase cascade involves a signaling complex containing B-Raf and HSP90. *J. Biol. Chem.*, **271**, 23626–23629.

271 Lovric, J., Bischof, O., and Moelling, K. (1994). Cell cycle-dependent association of Gag-Mil and hsp90. *FEBS Lett.*, **343**, 15–21.

272 Stancato, L. F., Chow, Y. H., Hutchison, K. A., Perdew, G. H., Jove, R., and Pratt, W. B. (1993). Raf exists in a native heterocomplex with hsp90 and p50 that can be reconstituted in a cell-free system. *J. Biol. Chem.*, **268**, 21711–21716.

273 Wartmann, M. and Davis, R. J. (1994). The native structure of the activated Raf protein kinase is a membrane-bound multi-subunit complex. *J. Biol. Chem.*, **269**, 6695–6701.

274 Cutforth, T. and Rubin, G. M. (1994). Mutations in Hsp83 and cdc37 impair signaling by the sevenless receptor tyrosine kinase in Drosophila. *Cell*, **77**, 1027–1036.

275 Donze, O. and Picard, D. (1999). Hsp90 binds and regulates Gcn2, the ligand-inducible kinase of the alpha subunit of eukaryotic translation initiation factor 2 [corrected]. *Mol. Cell Biol.*, **19**, 8422–8432.

276 Bernstein, S. L., Russell, P., Wong, P., Fishelevich, R., and Smith, L. E. (2001). Heat shock protein 90 in retinal ganglion cells: association with axonally transported proteins. *Vis. Neurosci.*, **18**, 429–436.

277 Lipsich, L. A., Cutt, J. R., and Brugge, J. S. (1982). Association of the transforming proteins of Rous, Fujinami, and Y73 avian sarcoma viruses with the same two cellular proteins. *Mol. Cell Biol.*, **2**, 875–880.

278 Ziemiecki, A., Catelli, M. G., Joab, I., and Moncharmont, B. (1986). Association of the heat shock protein hsp90 with steroid hormone receptors and tyrosine kinase oncogene products. *Biochem. Biophys. Res. Commun.*, **138**, 1298–1307.

279 Hutchison, K. A., Brott, B. K., De Leon, J. H., Perdew, G. H., Jove, R., and Pratt, W. B. (1992). Reconstitution of the multiprotein complex of pp60src, hsp90, and p50 in a cell-free system. *J. Biol. Chem.*, **267**, 2902–2908.

280 Oppermann, H., Levinson, W., and Bishop, J. M. (1981). A cellular protein that associates with the transforming protein of Rous sarcoma virus is also a heat-shock protein. *Proc. Natl. Acad. Sci. U.S.A*, **78**, 1067–1071.

281 Koyasu, S., Nishida, E., Kadowaki, T., Matsuzaki, F., Iida, K., Harada, F., Kasuga, M., Sakai, H., and Yahara, I. (1986). Two mammalian heat shock proteins, HSP90 and HSP100, are actin-binding proteins. *Proc. Natl. Acad. Sci. U.S.A*, **83**, 8054–8058.

282 Saleh, A., Srinivasula, S. M., Balkir, L., Robbins, P. D., and Alnemri, E. S. (2000). Negative regulation of the Apaf-1 apoptosome by Hsp70. *Nat. Cell Biol.*, **2**, 476–483.

283 Gusarova, V., Caplan, A. J., Brodsky, J. L., and Fisher, E. A. (2001). Apoprotein B degradation is promoted

by the molecular chaperones hsp90 and hsp70. *J. Biol. Chem.*, **276**, 24891–24900.

284 Kumar, R., Grammatikakis, N., and Chinkers, M. (2001). Regulation of the atrial natriuretic peptide receptor by heat shock protein 90 complexes. *J. Biol. Chem.*, **276**, 11371–11375.

285 Zhao, C. and Wang, E. (2004). Heat shock protein 90 suppresses tumor necrosis factor alpha induced apoptosis by preventing the cleavage of Bid in NIH3T3 fibroblasts. *Cell Signal.*, **16**, 313–321.

286 Nishida, E., Koyasu, S., Sakai, H., and Yahara, I. (1986). Calmodulin-regulated binding of the 90-kDa heat shock protein to actin filaments. *J. Biol. Chem.*, **261**, 16033–16036.

287 Bogatcheva, N. V., Ma, Y., Urosev, D., and Gusev, N. B. (1999). Localization of calponin binding sites in the structure of 90 kDa heat shock protein (Hsp90). *FEBS Lett.*, **457**, 369–374.

288 Uzawa, M., Grams, J., Madden, B., Toft, D., and Salisbury, J. L. (1995). Identification of a complex between centrin and heat shock proteins in CSF-arrested Xenopus oocytes and dissociation of the complex following oocyte activation. *Dev. Biol.*, **171**, 51–59.

289 Imai, J. and Yahara, I. (2000). Role of HSP90 in salt stress tolerance via stabilization and regulation of calcineurin. *Mol. Cell Biol.*, **20**, 9262–9270.

290 Loo, M. A., Jensen, T. J., Cui, L., Hou, Y., Chang, X. B., and Riordan, J. R. (1998). Perturbation of Hsp90 interaction with nascent CFTR prevents its maturation and accelerates its degradation by the proteasome. *EMBO J.*, **17**, 6879–6887.

291 Stemmann, O., Neidig, A., Kocher, T., Wilm, M., and Lechner, J. (2002). Hsp90 enables Ctf13p/Skp1p to nucleate the budding yeast kinetochore. *Proc. Natl. Acad. Sci. U.S.A*, **99**, 8585–8590.

292 Garcia-Cardena, G., Fan, R., Shah, V., Sorrentino, R., Cirino, G., Papapetropoulos, A., and Sessa, W. C. (1998). Dynamic activation of endothelial nitric oxide synthase by Hsp90. *Nature*, **392**, 821–824.

293 Ficker, E., Dennis, A. T., Wang, L., and Brown, A. M. (2003). Role of the cytosolic chaperones Hsp70 and Hsp90 in maturation of the cardiac potassium channel HERG. *Circ. Res.*, **92**, e87–100.

294 Hoshino, T., Wang, J., Devetten, M. P., Iwata, N., Kajigaya, S., Wise, R. J., Liu, J. M., and Youssoufian, H. (1998). Molecular chaperone GRP94 binds to the Fanconi anemia group C protein and regulates its intracellular expression. *Blood*, **91**, 4379–4386.

295 Inanobe, A., Takahashi, K., and Katada, T. (1994). Association of the beta gamma subunits of trimeric GTP-binding proteins with 90-kDa heat shock protein, hsp90. *J. Biochem. (Tokyo)*, **115**, 486–492.

296 Busconi, L., Guan, J., and Denker, B. M. (2000). Degradation of heterotrimeric Galpha(o) subunits via the proteosome pathway is induced by the hsp90-specific compound geldanamycin. *J. Biol. Chem.*, **275**, 1565–1569.

297 Vaiskunaite, R., Kozasa, T., and Voyno-Yasenetskaya, T. A. (2001). Interaction between the G alpha subunit of heterotrimeric G(12) protein and Hsp90 is required for G alpha(12) signaling. *J. Biol. Chem.*, **276**, 46088–46093.

298 Tahbaz, N., Carmichael, J. B., and Hobman, T. C. (2001). GERp95 belongs to a family of signal-transducing proteins and requires Hsp90 activity for stability and Golgi localization. *J. Biol. Chem.*, **276**, 43294–43299.

299 Mayama, J., Kumano, T., Hayakari, M., Yamazaki, T., Aizawa, S., Kudo, T., and Tsuchida, S. (2003). Polymorphic glutathione S-transferase subunit 3 of rat liver exhibits different susceptibilities to carbon tetrachloride: differences in their interactions with heat-shock protein 90. *Biochem. J.*, **372**, 611–616.

300 Venema, R. C., Venema, V. J., Ju, H., Harris, M. B., Snead, C., Jilling, T.,

Dimitropoulou, C., Maragoudakis, M. E., and Catravas, J. D. (2003). Novel complexes of guanylate cyclase with heat shock protein 90 and nitric oxide synthase. *Am. J. Physiol Heart Circ. Physiol*, **285**, H669–H678.

301 Herbertsson, H., Kuhme, T., and Hammarstrom, S. (1999). The 650-kDa 12(S)-hydroxyeicosatetraenoic acid binding complex: occurrence in human platelets, identification of hsp90 as a constituent, and binding properties of its 50-kDa subunit. *Arch. Biochem. Biophys.*, **367**, 33–38.

302 Schnaider, T., Oikarinen, J., Ishiwatari-Hayasaka, H., Yahara, I., and Csermely, P. (1999). Interactions of Hsp90 with histones and related peptides. *Life Sci.*, **65**, 2417–2426.

303 Joly, G. A., Ayres, M., and Kilbourn, R. G. (1997). Potent inhibition of inducible nitric oxide synthase by geldanamycin, a tyrosine kinase inhibitor, in endothelial, smooth muscle cells, and in rat aorta. *FEBS Lett.*, **403**, 40–44.

304 Agarraberes, F. A. and Dice, J. F. (2001). A molecular chaperone complex at the lysosomal membrane is required for protein translocation. *J. Cell Sci.*, **114**, 2491–2499.

305 Nakamura, T., Hinagata, J., Tanaka, T., Imanishi, T., Wada, Y., Kodama, T., and Doi, T. (2002). HSP90, HSP70, and GAPDH directly interact with the cytoplasmic domain of macrophage scavenger receptors. *Biochem. Biophys. Res. Commun.*, **290**, 858–864.

306 Kang, J., Kim, T., Ko, Y. G., Rho, S. B., Park, S. G., Kim, M. J., Kwon, H. J., and Kim, S. (2000). Heat shock protein 90 mediates protein-protein interactions between human aminoacyl-tRNA synthetases. *J. Biol. Chem.*, **275**, 31682–31688.

307 Peng, Y., Chen, L., Li, C., Lu, W., and Chen, J. (2001). Inhibition of MDM2 by hsp90 contributes to mutant p53 stabilization. *J. Biol. Chem.*, **276**, 40583–40590.

308 Kellermayer, M. S. and Csermely, P. (1995). ATP induces dissociation of the 90 kDa heat shock protein (hsp90) from F-actin: interference with the binding of heavy meromyosin. *Biochem. Biophys. Res. Commun.*, **211**, 166–174.

309 Hubert, D. A., Tornero, P., Belkhadir, Y., Krishna, P., Takahashi, A., Shirasu, K., and Dangl, J. L. (2003). Cytosolic HSP90 associates with and modulates the Arabidopsis RPM1 disease resistance protein. *EMBO J.*, **22**, 5679–5689.

310 Bender, A. T., Silverstein, A. M., Demady, D. R., Kanelakis, K. C., Noguchi, S., Pratt, W. B., and Osawa, Y. (1999). Neuronal nitric-oxide synthase is regulated by the Hsp90-based chaperone system in vivo. *J. Biol. Chem.*, **274**, 1472–1478.

311 Ishiwatari-Hayasaka, H., Maruya, M., Sreedhar, A. S., Nemoto, T. K., Csermely, P., and Yahara, I. (2003). Interaction of neuropeptide Y and Hsp90 through a novel peptide binding region. *Biochemistry*, **42**, 12972–12980.

312 Adinolfi, E., Kim, M., Young, M. T., Di Virgilio, F., and Surprenant, A. (2003). Tyrosine phosphorylation of HSP90 within the P2X7 receptor complex negatively regulates P2X7 receptors. *J. Biol. Chem.*, **278**, 37344–37351.

313 Momose, F., Naito, T., Yano, K., Sugimoto, S., Morikawa, Y., and Nagata, K. (2002). Identification of Hsp90 as a stimulatory host factor involved in influenza virus RNA synthesis. *J. Biol. Chem.*, **277**, 45306–45314.

314 Bruneau, N., Lombardo, D., and Bendayan, M. (1998). Participation of GRP94-related protein in secretion of pancreatic bile salt-dependent lipase and in its internalization by the intestinal epithelium. *J. Cell Sci.*, **111 (Pt 17)**, 2665–2679.

315 Banumathy, G., Singh, V., and Tatu, U. (2002). Host chaperones are recruited in membrane-bound complexes by Plasmodium falciparum. *J. Biol. Chem.*, **277**, 3902–3912.

316 Pai, K. S., Mahajan, V. B., Lau, A., and Cunningham, D. D. (2001). Thrombin receptor signaling to cytoskeleton requires Hsp90. *J. Biol. Chem.*, **276**, 32642–32647.

317 Tsubuki, S., Saito, Y., and Kawashima, S. (1994). Purification and characterization of an endogenous inhibitor specific to the Z-Leu-Leu-Leu-MCA degrading activity in proteasome and its identification as heat-shock protein 90. *FEBS Lett.*, **344**, 229–233.

318 Sakisaka, T., Meerlo, T., Matteson, J., Plutner, H., and Balch, W. E. (2002). Rab-alphaGDI activity is regulated by a Hsp90 chaperone complex. *EMBO J.*, **21**, 6125–6135.

319 Hu, Y. and Mivechi, N. F. (2003). HSF-1 interacts with Ral-binding protein 1 in a stress-responsive, multiprotein complex with HSP90 in vivo. *J. Biol. Chem.*, **278**, 17299–17306.

320 Gilmore, R., Coffey, M. C., and Lee, P. W. (1998). Active participation of Hsp90 in the biogenesis of the trimeric reovirus cell attachment protein sigma1. *J. Biol. Chem.*, **273**, 15227–15233.

321 Fortugno, P., Beltrami, E., Plescia, J., Fontana, J., Pradhan, D., Marchisio, P. C., Sessa, W. C., and Altieri, D. C. (2003). Regulation of survivin function by Hsp90. *Proc. Natl. Acad. Sci. U.S.A*, **100**, 13791–13796.

322 Abbas-Terki, T. and Picard, D. (1999). Alpha-complemented beta-galactosidase. An in vivo model substrate for the molecular chaperone heat-shock protein 90 in yeast. *Eur. J. Biochem.*, **266**, 517–523.

323 Muresan, Z. and Arvan, P. (1997). Thyroglobulin transport along the secretory pathway. Investigation of the role of molecular chaperone, GRP94, in protein export from the endoplasmic reticulum. *J. Biol. Chem.*, **272**, 26095–26102.

324 Sanchez, E. R., Redmond, T., Scherrer, L. C., Bresnick, E. H., Welsh, M. J., and Pratt, W. B. (1988). Evidence that the 90-kilodalton heat shock protein is associated with tubulin-containing complexes in L cell cytosol and in intact PtK cells. *Mol. Endocrinol.*, **2**, 756–760.

325 Melnick, J., Aviel, S., and Argon, Y. (1992). The endoplasmic reticulum stress protein GRP94, in addition to BiP, associates with unassembled immunoglobulin chains. *J. Biol. Chem.*, **267**, 21303–21306.

326 Hung, J. J., Chung, C. S., and Chang, W. (2002). Molecular chaperone Hsp90 is important for vaccinia virus growth in cells. *J. Virol.*, **76**, 1379–1390.

24
Small Heat Shock Proteins: Dynamic Players in the Folding Game

Franz Narberhaus and Martin Haslbeck

24.1
Introduction

The most poorly conserved molecular chaperone family comprises the vertebrate eye lens α-crystallins and numerous small heat shock proteins (sHsps). Although proteins belonging to this superfamily are diverse, most of them share six characteristic features: (1) a modestly conserved α-crystallin domain of about 90 amino acids, (2) a small molecular mass between 12 and 43 kDa, (3) induction by stress conditions, (4) chaperone-like activity in preventing other proteins from aggregation, (5) formation of large oligomers, and (6) a highly dynamic quaternary structure [1–9]. This chapter reports on the structural and functional properties of sHsps with an emphasis on their dynamic oligomeric nature.

24.2
α-Crystallins and the Small Heat Shock Protein Family: Diverse Yet Similar

The vertebrate eye lens α-crystallins are the founding members of a protein superfamily, the small heat shock proteins, sHsps or α-Hsps for short [2, 7, 8]. Two isoforms, αA- and αB-crystallin, which exhibit close to 60% amino acid sequence identity, usually occur in a molar ratio of three to one and makeup one-third of the eye lens [10]. As the name implies, functional α-crystallins are crucial for transparency of the eye lens, a crowded environment with an extremely high protein concentration. While αA-crystallin is mainly confined to the lenticular compartment, its close relative αB-crystallin was found to be a heat-inducible protein present in many other tissues [11–13]. Ignolia and Craig were the first to notice that four small heat shock proteins from *Drosophila* are related to mammalian α-crystallins [14]. Meanwhile, members of the sHsp family have been reported in all kingdoms from archaea and bacteria to plants and animals [2, 3, 7, 8]. A notable exception are at least 11 pathogenic bacteria, whose completed genome sequences do not provide any hint at the presence of α-Hsp genes [2, 15, 16]. On the other hand, many organisms encode multiple sHsps. Distinct sHsp classes residing in different cellular compartments can be distinguished in plants. For example, the *Arabidopsis* genome revealed a total of 13 sHsps belonging to six classes defined on the

Protein Folding Handbook. Part II. Edited by J. Buchner and T. Kiefhaber

ISBN: 3-527-30784-2

basis of their intracellular localization and sequence similarity. Six additional open reading frames potentially encode further family members [17]. A list of representative sHsps from prokaryotes and eukaryotes is presented in Table 24.1.

The molecular mass of the majority of sHsps is around 20 kDa giving rise to their designation small Hsp. However, the full range extends from 12 to 43 kDa. Unlike other molecular chaperones, sHsps are poorly conserved and often exhibit not more than 20% sequence identity in pairwise alignments. Common to all family members is a moderately conserved α-crystallin domain of about 90 amino acids flanked by a variable N-terminal region and a short C-terminal extension (Figure 24.1). Multiple sequence alignments including sHsps from all kingdoms revealed only very few consensus residues [2, 7, 8]. Not a single amino acid is invariant or shared by all family members. The most common denominators are two short motifs (FxRxxxL and AxxxxGVL) towards the end of the α-crystallin domain and an IxI sequence in the C-terminal extension (Figure 24.1). The N-terminal region varies substantially in both length and sequence. While Hsp12 proteins of *Caenorhabditis elegans* carry a short N-terminus of about 20 amino acids [18], Hsp42 from yeast owes its length to an extended N-terminal region [19]. Sequence similarities in the N-terminal domain are restricted to closely related proteins. Phylogenetic analyses suggest that sHsps diverged very early in evolution [7]. A complex history of repeated gene duplications and lateral gene transfer probably accounts for the presence of multiple sHsps in some organisms. Bacteria lacking these proteins might have lost the corresponding gene(s) during adaptation to specialized conditions. Since sHsps are generally less critical for survival than the major chaperones, weaker functional constraints might have allowed a much higher diversification as compared to Hsp60 and Hsp70.

Many sHsps are not expressed under normal physiological conditions but rapidly accumulate during heat stress by classical prokaryotic or eukaryotic heat shock control mechanisms [2, 20, 21]. Microarray-based gene expression profiling in *E. coli* revealed a 300-fold induction of *ibpAB*, which by far exceeds the induction of all other heat shock genes [22]. Massive heat induction of sHsps was also reported in animal and plant tissues [3, 23]. Interestingly, the presence of sHsps is not restricted to eye lenses and heat-stressed cells. In bacteria, plants, and animals, they have been discovered during development processes, metabolic transitions, environmental insults, and many other conditions [2, 3, 9, 23, 24]. The broad expression pattern of sHsps is consistent with both a general protective function on one hand and specialized tasks in certain cell types on the other.

24.3 Cellular Functions of α-Hsps

24.3.1 Chaperone Activity in Vitro

From in vitro studies in the early 1990s, it emerged that the primary function of sHsps might be to bind denatured proteins in order to prevent their aggregation.

Tab. 24.1. Representative members of the α-Hsp family.

Kingdom	*Organism*	*Protein*	*Complex*	*Chaperone*	*Comments*	*References*
Archaea	*M. jannaschii*	Hsp16.5	24mer	Yes	Crystal structure, hollow sphere	110, 156
	P. furiosus	Pfu-sHsp	Large	Yes		157, 158
Bacteria	*E. coli*	IbpA/B	Large	Yes	Associates with inclusion bodies	53, 159
	M. tuberculosis	Hsp16.3	9mer	Yes	Antigen, trimer of trimers	115, 160
	Synechocystis sp.	Hsp17 (16.6)	24mer	Yes	Stabilizes membranes	97, 122
Eukaryotes	Yeast	Hsp26	24mer	Yes	Temperature-regulated chaperone	114
		Hsp42	12mer	Yes		161
	C. elegans	Hsp12.6	Monomer	No		18
	Wheat	Hsp16.9	12mer	Yes	Crystal structure, two hexameric rings	113
	Pea	Hsp18.1	12mer	Yes		29
	Mouse	Hsp25	16mer	Yes	Concentration-dependent equilibrium among 16mer, tetramers, and dimers	46, 107, 162
	Vertebrates	αA/B-crystallin	Large	Yes	Variable quaternary structure, molar ratio of 3:1 in the eye lens	25, 163

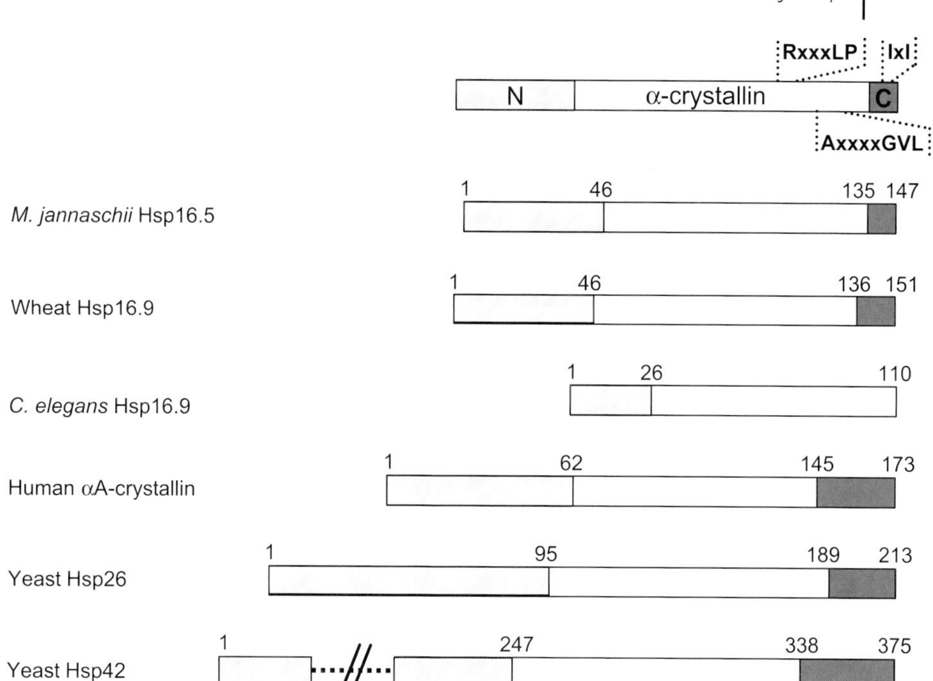

Fig. 24.1. Domain structure of sHsps. The overall domain structure is depicted on top of the figure. The N-terminal region, α-crystallin domain, C-terminal extension, and conserved motifs are indicated. Representative sHsps are shown below. Numbering is based on the assignment by Kim et al. [110].

Bovine α-crystallin was first reported to have chaperone-like activity [25]. It prevented thermal aggregation of various model substrates. Since then, it has been demonstrated that almost all sHsps are able to prevent the formation of thermally or chemically induced light-scattering aggregates (Table 24.1). Some additional examples are human Hsp27 [26], frog Hsp30C [27], shrimp p26 [28], pea Hsp17.7 [29], tobacco Hsp18 [30], two classes of sHsps from the soybean symbiont *Bradyrhizobium japonicum* [31], Lo18 from the lactic acid bacterium *Oenococcus oeni* [32], and Hsp16 and $sHsp_{Tm}$ from the thermophilic bacteria *Synechococcus vulcanus* and *Thermotoga maritima*, respectively [33, 34]. Only the smallest known representatives of the sHsp family, Hsp12.2, Hsp12.3, and Hsp12.6 of *C. elegans*, have failed to show chaperone-like properties [18, 35].

Since natural substrates are largely unknown, the model substrates chosen for in vitro studies usually are commercially available proteins, such as citrate synthase (CS) or malate dehydrogenase (MDH) from pig heart mitochondria, hog muscle lactate dehydrogenase, or bovine insulin. Some fairly robust chaperone assays will be described in the Appendix. The fact that such a wide spectrum of artificial substrates is protected from precipitation suggests that sHsps act as rather promis-

[65]. An R116G exchange αA-crystallin is responsible for autosomal dominant congenital cataract in humans [66]. The mutated protein was shown to have reduced chaperone activity [67–69]. An equivalent mutation in αB-crystallin (R120G) causes cataract formation and desmin-related myopathy, an inherited neuromuscular disorder [70]. The aggregation of desmin filaments is thought to result from the reported defect in chaperone activity [67, 71, 72]. Somewhat surprisingly, a mouse deleted of the αB-crystallin gene carried normal eye lenses and developed normally except for a reduced life span [73].

Interestingly, αB-crystallin is overexpressed in many neurological disorders such as Alzheimer [74], Creutzfeld-Jakob [75], and other diseases [76, 77]. The clinical importance of sHsps is further underlined by an ever-growing number of other reported activities, among them, interaction with actin filaments and microtubules [78], protection against oxidative stress and modulation of the intracellular redox state [79], interference with apoptosis [80], translation inhibition [81], and neuroprotective effects [82]. The involvement of Hsps in disease and important cellular processes is potentially applicable to the development of diagnostic tests and novel drugs [83, 84].

Specific subsets of the multiple sHsps in plants are induced by an amazing array of conditions including germination, embryo and pollen development, and fruit maturation [3, 24]. How the complex developmental expression pattern is coordinated remains to be determined. Osmotic shock, oxidative stress, cold storage, and heavy metals also trigger the induction of certain sHsp species. Again, it is tempting to speculate that protein misfolding is a common theme in all of these events and that the sHsps carry out a principle protective function by preventing deleterious protein aggregation. In fact, it has been demonstrated by an in vivo reporter system that expression of various sHsps protects firefly luciferase in *Arabidopsis* cell suspension cultures [85, 86]. Interestingly, the expression of various plant sHsps has been reported to enhance thermotolerance of *E. coli*, again emphasizing a general rather than specific protective effect of these chaperones [87, 88]. Much to the same effect, overexpression of endogenous sHsps increased stress tolerance of *E. coli* and *A. thaliana* [89, 90], whereas disruption of the *hsp30* gene in *Neurospora crassa* resulted in reduced thermotolerance [91]. In contrast to the Hsp60 and Hsp70 chaperones, one of which is often essential for survival in many cell types, sHsps are dispensable. Apparently, their lack can be offset by the other constituents of the multi-chaperone network. Growth defects associated with accumulation of aggregated proteins were observed only in an *E. coli ibpAB* mutant during extreme heat shock [92, 93]. The defect was more pronounced in combination with a *dnaK* mutant allele, suggesting that a functional interaction between sHsps and the Hsp70 system also exists in vivo [92].

24.3.3
Other Functions

Inactivation of the *hsp17* gene of *Synechocystis* sp. PCC6803 coding for the single sHsp member in this cyanobacterium resulted not only in decreased thermotolerance but also in reduced activity of the photosynthetic apparatus and disrupted integrity of thylakoid membranes [94, 95]. Hsp17 was found to be associated with thylakoid membranes, and transcription of *hsp17* responded to changes in membrane physical order, which led to its designation as "fluidity gene" [96]. Purified Hsp17 exhibited a dual function, acting either as chaperone by binding to denatured proteins or as stabilizer of hyperfluid lipid membranes by penetrating into the membrane hydrophobic core [97]. Preservation of membrane integrity by sHsps was postulated to be a general mechanism because α-crystallin stabilized synthetic and cyanobacterial membranes much like Hsp17 [98]. The observed membrane association of α-crystallin [99] and other sHsps [100–102] supports this hypothesis. Partitioning between soluble and membrane-bound states may be the key to whether chaperone function or membrane stabilization prevails.

Several reports have documented an interaction between sHsps and nucleic acids. Untranslated mRNAs have been detected in heat shock granules from plants and mammals [56, 103]. It is not clear, however, whether it is a fortuitous association mediated via other proteins that are integrated into the granules or whether the sHsps have a direct affinity towards nucleic acids. Some reports show that sHsps are indeed able to bind single-stranded or double-stranded DNA [104, 105], probably by helical structures [106]. Many interesting sHsp activities in maintaining the integrity of various cellular macromolecules may still remain to be explored.

24.4 The Oligomeric Structure of α-Hsps

Despite their sequence diversity, sHsps share secondary structure. Predictions that suggest a high β-sheet content in the α-crystallin domain are consistent with CD-spectra dominated by β-sheets in several sHsps [107, 108]. Calculation of the β-sheet content from far UV spectra of Hsp25, Hsp26, and α-crystallin revealed β-sheet contents between 20% and 30% [107, 109], correlating directly with the content of α-crystallin domain in the overall sequence and indicating high α-helical and unstructured parts in the N-terminal domain.

One of the most striking features of sHsps is their organization in large oligomeric structures, comprising nine to about 50 subunits. For three family members the structure of these complexes has been solved (Figure 24.3), revealing hollow, globule-like structures with outside diameters of 120 and 190 Å. While the quaternary structure of Hsp16.5 from *Methanococcus jannaschii* is well ordered [110], the oligomers formed by α-crystallin and Hsp27 show more structural variability and the ability to acquire and release subunits [111, 112]. Interestingly, wheat Hsp16.9 assembles into a dodecameric double disk. Each disk is organized as a trimer of dimers [113]. Hydrophobic patches become exposed upon disassembly of the complex into dimers. The structures of all sHsps solved so far support the idea that a dimer is the smallest exchangeable unit. This minimal dimeric building block seems to be conserved and may represent the functional unit concerning chaper-

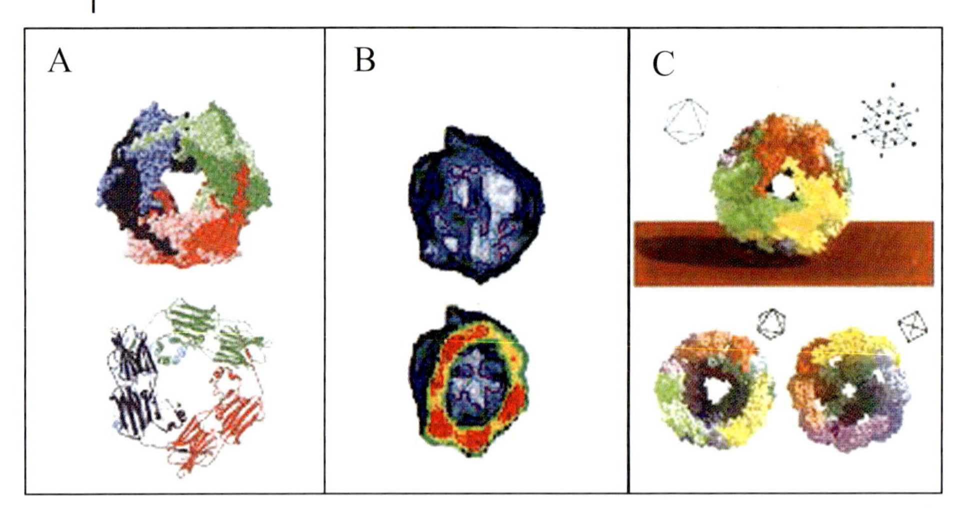

Fig. 24.3. Three-dimensional structure of sHsps. (A) Crystal structure of wheat Hsp16.9 [113]. The dodecamer arranged as two hexameric discs is 9.5 nm wide and 5.5 nm high. (B) Cryo-electron microscopic image of α-crystallin [111]. The outer diameter of the 32mer is 18 nm [111]. (C) Crystal structure of Hsp16.5 from *M. jannaschii*, which assembles into a 24mer with an outer diameter of 12 nm [110]. Reproduced with permission of the publisher and authors.

one properties [9, 109, 113, 114]. The necessary exception to this general picture might be Hsp16.3 from *Mycobacterium tuberculosis*, for which a trimeric substructure has been proposed. According to cryo-electron microscopy data, the triangular particle is composed of nine subunits organized as a trimer of trimers [115].

It is matter of controversy which parts of sHsps contribute to the association of the oligomeric structures and which are involved in the interaction with nonnative proteins [2, 4, 9]. A common property of all sHsp structures is the localization of the N-terminal regions in the interior of the oligomeric structures. In Hsp16.5 and α-crystallin, these disordered regions sequester inside the sphere; in the case of Hsp16.9, they are buried in the oligomeric structure. The liberated, hydrophobic N-terminal part might represent the substrate-binding site of disassembled sHsp particles [109, 113].

To asses the role of the N-terminal region, the four Hsp12 proteins from *C. elegans* have been investigated [18, 35]. They are the smallest naturally occurring representatives of the family and are reduced to the α-crystallin domain preceded by an exceptionally short N-terminal region of 25 or 26 residues. They form complexes from monomers to tetramers, but neither of them displays chaperone activity. Only 16.2 from *C. elegans*, a typical member with an N-terminal extension of 41 residues, builds a higher oligomer of 14 to 24 subunits and displays chaperone activity [116]. N-terminal deletions resulted in trimeric or tetrameric complexes lacking chaperone activity. In recent studies on yeast Hsp26, an N-terminally truncated construct turned out to be dimeric and inactive [109]. It was concluded that regions that are important for both the assembly of the 24mer and the interaction with nonnative proteins reside in the N-terminal part of the protein. This view is supported by thermal unfolding experiments that show that full-length Hsp26 exhibits two transitions, one in the heat shock temperature range and one at much higher temperatures. Only the high-temperature transition was observed in the case of Hsp26ΔN. Furthermore, these results suggest that less energy is required for dissociation of the 24mer than for the dissociation and unfolding of the dimer. Quantitative data for the changes in energy involved in these processes were obtained by analyzing urea-induced unfolding transitions at 25 °C and 43 °C. These data are in good agreement with the unfolding transitions and support the hypothesis of a thermolabile Hsp26 assembly that dissociates to stable dimers at elevated temperatures [109, 114].

Many efforts to identify individual amino acid residues involved in subunit interaction or chaperone activity by point mutagenesis have had limited success [52, 117–120]. Although some point mutations resulted in significantly enlarged assemblies, chaperone activity was barely affected. Point mutations in bacterial sHsps have been more telling because they pinpointed a number of residues in the α-crystallin domain and C-terminal extension that are critical for oligomerization and chaperone activity [42, 121–123]. In summary, the emerging picture is that residues in all three domains of sHsp are required for oligomerization [2]. While the α-crystallin domain is necessary for dimer formation and thus assembles the basic building block, both flanking regions promote the formation of higher-order structures.

The oligomeric state of mammalian sHsps can be altered by various posttranslational modifications, in particular by serine-specific phosphorylation [124–126]. Human Hsp27, for example, possesses three phosphorylation sites, S15, S78, and S82, whose modification via a MAP kinase cascade leads to eight possible isoforms [127]. Thr143 comprises an alternative phosphorylation site in Hsp27 that is phosphorylated by a cGMP-dependent protein kinase [128]. The only reported example of a phosphorylated plant sHsp is Hsp22 from maize mitochondria. Covalent modification occurs at a serine residue by an unspecified kinase activity [129]. Phosphorylation of human Hsp27 and αB-crystallin decreased its oligomeric size and reduced chaperone activity [130–132]. With the exception of αA-crystallin, phosphorylation of mammalian sHsps is regulated in response to stress, cytokines, and growth factors [133–135]. Phosphatase-mediated dephosphorylation regulates the phosphorylation state [136–138]. Apart from the ambient physical and chemical conditions, controlled covalent modification of sHsp provides an additional mechanism to modulate the structural and functional properties of sHsps according to the cellular demands.

24.5 Dynamic Structures as Key to Chaperone Activity

While the crystal structures of Hsp16.5 and Hsp16.9 might appear rather rigid and stable, sHsp complexes in fact are very dynamic, enabling them to exchange subunits and to form hetero-oligomeric assemblies. For example, *Triticum aestivum* Hsp16.9 has been shown to dissociate into sub-oligomeric species at higher temperatures. As a consequence, Hsp16.9 is able to interact and form complexes with its close homologue *Pisum sativum* Hsp18.1 [113]. Complex formation between both sHsps involves subunit exchange, as has been demonstrated elegantly by electrospray mass spectrometry [139].

The exchange of subunits between sHsps has also been demonstrated for other sHsps from plants and bacteria [31, 140]. Exchange occurred only if the proteins were from the same class of sHsps. In mammals, various sHsps frequently exchange subunits. In the vertebrate eye lens, for example, α-crystallin forms hetero-aggregates containing αA and αB subunits [141]. In addition, αA-crystallin exchanges subunits with Hsp27 [112]. Mixed polymers of αA-crystallin, αB-crystallin, and Hsp25 [107], and between HspB2 and HspB3, have been described [142]. It is not clear what the function of this intermolecular subunit exchange in sHsps might be. However, in the case of cells that express several sHsps in the same compartment, the formation of hetero-complexes most likely is relevant in vivo.

It is possible that mixed complexes are the result of a fortuitous interaction of closely related proteins maintaining them in an inactive storage form. Regardless of whether homo-oligomeric or hetero-oligomeric complexes have been formed, they need to dissociate in order to gain chaperone function (Figure 24.2). One interpretation of this dynamic behavior is that substrate-binding sites are buried in the complex but become exposed by dissociation. Major structural rearrangements upon substrate binding lead to the formation of even larger substrate-sHsp complexes that need to be taken apart with the help of other chaperones [54, 55]. The spontaneous dissociation-reassociation process might be some sort of sensing process monitoring the presence of nonnative proteins in the cellular environment.

24.6 Experimental Protocols

24.6.1 Purification of sHsps

Purification of Affinity-tagged sHsps

It is well documented that sHsps carrying a short tag that facilitates purification are functional chaperones. N-terminally histidine-tagged IbpA and IbpB of *E. coli* [143], C-terminally histidine-tagged class A and class B sHsps from *B. japonicum* [31], and C-terminally Strep-tagged pea Hsp18.1 and *Synechocystis* Hsp16.6 [144] were shown to suppress light-scattering of model substrates. Affinity-tagged sHsps can be purified according to standard protocols.

Purification of Hsp26 from *Saccharomyces cerevisiae*

The purification method described below allows the overexpression and purification of Hsp26 from *Saccharomyces cerevisiae* [114, 145]. A typical yield to be achieved is about 3–5 mg of purified protein from one liter of cell culture. One liter of cell culture amounted to an average of 4 g wet cells. The purity of the protein was estimated to be >95% by densitometric scanning. After each purification step, fractions were pooled according to the appearance of the respective band on SDS-PAGE. If the band pattern was ambiguous, Western blot analysis using specific antibodies against the sHsp was performed.

Protein Expression

High-level expression of Hsp26 was achieved using the shuttle-vector pJV517, which is a derivative of the 2μ plasmid pRS464. The high copy number ensured good expression levels. The plasmid contained the Ura3 gene for selection of yeast cells and the β-lactamase gene that allowed selection in *E. coli*. For protein expression, the Ura-yeast strain JT(DIP) GPD26(A) was used. The yeast strain and the plasmid were a generous gift from S. Lindquist (University of Chicago). Cells were grown at 30 °C to late logarithmic phase to an optical density of about 0.8 at 595 nm and harvested by centrifugation (2500 g, 5 min, 4 °C).

Purification

Purification of Hsp26 was accomplished by two different anion exchanges and a gel filtration column. During the purification, all buffers and reaction vessels were pre-cooled on ice and the purification steps were always carried out at 4 °C.

1. The cell pellet was washed with ice-cold buffer A (40 mM HEPES-KOH pH 7.5, 1 mM DTE, 1 mM EDTA) + 50 mM NaCl. The buffer contained a protease-inhibitor mix (1 μM Leupeptin, 2.5 mM *p*-amino benzoic acid (PABA), 1 mM Pefabloc, 1 μM Pepstatin) to avoid protease degradation during the purification procedure.
2. The cell pellet was resuspended 1:2 in buffer A + 50 mM NaCl + protease-inhibitor mix, and cells were broken with a Basic Z cell disrupter (Constant Systems).
3. The cell lysate was centrifuged (20 200 g, 45 min, 4 °C), fresh protease inhibitors were added, and the soluble extract was applied to a 50-mL DEAE Sephacel (Amersham Biosciences) ion exchange chromatography column equilibrated in buffer A + 50 mM NaCl. Hsp26 was eluted with a 500-mL NaCl gradient in buffer A ranging from 50 to 500 mM NaCl. Hsp26 containing fractions that eluted at 200–300 mM NaCl were pooled and dialyzed against buffer A + 50 mM NaCl.
4. The protein solution was loaded onto a 6-mL Resource-Q ion exchange column (Amersham Bisciences), equilibrated in buffer A + 50 mM NaCl. A linear NaCl gradient (50–500 mM) was used for elution. Hsp26 eluted in the range of 200–250 mM NaCl.
5. The Hsp26-containing fractions were pooled and further purified on a 90-mL Superdex S200 pg column (Amersham Biosciences) equilibrated in buffer B (40 mM HEPES-KOH pH 7.5, 1 mM DTE, 1 mM EDTA) + 200 mM NaCl. The

column was operated at a flow rate of 0.5 mL min^{-1}. Hsp26 eluted at about 60-mL buffer volume, indicating a large oligomeric species. Fractions containing pure Hsp26 were pooled, dialyzed against buffer A + 50 mM NaCl, and concentrated to approximately 3 mg mL^{-1} by ultrafiltration using an Amicon cell with a YM30 membrane and Centricon Microconcentrators with 30-kDa cutoff. The protein solution was centrifuged at 20 000 g at 4 °C, aliquoted, frozen in liquid nitrogen, and stored at −80 °C.

Purification of Recombinant Murine Hsp25 and Human Hsp27

The method described below allows the purification of about 10 mg of recombinant mammalian sHsp (mouse Hsp25, human Hsp27) per liter of cell culture [26, 46, 145, 146]. One liter of cell culture amounted to a wet cell pellet of about 3 g. The protocol typically led to a purity of ≥90%.

Protein Expression

For recombinant expression of Hsp25 and Hsp27, we used the *E. coli* strain BL21 (DE3) and the plasmids pAK3038Hsp25 [26] and pAK3038Hsp27 [146], respectively. A 50-mL pre-culture in LB medium containing 200 μg mL^{-1} ampicillin was inoculated with single colonies of *E. coli* BL21 (DE3) harboring the plasmid pAK3038Hsp25/27 grown overnight on an LB agar plate containing 100 μg mL^{-1} ampicillin. The pre-culture was grown at 37 °C for 135 min.

To inoculate each of six 1-L cultures (LB medium containing 0.4% glucose, 200 μg mL^{-1} ampicillin), 5 mL of the pre-culture was used. After about 4 h shaking at 37 °C, the expression of Hsp25/27 was induced by addition of IPTG (final concentration: 0.4 mM). The cells were incubated for another 2 h at 37 °C and subsequently harvested and pooled by centrifugation at 2600 g and 4 °C. The wet cell pellet can be stored at −80 °C if necessary.

Purification

The purification of recombinant mammalian sHsps was achieved by ammonium sulfate precipitation and anion exchange chromatography.

1. The cell pellet was resuspended in 50 mL lysis buffer (50 mM Tris/HCl, pH 8.0, 100 mM NaCl, 1 mM EDTA) and centrifuged for 10 min at 4 °C and 4000 g.
2. For lysis, 60 mL lysis buffer, 160 μL 50 mM PMSF (freshly dissolved in methanol, final concentration: 0.13 mM), and 2 mL lysozyme (25 mg mL^{-1} in lysis buffer) were added to the pellet, and re-dissolved cells were incubated on ice for 20 min. Subsequently, 80 mg solid sodium desoxycholate was added and cells were transferred to 37 °C. After 15 min incubation under constant stirring with a glass rod, about 1–2 mg of lyophilized DNase I was added and this step was repeated every 5 min until the viscosity of the solution decreased significantly (about 15 min after first addition of DNase I). The solution was then centrifuged for 15 min at 4 °C at 20 000 g.
3. The supernatant was transferred to room temperature, and a saturated solution of ammonium sulfate (pH 7.0) was added drop-wise under constant stirring un-

til a final saturation of 40% was reached. The stirring was continued for a further 30 min, and the sample was then centrifuged for 10 min at 20 400 g and 20 °C. The supernatant was discarded and the Hsp25/27-containing pellet was re-dissolved in 20 mL of buffer 1 (20 mM Tris/HCl, pH 7.6, 10 mM $MgCl_2$, 30 mM ammonium chloride, 0.5 mM dithiothreitol, 0.05 mM NaN_3, 2 μM PMSF). The re-dissolved pellet was dialyzed three times for at least 6 h at 4 °C against 600 mL of buffer 1 and then centrifuged again as described above.

4. Ion exchange chromatography was carried out on a 135-mL DEAE-Sepharose CL-6B (Amersham Biosciences) column. After equilibration of the column with 300 mL of buffer 1, the dialyzed sample was applied and the column was washed with 100 mL of buffer 1 and developed with a 500-mL gradient from 0 to 200 mM NaCl in buffer 1. Hsp25 eluted as a single peak at about 100 mM NaCl, whereas Hsp27 eluted at about 120 mM NaCl.
5. Pooled peak fractions containing Hsp25/27 were precipitated overnight at 4 °C after adding solid ammonium sulfate to a final saturation of 50%. The precipitated protein was pelleted by centrifugation for 10 min at 20 000 g and 20 °C.
6. The pellet was re-dissolved in 2 mL of buffer 1 and dialyzed three times against 600 mL of buffer 1 at 4 °C. The dialyzed sample was again centrifuged as above. The supernatant contains Hsp25/27, which can be stored in aliquots at −80 °C.

24.6.2 Chaperone Assays

Analysis of Chaperone Function

Progress in understanding the mechanism of chaperone function has been achieved mainly by a reductionist biochemical approach using purified chaperones and "model substrate proteins." This term implies that the substrate proteins do not necessarily represent natural, in vivo substrates. In many cases, the underlying rationale was that proteins that are recalcitrant folders in vitro may also have problems folding in the cell and are therefore good representatives of the unknown natural targets.

A number of different substrate proteins have been used to study the function of chaperones including RuBisCO [147], rhodanese [148], malate dehydrogenase [149, 150], and citrate synthase [145, 151]. Those substrates differ in their quaternary structure, their rate of folding, and their tendency to undergo irreversible side reactions during folding and unfolding [152].

Citrate Synthase Assay

Description of Enzyme

Enzyme: Citrate synthase (CS), mitochondrial, from pig heart
E.C.: 4.1.3.7.
MW: 48.969 kDa
$E^{280\ \text{nm}}$ 1.56, d = 1 cm
Sp. activity: ~150 U/mg at 25 °C, pH 8.0 using dithionitrobenzoic acid (DTNB)

Description of Assay

CS catalyzes the reaction of oxaloacetic acid (OAA) and acetyl-coenzymeA (Ac-CoA) to citrate and CoA. The enzyme activity is determined by a colorimetric test using DTNB. DTNB reacts with the free thiol groups of the reaction product CoA. This reaction can be easily followed in a spectrophotometer at 412 nm. CS can be used as a substrate for chaperones since it rapidly denatures and aggregates at heat shock temperatures >37 °C [145]. OAA stabilizes CS, shifting the midpoint of thermal transition from 43 °C to 66.5 °C.

Materials and Suppliers

Acetyl-CoA (Roche)
Oxaloacetic acid (OAA) (Sigma)
Dithionitrobenzoic acid (DTNB) (Sigma)
Disposable cuvettes, d = 1 cm
Spectrophotometer with thermostated cell holder

Solutions

CS stock solution: 15 μM in 50 mM TE buffer [153] or commercially available CS (Sigma), dialyzed against TE buffer.
TE buffer: 50 mM Tris/HCl, 2 mM EDTA, pH 8.0
OAA solution: 10 mM, dissolve in 50 mM Tris no pH ⇒ ~pH 8.0
DTNB solution: 10 mM, dissolve in TE buffer, pH 8.0, DTNB is poorly soluble, requires extended stirring to dissolve
Ac-CoA solution: 5 mM, dissolve in TE buffer, pH 8.0
Do not freeze and thaw the substrate solutions, store on ice

Assay

The reaction mixture for the activity assay consists of 930 μL TE, 10 μL OAA solution, 10 μL DTNB solution, and 30 μL Ac-CoA solution. The reaction is started by the addition of 20 μL of 0.15 μM CS (monomer), and the change in absorbance is followed for 1 min in a UV spectrophotometer at 412 nm. The specific activity can be calculated as follows:

$$\text{Specific activity (U mg}^{-1}) = \Delta E/\text{min} \times V/(\varepsilon d v c)$$

where V is the test volume (mL), ε is the molar extinction coefficient of DTNB (13 600 M^{-1} cm^{-1}), d is the path length of the cell (cm), v is the sample volume (mL), and c is the enzyme concentration (mg mL^{-1}). The specific activity of native CS from pig heart is 150 U mg^{-1}.

Thermal Inactivation and Reactivation

CS (15 μM) is diluted 100-fold with stirring into the corresponding buffer (usually 40 mM HEPES-KOH, pH 7.5 or 50 mM Tris-HCL, pH 8.0 (25 °C), which is pre-

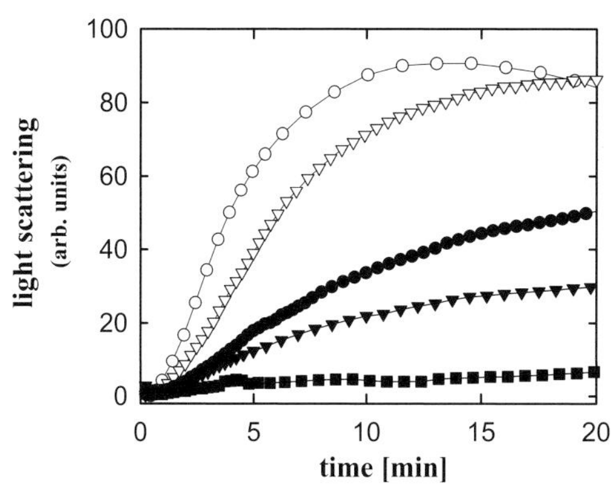

Fig. 24.4. Typical chaperone assay using CS as substrate. Influence of Hsp26 on the thermal aggregation of CS. CS (final concentration: 75 nM) was diluted into a thermostated solution of 37.5 nM (▽), 55 nM (●), 75 nM (τ), and 150 nM Hsp26 complex (■). Open circles represent the spontaneous aggregation of CS at 43 °C. The kinetics of aggregation was determined by measuring light scattering of the samples.

incubated at 25 °C in a 2-mL Eppendorf tube equipped with a small stirring bar. The activity is determined and set to 100%. The inactivation is started by placing the test tube in a 43 °C water bath. During the time course of inactivation, aliquots are withdrawn to determine activity.

Reactivation of CS can be initiated either by a temperature shift back to 25 °C or by addition of 1 mM OAA at 43 °C. During the time course of reactivation, aliquots are withdrawn to determine activity.

Thermal Aggregation of CS

Light scattering is used to examine the influence of sHsps on the thermal aggregation of CS (Figure 24.4). CS (15 μM) is diluted 1:50 in 40 mM HEPES/KOH, pH 7.5, and equilibrated at 43 °C in the presence and in the absence of sHsps. Aggregation kinetics is measured in a fluorescence spectrophotometer in a stirred and thermostated quartz cell at 400 nm. Alternatively, the aggregation of 2–4 μM (end concentration) of CS can be monitored in a VIS photometer at 400 nm.

Insulin Assay

Description of Enzyme

Insulin is a two-chain polypeptide hormone produced by the β-cells of pancreatic islets. The A (21 aa) and B (30 aa) chains are joined by two interchain disulfide bonds. The A chain contains an additional intrachain disulfide bond.
Insulin from bovine pancreas
MW: 5.8 kDa
$A_{280\ \mathrm{nm},0.1\%} = 1.06$
1.06 OD is 1 mg mL^{-1}

Assay Description

On reduction of the disulfide bonds of insulin with dithiothreitol (DTT), the insulin B chain will aggregate and precipitate, while the A chain remains in solution. This unfolding process can be monitored by measuring the apparent absorbance due to the increase in scattering.

This is a simple semi-quantitative assay to describe the ability of the protein under investigation to act as a chaperone [154]. The assay can be used at different temperatures and salt conditions but seems very sensitive to pH changes: the higher the pH, the weaker the aggregation reaction. The aggregation reaction is slightly faster at higher temperatures.

Materials and Suppliers

Insulin (Sigma)
UV cuvettes, d = 1 cm, 120 μL
Spectrophotometer with thermostated cell holder
240 mM stock solution of DTT

Separation of Stock Solution of Insulin

1. Resuspend 10 mg of insulin to a concentration of 7 mg mL^{-1} in buffer (10 mM $NaPO_4$, pH 7.0).
2. To dissolve the insulin, add concentrated HCl until the solution is clear, than increase the pH by adding concentrated NaOH until the solution gets slightly cloudy again.
3. Stock solution can be kept on ice or at −20 °C.

Assay

Conditions: 25–45 °C

Add different amounts of the potential chaperone to 50 μM insulin in a total volume of 115 μL and use the same amount of dialysis buffer as control. The reaction is started by adding this solution to a 120-μL cuvette already containing 10 μL of DTT solution. The aggregation is significantly slower when lower DTT concentrations are used. The cuvette is placed in the UV spectrophotometer and the kinetics of aggregation is followed at 400 nm.

24.6.3 Monitoring Dynamics of sHsps

Size-exclusion Chromatography

The dynamics of sHsps has been studied by various techniques, among them electron microscopy [111, 114], native PAGE [114], affinity co-purification [42], fluorescence resonance energy transfer [44], and mass spectrometry [139, 155]. The most frequently used method is size-exclusion HPLC (SEC). A critical aspect in such gel filtration experiments is to use columns that can be operated at varying temperatures or buffer conditions. Here, TosoHaas TSK 3000 PW (30 cm × 0.75 cm; sepa-

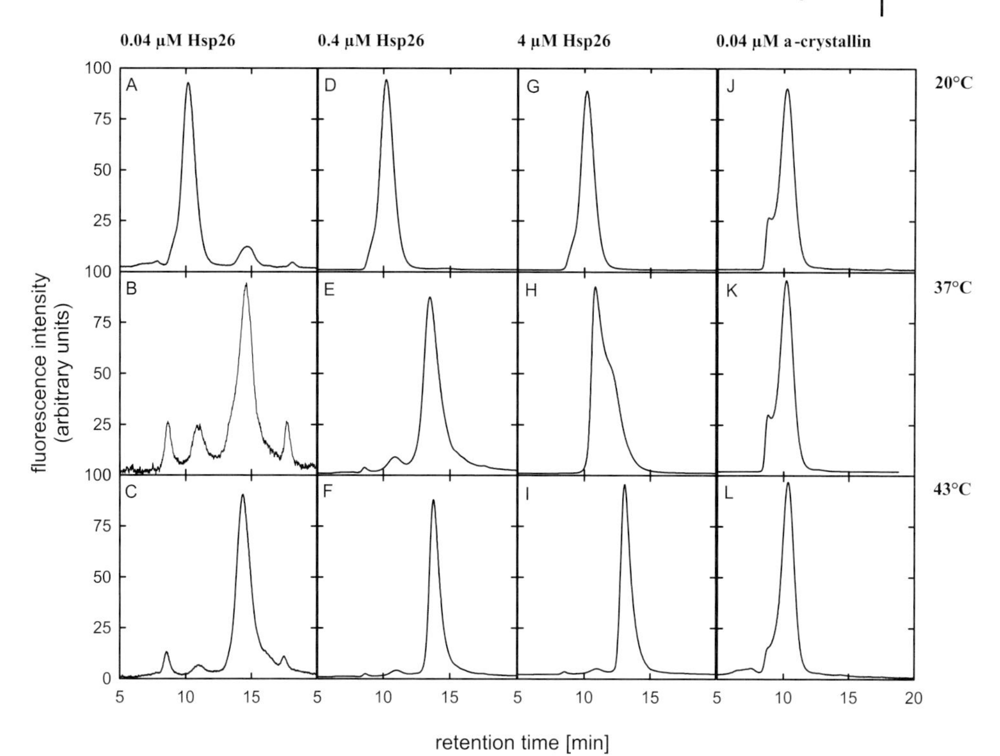

Fig. 24.5. Influence of elevated temperature and concentration on Hsp26 and α-crystallin quaternary structure. SEC was performed using a TosoHaas TSK 4000 PW column at different temperatures with a flow rate of 0.75 mL min^{-1} in 40 mM HEPES/KOH, 150 mM KCl, pH 7.4. Hsp26 and α-crystallin were incubated for 20 min at the indicated concentrations and temperatures and then applied to the column, which was kept at the same temperatures. The signal intensity was normalized.

ration range 0.5–800 kDa) or TSK 4000 PW columns (30 cm × 0.75 cm; separation range 0.5–1200 kDa) proved to be useful. Chromatography usually is carried out at 10–50 °C in buffers containing 150–300 mM salt at physiological pH conditions, with a flow rate of 0.5–0.75 mL min^{-1} (Figure 24.5). To analyze the dynamics of sHsps, it is important to perform the experiments at different concentrations of sHsps, since the concentration often has a high impact on the oligomeric state and stability of sHsp complexes.

Acknowledgements

We thank Hauke Hennecke and Johannes Buchner for generous support. Work in our laboratories was financed by grants from the Swiss National Foundation for Scientific Research and the Swiss Federal Institute of Technology to F.N. and by a grant from the Deutsche Forschungsgemeinschaft (SFB 594) to J.B. and M.H.

References

1 MacRae, T. H. (2000) Structure and function of small heat shock/α-crystallin proteins: established concepts and emerging ideas. *Cell. Mol. Life Sci.* **57**, 899–913.

2 Narberhaus, F. (2002) α-Crystallin-type heat shock proteins: Socializing minichaperones in the context of a multichaperone network. *Microbiol. Mol. Biol. Rev.* **66**, 64–93.

3 Waters, E. R., Lee, G. J., and Vierling, E. (1996) Evolution, structure and function of the small heat shock proteins in plants. *J. Exp. Bot.* **47**, 325–338.

4 Haslbeck, M., and Buchner, J. (2002) Chaperone function of sHsps. *Prog. Mol. Subcell. Biol.* **28**, 37–59.

5 Haslbeck, M. (2002) sHsps and their role in the chaperone network. *Cell. Mol. Life Sci.* **59**, 1649–1657.

6 Ehrnsperger, M., Buchner, J., and Gaestel, M. (1998) Structure and function of small heat-shock proteins. In *Molecular chaperones in the life cycle of proteins: Structure, function and mode of action* (Fink, A. L., and Goto, Y., eds), pp. 533–575, Marcel Dekker, Inc., New York.

7 De Jong, W. W., Caspers, G. J., and Leunissen, J. A. M. (1998) Genealogy of the α-crystallin – small heat-shock protein superfamily. *Int. J. Biol. Macromol.* **22**, 151–162.

8 Caspers, G. J., Leunissen, J. A. M., and de Jong, W. W. (1995) The expanding small heat-shock protein family, and structure predictions of the conserved 'α-crystallin domain'. *J. Mol. Evol.* **40**, 238–248.

9 Van Montfort, R., Slingsby, C., and Vierling, E. (2001) Structure and function of the small heat shock protein/α-crystallin family of molecular chaperones. *Adv. Protein Chem.* **59**, 105–156.

10 Horwitz, J. (2003) Alpha-crystallin. *Exp. Eye Res.* **76**, 145–153.

11 Klemenz, R., Fröhli, E., Steiger, R. H., Schäfer, R., and Aoyama, A. (1991) αB-crystallin is a small heat shock protein. *Proc. Natl. Acad. Sci. USA* **88**, 3652–3656.

12 Dubin, R. A., Wawrousek, E. F., and Piatigorsky, J. (1989) Expression of the murine αB-crystallin gene is not restricted to the lens. *Mol. Cell. Biol.* **9**, 1083–1091.

13 Bhat, S. P., and Nagineni, C. N. (1989) αB subunit of lens-specific protein α-crystallin is present in other ocular and non-ocular tissues. *Biochem. Biophys. Res. Commun.* **158**, 319–325.

14 Ingolia, T. D., and Craig, E. A. (1982) Four small *Drosophila* heat shock proteins are related to each other and to mammalian α-crystallin. *Proc. Natl. Acad. Sci. USA* **79**, 2360–2364.

15 Münchbach, M., Nocker, A., and Narberhaus, F. (1999) Multiple small heat shock proteins in rhizobia. *J. Bacteriol.* **181**, 83–90.

16 Kappé, G., Leunissen, J. A., and de Jong, W. W. (2002) Evolution and diversity of prokaryotic small heat shock proteins. *Prog. Mol. Subcell. Biol.* **28**, 1–17.

17 Scharf, K. D., Siddique, M., and Vierling, E. (2001) The expanding family of *Arabidopsis thaliana* small heat stress proteins and a new family of proteins containing α-crystallin domains (Acd proteins). *Cell Stress Chaperones* **6**, 225–237.

18 Leroux, M. R., Ma, B. J., Batelier, G., Melki, R., and Candido, E. P. M. (1997) Unique structural features of a novel class of small heat shock proteins. *J. Biol. Chem.* **272**, 12847–12853.

19 Wotton, D., Freeman, K., and Shore, D. (1996) Multimerization of Hsp42p, a novel heat shock protein of *Saccharomyces cerevisiae*, is dependent on a conserved carboxyl-terminal sequence. *J. Biol. Chem.* **271**, 2717–2723.

20 Nover, L., and Scharf, K. D. (1997) Heat Stress Proteins and Transcription Factors. *Cell. Mol. Life Sci.* **53**, 80–103.

21 Morimoto, R. I. (1998) Regulation of the heat shock transcriptional response: cross talk between a family of heat shock factors, molecular chaperones, and negative regulators. *Genes Dev.* **12**, 3788–3796.
22 Richmond, C. S., Glasner, J. D., Mau, R., Jin, H., and Blattner, F. R. (1999) Genome-wide expression profiling in *Escherichia coli* K-12. *Nucleic Acids Res.* **27**, 3821–3835.
23 Arrigo, A. P., and Landry, J. (1994) Expression and function of the low-molecular-weight heat shock proteins. In *The biology of heat shock proteins and molecular chaperones* (Morimoto, R. I., Tissiéres, A., and Georgopoulos, C., eds), pp. 335–373, Cold Spring Harbor Laboratory Press, Cold Spring Harbor, N.Y.
24 Sun, W., Van Montagu, M., and Verbruggen, N. (2002) Small heat shock proteins and stress tolerance in plants. *Biochim. Biophys. Acta* **1577**, 1–9.
25 Horwitz, J. (1992) α-crystallin can function as a molecular chaperone. *Proc. Natl. Acad. Sci. USA* **89**, 10449–10453.
26 Jakob, U., Gaestel, M., Engel, K., and Buchner, J. (1993) Small heat shock proteins are molecular chaperones. *J. Biol. Chem.* **268**, 1517–1520.
27 Fernando, P., and Heikkila, J. J. (2000) Functional characterization of *Xenopus* small heat shock protein, Hsp30C: the carboxyl end is required for stability and chaperone activity. *Cell Stress Chaperones* **5**, 148–159.
28 Liang, P., Amons, R., Macrae, T. H., and Clegg, J. S. (1997) Molecular characterization of a small heat shock/alpha-crystallin protein in encysted Artemia embryos. *Eur. J. Biochem.* **243**, 225–232.
29 Lee, G. J., Pokala, N., and Vierling, E. (1995) Structure and in vitro molecular chaperone activity of cytosolic small heat shock proteins from pea. *J. Biol. Chem.* **270**, 10432–10438.
30 Smýkal, P., Masin, J., Hrdý, I., Konopásek, I., and Zárský, V. (2000) Chaperone activity of tobacco HSP18, a small heat-shock protein, is inhibited by ATP. *Plant J.* **23**, 703–713.
31 Studer, S., and Narberhaus, F. (2000) Chaperone activity and homo- and hetero-oligomer formation of bacterial small heat shock proteins. *J. Biol. Chem.* **275**, 37212–37218.
32 Delmas, F., Pierre, F., Coucheney, F., Divies, C., and Guzzo, J. (2001) Biochemical and physiological studies of the small heat shock protein Lo18 from the lactic acid bacterium *Oenococcus oeni*. *J. Mol. Microbiol. Biotechnol.* **3**, 601–610.
33 Michelini, E. T., and Flynn, G. C. (1999) The unique chaperone operon of *Thermotoga maritima*: Cloning and initial characterization of a functional Hsp70 and small heat shock protein. *J. Bacteriol.* **181**, 4237–4244.
34 Roy, S. K., Hiyama, T., and Nakamoto, H. (1999) Purification and characterization of the 16-kDa heat-shock-responsive protein from the thermophilic cyanobacterium *Synechococcus vulcanus*, which is an α-crystallin-related, small heat shock protein. *Eur. J. Biochem.* **262**, 406–416.
35 Kokke, B. P. A., Leroux, M. R., Candido, E. P. M., Boelens, W. C., and deJong, W. W. (1998) *Caenorhabditis elegans* small heat-shock proteins Hsp12.2 and Hsp12.3 form tetramers and have no chaperone-like activity. *FEBS Lett.* **433**, 228–232.
36 Santhoshkumar, P., and Sharma, K. K. (2002) *Caenorhabditis elegans* small heat-shock proteins Hsp12.2 and Hsp12.3 form tetramers and have no chaperone-like activity. *Biochim. Biophys. Acta* **1598**, 115–121.
37 Bhattacharyya, J., Santhoshkumar, P., and Sharma, K. K. (2003) A peptide sequence-YSGVCHTDLHA-WHGDWPLPVK [40–60]-in yeast alcohol dehydrogenase prevents the aggregation of denatured substrate proteins. *Biochem. Biophys. Res. Commun.* **307**, 1–7.
38 Lee, G. J., Roseman, A. M., Saibil, H. R., and Vierling, E. (1997) A

small heat shock protein stably binds heat denatured model substrates and can maintain a substrate in a folding competent state. *EMBO J.* **16**, 659–671.

39 Sharma, K. K., Kumar, G. S., Murphy, A. S., and Kester, K. (1998) Identification of 1,1′-Bi(4-anilino)naphthalene-5,5′-disulfonic acid binding sequences in α-crystallin. *J. Biol. Chem.* **273**, 15474–15478.

40 Sharma, K. K., Kumar, R. S., Kumar, G. S., and Quinn, P. T. (2000) Synthesis and characterization of a peptide identified as a functional element in αA-crystallin. *J. Biol. Chem.* **275**, 3767–3771.

41 Sharma, K. K., Kaur, H., and Kester, K. (1997) Functional elements in molecular chaperone α-crystallin: identification of binding sites in αB-crystallin. *Biochem. Biophys. Res. Commun.* **239**, 217–222.

42 Lentze, N., Studer, S., and Narberhaus, F. (2003) Structural and functional defects caused by point mutations in the α-crystallin domain of a bacterail α-heat shock protein. *J. Mol. Biol.* **328**, 927–937.

43 Lindner, R. A., Kapur, A., Mariani, M., Titmuss, S. J., and Carver, J. A. (1998) Structural alterations of α-crystallin during its chaperone action. *Eur. J. Biochem.* **258**, 170–183.

44 Bova, M. P., Ding, L. L., Horwitz, J., and Fung, B. K. (1997) Subunit exchange of a α-crystallin. *J. Biol. Chem.* **272**, 29511–29517.

45 Hartl, F. U. (1996) Molecular chaperones in cellular protein folding. *Nature* **381**, 571–580.

46 Ehrnsperger, M., Gräber, S., Gaestel, M., and Buchner, J. (1997) Binding of non-native protein to Hsp25 during heat shock creates a reservoir of folding intermediates for reactivation. *EMBO J.* **16**, 221–229.

47 Rajaraman, K., Raman, B., and Rao, C. M. (1996) Molten-globule state of carbonic anhydrase binds to the chaperone-like α-crystallin. *J. Biol. Chem.* **271**, 27595–27600.

48 Rawat, U., and Rao, M. (1998) Interactions of chaperone alpha-crystallin with the molten globule state of xylose reductase. Implications for reconstitution of the active enzyme. *J. Biol. Chem.* **273**, 9415–9423.

49 Lindner, R. A., Kapur, A., and Carver, J. A. (1997) The interaction of the molecular chaperone, α-crystallin, with molten globule states of bovine α-lactalbumin. *J. Biol. Chem.* **272**, 27722–27729.

50 Wang, K., and Spector, A. (2001) ATP causes small heat shock proteins to release denatured protein. *Eur. J. Biochem.* **268**, 6335–6345.

51 Muchowski, P. J., and Clark, J. I. (1998) ATP-enhanced molecular chaperone functions of the small heat shock protein human αB crystallin. *Proc. Natl. Acad. Sci. USA* **95**, 1004–1009.

52 Muchowski, P. J., Hays, L. G., Yates, J. R., and Clark, J. I. (1999) ATP and the core "α-crystallin" domain of the small heat-shock protein αB-crystallin. *J. Biol. Chem.* **274**, 30190–30195.

53 Allen, S. P., Polazzi, J. O., Gierse, J. K., and Easton, A. M. (1992) Two novel heat shock genes encoding proteins produced in response to heterologous protein expression in *Escherichia coli*. *J. Bacteriol.* **174**, 6938–6947.

54 Stromer, T., Ehrnsperger, M., Gaestel, M., and Buchner, J. (2003) Analysis of the interaction of small heat shock proteins with unfolding proteins. *J. Biol. Chem.* **278**, 18015–18021.

55 Mogk, A., Schlieker, C., Friedrich, K. L., Schönfeld, H. J., Vierling, E., and Bukau, B. (2003) Refolding of substrates bound to small Hsps relies on a disaggregation reaction mediated most efficiently by ClpB/Dank. *J. Biol. Chem.* **278**, 31033–31042.

56 Nover, L., Scharf, K. D., and Neumann, D. (1989) Cytoplasmic heat shock granules are formed from precursor particles and are associated with a specific set of mRNAs. *Mol. Cell. Biol.* **9**, 1298–1308.

57 Kirschner, M., Winkelhaus, S., Thierfelder, J. M., and Nover, L.

(2000) Transient expression and heat-stress-induced co-aggregation of endogenous and heterologous small heat-stress proteins in tobacco protoplasts. *Plant J.* **24**, 397–411.

58 Wang, K., and Spector, A. (1994) The chaperone activity of bovine α-crystallin. Interaction with other lens crystallins in native and denatured states. *J. Biol. Chem.* **269**, 13601–13608.

59 Smýkal, P., Hrdý, I., and Pechan, P. M. (2000) High-molecular-mass complexes formed in vivo contain smHSPs and HSP70 and display chaperone-like activity. *Eur. J. Biochem.* **267**, 2195–2207.

60 Veinger, L., Diamant, S., Buchner, J., and Goloubinoff, P. (1998) The small heat-shock protein IbpB from *Escherichia coli* stabilizes stress-denatured proteins for subsequent refolding by a multichaperone network. *J. Biol. Chem.* **273**, 11032–11037.

61 Lee, G. J., and Vierling, E. (2000) A small heat shock protein cooperates with heat shock protein 70 systems to reactivate a heat-denatured protein. *Plant Physiol.* **122**, 189–197.

62 Fagerholm, P. P., Philipson, B. T., and Lindström, B. (1981) Normal human lens – the distribution of proteins. *Exp. Eye Res.* **33**, 615–620.

63 Rao, P. V., Huang, Q. L., Horwitz, J., and Zigler, J. S., Jr. (1995) Evidence that *a*-crystallin prevents non-specific protein aggregation in the intact eye lens. *Biochim. Biophys. Acta* **1245**, 439–447.

64 Carver, J. A., Nicholls, K. A., Aquilina, J. A., and Truscott, R. J. (1996) Age-related changes in bovine α-crystallin and high-molecular-weight protein. *Exp. Eye Res.* **63**, 639–647.

65 Brady, J. P., Garland, D., Duglas-Tabor, Y., Robison, W. G. J., Groome, A., and Wawrousek, E. F. (1997) Targeted disruption of the mouse αA-crystallin gene induces cataract and cytoplasmic inclusion bodies containing the small heat shock protein αB-crystallin. *Proc. Natl. Acad. Sci. USA* **94**, 884–889.

66 Litt, M., Kramer, P., LaMorticella, D. M., Murphey, W., Lovrien, E. W., and Weleber, R. G. (1998) Autosomal dominant congenital cataract associated with a missense mutation in the human alpha crystallin gene CRYAA. *Hum. Mol. Genet.* **7**, 471–474.

67 Kumar, L. V., Ramakrishna, T., and Rao, C. M. (1999) Structural and functional consequences of the mutation of a conserved arginine residue in αA and αB crystallins. *J. Biol. Chem.* **274**, 24137–24141.

68 Cobb, B. A., and Petrash, J. M. (2000) Characterization of α-crystallin-plasma membrane binding. *Biochemistry* **39**, 15791–15798.

69 Shroff, N. P., Cherian-Shaw, M., Bera, S., and Abraham, E. C. (2000) Mutation of R116C results in highly oligomerized αA-crystallin with modified structure and defective chaperone-like function. *Biochemistry* **39**, 1420–1426.

70 Vicart, P., Caron, A., Guicheney, P., Li, Z., Prevost, M. C., Faure, A., Chateau, D., Chapon, F., Tome, F., Dupret, J. M., Paulin, D., and Fardeau, M. (1998) A missense. mutation in the αB-crystallin chaperone gene causes a desmin-related myopathy. *Nat. Genet.* **20**, 92–95.

71 Perng, M. D., Muchowski, P. J., van den Ijssel, P., Wu, G. J., Hutcheson, A. M., Clark, J. I., and Quinlan, R. A. (1999) The cardiomyopathy and lens cataract mutation in αB-crystallin alters its protein structure, chaperone activity, and interaction with intermediate filaments in vitro. *J. Biol. Chem.* **274**, 33235–33243.

72 Bova, M. P., Yaron, O., Huang, Q. L., Ding, L. L., Haley, D. A., Stewart, P. L., and Horwitz, J. (1999) Mutation R120G in αB-crystallin, which is linked to a desmin-related myopathy, results in an irregular structure and defective chaperone-like function. *Proc. Natl. Acad. Sci. USA* **96**, 6137–6142.

73 Brady, J. P., Garland, D. L., Green, D. E., Tamm, E. R., Giblin, F. J., and Wawrousek, E. F. (2001) αB-crystallin

in lens development and muscle integrity: a gene knockout approach. *Invest. Ophthalmol. Vis. Sci.* **42**, 2924–2934.

74 Renkawek, K., Voorter, C. E., Bosman, G. J., van Workum, F. P., and de Jong, W. W. (1994) Expression of αB-crystallin in Alzheimer's disease. *Acta Neuropathol.* **87**, 155–160.

75 Renkawek, K., de Jong, W. W., Merck, K. B., Frenken, C. W., van Workum, F. P., and Bosman, G. J. (1992) αB-crystallin is present in reactive glia in Creutzfeldt-Jakob disease. *Acta Neuropathol.* **83**, 324–327.

76 Lowe, J., Landon, M., Pike, I., Spendlove, I., McDermott, H., and Mayer, R. J. (1990) Dementia with beta-amyloid deposition: involvement of alpha B-crystallin supports two main diseases. *Lancet* **336**, 515–516.

77 Head, M. W., Corbin, E., and Goldman, J. E. (1993) Overexpression and abnormal modification of the stress proteins αB-crystallin and HSP27 in Alexander disease. *Am. J. Pathol.* **143**, 1743–1153.

78 Mounier, N., and Arrigo, A. P. (2002) Actin cytoskeleton and small heat shock proteins: how do they interact? *Cell Stress Chaperones* **7**, 167–176.

79 Arrigo, A. P., Paul, C., Ducasse, C., Sauvageot, O., and Kretz-Remy, C. (2002) Small stress proteins: modulation of intracellular redox state and protection against oxidative stress. *Prog. Mol. Subcell. Biol.* **28**, 171–184.

80 Arrigo, A. P., Paul, C., Ducasse, C., Manero, F., Kretz-Remy, C., Virot, S., Javouhey, E., Mounier, N., and Diaz-Latoud, C. (2002) Small stress proteins: novel negative modulators of apoptosis induced independently of reactive oxygen species. *Prog. Mol. Subcell. Biol.* **28**, 185–204.

81 Cuesta, R., Laroia, G., and Schneider, R. J. (2000) Chaperone hsp27 inhibits translation during heat shock by binding eIF4G and facilitating dissociation of cap-initiation complexes. *Genes Dev.* **14**, 1460–1470.

82 Akbar, M. T., Lundberg, A. M., Liu, K., Vidyadaran, S., Wells, K. E., Dolatshad, H., Wynn, S., Wells, D. J., Latchman, D. S., and de Belleroche, J. (2003) The neuroprotective effects of heat shock protein 27 overexpression in transgenic animals against kainate-induced seizures and hippocampal cell death. *J. Biol. Chem.* **278**, 19956–19965.

83 Clark, J. I., and Muchowski, P. J. (2000) Small heat-shock proteins and their potential role in human disease. *Curr. Opin. Struct. Biol.* **10**, 52–59.

84 Crabbe, M. J. C., and Hepburne-Scott, H. W. (2001) Small heat shock proteins (sHsps) as potential drug targets. *Curr. Pharm. Biotech.* **2**, 77–111.

85 Löw, D., Brändle, K., Nover, L., and Forreiter, C. (2000) Cytosolic heat-stress proteins Hsp17.7 class I and Hsp17.3 class II of tomato act as molecular chaperones in vivo. *Planta* **211**, 575–582.

86 Forreiter, C., Kirschner, M., and Nover, L. (1997) Stable transformation of an Arabidopsis cell suspension culture with firefly luciferase providing a cellular system for analysis of chaperone activity in vivo. *Plant Cell* **9**, 2171–2181.

87 Yeh, C. H., Chang, P. F. L., Yeh, K. W., Lin, W. C., Chen, Y. M., and Lin, C. Y. (1997) Expression of a gene encoding a 16.9-kDa heat-shock protein, Oshsp16.9, in *Escherichia coli* enhances thermotolerance. *Proc. Natl. Acad. Sci. USA* **94**, 10967–10972.

88 Soto, A., Allona, I., Collada, C., Guevara, M. A., Casado, R., Rodriguez-Cerezo, E., Aragoncillo, C., and Gomez, L. (1999) Heterologous expression of a plant small heat-shock protein enhances *Escherichia coli* viability under heat and cold stress. *Plant Physiol.* **120**, 521–528.

89 Sun, W., Bernard, C., van de Cotte, B., Van Montagu, M., and Verbruggen, N. (2001) At-HSP17.6A, encoding a small heat-shock protein in *Arabidopsis*, can enhance

osmotolerance upon overexpression. *Plant J.* **27**, 407–415.

90 Kitagawa, M., Matsumura, Y., and Tsuchido, T. (2000) Small heat shock proteins, IbpA and IbpB, are involved in resistances to heat and superoxide stresses in *Escherichia coli*. *FEMS Microbiol. Lett.* **184**, 165–171.

91 Plesofsky-Vig, N., and Brambl, R. (1995) Disruption of the gene for *hsp30*, an α-crystallin-related heat shock protein of *Neurospora crassa*, causes defects in thermotolerance. *Proc. Natl. Acad. Sci. USA* **92**, 5032–5036.

92 Thomas, J. G., and Baneyx, F. (1998) Roles of the *Escherichia coli* small heat shock proteins IbpA and IbpB in thermal stress management: Comparison with ClpA, ClpB, and HtpG in vivo. *J. Bacteriol.* **180**, 5165–5172.

93 Kuczynska-Wisnik, D., Kedzierska, S., Matuszewska, E., Lund, P., Taylor, A., Lipinska, B., and Laskowska, E. (2002) The *Escherichia coli* small heat-shock proteins IbpA and IbpB prevent the aggregation of endogenous proteins denatured in vivo during extreme heat shock. *Microbiology* **148**, 1757–1765.

94 Lee, S., Owen, H. A., Prochaska, D. J., and Barnum, S. R. (2000) HSP16.6 is involved in the development of thermotolerance and thylakoid stability in the unicellular cyanobacterium, *Synechocystis* sp. PCC 6803. *Curr. Microbiol.* **40**, 283–287.

95 Lee, S. Y., Prochaska, D. J., Fang, F., and Barnum, S. R. (1998) A 16.6-kilodalton protein in the cyanobacterium *Synechocystis* sp. PCC 6803 plays a role in the heat shock response. *Curr. Microbiol.* **37**, 403–407.

96 Horváth, I., Glatz, A., Varvasovszki, V., Török, Z., Páli, T., Balogh, G., Kovács, E., Nádasdi, L., Benkö, S., Joó, F., and Vigh, L. (1998) Membrane physical state controls the signaling mechanism of the heat shock response in *Synechocystis* PCC 6803: identification of *hsp17* as a "fluidity gene". *Proc. Natl. Acad. Sci. USA* **95**, 3513–3518.

97 Török, Z., Goloubinoff, P., Horváth, I., Tsvetkova, N. M., Glatz, A., Balogh, G., Varvasovszki, V., Los, D. A., Vierling, E., Crowe, J. H., and Vigh, L. (2001) *Synechocystis* HSP17 is an amphitropic protein that stabilizes heat-stressed membranes and binds denatured proteins for subsequent chaperone-mediated refolding. *Proc. Natl. Acad. Sci. USA* **98**, 3098–3103.

98 Tsvetkova, N. M., Horvath, I., Török, Z., Wolkers, W. F., Balogi, Z., Shigapova, N., Crowe, L. M., Tablin, F., Vierling, E., Crowe, J. H., and Vigh, L. (2002) Small heat-shock proteins regulate membrane lipid polymorphism. *Proc. Natl. Acad. Sci. USA* **99**, 13504–13509.

99 Cobb, B. A., and Petrash, J. M. (2000) Characterization of α-crystallin-plasma membrane binding. *J. Biol. Chem.* **275**, 6664–6672.

100 Lee, B. Y., Hefta, S. A., and Brennan, P. J. (1992) Characterization of the major membrane protein of virulent *Mycobacterium tuberculosis*. *Infect. Immun.* **60**, 2066–2074.

101 Lünsdorf, H., Schairer, H. U., and Heidelbach, M. (1995) Localization of the stress protein SP21 in indole-induced spores, fruiting bodies, and heat-shocked cells of *Stigmatella aurantiaca*. *J. Bacteriol.* **177**, 7092–7099.

102 Jobin, M. P., Delmas, F., Garmyn, D., Diviès, C., and Guzzo, J. (1997) Molecular characterization of the gene encoding an 18-kilodalton small heat shock protein associated with the membrane of *Leuconostoc oenos*. *Appl. Environ. Microbiol.* **63**, 609–614.

103 Kedersha, N. L., Gupta, M., Li, W., Miller, I., and Anderson, P. (1999) RNA-binding proteins TIA-1 and TIAR link the phosphorylation of eIF-2α to the assembly of mammalian stress granules. *J. Cell. Biol.* **147**, 1431–1442.

104 Singh, K., GrothVasselli, B., and Farnsworth, P. N. (1998) Interaction of DNA with bovine lens α-crystallin: its functional implications. *Int. J. Biol. Macromol.* **22**, 315–320.

105 Pietrowski, D., Durante, M. J., Liebstein, A., Schmitt-John, T., Werner, T., and Graw, J. (1994) α-Crystallins are involved in specific interactions with the murine gamma D/E/F-crystallin-encoding gene. *Gene* **144**, 171–178.

106 Bloemendal, M., Toumadje, A., and Johnson, W. C. (1999) Bovine lens crystallins do contain helical structures: a circular dichroism study. *Biochim. Biophys. Acta* **1432**, 234–238.

107 Merck, K. B., Groenen, P. J. T. A., Voorter, C. E. M., de Haard-Hoekman, W. A., Horwitz, J., Bloemendal, H., and de Jong, W. W. (1993) Structural and functional similarities of bovine α-crystallin and mouse small heat-shock protein – A family of chaperones. *J. Biol. Chem.* **268**, 1046–1052.

108 Dudich, I. V., Zav'yalov, V. P., Pfeil, W., Gaestel, M., Zav'yalova, G. A., Denesyuk, A. I., and Korpela, T. (1995) Dimer structure as a minimum cooperative subunit of small heat-shock proteins. *Biochim. Biophys. Acta* **1253**, 163–168.

109 Stromer, T., Fischer, E., Richter, K., Haslbeck, M., and Buchner, J. (2004) Analysis of the regulation of the molecular chaperone Hsp26 by temperature-induced dissociation: the N-terminal domail is important for oligomer assembly and the binding of unfolding proteins. *J. Biol. Chem.* **278**, 25970–25976.

110 Kim, K. K., Kim, R., and Kim, S. H. (1998) Crystal structure of a small heat-shock protein. *Nature* **394**, 595–599.

111 Haley, D. A., Horwitz, J., and Stewart, P. L. (1998) The small heat-shock protein, α-B-crystallin, has a variable quaternary structure. *J. Mol. Biol.* **277**, 27–35.

112 Bova, M. P., McHaourab, H. S., Han, Y., and Fung, B. K. K. (2000) Subunit exchange of small heat shock proteins – Analysis of oligomer formation of αA-crystallin and Hsp27 by fluorescence resonance energy transfer and site-directed truncations. *J. Biol. Chem.* **275**, 1035–1042.

113 van Montfort, R. L., Basha, E., Friedrich, K. L., Slingsby, C., and Vierling, E. (2001) Crystal structure and assembly of a eukaryotic small heat shock protein. *Nat. Struct. Biol.* **8**, 1025–1030.

114 Haslbeck, M., Walke, S., Stromer, T., Ehrnsperger, M., White, H. E., Chen, S. X., Saibil, H. R., and Buchner, J. (1999) Hsp26: a temperature-regulated chaperone. *EMBO J.* **18**, 6744–6751.

115 Chang, Z. Y., Primm, T. P., Jakana, J., Lee, I. H., Serysheva, I., Chiu, W., Gilbert, H. F., and Quiocho, F. A. (1996) *Mycobacterium tuberculosis* 16-kDa antigen (Hsp16.3) functions as an oligomeric structure in vitro to suppress thermal aggregation. *J. Biol. Chem.* **271**, 7218–7223.

116 Leroux, M. R., Melki, R., Gordon, B., Batelier, G., and Candido, E. P. M. (1997) Structure-function studies on small heat shock protein oligomeric assembly and interaction with unfolded polypeptides. *J. Biol. Chem.* **272**, 24646–24656.

117 Derham, B. K., van Boekel, M. A., Muchowski, P. J., Clark, J. I., Horwitz, J., Hepburne-Scott, H. W., de Jong, W. W., Crabbe, M. J., and Harding, J. J. (2001) Chaperone function of mutant versions of αA- and αB-crystallin prepared to pinpoint chaperone binding sites. *Eur. J. Biochem.* **268**, 713–721.

118 Horwitz, J., Bova, M., Huang, Q. L., Ding, L., Yaron, O., and Lowman, S. (1998) Mutation of αB-crystallin: effects on chaperone-like activity. *Int. J. Biol. Macromol.* **22**, 263–269.

119 Koteiche, H. A., and Mchaourab, H. S. (1999) Folding pattern of the α-crystallin domain in αA-crystallin determined by site-directed spin labeling. *J. Mol. Biol.* **294**, 561–577.

120 Smulders, R. H. P. H., Carver, J. A., Lindner, R. A., van Boekel, M. A. M., Bloemendal, H., and de Jong, W. W. (1996) Immobilization of the C-terminal extension of bovine αA-crystallin reduces chaperone-like activity. *J. Biol. Chem.* **271**, 29060–29066.

121 Studer, S., Obrist, M., Lentze, N., and Narberhaus, F. (2002) A critical motif for oligomerization and chaperone activity of bacterial α-heat shock proteins. *Eur. J. Biochem.* **269**, 3578–3586.

122 Giese, K. C., and Vierling, E. (2002) Changes in oligomerization are essential for the chaperone activity of a small heat shock protein in vivo and in vitro. *J. Biol. Chem.* **277**, 46310–46318.

123 Berengian, A. R., Parfenova, M., and Mchaourab, H. S. (1999) Site-directed spin labeling study of subunit interactions in the α-crystallin domain of small heat-shock proteins: Comparison of the oligomer symmetry in αA-crystallin, Hsp27 and Hsp16.3. *J. Biol. Chem.* **274**, 6305–6314.

124 Kantorow, M., and Piatigorsky, J. (1998) Phosphorylations of alpha A- and alpha B-crystallin. *Int. J. Biol. Macromol.* **22**, 307–314.

125 Gaestel, M. (2002) sHsp-phosphorylation: enzymes, signaling pathways and functional implications. *Prog. Mol. Subcell. Biol.* **28**, 151–169.

126 Derham, B. K., and Harding, J. J. (1999) α-crystallin as a molecular chaperone. *Prog. Retin. Eye Res.* **18**, 463–509.

127 Kemp, B. E., and Pearson, R. B. (1990) Protein kinase recognition sequence motifs. *Trends Biochem. Sci.* **15**, 342–346.

128 Butt, E., Immler, D., Meyer, H. E., Kotlyarov, A., Laass, K., and Gaestel, M. (2001) Heat shock protein 27 is a substrate of cGMP-dependent protein kinase in intact human platelets: phosphorylation-induced actin polymerization caused by HSP27 mutants. *J. Biol. Chem.* **276**, 7108–7113.

129 Lund, A. A., Rhoads, D. M., Lund, A. L., Cerny, R. L., and Elthon, T. E. (2001) In vivo modifications of the maize mitochondrial small heat stress protein, HSP22. *J. Biol. Chem.* **276**, 29924–29929.

130 Ito, H., Kamei, K., Iwamoto, I., Inaguma, Y., Nohara, D., and Kato, K. (2001) Phosphorylation-induced Change of the Oligomerization State of αB-crystallin. *J. Biol. Chem.* **276**, 5346–5352.

131 Kato, K., Hasegawa, K., Goto, S., and Inaguma, Y. (1994) Dissociation as a result of phosphorylation of an aggregated form of the small stress protein, hsp27. *J. Biol. Chem.* **269**, 11274–11278.

132 Rogalla, T., Ehrnsperger, M., Preville, X., Kotlyarov, A., Lutsch, G., Ducasse, C., Paul, C., Wieske, M., Arrigo, A. P., Buchner, J., and Gaestel, M. (1999) Regulation of Hsp27 oligomerization, chaperone function, and protective activity against oxidative stress/tumor necrosis factor alpha by phosphorylation. *J. Biol. Chem.* **274**, 18947–18956.

133 Ito, H., Okamoto, K., Nakayama, H., Isobe, T., and Kato, K. (1997) Phosphorylation of αB-crystallin in response to various types of stress. *J. Biol. Chem.* **272**, 29934–29941.

134 van den IJssel, P. R., Overkamp, P., Bloemendal, H., and de Jong, W. W. (1998) Phosphorylation of αB-crystallin and HSP27 is induced by similar stressors in HeLa cells. *Biochem. Biophys. Res. Commun.* **247**, 518–523.

135 Voorter, C. E., de Haard-Hoekman, W. A., Roersma, E. S., Meyer, H. E., Bloemendal, H., and de Jong, W. W. (1989) The in vivo phosphorylation sites of bovine αB-crystallin. *FEBS Lett.* **259**, 50–52.

136 Gaestel, M., Benndorf, R., Hayess, K., Priemer, E., and Engel, K. (1992) Dephosphorylation of the small heat shock protein hsp25 by calcium/calmodulin-dependent (type 2B) protein phosphatase. *J. Biol. Chem.* **267**, 21607–21611.

137 Cairns, J., Qin, S., Philp, R., Tan, Y. H., and Guy, G. R. (1994) Dephosphorylation of the small heat shock protein Hsp27 in vivo by protein phosphatase 2A. *J. Biol. Chem.* **269**, 9176–9183.

138 Moroni, M., and Garland, D. (2001) In vitro dephosphorylation of α-crystallin is dependent on the state of

oligomerization. *Biochim. Biophys. Acta* **1546**, 282–290.

139 SOBOTT, F., BENESCH, J. L., VIERLING, E., and ROBINSON, C. V. (2002) Subunit exchange of multimeric protein complexes. Real-time monitoring of subunit exchange between small heat shock proteins by using electrospray mass spectrometry. *J. Biol. Chem.* **277**, 38921–38929.

140 HELM, K. W., LEE, G. J., and VIERLING, E. (1997) Expression and native structure of cytosolic class II small heat shock proteins. *Plant Physiol.* **114**, 1477–1485.

141 VAN DEN OETELAAR, P. J., VAN SOMEREN, P. F., THOMSON, J. A., SIEZEN, R. J., and HOENDERS, H. J. (1990) A dynamic quaternary structure of bovine alpha-crystallin as indicated from intermolecular exchange of subunits. *Biochemistry* **10**, 3488–3493.

142 SUGIYAMA, Y., SUZUKI, A., KISHIKAWA, M., AKUTSU, R., HIROSE, T., WAYE, M. M., TSUI, S. K., YOSHIDA, S., and OHNO, S. (2000) Muscle develops a specific form of small heat shock protein complex composed of MKBP/HSPB2 and HSPB3 during myogenic differentiation. *J. Biol. Chem.* **275**, 1095–1104.

143 KITAGAWA, M., MIYAKAWA, M., MATSUMURA, Y., and TSUCHIDO, T. (2002) *Escherichia coli* small heat shock proteins, IbpA and IbpB, protect enzymes from inactivation by heat and oxidants. *Eur. J. Biochem.* **269**, 2907–2917.

144 FRIEDRICH, K. L., GIESE, K. C., BUAN, N. R., and VIERLING, E. (2004) Interactions between small heat shock protein subunits and substrate in small heat shock protein/substrate complexes. *J. Biol. Chem.* **279**, 1080–1089.

145 BUCHNER, J., GRALLERT, H., and JAKOB, U. (1998) Analysis of chaperone function using citrate synthase as nonnative substrate protein. *Methods Enzymol.* **290**, 323–338.

146 GAESTEL, M., GROSS, B., BENNDORF, R., STRAUSS, M., SCHUNK, W. H., KRAFT, R., OTTO, A., BOHM, H., STAHL, J., DRABSCH, H., and et al. (1989) Molecular cloning, sequencing and expression in *Escherichia coli* of the 25-kDa growth-related protein of Ehrlich ascites tumor and its homology to mammalian stress proteins. *Eur. J. Biochem.* **179**, 209–213.

147 GOLOUBINOFF, P., CHRISTELLER, J. T., GATENBY, A. A., and LORIMER, G. H. (1989) Reconstitution of active dimeric ribulose bisphosphate carboxylase from an unfoleded state depends on two chaperonin proteins and Mg-ATP. *Nature* **342**, 884–889.

148 MENDOZA, J. A., ROGERS, E., LORIMER, G. H., and HOROWITZ, P. M. (1991) Chaperonins facilitate the in vitro folding of monomeric mitochondrial rhodanese. *J. Biol. Chem.* **266**, 13044–13049.

149 RANSON, N. A., BURSTON, S. G., and CLARKE, A. R. (1997) Binding, encapsulation and ejection: substrate dynamics during a chaperonin-assisted folding reaction. *J. Mol. Biol.* **266**, 656–664.

150 CHEN, S., ROSEMAN, A. M., HUNTER, A. S., WOOD, S. P., BURSTON, S. G., RANSON, N. A., CLARKE, A. R., and SAIBIL, H. R. (1994) Location of a folding protein and shape changes in GroEL-GroES complexes imaged by cryo-electron microscopy. *Nature* **371**, 261–264.

151 BUCHNER, J., SCHMIDT, M., FUCHS, M., JAENICKE, R., RUDOLPH, R., SCHMID, F. X., and KIEFHABER, T. (1991) GroE facilitates refolding of citrate synthase by suppressing aggregation. *Biochemistry* **30**, 1586–1591.

152 JAENICKE, R. (1993) What does protein refolding in vitro tell us about protein folding in the cell? *Philos. Trans. R. Soc. Lond. B Biol. Sci.* **339**, 287–294.

153 HASLBECK, M., SCHUSTER, I., and GRALLERT, H. (2003) GroE-dependent expression and purification of pig heart mitochondrial citrate synthase in Escherichia coli. *J. Chromatogr. B Analyt. Technol. Biomed. Life Sci.* **786**, 127–136.

154 FARAHBAKHSH, Z. T., HUANG, Q. L.,

Ding, L. L., Altenbach, C., Steinhoff, H. J., Horwitz, J., and Hubbell, W. L. (1995) Interaction of α-crystallin with spin-labeled peptides. *Biochemistry* **34**, 509–516.

155 Aquilina, J. A., Benesch, J. L., Bateman, O. A., Slingsby, C., and Robinson, C. V. (2003) Polydispersity of a mammalian chaperone: mass spectrometry reveals the population of oligomers in αB-crystallin. *Proc. Natl. Acad. Sci. USA* **100**, 10611–10616.

156 Kim, R., Kim, K. K., Yokota, H., and Kim, S. H. (1998) Small heat shock protein of *Methanococcus jannaschii*, a hyperthermophile. *Proc. Natl. Acad. Sci. USA* **95**, 9129–9133.

157 Laksanalamai, P., Jiemjit, A., Bu, Z., Maeder, L., and Robb, T. (2003) Multi-subunit assembly of the *Pyrococcus furiosus* small heat shock protein is essential for cellular protection at high temperature. *Extremophiles* **7**, 79–83.

158 Laksanalamai, P., Maeder, D. L., and Robb, F. T. (2001) Regulation and mechanism of action of the small heat shock protein from the hyperthermophilic archaeon *Pyrococcus furiosus. J. Bacteriol.* **183**, 5198–5202.

159 Shearstone, J. R., and Baneyx, F. (1999) Biochemical characterization of the small heat shock protein IbpB from *Escherichia coli. J. Biol. Chem.* **274**, 9937–9945.

160 Young, D., Lathriga, R., Hendrix, R., Sweetser, D., and Young, R. A. (1988) Stress proteins are immune targets in leprosy and tuberculosis. *Proc. Natl. Acad. Sci. USA* **85**, 4267–4270.

161 Haslbeck, M., Braun, N., Stromer, T., Richter, B., Model, N., Weinkauf, S., and Buchner, J. (2004) Hsp42 is the general small heat shock protein in the cytosol of Saccharomyces cerevisiae. *EMBO J.* **23**, 638–649.

162 Ehrnsperger, M., Lilie, H., Gaestel, M., and Buchner, J. (1999) The dynamics of Hsp25 quaternary structure – Structure and function of different oligomeric species. *J. Biol. Chem.* **274**, 14867–14874.

163 Haley, D. A., Bova, M. P., Huang, Q. L., McHaourab, H. S., and Stewart, P. L. (2000) Small heat-shock protein structures reveal a continuum from symmetric to variable assemblies. *J. Mol. Biol.* **298**, 261–272.

25
Alpha-crystallin: Its Involvement in Suppression of Protein Aggregation and Protein Folding

Joseph Horwitz

25.1
Introduction

Alpha-crystallins are one of the major protein components of the mammalian eye lens. There are two alpha-crystallin genes, alpha A and alpha B. In humans, the alpha A gene is found on chromosome 21 and encodes for a polypeptide containing 173 amino acid residues. The alpha B gene is on chromosome 11 and encodes for a polypeptide containing 175 amino acid residues. The amino acid sequence homology between alpha A and alpha B is about 57%. Alpha-crystallins are members of the small heat shock protein family [1]. All of the small heat shock proteins contain an "alpha-crystallin domain," which is a stretch of about 90 amino acids in the C-terminal domain that shares some degree of homology among the various members [1]. The properties of alpha-crystallins as well as of other small heat shock protein have been reviewed extensively [2–9]. A key property of the small heat shock proteins is their ability to interact with unfolded proteins [10–11].

25.2
Distribution of Alpha-crystallin in the Various Tissues

By far the organ with the highest concentration of alpha-crystallin is the eye lens. The concentration of alpha-crystallin in many mammalian lenses is estimated to be between 25% and 40% of the total soluble lens proteins. In fish lenses, the amount of alpha-crystallin is generally less than 10%. In the eye lens, the alpha A species is dominant. In most mammalian lenses, the weight ratio of alpha A to alpha B is about 3 to 1. Alpha-crystallin is not found in any of the invertebrate lenses that have been examined thus far. In 1989 it was first reported that alpha B crystallin exists in many organs and cells outside the lens [5, 12–14]. Shortly thereafter, it was shown to be a bona fide small heat shock protein [15]. In the heart, for example, alpha B accounts for about 2% of the total protein mass. In some of the neurological diseases in which alpha B is known to be overexpressed, the concentration of alpha B can be extremely high [16].

Protein Folding Handbook. Part II. Edited by J. Buchner and T. Kiefhaber

ISBN: 3-527-30784-2

25.3 Structure

When isolated from organs or cells, alpha-crystallin is always found as a heterogeneous multimeric assembly with a molecular weight distribution ranging from approximately 300,000 to over one million. The quaternary structure of the small heat shock proteins has been the subject of many studies (reviewed in Refs. [5, 6, 9]). At present, the crystal structures of only two members of the small heat shock protein family are known: MjHsp16.5 from *Methanococcus jannaschii* [17] and Hsp16.9 from wheat [18]. It is important to note that MjHsp16.5 and Hsp16.9 are both homogeneous in nature and possess a discrete number of subunits, whereas alpha-crystallin Hsp27 and many other small heat shock proteins are innately polydispersed. This can be seen in Figure 25.1, where the gel filtration elution profile and molecular weight distribution of a preparation of recombinant alpha B, Hsp27, and MjHsp16.5 are shown. This data was obtained with a system that utilizes size-exclusion chromatography with online laser light-scattering, absorbance, and refractive index detectors [19–21]. As seen in Figure 25.1, the molecular weight distribution of MjHsp16.5 does not change much across the elution profile. This profile is what one would expect from a homogeneous sample. In contrast, the heterogeneity of alpha B crystallin and Hsp27 is evident (Figure 25.1).

For Hsp27 the molecular weight obtained at the peak is 460 000. At the half-bandwidth point of the leading elution profile (~12.3 mL), the molecular weight is about 550 000, whereas the molecular weight at the trailing edge (~14 mL) is 360 000. The molecular weight of the monomeric Hsp27 is ~22 800, meaning that at the peak the complex is made up of 20 subunits, whereas at the half-bandwidth point, the complexes are made up of ~16–24 subunits. Similarly, for alpha B, the native oligomeric structure shown covers a range of 26–34 subunits. This may explain in part its inability to crystallize native alpha-crystallin or Hsp27.

The oligomeric structure of alpha-crystallin is not rigid. In 1990, Van Der Oetelaar et al. were the first to observe that there is an intermolecular exchange of subunits in alpha-crystallins, pointing to a dynamic quaternary structure [22]. More recently, Bova et al. [23] reexamined the subunit exchange properties of alpha A crystallin using fluorescence resonance energy transfer (FRET). These measurements proved that the subunit exchange strongly depends on temperature. Subunit exchange is a property that occurs in many if not all members of the small heat shock protein family [6]. Even the extremely stable 24mer small heat shock protein Hsp16.5 isolated from the hyperthermophilic Archaea *Methanococcus jannaschii* exhibits subunit exchange at temperatures that are physiologically relevant [24]. The dynamic structure of alpha-crystallin and other small heat shock proteins is believed to be a key property for its chaperone activity (reviewed in Ref. [6]).

In recent years, major advances into the secondary and tertiary structure of alpha-crystallin were achieved especially by using site-directed spin labeling [25, 26]. However, in the absence of x-ray crystal structure, we are forced to rely mostly on modeling [5]. The recent models rely mostly on the x-ray structure of Hsp16.9 and MjHsp16.5, which both possess the "alpha-crystallin domain" [5,

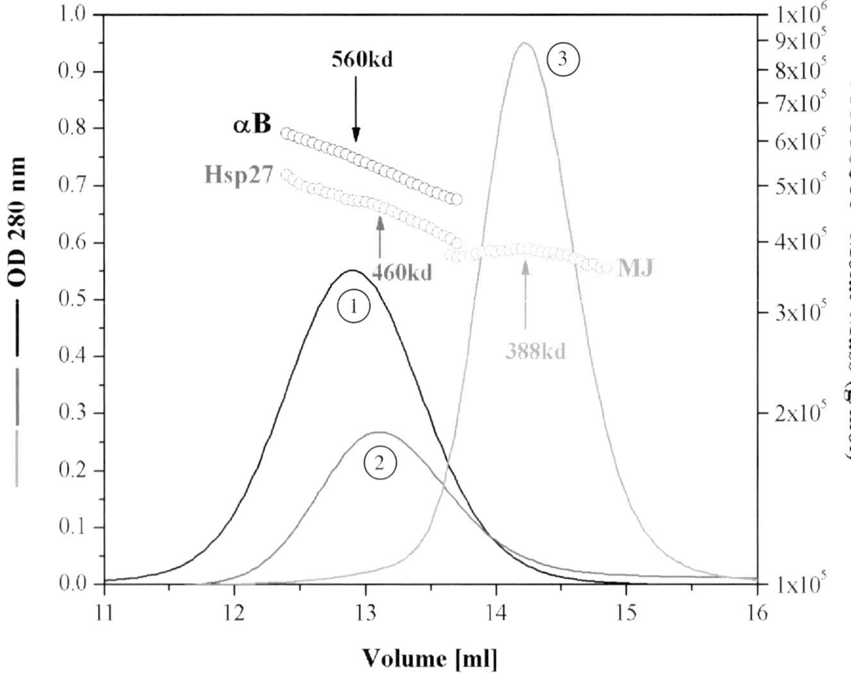

Fig. 25.1. Molar mass distribution and absorption spectra of recombinant alpha B crystallin, recombinant Hsp27, and recombinant Hsp16.5 from *Methanococcus jannaschii*. Size-exclusion chromatography was carried out on a Pharmacia Superose HR 6 column (1 × 30 cm) driven by a Pharmacia AKTA Basic system with a UV 900 monitor connected in line with an 18-angle laser light-scattering detector and a refractive index detector (DAWN EOS, Optilab DSP; Wyatt-Technology Corp. Santa Barbara, CA, USA). The solid lines represent the absorption profile of the eluted protein. The circles represent the molecular weight obtained as a function of the elution volume. The molecular weight is determined continuously every 20 microliters. Line spectra (1) obtained from 0.12 mg of alpha B. Line spectra (2) obtained from 0.1 mg of Hsp27. Line spectra (3) obtained from 0.15 mg of MjHsp16.5. All samples were dissolved in 50 mM phosphate buffer and 0.1 M NaCl, pH 7. Elution was performed at a rate of 0.5 mL min^{-1} and a temperature of 25 °C.

27]. Advances in cryo-electron microscopy provide a low-resolution picture of alpha-crystallin. These studies suggest that alpha-crystallin possesses a variable quaternary structure and a central cavity [28, 29]. The crystal structure of wheat Hsp16.9 as well as a cryo-electron microscopic image of alpha-crystallin and the crystal structure of Hsp16.5 from *M. jannaschii* are shown in Figure 24.3.

25.4
Phosphorylation and Other Posttranslation Modification

Alpha-crystallin is known to undergo many posttranslational modifications such as age-dependent truncation, glycation, deamidation, racemization, etc. [3, 30]. Of particular interest are the effects of phosphorylation on the function and structure of alpha-crystallin. The effects of phosphorylation on alpha-crystallin and other small heat shock proteins have been reviewed previously [5, 31, 32]. The effects of phosphorylation on the function of alpha-crystallin are controversial at present [9]. Recently, using site-directed mutagenesis to mimic phosphorylation, Ito et al. [33] reported a reduction in the size of the alpha B crystallin oligomer from 500 000 for native protein to 300 000 for the mimic-phosphorylated protein. Using lactate dehydrogenase at 50 °C as a target protein, the authors concluded that the phosphorylated mimic has a significantly reduced chaperone activity. Koteiche and Mchaourab [34], using T4 lysozyme as a target protein, showed that phosphorylation enhances the chaperone activity. In our laboratory we observed that phosphorylation of alpha B did not significantly affect its chaperone properties when lysozyme or lactalbumin were unfolded at 37 °C by reducing the disulfide bonds (unpublished data). These results underscore the importance of choosing an appropriate target protein as well as the unfolding conditions when assaying for the chaperone properties of alpha-crystallin.

25.5
Binding of Target Proteins to Alpha-crystallin

A common property of all the small heat shock proteins is to interact with unfolded or denatured target proteins and to prevent their nonspecific aggregation and precipitation (reviewed in Refs. [2, 6] and in Chapter 24). Alpha-crystallin will not interact with native protein but will recognize and interact with unfolded structures that are in a molten-globule-like state [35–38]. It will react more efficiently with slowly aggregating target proteins. This point was clearly demonstrated by Carver et al., who showed that alpha-crystallin was more efficient in arresting the aggregation of apo-α-lactalbumin as compared to holo-α-lactalbumin, which tends to aggregate much faster when the disulfide bonds are reduced [39]. Another example is comparing the interaction of alpha-crystallin with reduced lysozyme and apo-α-lactalbumin. Figure 25.2 shows the unfolding and aggregation of lysozyme and α-lactalbumin when the disulfide bonds are reduced with TCEP. To achieve complete suppression of the aggregation at 37 °C, the weight ratio of native bovine alpha-crystallin to lactalbumin is about 1:1, whereas for lysozyme the ratio of alpha-crystallin to lysozyme has to be 5:1. Alpha-lactalbumin and lysozyme are structurally closely related. However, they differ greatly in their unfolding behavior: lysozyme unfolds in a classic two-state model for protein denaturation, while alpha-lactalbumin unfolds through an intermediate molten-globule state [40, 41]. As can be seen in Figure 25.2, the kinetics of aggregation of lysozyme is significantly faster than that of alpha-lactalbumin. The lag time typically observed in this kind of experiment occurs because time is needed for nucleation and growth of the aggregate before scattering can be observed. Note that the lag-time for lysozyme is significantly shorter. The fact that kinetic considerations are extremely important in studying the interaction of alpha-crystallin with a target protein can also be exemplified with insulin. Insulin is another common target protein used to study

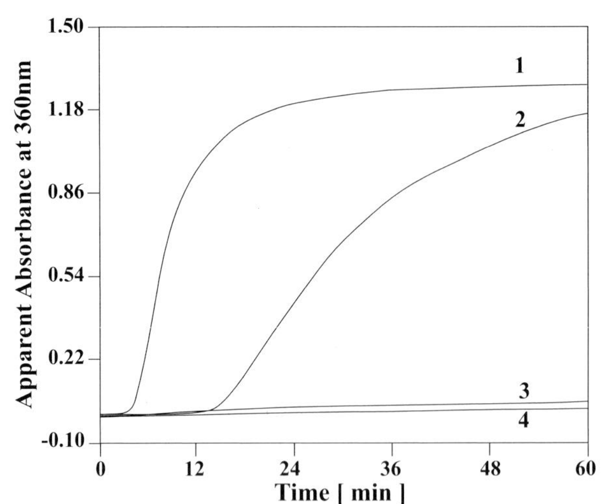

Fig. 25.2. Kinetics of the aggregation of lysozyme and bovine α-lactalbumin upon reduction of the disulfide bonds with Tris (2-carboxyethyl) phosphine (TCEP). Curve 1, 0.1 mg of lysozyme in 0.4 mL of buffer, pH 7.0, with 1 mM TCEP; Curve 2, 0.25 mg of α-lactalbumin in 0.4 mL of buffer, pH 6.7, with 5 mM of TCEP; Curve 3, same as curve 1, but with 0.5 mg of native bovine alpha-crystallin; Curve 4, same as curve 2, but with 0.25 mg of native bovine alpha-crystallin. Measurements were performed at 37 °C in a Shimadzu UV-2401PC absorption spectrophotometer. The temperature of the samples was controlled with a Peltier system; the path length was 1.0 cm.

the properties of small heat shock proteins [42, 43]. Upon reduction of the disulfide bonds, the B-chain of insulin will aggregate. The aggregation can be monitored by simple turbidometric measurements [42–44]. Holmgren found that thioredoxin catalyzes the reduction of insulin by DTT. He showed that thioredoxin at 5 μM concentration accelerated the reaction of 0.130 mM insulin with 1.0 mM DTT at pH 7 and 23 °C by about 20-fold [44]. The amount of native bovine alpha-crystallin needed to completely suppress the aggregation of 0.1 mg insulin dissolved in 0.4 mL of buffer is about 0.5 mg. In the presence of 5 μM of thioredoxin, 2 mg of native alpha-crystallin was needed to completely suppress the aggregation (data not shown).

Relatively little is known about the binding site and the mechanisms of the chaperone function of alpha-crystallin. Sharma and his coworkers have identified a region in the alpha A crystallin domain (residues 70–88) that binds to the target protein that they were using [45]. Interestingly, a synthetic peptide corresponding to this region can bind and suppress the aggregation of denatured protein, albeit with a much reduced efficiency [45, 46]. Other works suggest that the N-terminus of alpha-crystallin may also be involved in the binding of the target protein [47]. Recently, using site-directed spin labeling and electron paramagnetic resonance spectroscopy on a well-defined target protein (T4 lysozyme), Mchaourab et al. were able to show through careful thermodynamic analysis that alpha-crystallin possesses two modes of binding. One mode is a low-capacity mode that binds compact, native-like states. A second mode is a high-capacity mode that binds globally unfolded states. These two modes have different affinities and different numbers of binding sites [34, 48].

25.6 The Function of Alpha-crystallin

The various functions of the small heat shock proteins have been reviewed extensively ([5–7]; see also Chapter 24). One of the functions of the small heat shock proteins is to trap aggregation-prone denatured proteins and keep them in a refoldable conformation [49, 50]. The complex of the sHsp target protein can then interact with another chaperone system, such as Hsp70, in an ATP-dependent process to refold the target protein (see Figure 24.2). In the eye lens, however, the functions of alpha-crystallin are different. The first function of alpha-crystallin is, by virtue of its relatively high concentration, to contribute the necessary refractive index that the lens needs. The chaperone properties of alpha-crystallin are needed to control the unavoidable protein denaturation that takes place in the lens as a result of normal aging (reviewed in Refs. [8] and [9]). Because of the unique growth pattern of the lens, there is no protein turnover in the center of the lens. Thus, the center of a 60-year-old lens, for example, contains proteins that were synthesized during embryogenesis. These proteins undergo major posttranslational modification that result in aggregation and scattering, thereby compromising lens transparency. The data to date suggest that alpha-crystallin complexes with the old, denatured proteins and controls the unavoidable age-dependent aggregation processes [51–56].

While alpha-crystallin by itself is not a very efficient chaperone for refolding denatured proteins, there are many examples in the literature showing that it can protect in vivo other proteins and enzymes from various insults [3, 57–62].

25.7 Experimental Protocols

25.7.1 Preparation of Alpha-crystallin

A: Native Lens Alpha-crystallin

The preparation and purification of native eye lens alpha-crystallin have been described in detail previously [43]. Most investigators in the field use calf or cow lenses, as these are readily available from local slaughterhouses. The molar ratio of alpha A to alpha B in this preparation is about 3 to 1. As was mentioned previously, native lens crystallins undergo major posttranslational modification [30]. Thus, the age of the lens used may affect the quality and the properties of the alpha-crystallin obtained [43].

To separate the alpha A and alpha B isoforms from native alpha-crystallin, ion-exchange chromatography in the presence of a high concentration of urea (≥ 6 M) is generally used [63]. It should be emphasized, however, that after treatment of alpha-crystallin with urea the quaternary structure obtained following the removal of urea is different from the native state [64]. Thus, treatment with urea is not completely reversible. Whenever a high concentration of urea is used, care should be taken that the cyanate being generated does not carbamylate the alpha-crystallin. The native alpha A and alpha B subunits that are isolated from the native alpha-crystallin are also phosphorylated to various degrees [63].

B: Recombinant Alpha B and Alpha A-crystallin

Recombinant alpha-crystallin can be easily obtained according to standard procedures. A detailed procedure for the preparation of recombinant alpha B crystallin is described elsewhere [43]. The recombinant alpha-crystallin can be produced at relatively high yields and without the need to use urea or other chaotropic salts for solubilization. The purity of the preparation can be assessed by standard SDS-polyacrylamide gel electrophoresis. However, bacterial nucleotides that commonly bind the recombinant alpha-crystallin could be a major source of contamination. This contamination will not be visible on the SDS gels. Therefore, it is important to record the near-UV absorption spectra of the recombinant alpha-crystallin preparations. For highly purified preparations, a ratio of A280/A260 ≥ 1.5 should be obtained [43].

While native alpha-crystallin from eye lens as well as recombinant alpha A and alpha B possess many common properties, there are also major differences in some of their critical properties, such as conformational stability, quaternary structure, and ability to interact and suppress the aggregation of various target proteins [65–69]. Thus, it is important to choose the relevant species of alpha-crystallin for the specific question that is being studied.

C: Molecular Weight Determination

As stated previously, alpha-crystallin preparations in their native state are always heterogeneous. Mutations of some key residues in alpha-crystallin have a significant effect on the weight average and the heterogeneity of the alpha-crystallin complex. In addition, there is sometimes the need to know the molecular weight of the complex of alpha-crystallin with its target protein. The most frequently used method for determining the average molecular weight is size-exclusion chromatography (see Chapter 24). Other standard methods such as analytical ultracentrifugation and various light-scattering techniques are also being used. It is well established that standard size-exclusion chromatography may sometimes yield erroneous results because of the shape of the protein in question, possible interactions of the protein with the column, and the dependence on standard proteins for calibration of the column. As was mentioned earlier in Section 25.3 and shown in Figure 25.1, size-exclusion chromatography with an online multi-angle laser light-scattering detector and a refractive index detector can greatly improve the accuracy of molecular weight determination [19–21].

One of the major advantages of such a system is the ability to obtain reliable data at elevated temperatures. In general, the absorbance detectors that are used in most commercial liquid chromatography systems cannot operate at temperatures above 50 °C. The same problem exists with the commercially available analytical ultracentrifuges, where it is impossible to obtain data at high temperatures. The chromatographic and detection systems used in Figure 25.1 can be used at temperatures even higher than 100 °C. Thus, we were able to show that the small heat shock protein 16.5 from *Methanococcus jannaschii* retains its multimeric structure and subunit organization at 70 °C (see Figure 2 in Ref. [24]).

Another example of the advantage of the multi-angle laser light-scattering system is shown below in Figure 25.3, where the effects of phosphorylation mimics on the quaternary structure of alpha B crystallin are shown. Figure 25.3 compares the molecular weight distribution of recombinant alpha B crystallin with alpha B crystallin in which all three of the phosphorylation sites were mutated to aspartate to mimic phosphorylation. Ito et al. [33] were the first to study the effects of these mutations on the quaternary structure of alpha B crystallin. They fractionated wild-type alpha B and the mutated alpha B using sucrose density-gradient

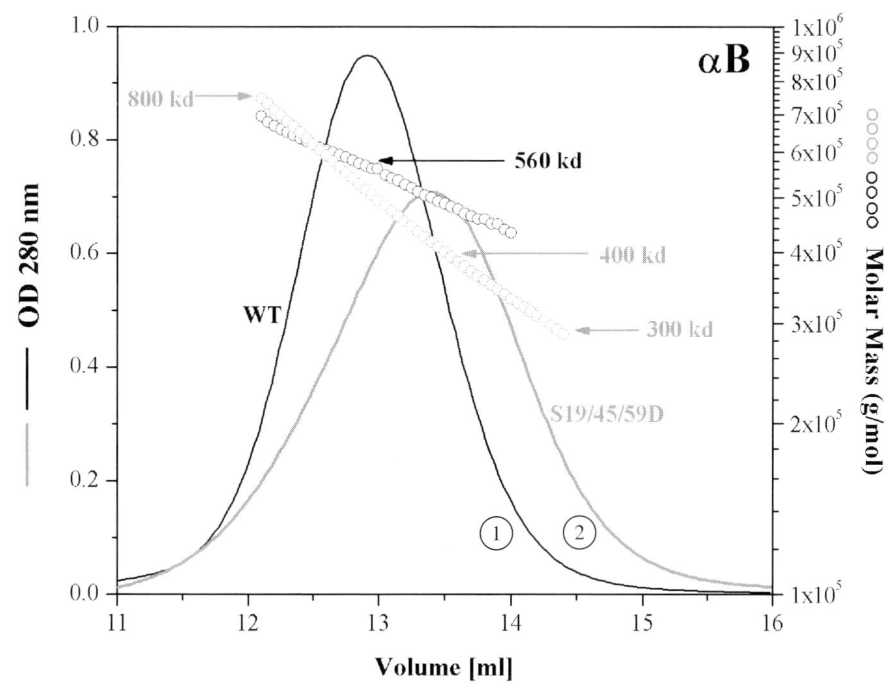

Fig. 25.3. Molar mass distribution and absorption spectra of recombinant alpha B and alpha B, where serine 19, 45, and 59 were mutated to aspartate. Curve (1), recombinant alpha B crystallin. Curve (2), S19/45/59 alpha B crystallin. All other conditions are the same as in Figure 25.1.

centrifugation. The fractions were than subjected to SDS-PAGE analysis followed by immunostaining with antibodies against the C-terminal of alpha B crystallin. Quantitation of these Western blot analyses suggested that the main peak of wild-type alpha B corresponds to a molecular weight of 500 000, whereas the molecular weight of the tripled phosphorylated mimics corresponds to 300 000. Using gel filtration analysis, they reported a molecular weight of 550 000 for wild-type alpha B and of 390 000 for the S19/45/59D mimics [33]. The data shown in Figure 25.3 is in excellent agreement with their gel filtration data. It should be noted, however, that in one single run a complete picture of the polydisperse nature of alpha-crystallin is observed with an accurate molecular weight of the whole elution profile. As can be noted from the slope of the line representing the molecular weight, the phosphorylation mimics not only decreased the weight average of alpha B but also significantly increased its polydispersity in agreement with the observation of Ito et al. [33]. The data presented in Figure 25.3 take less than an hour to obtain.

The Use of Mass Spectrometry for the Determination of the Quaternary Structure of Alpha-crystallin and Other Small Heat Shock Proteins

Recent advances in mass spectrometry make it possible to analyze high-molecular-weight protein complexes [70]. This powerful technology was recently utilized for real-time monitoring of subunit exchange between small heat shock proteins [71]. It was also used successfully to study the polydisperse nature of alpha B crystallin [73]. It can be expected that this approach will yield valuable information on the function and structure of alpha-crystallin as well as of other small heat shock proteins.

D: Chaperone Assays for Alpha-crystallin

The most commonly used method for assaying the chaperone properties of alpha-crystallin is monitoring its ability to suppress the aggregation of nonnative forms of proteins during the unfolding processes. Aggregation can be easily monitored by measuring the increase in light scattering [73–75]. Standard spectrophotometers or fluorometers are generally used [43, 76]. Typical data obtained is shown in Figure 25.2. Because the aggregation process is highly temperature-dependent, it is important that the temperature of the cell be controlled. Many proteins can serve as a target for the aggregation assay. Some of the more common target proteins, which cover a wide range of molecular weights, are given in Table 25.1.

E: Comments on Specific Proteins

Unfolding of Proteins by Breaking Disulfide Bonds Using Dithiothreitol or Tris (2-carboxyethyl) Phosphine

For target proteins such as insulin, lysozyme, α-lactalbumin, ovotransferrin [77], or abrin [79], in which unfolding is achieved by reducing the disulfide bonds, dithiothreitol (DTT) is generally used. In recent years, a new reagent for the specific reduction of disulfide bonds has become more popular, namely, Tris (2-carboxyethyl) phosphine (TCEP). This compound has major advantages over DTT [80, 81]. TCEP

Tab. 25.1. Target proteins commonly used for chaperone assay of alpha-crystallin.

Protein (MW)	*Unfolded by:*	*Assay temperature*	*References*
Insulin (6000)	Reduction with DTT, TCEP	25–37 °C	42, 43, Chapter 24
Lysozyme (14 400)	Reduction with DTT, TCEP	25–37 °C	77, 78
α-lactalbumin (14 400)	Reduction with DTT, TCEP	25–37 °C	43
Alcohol dehydrogenase (yeast) (4 × 35 000)	Heat to 37 °C	37 °C	43
Malate dehydrogenase (2 × 35 000)	Heat to 45 °C	45 °C	
Citrate synthase (mitochondrial pig heart) (49 000)	Heat to 43 °C 6 M guanidinium chloride at 25 °C	43 °C 25 °C	Chapter 24, 76

is more stable than DTT at a pH above 7.5 and in general is a faster and stronger reductant than DTT [81]. For example, instead of using 20 mM DTT for reducing insulin in a chaperone assay, we can use only 4 mM or less TCEP for a typical aggregation assay. Another advantage is that TCEP does not have any absorbance at 280 nm, whereas the molar absorptivity of oxidized DTT or DTE is $\varepsilon_{283} = 273$ M^{-1} cm^{-1}. Thus, in many cases oxidized DTT interferes with protein measurements at 280 nm. TCEP is commercially available from Pierce Chemical Company or Molecular Probes.

Insulin Assay

Details for the assay are given in Refs. [42] and [43] and in Chapter 24. Insulin from bovine pancreas or human recombinant insulin can be obtained from Sigma and Roche.

Alpha-lactalbumin Assay

Details are given in Ref. [43]. The aggregation assay is very sensitive to pH [82]. For α-lactalbumin at concentration of 0.5 mg mL^{-1} or lower, the pH should be below 7.0. The assay works well in the pH range of 6.5–6.8. Calcium-depleted α-lactalbumin is used.

Assay conditions:

Bovine α-lactalbumin (calcium-depleted) obtained from Sigma Chemical Co. is dissolved in a buffer containing 50 mM sodium phosphate, 0.1 M NaCl, and 2 mM ethylenediaminetetraacetic acid (EDTA). The final pH should be 6.7. Stock solutions should be kept over ice, and fresh solutions should be made daily.

For an aggregation assay using a standard absorption spectrophotometer with a 1.0-cm path length cell, a concentration of ~0.5 mg mL^{-1} lactalbumin will produce a good signal. Aggregation is initiated by adding DTT or TCEP to the sample.

Assay temperature: 25–50 °C.

For bovine α-lactalbumin: $E_{280}^{0.1\%} = 2$

Lysozyme Assay

Lysozyme assay is performed under conditions similar to those described for α-lactalbumin. Lysozyme is not as sensitive to pH as α-lactalbumin, and the assay works well at neutral pH.

Assay conditions:

Lysozyme obtained from Roche or Sigma is dissolved in a buffer containing 50 mM sodium phosphate and 0.1 M NaCl, pH 7.

For an aggregation assay using a standard absorption spectrophotometer with a 1.0-cm path length cell, a concentration of ~0.2 mg mL^{-1} will produce a good signal. Aggregation is initiated by adding 20 mM DTT or 2–5 mM TCEP.

Assay temperature: 25–50 °C.

For lysozyme: $E_{280}^{0.1\%} = 2.6$

Assays Involving Aggregation of Proteins by Heat Denaturation

Heat denaturation is commonly used for studying the chaperone properties of small heat shock proteins. Some of the commonly used target proteins include alcohol dehydrogenase, malate dehydrogenase, citrate synthase, luciferase, rhodanese, and α-glucosidase. The advantage of using these proteins as a target protein for alpha-crystallin is that all of them aggregate at temperatures below 50 °C, where alpha-crystallin still retains its native structure [83–85].

Alcohol Dehydrogenase

Yeast alcohol dehydrogenase will unfold and aggregate at 37 °C under the conditions described in Ref. [43].

Assay conditions:

Alcohol dehydrogenase (yeast) obtained from Roche or Sigma is dissolved in a buffer containing 50 mM sodium phosphate containing 0.1 M NaCl and 1 mM 1,10-phenanthroline. Stock solutions should be kept over ice, and fresh solutions should made daily.

For a standard aggregation assay as described above, a concentration of ~0.5 mg mL^{-1} is used. Aggregation is initiated by increasing the cell temperature to 37 °C.

Assay temperature: 37–40 °C.

For yeast alcohol dehydrogenase: $E_{280}^{0.1\%} = 1.3$

Note: For a fluorometer-based scattering analysis, a concentration as low as 15 μg mL^{-1} alcohol dehydrogenase can be used.

Malate Dehydrogenase

Malate dehydrogenase (pig heart, mitochondrial) is suitable for aggregation studies at temperatures around 45 °C.

Assay conditions:

Malate dehydrogenase (pig heart, mitochondrial) obtained from Roche is dissolved in a buffer containing 50 mM sodium phosphate and 0.1 M NaCl, pH 7. Stock solutions should be kept over ice.

For a standard aggregation assay as described above, a concentration of ~0.5 mg mL^{-1} is generally used. Aggregation is initiated by heating the enzyme to 45 °C.

Assay temperature: 45–48 °C.

For malate dehydrogenase: $E_{280}^{0.1\%} = 0.29$

Citrate Synthase

Citrate synthase is widely used for many chaperone studies. Details are given in Ref. [76] and in Chapter 24. Citrate synthase is available from Roche.

F: Interpretation of Light-scattering and Turbidometric Data Obtained During Protein Aggregation and During Suppression of the Aggregation with Alpha-crystallin

As was mentioned previously, the most common technique used to monitor protein aggregation and the effects of chaperones on the aggregation is light-scattering or turbidometric measurements in an absorption spectrophotometer or a fluorometer. The kinetics of aggregation, such as those shown in Figure 25.2 or Figure 24.4, is widely found in the published literature. However, to get meaningful kinetic parameters from such data is not a trivial matter. It should be emphasized that all of these aggregation-based assays are done under non-equilibrium conditions. The aggregation in most cases is an irreversible process. Thus, it is not simple to obtain kinetic parameters [48, 70, 86]. In addition to being strongly dependent on protein concentration and temperature, other conditions may have a significant effect on the kinetic parameters obtained from the light-scattering data, e.g., mixing the reaction solution during the aggregation measurements. It is obvious from the published data that most investigators who use turbidometric techniques do not mix the reactant during the measurements. In many cases, the published data show that there is a sudden drop in the intensity of the apparent absorbance or the light-scattering signal. This happens simply because the protein aggregates become so large that they sink to the bottom of the cell, thereby decreasing the apparent absorbance or the intensity of the light scattering. Thus, the distribution of aggregates in the cell may not be homogenous. Mixing during the reaction may have its own problems. For example, we have observed that with lysozyme, visibly large aggregates of various sizes occupy the entire cuvette during the unfolding with TCEP (Figure 25.2). If the solution is mixed during the reaction with a small magnetic stirrer, the aggregates observed are very fine. However, they tend to adsorb with time to the quartz cuvette wall, causing a slow decrease in the apparent absorbance measured as a function of time. The kinetics of aggregation observed under these two conditions is not the same. Similar results were obtained with malate dehydrogenase that was denatured by heating to 45 °C (data not shown).

Recently Kurganov [87, 88] studied the kinetics of irreversible protein aggregation in a system that utilizes measuring the increase in apparent absorbance as a function of time. This study shows that if the aggregation of the protein substrate follows first-order kinetics, then useful parameters may be obtained. This approach may be useful for characterizing the chaperone activity of alpha-crystallin and other small heat shock proteins.

G: Chaperone Assays Involving Protection of Enzyme Activities by Alpha-crystallin

It has been shown by several investigators that alpha-crystallin could protect in vitro some enzymes from inactivation when they are subjected to various stresses such as heat, ultraviolet light, or glycation [3, 57–62]. Enzyme activity assays can be more informative then the commonly used aggregation-based assays because protection from aggregation does not necessarily mean protection of enzyme activity. Some of the enzymes that were used included sorbitol dehydrogenase [58], several restriction enzymes [57, 59], Na/K-ATPase [62], catalase, glutathione reductase, superoxidase dismutase, and various enzymes from the glycolytic pathway [3, 61]. A commonly used enzyme is citrate synthase. Details about using citrate synthase for chaperone assays are given in Ref. [76] and in Chapter 24.

Acknowledgements

I thank Linlin Ding and Qingling Huang for continuous support of my research program. This work was supported by the National Eye Institute Merit R-37 EY3897 and by the National Eye Institute core grant EY 000331.

References

1 W. W. de Jong, G. J. Caspers and J. A. M. Leunissen, Genealogy of the alpha-crystallin-small heat-shock protein superfamily. *Int. J. Biol. Macromol.* **1998**, *22*, 151–162.
2 M. Ehrnsperger, M. Gaestel and J. Buchner, Molecular Chaperones in the Life Cycle of Proteins, **1998**, 553–577.
3 B. K. Derham and J. J. Harding, *Progress in Retinal and Eye Research* Vol. 18, Pergamon: Elsevier EB, **1999**, pp. 431–462.
4 R. Jaenicke and C. Slingsby, Lens crystallins and their microbial homologs: structure, stability, and function. *Crit. Rev. Biochem. Molec. Biol.* **2001**, *36*, 435–499.
5 R. van Montfort, C. Slinsgsby and E. Vierling, *Adv. Protein Chem. 5a*, Academic Press, **2002**, 105–156.
6 F. Narberhaus, Alpha-crystallin-type heat shock proteins: socializing minichaperones in the context of a multichaperone network. *Microbiol. Mol. Biol. Rev.* **2002**, *66*, 64–93.

7 A. P. Arrigo and W. E. G. Müller, *Small Stress Proteins*, **2002**, Springer-Verlag, Berlin.

8 J. Horwitz, The function of alpha-crystallin in vision. *Sem. Cell and Devel. Biol.* **2000**, *11*, 53–60.

9 J. Horwitz, Alpha-crystallin. *Exp. Eye Research*, **2003**, *76*, 145–153.

10 J. Horwitz, Alpha-crystallin can function as a molecular chaperone. *Proc. Natl. Acad. Sci. USA*, **1992**, *89*, 10449–10453.

11 U. Jakob, M. Gaestel, K. Engel and J. Buchner, Small heat shock proteins are molecular chaperones. *J. Biol. Chem.* **1993**, *268*, 1517–1520.

12 S. P. Bhat and C. N. Nagineni, Alpha B subunit of lens-specific protein alpha-crystallin is present in other ocular and non-ocular tissues. *Biochem. Biophys. Res. Commun.* **1989**, *158*, 319–325.

13 R. A. Dubin, E. F. Wawrousek, J. Piatigorsky, Expression of the murine alpha B-crystallin gene is not restricted to the lens. *Mol Cell Biol.* **1989**, *3*, 1083–1091.

14 T. Iwaki, A. Kume-Iwaki, R. K. Liem and J. E. Goldman, Alpha B-crystallin is expressed in non-lenticular tissues and accumulates in Alexander's disease brain. *Cell*, **1989**, *57*, 71–78.

15 R. Klemenz, E. Frohli, R. H. Steiger, R. Schafer and A. Aoyama, Alpha B-crystallin is a small heat shock protein. *Proc. Nat. Acad. Sci. USA*, **1991**, *88*, 3652–3656.

16 M. W. Head, G. E. Goldman, Small heat shock proteins, the cytoskeleton, and inclusion body formation. *Neuropathol Appl Neurobiol.* **2000**, *26*, 304–312.

17 K. K. Kim, R. Kim and S. H. Kim, Crystal structure of a small heat-shock protein. *Nature*, **1998**, *394*, 595–599.

18 R. L. M. van Montfort, E. Basha, K. L. Friedrich, C. Slingsby and E. Vierling, Crystal structure and assembly of a eukaryotic small heat shock protein. *Nature Struct. Biol.* **2001**, *8*, 1025–1030.

19 P. J. Wyatt, Light scattering and the absolute characterization of macromolecules. *Anal. Chim. Acta.* **1993**, *272*, 1–40.

20 J. Wen, T. Arakawa and J. S. Philo, Size-exclusion chromatography with on-line light-scattering, absorbance, and refractive index detectors for studying proteins and their interactions. *Anal. Biochem.* **1996**, *240*, 155–166.

21 E. Folta-Stogniew and K. R. Williams, Determination of molecular masses of proteins in solution: implementation of an HPLC size exclusion chromatography and laser light scattering service in a core laboratory. *J Biomolec. Tech.* **1999**, *10*, 51–63.

22 P. J. van den Oetelaar, P. F. van Someren, J. A. Thomson, R. J. Siezen, H. J. Hoenders, A dynamic quaternary structure of bovine alpha-crystallin as indicated from intermolecular exchange of subunits. *Biochemistry*, **1990**, *14*, 3488–3493.

23 M. P. Bova, L. L. Ding, J. Horwitz, B. K. Fung, Subunit exchange of αA-crystallin. *J Biol Chem.* **1997**, *272*, 29511–29517.

24 M. P. Bova, O. Huang, L. Ding, J. Horwitz, Subunit exchange, conformational stability, and chaperone-like function of the small heat shock protein 16.5 from *Methanococcus jannaschii*. *J Biol Chem.* **2002**, *277*, 38468–38475.

25 H. A. Koteiche and H. S. Mchaourab, Folding pattern of the alpha-crystallin domain in alphaA-crystallin determined by site-directed spin labeling. *J. Mol. Biol.* **1999**, *294*, 561–577.

26 H. A. Koteiche and H. S. Mchaourab, The determinants of the oligomeric structure in Hsp16.5 are encoded in the alpha-crystallin domain. *FEBS Lett.* **2002**, *519*, 16–22.

27 K. Guruprasad and K. Kumari, Three-dimensional models corresponding to the C-terminal domain of human αA and αB-crystallins based on the crystal structure pf the small heat-shock protein HSP16.9 from wheat. *Int. J. Biol. Macromol.* **2003**, *33*, 107–112.

28 D. A. Haley, J. Horwitz and P. L. Stewart, The small heat-shock protein, alphaB-crystallin, has a variable quaternary structure. *J. Mol. Biol.* **1998**, *277*, 27–35.

29 D. A. Haley, M. P. Bova, Q. L. Huang, H. S. Mchaourab and P. L. Stewart, Small heat-shock protein structures reveal a continuum from symmetric to variable assemblies. *J. Mol. Biol.* **2000**, *298*, 261–272.

30 P. J. Groenen, K. B. Merck, W. W. de Jong and H. Bloemendal, Structure and modifications of the junior chaperone alpha-crystallin. From lens transparency to molecular pathology. *Eur J Biochem.* **1994**, *225*, 1–19.

31 M. Gaestel, sHsp-phosphorylation: enzymes, signaling pathways and functional implications. In: A. P. Arrigo and W. E. G. Müller, Editors, *Small Stress Proteins* **2002**, *28*, 151–184.

32 K. Kato, H. Ito and Y. Inaguma, Expression and phosphorylation of mammalian small heat shock proteins. In: A. P. Arrigo and W. E. G. Müller, Editors, *Small Stress Proteins* **2002**, *28*, 129–150.

33 K. Kamei, I. Iwamoto, Y. Inaguma, D. Nohara and K. Kato, H. Ito, Phosphorylation-induced change of the oligomerization state of αB-crystallin. *J. Biol. Chem.* **2001**, *276*, 5346–5352.

34 H. A. Koteiche and H. S. Mchaourab, Mechanism of chaperone function in small heat-shock proteins. *J Biol Chem.* **2003**, *278*, 10361–10367.

35 K. P. Das, J. M. Petrash and W. K. Surewicz, Conformational properties of substrate proteins bound to a molecular chaperone α-crystallin. *J Biol Chem.* **1996**, *271*, 10449–10452.

36 K. P. Das, L. P. Choo-sith, J. M. Petrash and W. K. Surewicz, Insight into the secondary structure of non-native proteins bound to a molecular chaperone α-crystallin. *J Biol Chem.* **1999**, *274*, 33209–33212.

37 R. A. Lindner, A. Kapur and J. A. Carver, The interaction of the molecular chaperone, α-crystallin, with molten globule states of bovine α-lactalbumin. *J Biol Chem.* **1997**, *272*, 27722–27729.

38 K. Rajaraman, B. Raman, T. Ramakrishna and C. M. Rao, The chaperone-like alpha-crystallin forms a complex only with the aggregation-prone molten globule state of alpha-lactalbumin. *Biochem. Biophys. Res. Commun.* **1998**, *249*, 917–921.

39 J. A. Carver, R. A. Lindner, C. Lyon, D. Canet, H. Hernandez, C. M. Dobson and C. Redfield, The interaction of the molecular chaperone alpha-crystallin with unfolding alpha-lactalbumin: a structural and kinetic spectroscopic study. *J Mol Biol.* **2002**, *318*, 815–827.

40 C. M. Dobson, P. A. Evans, S. E. Radford, Understanding how proteins fold: the lysozyme story so far. *Trends Biochem Sci.* **1994**, *19*, 31–37.

41 S. E. Radford and C. M. Dobson, Insights into protein folding using physical techniques: studies of lysozyme and alpha-lactalbumin. *Philos Trans R Soc Lond B Biol Sci.* **1995**, *348*, 17–25.

42 Z. T. Farahbakhsh, Q. L. Huang, L. Ding, C. Altenbach, H. J. Steinhoff, J. Horwitz and W. L. Hubbell, Interaction of α-crystallin with spin-labeled peptides. *Biochemistry*, **1995**, *34*, 509–516.

43 J. Horwitz, Q. L. Huang, L. Ding and M. P. Bova, Lens alpha-crystallin: chaperone-like properties. *Methods in Enzymology*, **1998**, *290*, 365–383.

44 A. Holmgren, Thioredoxin catalyzes the reduction of insulin disulfides by dithiothreitol and dihydrolipoamide. *J Biol Chem.* **1979**, *254*, 9627–9632.

45 K. K. Sharma, R. S. Kumar, G. S. Kumar, and P. T. Quinn, Synthesis and characterization of a peptide identified as a functional element in αA-crystallin. *J Biol Chem*, **2000**, *275*, 3767–3771.

46 J. Bhattacharyya and K. K. Sharma, Conformational specificity of mini-alphaA-crystallin as a molecular

chaperone. *J Pept Res.* **2001**, *57*, 428–434.
47 J. B. Smith, Y. Lin and D. L. Smith, Identification of possible regions of chaperone activity in lens alpha-crystallin. *Exp. Eye Res.* **1996**, *63*, 125–128.
48 H. S. Mchaourab, E. K. Dodson, and H. A. Koteiche, Mechanism of chaperone function in small heat shock proteins. Two-mode binding of the excited states of T4 lysozyme mutants by αA-crystallin. *J Biol Chem*, **2002**, *277*, 40557–40566.
49 M. Ehrnsperger, S. Graber, M. Gaestel and J. Buchner, Binding of non-native protein to Hsp25 during heat shock creates a reservoir of folding intermediates for reactivation. *EMBO J.* **1997**, *16*, 221–229.
50 G. J. Lee, A. M. Roseman, H. R. Saibil and E. Vierling, A small heat shock protein stably binds heat-denatured model substrates and can maintain a substrate in a folding-competent state. *EMBO J.* **1997**, *16*, 659–671.
51 K. Wang and A. Spector, The chaperone activity of bovine alpha crystallin. Interaction with other lens crystallins in native and denatured states. *J. Biol. Chem.* **1994**, *269*, 13601–13608.
52 D. Boyle and L. Takemoto, Characterization of the alpha-gamma and alpha-beta complex: evidence for an in vivo functional role of alpha-crystallin as a molecular chaperone. *Exp. Eye Res.* **1994**, *58*, 9–16.
53 P. V. Rao, Q. L. Huang, J. Horwitz and S. J. Zigler, Evidence that alpha-crystallin prevents non-specific protein aggregation in the intact eye lens. *Biochem. Biophys. Acta.* **1995**, *1245*, 439–447.
54 J. A. Carver, K. A. Nicholls, J. A. Aquilina and R. J. W. Truscott, Age-related changes in bovine alpha-crystallin and high-molecular-weight protein. *Exp. Eye Res.* **1996**, *63*, 639–647.
55 Y. C. Chen, G. E. Reid, R. J. Simpson and R. J. W. Truscott, Molecular evidence for the involvement of alpha crystallin in th colouration/crosslinking of crystallins in age-related nuclear cataract. *Exp. Eye Res.* **1997**, *65*, 835–840.
56 R. J. W. Truscott, Y. C. Chen and D. C. Shaw, Evidence for the participation of alpha B-crystallin in human age-related nuclear cataract. *Int. J. Biol. Macromol.* **1998**, *22*, 321–330.
57 J. F. Hess and P. G. FitzGerald, Protection of a restriction enzyme from heat inactivation by [alpha]-crystallin. *Mol. Vision.* **1998**, *4*, 29–32.
58 I. Marini, R. Moschini, A. Del Corso and U. Mura, Complete protection by alpha-crystallin of lens sorbitol dehydrogenase undergoing thermal stress. *J. Biol. Chem.* **2000**, *275*, 32559–32565.
59 P. Santhoshkumar and K. K. Sharma, Analysis of alpha-crystallin chaperone function using restriction enzymes and citrate synthase. *Mol. Vision.* **2001**, *7*, 172–177.
60 G. B. Reddy, P. Y. Reddy and P. Suryanarayana, AlphaA- and alphaB-crystallins protect glucose-6-phosphate dehydrogenase against UVB irradiation-induced inactivation. *Biochem Biophys Res Commun.* **2001**, *282*, 712–716.
61 B. K. Derham and J. J. Harding, Enzyme activity after resealing within ghost erythrocyte cells, and protection by alpha-crystallin against fructose-induced inactivation. *Biochem J.* **2002**, *368*, 865–874.
62 B. K. Derham, J. C. Ellory, A. J. Bron and J. J. Harding, The molecular chaperone alpha-crystallin incorporated into red cell ghosts protects membrane Na/K-ATPase against glycation and oxidative stress. *Eur J Biochem.* **2003**, *270*, 2605–2611.
63 H. Bloemendal and G. Groenewoud, One-step separation of the subunits of alpha-crystallin by chromatofocusing in 6 M urea. *Anal Biochem.* **1981**, *117*, 327–329.
64 E. W. Doss-Pepe, E. L. Carew and J. F. Koretz, Studies of the denaturation patterns of bovine alpha-crystallin using an ionic denaturant, guanidine

hydrochloride and a non-ionic denaturant, urea. *Exp Eye Res.* **1998**, *67*, 657–679.

65 M. A. van Boekel, F. de Lange, W. J. de Grip, W. W. de Jong, Eye lens alphaA- and alphaB-crystallin: complex stability versus chaperone-like activity. Biochim Biophys Acta. **1999**, *1434*, 114–123.

66 J. Horwitz, M. P. Bova, L. L. Ding, D. A. Haley, P. L. Stewart, Lens alpha-crystallin: function and structure. Eye. **1999**, *13*, 403–408.

67 T. X. Sun, N. J. Akhtar, J. J. Liang, Thermodynamic stability of human lens recombinant alphaA- and alphaB-crystallins. *J Biol Chem.* **1999**, *274*, 34067–34071.

68 S. Abgar, J. Backmann, T. Aerts, J. Vanhoudt, J. Clauwaert, The structural differences between bovine lens alphaA- and alphaB-crystallin. *Eur J Biochem.* **2000**, *267*, 5916–5925.

69 J. J. Liang, T. X. Sun, N. J. Akhtar, Heat-induced conformational change of human lens recombinant alphaA- and alphaB-crystallins. *Mol Vis.* **2000**, *6*, 10–14.

70 F. Sobott, C. V. Robinson, Protein complexes gain momentum. *Curr Opin Struct Biol.* **2002**, *12*, 729–734.

71 F. Sobott, J. L. Benesch, E. Vierling, C. V. Robinson, Subunit exchange of multimeric protein complexes. Real-time monitoring of subunit exchange between small heat shock proteins by using electrospray mass spectrometry. *J Biol Chem.* **2002**, *277*, 38921–38929.

72 J. A. Aquilina, J. L. Benesch, O. A. Bateman, C. Slingsby, C. V. Robinson, Polydispersity of a mammalian chaperone: mass spectrometry reveals the population of oligomers in alphaB-crystallin. *Proc Natl Acad Sci USA.* **2003**, *100*, 10611–10616.

73 G. Zettlmeissl, R. Rudolph, R. Jaenicke, Reconstitution of lactic dehydrogenase. Noncovalent aggregation vs. reactivation. 1. Physical properties and kinetics of aggregation. *Biochemistry.* **1979**, *18*, 5567–5571.

74 J. Buchner, M. Schmidt, M. Fuchs, R. Jaenicke, R. Rudolph, F. X. Schmid, T. Kiefhaber, GroE facilitates refolding of citrate synthase by suppressing aggregation. *Biochemistry.* **1991**, *30*, 1586–1591.

75 J. Buchner, Supervising the fold: functional principles of molecular chaperones. *FASEB J.* **1996**, *10*, 10–19.

76 J. Buchner, H. Grallert, U. Jakob, Analysis of chaperone function using citrate synthase as nonnative substrate protein. *Methods Enzymol.* **1998**, *290*, 323–338.

77 S. Abgar, N. Yevlampieva, T. Aerts, J. Vanhoudt, J. Clauwaert, Chaperone-like activity of bovine lens alpha-crystallin in the presence of dithiothreitol-destabilized proteins: characterization of the formed complexes. *Biochem Biophys Res Commun.* **2000**, *276*, 619–625.

78 S. Abgar, J. Vanhoudt, T. Aerts, J. Clauwaert, Study of the chaperoning mechanism of bovine lens alpha-crystallin, a member of the alpha-small heat shock superfamily. *Biophys J.* **2001**, *80*, 1986–1995.

79 G. B. Reddy, S. Narayanan, P. Y. Reddy, I. Surolia, Suppression of DTT-induced aggregation of abrin by alphaA- and alphaB-crystallins: a model aggregation assay for alpha-crystallin chaperone activity in vitro. *FEBS Lett.* **2002**, *522*, 59–64.

80 E. B. Getz, M. Xiao, T. Chakrabarty, R. Cooke, P. R. Selvin, A comparison between the sulfhydryl reductants tris(2-carboxyethyl)phosphine and dithiothreitol for use in protein biochemistry. *Anal Biochem.* **1999**, *273*, 73–80.

81 J. C. Han, G. Y. Han, A procedure for quantitative determination of tris(2-carboxyethyl)phosphine, an odorless reducing agent more stable and effective than dithiothreitol. *Anal Biochem.* **1994**, *220*, 5–10.

82 K. Rajaraman, B. Raman, T. Ramakrishna, C. M. Rao, The chaperone-like alpha-crystallin forms a complex only with the aggregation-

prone molten globule state of alpha-lactalbumin. *Biochem Biophys Res Commun.* **1998**, *249*, 917–921.

83 A. Tardieu, D. Laporte, P. Licinio, B. Krop, M. Delaye, Calf lens alpha-crystallin quaternary structure. A three-layer tetrahedral model. *J Mol Biol.* **1986**, *192*, 711–724.

84 W. K. Surewicz, P. R. Olesen, On the thermal stability of alpha-crystallin: a new insight from infrared spectroscopy. *Biochemistry.* **1995**, *34*, 9655–9660.

85 K. P. Das, W. K. Surewicz, Temperature-induced exposure of hydrophobic surfaces and its effect on the chaperone activity of alpha-crystallin. *FEBS Lett.* **1995**, *369*, 321–325.

86 M. A. Speed, J. King, D. I. C. Wang, Polymerization mechanism of polypeptide chain aggregation. *Biotechnology and Bioengineering.* **1997**, *54*, 333–343.

87 B. I. Kurganov, Kinetics of protein aggregation. Quantitative estimation of the chaperone-like activity in test-systems based on suppression of protein aggregation. *Biochemistry (Moscow)*. **2002**, *67*, 409–422.

88 K. Wang, B. I. Kurganov, Kinetics of heat- and acidification-induced aggregation of firefly luciferase. *Biophys Chem.* **2003**, *106*, 97–109.

26
Transmembrane Domains in Membrane Protein Folding, Oligomerization, and Function

Anja Ridder and Dieter Langosch

26.1
Introduction

26.1.1
Structure of Transmembrane Domains

Integral membrane proteins traverse the membrane with one or more transmembrane segments (TMSs). The basic architecture of a membrane consists of a hydrophobic region of about 30 Å, formed by the lipid acyl chains, and two more polar interface regions that are both about 15 Å thick [1]. In order to fulfill the hydrogen bond–forming capacity of the polypeptide, proteins generally span the membrane either with α-helices or β-strands (Figure 26.1).

Since in β-strands hydrogen bonds are formed between residues in adjacent segments, proteins have to form cylindrical β-barrels to satisfy all their hydrogen bonds. In α-helices, hydrogen bonds are formed between residues within the same segment. Therefore, proteins can either span the membrane with a single hydrophobic α-helical transmembrane segment (bitopic proteins) or form a bundle of helices (polytopic proteins).

As an increasing number of polytopic membrane proteins are structurally characterized with high-resolution methods, and as oligomerization of more and more bitopic membrane proteins is studied with biochemical, biophysical, and genetic techniques, it is becoming increasingly clear that folding of membrane-integral domains and biological function is strongly dependent on interactions between α-helical TMSs. A number of excellent reviews have been published recently covering membrane protein folding and TMS-TMS interactions from different perspectives [1–12], and a related chapter on membrane protein folding by Tamm is available in Volume 1 of this handbook. Naturally, the structure of the membrane-spanning domains of the β-barrel type, as observed with many bacterial outer-membrane proteins, is also dependent on interactions between their TMSs [13, 14]. Because these interactions are less well understood than those between α-helical TMSs, they will be excluded from the present review.

Protein Folding Handbook. Part II. Edited by J. Buchner and T. Kiefhaber

ISBN: 3-527-30784-2

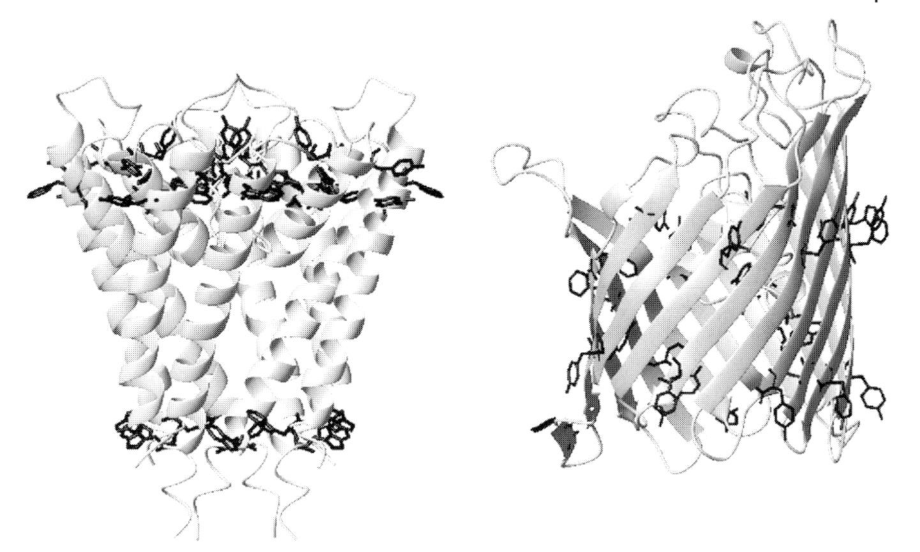

Fig. 26.1. Structures of two membrane proteins. On the left is the α-helical potassium channel from *Streptomyces lividans* [272] (PDB access code 1BL8). On the right, the β-barrel protein PhoE from *Escherichia coli* is shown [273] (PDB access code 1PHO). The side chains of tryptophan and tyrosine residues in the proteins are shown in black.

26.1.2
The Biosynthetic Route towards Folded and Oligomeric Integral Membrane Proteins

It appears that some proteins or peptides have the ability to insert into membranes spontaneously, i.e., without the use of proteinaceous factors. In the thylakoid membrane of chloroplasts, several proteins seem to insert as helical hairpins, depending on the hydrophobicity of the TMSs [15–17]. Small proteins with a single hydrophobic segment and small, if any, flanking regions also seem to be able to insert spontaneously into lipid bilayers [18], a situation that may also apply to some toxins or antimicrobial peptides.

However, most proteins become integrated into biological membranes via proteinaceous machineries. Co-translational membrane integration is a process that is similar in the eukaryotic endoplasmic reticulum (ER) membrane and bacterial inner membranes [19–22]. In the first step, a signal recognition particle (SRP) binds to the N-terminal signal sequence or to the first TMS of the protein when it emerges from the ribosome. These ribosome–nascent chain complexes are targeted to the SRP receptor in the membrane (performed by FtsY in bacteria) and subsequently transferred to the translocon. In eukaryotic ER membranes, this is the Sec61p complex, consisting of α-, β-, and γ-subunits, while in *E. coli* the homologous proteins are called SecYEG. The protein is then further translated and TMSs exit the translocation channel laterally into the lipid bilayer. For some proteins this occurs during translation, while other membrane proteins are transferred into the lipids only after the complete protein is synthesized. In the latter case, interactions between TMSs can occur within the translocon. It is commonly thought that TMSs are initially oriented and integrated into the membrane as independent units. However, recent results show that intraprotein interactions, stop-transfer effector proteins, and co-translational modifications can also play a role in determining the final topology of proteins within the membrane [21].

The membrane insertion of individual helices is the first stage in the widely accepted two-stage model of membrane protein folding, while in the second stage the helices come together through helix-helix interactions to produce the final folded conformation of the protein [7, 23]. It has been shown early on for a number of cases that membrane protein fragments produced by proteolytic cleavage of the loops or by co-expression of separate domains can be reconstituted to functional proteins. The most prominent example is bacteriorhodopsin, where a large number of different TMS fragments were produced and reassembled and where none of the connecting loops appears to be absolutely required for function [24–26]. Other examples are reviewed elsewhere [10, 12]. These results demonstrate that the information directing folding of functional membrane proteins is at least partly contained within the structure and sequence of the TMSs.

26.1.3 Structure and Stability of TMSs

26.1.3.1 Amino Acid Composition of TMSs and Flanking Regions

In contrast to the wealth of structural information obtained for soluble proteins, very few detailed structures of membrane proteins have been obtained. Nevertheless, much can be deduced about the structure of membrane proteins by examining their primary amino acid sequences.

The average length of a transmembrane (TM) helix is 20–25 amino acids, which is enough to span the hydrophobic part of a lipid bilayer [27–29]. Due to the need to be accommodated in the hydrophobic part of a lipid bilayer, TMSs are enriched in the hydrophobic amino acids Phe, Ile, Leu, Met, and Val [28, 30–33]. Single-spanning membrane protein TMSs tend to be more hydrophobic than TMSs from multi-spanning proteins and contain fewer polar amino acids [27, 32, 33]. The nucleotide bias of a particular organism affects the amino acid composition of TMSs such that organisms with a higher GC content in their genomes have more Val and Ala in their TMSs, whereas organisms with a lower GC content contain more Ile and Phe residues [34]. It is presently unknown whether the resulting difference in hydrophobicity is compensated for by the length of the TMSs.

In the membrane-flanking regions of TMSs, positively charged residues are more abundant at the cytoplasmic side of the membrane than at the periplasmic side, while there is no such bias for negatively charged residues [31–33]. Positive charges are more difficult to translocate across the membrane and thus control the topology of membrane proteins; this is called the "positive-inside rule" [35].

In most of the structures of membrane proteins that have been determined, a remarkable distribution of aromatic amino acids is observed. In β-barrel as well as

α-helical proteins, these residues form aromatic belts at the membrane-water interface (see Figure 26.1). This interfacial enrichment of aromatic residues has been confirmed by statistical analysis of membrane proteins [18, 27, 31, 36]. The role of these belts is not completely clear, but it is probably related to the strong affinity of Trp and Tyr residues for the interfacial region of lipid bilayers [37–40]. This may anchor proteins within the membrane and influence their precise positioning [40–42].

Using the information described above, it can be quite reliably predicted which proteins are membrane proteins and where TMSs are located in the sequence. Most of these prediction programs are based on hydrophobicity plots and scan an amino acid sequence for hydrophobic stretches of about 20 residues. Very often used are the Kyte and Doolittle [43] or GES [44] hydrophobicity scales. If the hydrophobicity of a segment is higher than a certain threshold value, it is considered to be a TMS. Other methods are based on comparisons with TMSs of well-characterized membrane proteins [45, 46] or use evolutionary information [47] to determine which amino acids are likely to be in the membrane [48]. When the TMSs have been determined, the transmembrane topology of a protein can subsequently be predicted by considering the preference of certain amino acids for the *cis* or *trans* side of the membrane [49–51]. Until now, the nucleotide bias of a particular organism usually has not been taken into account in these predictions [34].

26.1.3.2 Stability of Transmembrane Helices

In general, the stability of an α-helix depends both on intrahelical $N{-}H \cdots O{=}C$ hydrogen bonds between backbone atoms of successive helical turns and on intrahelical interactions between adjacent amino acid side chains [52]. As all of these interactions are essentially electrostatic in nature (see Section 26.2.1.1), it follows that their strength increases when a polypeptide chain is transferred from polar aqueous solution (dielectricity constant $\varepsilon = 80$) into the apolar part of a lipid bilayer. The magnitude of this increase depends on amino acid sequence and on the dielectric constant of the bilayer. That TM helices in apolar environments are indeed more stable than soluble helices is experimentally supported by the finding that denaturation of membrane proteins in detergent solution by heat or denaturants often causes only minimal loss of helical structure [3].

In soluble proteins, Leu and Ala residues rank among the best helix promoters, while Ile and Val destabilize α-helices and are over-represented in β-sheets; Gly strongly destabilizes soluble helices, and Pro is known to be a helix breaker [53–57]. Using a host-guest approach, different residue types were placed into the invariant framework of a model peptide and the latter's structure in media of different polarity was determined. It was found that the helix-destabilizing effect of Gly, Val, and Ile was much smaller upon insertion of the peptide into the apolar membrane-mimetic environment of detergent or lipid micelles than in aqueous buffer [58]. In a later study, the same authors showed that Ile, Leu, and Val were the residue types with the highest propensity of α-helices in apolar media, whereas Gly and Pro had the most destabilizing effects [59, 60]. That residues with β-branched side chains are effective helix formers in membranes is supported by the observation that Ile and/or Val make up about half of the α-helical TMSs of the bacteriophage M13 major coat protein [61] and of the pulmonary surfactant–associated polypeptide SP-C [62].

As in soluble helices, Pro residues have a destabilizing effect on TM-helices for two reasons: (1) the imide within its cyclic side chain cannot hydrogen bond to the carbonyl oxygen of the peptide bond at the i-3 or i-4 position, and (2) a steric clash between the Pro ring and the backbone carbonyl group at the i-4 position is produced [63]. Despite these destabilizing effects, Pro is well tolerated in TM helices due to their greater overall stability; in addition, the missing hydrogen bond may partially be replaced by a weak hydrogen bond involving the proton linked to the δ carbon of the Pro ring and the carbonyl group at i-4 [64]. Nevertheless, Pro residues in TMSs are frequently, but not always, associated with local bends or distortions of helical structure that may be relevant for protein function [63, 65]. A similar situation may apply to Gly residues that occur with high frequency in TMSs. Gly is thought to locally disrupt intrahelical side chain packing and appears to have a greater destabilizing effect on helices in a polar environment than in apolar media [58]. Interestingly, bends in TM helices are frequently seen when Pro and Gly residues are spaced four residues apart [66].

26.2
The Nature of Transmembrane Helix-Helix Interactions

26.2.1
General Considerations

A complete thermodynamic description of TMS-TMS interactions is a complex task since partitioning of the helices into the lipid bilayer, stability of and attractive forces between the helices, entropic factors, competition between protein-protein, protein-lipid, and lipid-lipid interactions, and environmental constraints have to be taken into account. At present, the degrees by which these factors define a given interaction are not understood at a quantitative level and we therefore limit the discussion to qualitative descriptions. We will summarize first how TMS-TMS interactions result from different attractive forces and entropic factors. Second, we will discuss how these concepts evolved from the study of polytopic and bitopic membrane proteins. Finally, we discuss how TMS-TMS interactions may be influenced by protein-lipid interactions.

26.2.1.1 Attractive Forces within Lipid Bilayers

Of the four basic attractive forces – i.e., weak and strong nuclear forces, gravitation, and electrostatic interactions – the latter account for folding of and interactions between proteins. Electrostatic interactions are commonly classified into three main categories. Ionic, or coulomb, interactions attract stable point charges such as ionized amino acid side chains. These are strong and decrease proportionally with the distance between them. Hydrogen bonds are based on protons that are shared by appropriately positioned electronegative atoms and are of primarily electrostatic nature. Dipole-dipole, or van der Waals, forces develop between permanent dipoles, between permanent and induced dipoles, or between induced and thus fluctuating dipoles (dispersion forces). In general, van der Waals forces are weaker than ionic interactions or hydrogen bonds; their strength depends on the polarity and polarizability of the binding partners and decreases with the sixth power of distance [52]. It has frequently been noted that the strength of ionic interactions and hydrogen bonds strongly increases when the polarity of the environment decreases, since electrostatic forces are inversely related to the dielectric constant ε. It is often overlooked, however, that the same applies to van der Waals interactions [67, 68]. These considerations imply that all attractive forces mediating protein-protein interaction are generally stronger in the low dielectric environment of the lipid bilayer acyl chain region than in aqueous solution.

26.2.1.2 **Forces between Transmembrane Helices**

All of the forces discussed in Section 26.2.1.1 – that is, ionic interactions, hydrogen bonds, and van der Waals interactions – can contribute to TMS-TMS interactions. Their relative contributions to the total enthalpy of interaction largely depend on amino acid composition and sequence. Since TMSs are built mostly from apolar amino acids, van der Waals interactions between their side chains appear to be a major and universal driving force in TMS assembly. Albeit weak, unidirectional, and strongly dependent on distance, van der Waals interactions apply to any type of side chain atom and may accumulate over entire interfacial regions. This is of particular relevance where an interface forms between helix surfaces that are complementary in shape and therefore form extensive interfacial areas. Interhelical hydrogen bonds appear to come in a variety of flavors, distinguished by the electronegativity and proximity of donors and acceptors that determine their strength. Hydrogen bonds can form (1) between the side chains of polar, yet non-ionizable, residues; (2) between the hydrogens connected to C_α carbon atoms and main-chain carbonyl groups or polar side chains; (3) between N–H groups of the main chain and polar side chains; and (4) between isolated ionizable residues that are likely to be uncharged, yet polar, in the membrane. There is experimental evidence for all of these alternatives, as is described in Section 26.2.2. Since hydrogen bonds exhibit a greater degree of directionality than van der Waals interactions, they are likely to complement the latter and to enhance specificity and stability of membrane protein folding and interaction [5]. Whether interactions between ionized side chains occur between TMSs is an unsettled question. The considerable enthalpic cost of transferring a point charge into an apolar environment makes the existence of isolated ionized side chains in membranes quite unlikely [69]. On the other hand, this cost would drop substantially provided that two charges of opposite sign interact and thus neutralize each other. Formation of strong ionic bonds between TMSs is therefore a realistic scenario.

As outlined in Section 26.2.1.1, electrostatic forces are strengthened in apolar environments. Whereas standard textbooks frequently depict the acyl chain region of a lipid bilayer as a slab of uniformly low polarity with dielectric constants $\varepsilon \sim 2$, the situation is more complex in reality. Nitroxide radicals, whose spin density depends on the polarity of the surrounding medium, were coupled to different positions of phospholipids, fatty acids, or cholesterol in order to report the polarity of model bilayers at different depths of penetration by electron paramagnetic resonance spectroscopy. The results revealed steep polarity gradients within the acyl chain region. Specifically, the apolar character of membranes composed of monounsaturated phospholipids increased from the headgroup region towards the middle of the acyl chain region. Inclusion of cholesterol reduced polarity in the central regions of saturated membranes and broadened the apolar central region in unsaturated membranes [70]. A similar picture emerged from neutron diffraction studies, where it was found that the concentration of the strongly hydrophobic molecule hexane dissolved in a bilayer peaked at its center and decreased towards its boundaries [71]. Accordingly, TMSs may experience the lowest polarity, and therefore strongest interactions, at their central regions. The same argument would apply to TMSs in detergent micelles, since their terminal regions would be close to the charged surface of the micelle, whereas the central regions would be embedded in its strongly hydrophobic core.

26.2.1.3 Entropic Factors Influencing Transmembrane Helix-Helix Interactions

Apart from enthalpic contributions, the free energy of TMS-TMS interactions is likely to be influenced by entropic factors in ways that are significantly different from the situation encountered with soluble proteins. First, TMSs are pre-oriented in the membrane upon biosynthetic insertion, with their long axes roughly perpendicular to the plane of the bilayer. Therefore, the loss of backbone entropy that is associated with any protein-protein interaction is expected to be smaller upon assembly of TMSs than of soluble helices tumbling freely in isotropic solution [72]. Second, helix-helix interactions are also influenced by the loss of side chain entropy upon association, provided that residues buried within interfaces exhibit only one out of several alternative rotameric states. On first approximation, this phenomenon would be of equal importance for soluble and membrane-embedded helices. On the other hand, the number of accessible rotameric states available to many side chains drops when a helix folds from the denatured state. Therefore, association of soluble helices that are usually in equilibrium with denatured states would be accompanied with a greater entropic penalty than assembly of preformed TM helices. Loss of side chain entropy is absent for Gly and minimal for those residues that adopt only one (Val) or three (Leu and Ile) rotamers in helices [10]; this would favor these amino acids in TMS-TMS interfaces. Furthermore, the hydrophobic effect is considered to be a major driving force in soluble protein folding. This effect is thought to evolve when apolar protein surfaces interact and thereby release ordered networks of water molecules associated with them [52]. It may be more important for inducing hydrophobic collapse than for stabilization of the folded structure, since many enzymes remain functional in apolar organic solvents [73]. Due to the scarcity of water molecules in the hydrophobic part of the membrane, the hydrophobic effect is thought to be of little or no importance for TMS-

TMS interactions [2, 10]. It has been argued, however, that other solvophobic factors based on lipids play a role in the membrane. Accordingly, the entropy of a membrane may increase as soon as lipids are removed from interacting protein surfaces [9].

26.2.2
Lessons from Sequence Analyses and High-resolution Structures

Careful analyses of the amino acid sequences and crystal structures of membrane proteins can provide insight into how TMSs pack together. Transmembrane helices of polytopic proteins that are adjacent in sequence always pack against each other in an antiparallel mode in the folded structure due to the topological constraints of the connecting loops. TMSs farther apart in sequence or TMSs from interacting subunits may pack in either parallel or antiparallel mode [74–76]. Thus, loops are at least partially responsible for the sequential order and relative orientation of the TMSs in a folded structure. An antiparallel arrangement of helices seems to allow for tighter packing than a parallel one [77]. A role for the helix dipole moment in antiparallel packing of TMSs does not seem likely [78].

Similar to the situation with soluble globular proteins [79], interacting TM helices within polytopic membrane proteins rarely display exactly parallel long axes but rather adopt either positive or negative crossing angles Ω, i.e., they form left- or right-handed pairs, respectively [74, 80] (Figure 26.1). The sign of the crossing angles depends on the geometry of side chain packing in the common interface. Packing of TM helices is predominantly left-handed [76, 77, 80], with crossing angles at around $+20^\circ$ [74], although this result may be biased by the limited number of available high-resolution structures. Pairs of helices with positive crossing angles have a larger average interaction surface than those with negative crossing angles [77].

TMS-TMS interfaces adopting positive crossing angles follow a repeated heptad [*abcdefg*] pattern of amino acids whose side chains interact via a "knobs-into-holes" packing reminiscent of soluble leucine zipper interaction domains [80]. There, residues located at the *a* or *d* positions protrude into cavities formed by *a*-, *d*-, *e*-, and g-type residues of the partner helix. An analysis of three membrane protein structures showed that the positions in such heptad repeat patterns are predominantly occupied by hydrophobic amino acids. Their composition is thus similar to the general composition of TMSs, although Trp seems to be over-represented. In that study, no positional specificity of particular amino acids was found [80].

The other general mode of TMS-TMS packing, generating negative crossing angles, is described more appropriately by "ridges-into-grooves" packing, where ridges are formed from protruding side chains and grooves correspond to the valleys between them [2, 79].

Sequence analyses showed that amino acid substitutions in general are more likely to occur at residues facing the lipid bilayer than at helix-helix interfaces, due to the need to conserve specific packing [81]. In support of this, single-spanning transmembrane proteins are more tolerant to mutation than multi-spanning proteins, indicating that they experience less sequence constraints [32, 33]. Surprisingly, the oligomerizing surfaces of polytopic transmembrane proteins are not well conserved [81]. However, this observation is probably caused by the fact that only membrane proteins with large, extended contacting surfaces located on different helices were analyzed.

In contrast to soluble helices, Gly and Pro are relatively common in TMSs and by themselves do not necessarily disrupt α-helical structure [66, 75], although disruption of helices is frequently observed when two Pro or a Pro and a Gly together are present within the same helix [66]. Conserved motifs in TMSs of membrane protein families often contain Gly or Pro residues [82]. This points to special roles for these potentially helix-breaking residues in transmembrane helices. Pro residues are enriched in the middle of TM helices compared to soluble ones [32, 33] and are usually oriented towards the protein interior where they contribute to tight packing [65, 75, 77, 83, 84]. A likely reason for this is that it would be unfavorable to direct the unsatisfied hydrogen bond of the carbonyl group of the residue at i-4 towards the lipids, whereas in the protein interior it could hydrogen bond to another helix or to prosthetic groups. Another function for Pro residues in TMSs was recently proposed [85]. In the cystic fibrosis transmembrane conductance regulator, a Pro prevents aggregation of the protein in the aqueous phase or in the translocation pore by hindering the formation of β-sheet structure. By counteracting the β-sheet propensities of surrounding amino acids in the TMS, Pro may thus destabilize misfolded states of the protein [85].

Gly residues in TMSs frequently occur in pairs that are spaced four amino acids apart [27] and thus localize to the same face of the helix. Indeed, the GxxxG motif is the single most enriched pairwise motif found in membrane-spanning segments, where it frequently occurs with β-branched residues at neighboring positions [28]. In addition, pairs of small residues with spacings of four or seven amino acids are conspicuous in light of their high degree of conservation within homologous TMSs in membrane protein families [85] and are therefore thought to play essential roles in the packing of transmembrane helices. In support of this, the small residues Gly, Ala, Ser, and Thr show a large preference for being oriented towards the interior of membrane proteins, where they do not create voids or pockets but pack tightly with other residues [66, 75, 84–86].

In contrast, large hydrophobic amino acids seem to be more loosely packed in membrane proteins than these small residues and are more often oriented to the lipids [75, 86]. Nevertheless, they are important for TMS packing, since in crystallized membrane protein structures 15 of the most frequent 20 pairs of neighboring residues in TM regions contain at least one L, I, or V residue [84]. Hydrophobic residues appear to be highly mutable in TMSs, indicating that they are relatively interchangeable. However, Leu was found to be only half as mutable as the other hydrophobic residues, which could indicate a special role for Leu in the packing of TMSs as leucine zippers [32, 33].

The more polar faces of transmembrane helices are enriched in aromatic amino acids, indicating that these residues are generally buried within the protein interior [83, 87]. Phe appears to be very versatile, since it can pack with a large variety of other residues [77, 84]. Trp and Tyr residues show a tendency to cluster together, which is probably due to their enrichment in the membrane-water interface, where they can interact with the lipid headgroups [27] (see Section 26.1.3.1).

Polar residues tend to be fairly conserved in membrane proteins, suggesting special roles in packing and/or function [77], e.g., in the formation of hydrogen bonds, ionic interactions, and/or cofactor binding.

Comparison of transmembrane sequences of mesophilic and thermophilic organisms can provide insight into the importance of certain sequence motifs in the packing and stability of membrane proteins [29]. Thermophilic organisms contain more Asp and Glu residues in their TMSs, which can form stronger hydrogen bonds than the Asn and Gln residues more often found in mesophilic organisms. A striking reduction in the occurrence of Cys residues was found in thermophiles, most likely because their reactivity could be harmful, especially at higher temperatures [29]. The fact that it can readily be replaced by other residues indicates that Cys in most cases has no specific function. Indeed, Cys residues rarely contact each other in TMSs [84], suggesting that S–S bridges almost never occur within the membrane. In thermophilic organisms, more of the small residues Gly, Ala, and Ser are found in pair motifs, which could allow for tighter packing between helices and thus for higher thermal stability than in mesophilic organisms [29].

Almost all transmembrane helices form hydrogen bonds with other helices, and the corresponding interfaces are packed more closely than those without [88]. The modes of interhelical hydrogen bonding appear to be more diverse in the membrane than between α-helices from soluble proteins [84] (see Section 26.2.1.2).

Asn and Gln show a high propensity to form strong hydrogen bonds in TMSs [84]. Consequently, these polar residues are often conserved in transmembrane helices, where they interact with other polar amino acids [75]. When located in the middle of TMSs, they tend to be buried within the TMS-TMS interfaces, while they do not show this preferential orientation in the lipid headgroup region [86].

Ser and Thr residues can form hydrogen bonds with the carbonyl atoms at positions i-4 or i-3 on the same helix [76] or can form interhelical hydrogen bonds, e.g., with backbone nitrogens [75].

Interhelical hydrogen bonds and backbone contacts occur more frequently between helix pairs crossing each other at positive packing angles [76]. Indeed, such interactions are often found where helices cross each other via GxxxG motifs, since Gly allows the helices to come into close contact [76, 84]. In addition, Gly has not one but two H atoms that can participate in C_α–H $\cdots$ O hydrogen bonds.

Multiple interhelical hydrogen bonds appear to be arranged in two kinds of spatial motifs [84]. In the "serine zipper," the side chains of two amino acids located on opposing helices form hydrogen bonds with the peptide backbone at the opposite residue. This is mostly seen for Ser-Ser pairs, and the seven-residue spacing is reminiscent of the leucine-zipper motif [84]. Indeed, such a serine zipper can often be found in combination with a leucine zipper, forming a mixed serine–leucine-zipper interface [89]. The second spatial arrangement of hydrogen bonds, the "polar clamp," is formed by three amino acids on two different helices, with two hydrogen bonds between them, so that a polar residue (Glu, Lys, Asn, Gln,

Arg, Ser, Thr) is clamped by hydrogen bonds to either backbone atoms or other side chains [84].

26.2.3
Lessons from Bitopic Membrane Proteins

Although the atomic structure of only one bitopic dimeric membrane protein TMS is currently known [90, 91], residue patterns of a number of TMS-TMS interfaces have been mapped by mutational analysis (see protocols, Section 26.4). Accordingly and in analogy to polytopic proteins (see Section 26.2.2), bitopic membrane proteins may be grouped into two broad categories. One exhibits interfacial $[abcd]_n$ repeats, where *a* and *b* correspond to interfacial residues and the TMSs appear to form right-handed pairs, i.e., they cross each other at negative angles. The other one is based on the $[abcdefg]_n$ heptad repeat motif of left-handed pairs characterized by positive crossing angles.

26.2.3.1 Transmembrane Segments Forming Right-handed Pairs

The homodimeric glycophorin A, an erythrocyte protein of unknown function, has long served as the paradigm for TMSs interacting at a negative angle, thus forming a right-handed pair. The interface between its TMSs corresponds to the pattern LI..GV..GV..T, as originally shown by Engelman and coworkers and others by mutational analysis (see Section 26.4) in detergent solution [92–94] and in membranes [95–97]. The structure of this interface gives rise to a negative crossing angle as implied by molecular modeling [98, 99] and confirmed by nuclear magnetic resonance studies in detergent [90] and membranes [91]. This interaction is dominated by a GxxxG motif that is central to the TMS-TMS interface, as mutation of these Gly residues to Ala destabilizes the interaction much more than mutation of the other interfacial residues [93, 95, 100, 101]. The glycophorin A TMS-TMS interaction is apparently driven by a complex mixture of attractive forces and entropic factors. The GxxxG motif may drive assembly by formation of a flat helix surface that allows multiple van der Waals interactions to form. In addition, the entropy loss upon association is considered minimal for Gly and the neighboring Val residues [102]. Moreover, the Gly residues reduce the distance between the helix axes and thus may facilitate hydrogen bond formation between their C_α-hydrogens and the backbone of the partner helix [76]. That the GxxxG motif is of prime importance is supported by the observation that changing the residue spacing between both Gly residues affects dimerization [103, 104] and that a GxxxG pair induces self-interaction of oligo-methionine or oligo-valine helices in membranes [103]. Interestingly, GxxxG motifs appear also to drive homophilic interactions of TMSs that are otherwise unrelated in sequence and are derived from membrane proteins of diverse function, including M13 major coat protein [105] and syndecan 3 [106]. Further, degenerate versions of this motif have been proposed to mediate TMS-TMS interactions of erbB receptors [107]. Apart from their occurrence in TMSs, GxxxG motifs seem to be prevalent in helix-helix interfaces of soluble proteins where they are also supported by $C_\alpha H \cdots O$ hydrogen bonds [108].

A traditional interhelical hydrogen bond between the Thr residues of the glycophorin TMS is not seen in detergent [90] but is suggested by the structure determined in membranes [91]. It should also be noted that the distance between the dimerization motif and the C-terminal flanking charged residues influences glycophorin TMS-TMS packing for reasons that are not entirely clear [109].

A residue pattern similar to that of glycophorin A accounts for interactions between the TMSs of SNARE (soluble NSF-attachment protein receptor) proteins. SNAREs form an evolutionarily conserved family of proteins that are essential for all types of intracellular membrane fusion events [110]. These proteins form a stable ternary complex that appears to bridge apposed membranes prior to their actual fusion [111]. While crystallographic analysis [112] has shown that a soluble version of the complex corresponds to a coiled-coil structure, the crystal structure of the complete SNARE complex including the TMSs is not known. In vivo, a fraction of the SNARE synaptobrevin II not participating in SNARE complex formation appears to exist in association with the synaptic vesicle protein synaptophysin or as a homodimer, as revealed by cross-linking experiments performed on brain fractions [113–115] or visualization of fluorescently tagged molecules in live cells [116]. A homodimeric form also develops from recombinant full-length synaptobrevin II in detergent solution [117], liposomes [118], or bacterial membranes [119]. Alanine-scanning mutagenesis indicated that synaptobrevin homodimerization depends on a specific residue pattern within its TMS: LxxICxxxLxxII. Grafting these six residues onto an inert oligo-alanine host sequence restored homodimerization in detergent solution; two additional residues were required to complete the motif when assayed in membranes: ILxxICxxILxxII [119]. These amino acids form a contiguous patch on the synaptobrevin TM helix and are thus supposed to constitute the TMS-TMS interface (Figure 26.2). The spacing of these interfacial residues suggests that the self-interacting TMSs adopt a negative crossing angle [117, 119]. These experimental results are supported by a computational study revealing that the critical residues form a tightly packed interface between α-helices that approach each other most closely at Cys103 and cross each other at a packing angle of -38° [120]. In the model, the synaptobrevin TMS-TMS interface comprises a much smaller molecular surface area (3.96 nm^2) than that of glycophorin A (5.50 nm^2), is based mainly on van der Waals interactions, and is not stabilized by intermolecular hydrogen bonds. Although the loss of side chain rotamer entropy upon burying Leu and Ile side chains within interfaces is considered to be small (see Section 26.2.1.3), it appears to exceed that of glycophorin TMS-TMS dimerization [120]. These differences are most likely due to the fact that the synaptobrevin TMS does not harbor a GxxxG motif. Interestingly, the interfacial residues of the synaptobrevin II TMS are almost completely conserved within the TMS of syntaxin 1A, the natural binding partner of synaptobrevin II [119]. This suggested that the TMS also mediates self-assembly of syntaxin or its heterophilic association with synaptobrevin, and this prediction was experimentally confirmed [118, 119]. A number of hypotheses concerning the functional relevance of synaptobrevin TMS-TMS interaction have been forwarded, including stabilization [117] or multimerization [119, 121] of the SNARE complex and/or propagating the "zippering up" of the

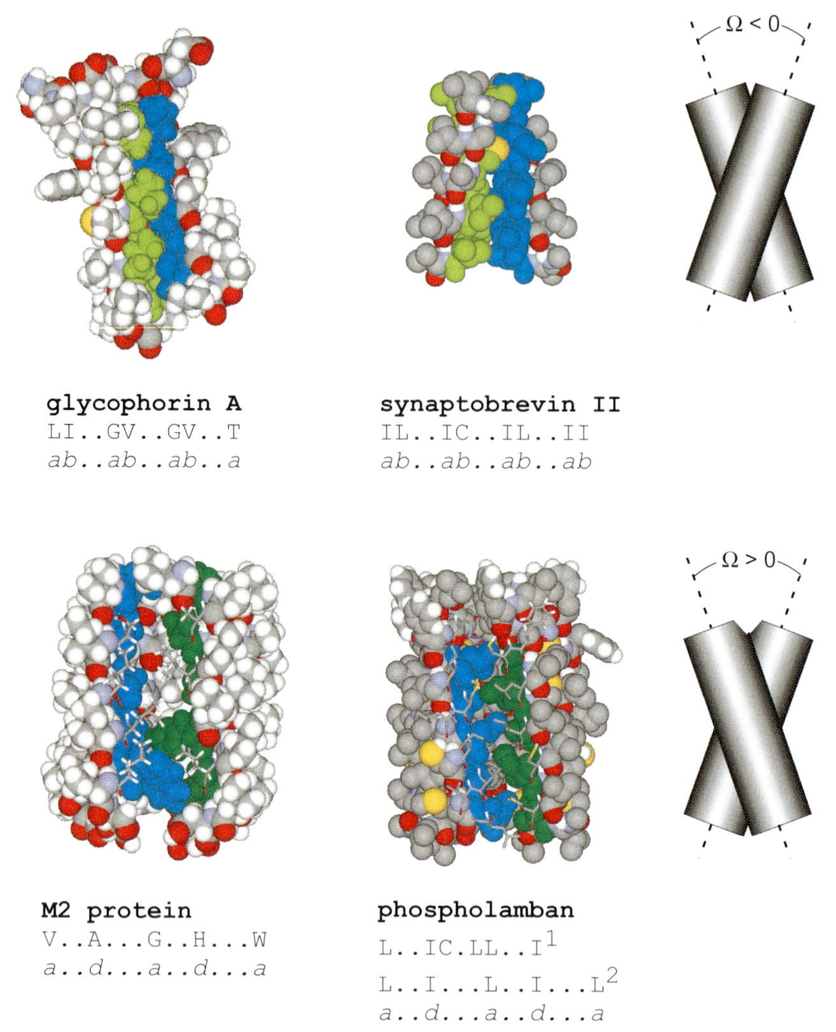

Fig. 26.2. Models of self-assembling TMSs from bitopic membrane proteins. The TMSs of glycophorin A (PDB code 1afo; only residues 70–90 are shown) and synaptobrevin II (PDB coordinates kindly provided by Dr. Karen Fleming) form right-handed homodimers with crossing angles $\Omega < 0°$, whereas the TMSs from the tetrameric influenza M2 protein (PDB coordinates kindly provided by Dr. William DeGrado) or from the pentameric phospholamban (PDB coordinates kindly provided by Dr. Isaiah Arkin) form left-handed pairs with $\Omega > 0°$. For easier visualization of TMS-TMS packing, the interfacial residues from neighboring helices are colored green and blue, respectively (the sulfur atom of synaptobrevin Cys 103 is in yellow). In the M2 protein and phospholamban, part of the residues is shown not in CPK but in stick representation to allow for better visual access to the interface. The patterns of the dominant interfacial residues as given below the models were determined by NMR (glycophorin A [90]) and/or mutational or functional analyses (glycophorin A [93]; synaptobrevin II [117]; M2 protein [123]; phospholamban [127][1] and [128][2]). Lowercase letters underneath the interfacial residues denote the interfacial residue positions according to the consensus patterns for negative ($[ab..]_n$) or positive ($[a..d...]_n$) helix-helix crossing angels.

cytoplasmic coiled-coil domains into the membrane at the onset of membrane fusion [118, 119]. In addition, the ability of SNARE proteins to enter the SNARE complex may be influenced by TMS-mediated homodimerization.

26.2.3.2 Transmembrane Segments Forming Left-handed Assemblies

The M2 proton channel from influenza A virus is one example of a TMS that is likely to adopt positive crossing angles. M2 forms a tetramer that is stabilized by intermolecular disulfide bonds. Since these are not essential for tetramerization, the protein can assemble by way of noncovalent TMS-TMS interactions. Its function is to form a proton channel in the envelope of the virus that is blocked by the antiviral drug amantadine [122]. Channel function can be reconstituted in mRNA-injected *Xenopus laevis* oocytes and proton translocation can be determined electrophysiologically. DeGrado, Pinto, and coworkers have determined the degree of amantadine blockage for a range of point mutants generated by cysteine-scanning mutagenesis in comparison with the wild-type protein. A pattern of sensitive residues (V..A...G..H...W) emerged with the periodicity of *a* and *d* positions in a left-handed helix-helix pair, suggesting a positive interhelical crossing angle [123]. This model was supported by direct analysis of oligomeric states upon disulfide oxidation [124] and refined by molecular modeling [125] and spectroscopic analysis ([10] and references cited therein).

Phospholamban TMSs also appear to cross each other at a positive packing angle upon homophilic assembly. This protein is part of the sarcoplasmatic reticulum membrane, where it functions as a regulator of Ca^{2+}-ATPase function [126]. Recombinant phospholamban forms a mixture of different oligomeric species in detergent solution up to a predominant pentamer, and the relevance of the individual TMS residues for assembly has been determined by saturation [127] and scanning [128] mutagenesis. Although the patterns of sensitive residues slightly deviated in both studies (L..IC.LL..I [127]; L..I...L..I...L [128]), it became apparent that the interfacial residues follow a heptad periodicity indicative of a positive crossing angle. Molecular modeling supported this view [129–131]. The exclusive participation of Leu, Ile, and Cys residues in interface formation is reminiscent of the situation of SNARE proteins as described in the previous section. It suggests that the interaction is driven mainly by van der Waals interactions and may be supported by minimal entropy loss upon association.

A leucine-zipper type of TMS-TMS interaction was also proposed by us to contribute to homodimerization of erythropoietin receptors. Whereas these cytokine receptors were originally thought to be activated by ligand-mediated dimerization, more recent evidence has revealed the existence of preformed dimers whose active conformation is stabilized by ligand binding [132, 133]. That the TMS may participate in receptor assembly was suggested by its ability to self-interact in membranes [134] and was confirmed by showing that the wild-type TMS sequence is important for receptor assembly [135] and function [136]. Thus, TMS-TMS assembly may support formation of a ligand-independent receptor dimer. The relevant interface was mapped by asparagine-scanning mutagenesis (see Section 26.4.3), which confirmed that it corresponds to a heptad repeat pattern [274]. Interestingly, the same face of the mEpoR TM helix may alternatively mediate homophilic interaction or heterophilic binding to the TMS of the viral membrane protein gp55-P in cells infected with polycythemic Friend spleen focus-forming virus. The disulfide-linked, single-span, homodimeric gp55-P binds to the erythropoietin receptor in a way that depends on the TMS sequences of both proteins and thereby induces constitutive activation [137]. Computational searching of low-energy structures and model building provided a 3-D model of this TMS-TMS interface and predicted that it corresponds to a left-handed leucine zipper stabilized by van der Waals interactions [137]. Since the residues that account for homophilic assembly of the erythropoietin receptor TMS correspond precisely to the residues contacting the gp55-P TMS in the model, homophilic interaction may compete with heterophilic interaction in the membrane of an infected erythroid cell.

A TMS-TMS interaction of the leucine zipper type was also proposed for cadherins that are calcium-dependent, cell-cell adhesion molecules whose function requires lateral clustering within the cell's plasma membrane [134]. Lateral cadherin clustering involves interactions between extracellular and cytoplasmic domains [138]. In addition, their TMSs appear to contribute to lateral assembly, since mutations reducing TMS-TMS interactions also reduced adhesiveness of E-cadherin in transfected cells [139].

Apart from these natural sequences, model systems have been developed in order to study the forces driving TMS-TMS assembly. For example, an oligo-Leu sequence that functions as an artificial TMS has been shown to self-interact in membranes and in detergent solution, where it forms oligomeric structures of mixed stoichiometry ([134, 275]). As self-interaction was preserved with a heptad-repeat pattern of Leu residues, the oligo-Leu sequence is regarded as a prototypical membrane-spanning leucine-zipper interaction domain [134] whose assembly is presumably driven by van der Waals interactions. Interestingly, the affinity of the oligo-Leu sequence was dramatically enhanced when Engelman and coworkers placed an Asn residue within the oligo-Leu TMS. A nuclear magnetic resonance study revealed that an intermolecular hydrogen bond had formed between the carboxamide side chains of Asn [140]. In a different study, DeGrado and coworkers investigated the effect of Asn on self-assembly of the GCN4 leucine-zipper domain that was reengineered to a membrane-soluble molecule (termed MS1) by exchanging most hydrophilic residues to apolar ones [141]. In the original GCN4 molecule, the leucine-zipper domain mediates high-affinity homodimerization and its structural features are known in great detail [142–145]. The *a* and *d* positions of this domain are occupied by Leu and Val residues, respectively, except for one *d* position that corresponds to an Asn. In the water-soluble form, this Asn residue is destabilizing relative to a hydrophobic interaction but specifies the dimeric state in favor of other oligomeric forms [144, 146]. In the membrane-soluble MS1 peptide, however, Asn strongly increased thermodynamic stability [141], whereas the Leu and Val residues at the other *a* and *d* positions contributed little to the free energy of association [10]. Thus, hydrogen bond formation between Asn side chains appears to provide a much stronger attractive force within the apolar milieu of detergent micelles or membranes than in aqueous solution. Later, it was shown that other polar amino acids such as Asp, Gln, Glu, and His also strongly enhanced self-assembly of the membrane-spanning leucine-zipper models [147, 148]. We recently used the effect exerted by Asn to probe the interface of the original oligo-Leu model. When all positions of the oligo-Leu sequence were systematically mutated in asparagine-scanning mutagenesis, the most sensitive positions correspond to a heptad repeat pattern of residues, confirming a leucine-zipper type of side chain packing. Further, asparagines had a much stronger impact on self-assembly when located at the center of the oligo-leucine sequence than at the termini, and a similar phenomenon was observed upon asparagine mutagenesis of the erythropoietin receptor TMS [274]. We believe that the dependence of the strength of the hydrogen bond between Asn residues on their depth of bilayer penetration is related to polarity gradients in lipid bilayers (see Section 26.2.1.2). This conclusion is supported by a recent study addressing the effect of asparagine on self-assembly of the MS1 peptide [86]. By analytical ultracentrifugation of corresponding synthetic peptides in detergent micelles, these authors showed that asparagine residues located within the apolar region of the peptide provide a significantly larger driving force on helix-helix interaction than an asparagine at a position near the apolar/polar interface of the micelle.

It has been argued that the existence of polar residues forming hydrogen bonds between TMSs could actually be dangerous since they might induce unspecific assembly in the membrane [140]. That this danger is real is highlighted by a number of growth factor receptors that are constitutively activeted by TMS mutations in different hereditary diseases. For example, the *neu* tyrosine kinase receptor is activated by a substitution of V664 within its TMS for a glutamic acid residue [149] that appears to induce permanent receptor dimerization by interhelical hydrogen bond formation [69]. The tyrosine kinase activity of fibroblast growth factor receptor 3 is activated by a glycine-to-arginine exchange within its TMS in patients suffering from achondroplasia [150]. Further, mutating S498 of the thrombopoietin receptor TMS to asparagine rendered this receptor constitutively active [151], and mutation of T617 to asparagine within the granulocyte colony-stimulating factor receptor TMS as found in patients with acute myeloid leukemia conferred growth-factor independence in expressing cells [152].

In addition to hydrogen bonds, it is quite likely that ionic interactions between charged residues take place in the apolar environment. Conceptually, charge-charge interactions between TMSs are difficult to envision since ionizable residues are likely to be integrated into membranes in an uncharged state. The T cell receptor is a model where this issue has been examined in considerable detail. This receptor is composed of six different chains that form the $\alpha\beta$ heterodimer responsible for ligand recognition plus the CD3$\gamma\varepsilon$, CD3$\delta\varepsilon$, and $\zeta\zeta$ signaling modules [153]. Three basic residues are found in the TM domains of the T cell receptor $\alpha\beta$ heterodimer, while a pair of acidic residues is present in each of the three associated signaling dimers. It was shown earlier that interactions between TMS residues of opposite charge contribute to association of certain receptor subunits [154–156], while a recent study addressed the role of charged residues in formation of the complete receptor complex [153]. According to this model, each of the nine ionizable residues is essential for receptor assembly. Each assembly step is promoted by interaction between the basic residues of the $\alpha\beta$ heterodimer and the acidic residues of the different signaling modules. Whether these residues are truly ionized prior to interaction is not clear. It was shown previously that unassembled α and β chains have a short half-life in cell membranes due to the basic residues [157], and model proteins with basic residues remain associated with the Sec61p channel of the endoplasmic reticulum rather than diffusing into the lipid [158]. Therefore, it was argued that newly synthesized subunits might stay at the vicinity of the translocon where their ionizable residues could be in the charged state before associating with cognate subunits [153]. This model of T cell receptor assembly is an interesting case, as it provides evidence that certain types of TMS-TMS interactions may depend on chaperoning by accessory proteins.

The issue of ionic interactions between TMSs was also addressed recently by an in vitro study based on synthetic hydrophobic peptides with non-natural sequences. Upon reconstitution into liposomal membranes, self-interaction of these oligo-Leu-based peptides was not enhanced by the presence of ionizable residues that were of opposite signs. It was thus proposed that ionic interactions would not contribute to TM helix-helix interactions [159]. The reason for the obvious discrepancy in the results obtained with the T cell receptor [153] is not known. Clearly, no chaperoning factors were present in the liposome system, and it is not certain that these peptides were inserted correctly into the plane of the bilayer.

26.2.4
Selection of Self-interacting TMSs from Combinatorial Libraries

Another tool for exploring the mechanisms underlying TMS-TMS interactions is the selection of transmembrane sequences capable of self-interaction from combinatorial libraries [102, 160, 161]. These libraries are created by randomization of amino acid motifs characteristic for TMS-TMS interfaces. In these systems, TMS-TMS interaction results in a phenotype that can be selected for. For this purpose, the ToxR transcription activator system is used, which exists in two versions: TOXCAT [96] and POSSYCCAT [160] (see Section 26.4).

Randomization of the LIxxGVxxGVxxT pattern that mediates the high-affinity dimerization of glycophorin A followed by selection of self-interacting sequences using TOXCAT yielded GxxxG motifs in over 80% of all isolates [102]. As discussed previously, Gly residues reduce the distance between the helix axes and thus may facilitate hydrogen bond formation between C_α-hydrogens and the backbone of the partner helix [76]. In addition, β-branched residues (Ile, Val) were apparently preferred at positions adjacent to the glycines. Such motifs might drive TMS-TMS packing by formation of a flat helix surface. Moreover, the side chains of Ile and Val prefer a single rotameric state in an α-helix; this limits the entropy loss that is normally associated with the freezing of side chain conformations upon protein-protein interaction [102]. When the glycophorin motif was randomized with a set of hydrophobic residues from which Gly was excluded, clusters of Ser and Thr residues were enriched in the selected self-interacting TMSs [161]. Although single Ser or Thr residues did not support TMS-TMS interactions in previous studies

[147, 148], multiples of them apparently do. This suggests that multiple weak interhelical hydrogen bonds – formed either between hydroxylated side chains or between them and main-chain atoms – may act cooperatively to drive TMS-TMS assembly.

The heptad repeat motif *ga..de.ga..de.ga* characteristic of TMS-TMS interfaces adopting positive crossing angles was randomized with three different sets of mostly hydrophobic residues, and self-interacting sequences were selected with the POSSYCCAT system [160]. A set of only five hydrophobic residues (Leu, Ile, Val, Met, and Phe) yielded heptad patterns that were capable of self-interaction in a rather sequence-independent way, although Val was over-represented in those TMSs with the highest affinities. Randomizing with more complex amino acid mixtures drastically decreased the fraction of self-interacting TMSs and resulted in an enrichment of Ile and Leu in high-affinity sequences.

It is likely that these aliphatic side chains contribute to TMS-TMS interaction by virtue of their relatively large contribution to van der Waal's interactions in a well-packed interface. In addition, loss of side chain entropy may be minimal for Leu, Ile, and Val. In contrast to the results obtained upon randomizing the motif associated with negative crossing angles [102, 161], no enrichment of GxxxG motifs or hydroxylated amino acids was seen, and the content of Pro residues was negatively correlated to self-interaction of the heptad pattern [160]. On the other hand, Pro was abundant in high-affinity sequences assumed to adopt negative crossing angles, where the free i-4 carbonyl was proposed to engage in interhelical hydrogen bonding to a Ser or Thr residue of the partner helix [161]. These differences suggest that weak interhelical hydrogen bonding may be more relevant for right-handed TMS pairs than for left-handed ones. This is in agreement with the observation that interhelical hydrogen bonds involving C_α-hydrogens were primarily associated with TM helix pairs crossing each other at negative packing angles [76].

26.2.5
Role of Lipids in Packing/Assembly of Membrane Proteins

The packing of transmembrane helices is likely to be influenced by the surrounding lipids. Excellent reviews of lipid-protein interactions in biological membranes have recently appeared, with an extensive treatment of high-resolution structures of membrane proteins that contain tightly bound lipid molecules even after detergent solubilization [162–164]. In a number of cases, these lipids are found at or between protein interaction surfaces and thus might directly support oligomerization by stabilizing TMS-TMS interactions. In the crystal structure of bacteriorhodopsin, which normally occurs as a trimer in its native purple membrane, six glycolipids were found in the central compartment of the trimer. Mutations in the binding sites for these lipids disrupted the trimeric organization of the protein [165]. For a review about lipid-protein interactions in the purple membrane, see Ref. [166]. When tightly bound cardiolipin molecules were removed from bovine cytochrome *c* oxidase by digestion with phospholipase, two subunits dissociated from the complex [167]. The structure of cytochrome *bc*1 also contains several phospholipids, which are suggested to play a role in stabilization of the multisubunit complex [168]. Moreover, a phosphatidylcholine (PC) molecule has been found at the interface between the L and M subunits of the photosynthetic reaction center [169], and three cardiolipin molecules are present at the trimer interface of formate dehydrogenase [170].

Apart from this direct role, lipids may influence TMS-TMS interactions in more indirect ways. First, interactions between proteins and lipids in a membrane are thermodynamically coupled. Thus, the free energy of helix-helix association in lipid bilayers can be described as:

$$\Delta G_a = \Delta G_{HH} + n/2\Delta G_{LL} - n\Delta G_{HL} \qquad (1)$$

where $\Delta G_{HH}, \Delta G_{LL}$, and ΔG_{HL} are the free energies of helix-helix, lipid-lipid, and helix-lipid interactions, respectively, and it is assumed that n lipid molecules are displaced from the helices upon interaction [1]. Therefore, oligomerization of transmembrane helices will be facilitated when helix-helix or lipid-lipid interactions are more favorable than helix-lipid interactions. Molecular dynamics simulations of glycophorin A in lipid bilayers suggest that these helix-helix and helix-lipid interactions can be quite similar [171]. Studies with transmembrane model peptides indicate that the free energy of helix-helix interaction increases with increasing lipid acyl chain length [172], presumably because the lipid-lipid interactions become more favorable. Other theoretical descriptions of how lipid properties can influence transmembrane helix-helix interactions have recently been reviewed [173–175]. Also, entropic factors play a role, since upon helix-helix association within the lipid bilayer, the displacement of protein-bound lipids to the bulk lipid increases overall entropy. In other words, reduction of the solvent-exposed surface would be associated with an increase in overall entropy. This is analogous to the hydrophobic effect, which is one force driving the folding of soluble proteins in water [9].

Second, oligomerization or aggregation of membrane proteins has been proposed to be one of the possible consequences of hydrophobic mismatch with the membrane. Hydrophobic mismatch occurs when the length of a hydrophobic TMS is not equal to the thickness of the apolar part of the lipid bilayer. The resulting exposure of hydrophobic surfaces to a hydrophilic environment is unfavorable, and a number of possible ways of relieving potential mismatch have been proposed [176, 177]. The area of the protein that is exposed to the lipid bilayer can be reduced by adaptation of the lipids, tilting of transmembrane helices, or oligomerization of the protein. This indeed occurs with bacteriorhodopsin, which was monomeric when reconstituted into phosphatidylcholines with chain lengths from 12 to 22 carbon atoms but aggregated in thinner or thicker bilayers [178]. This effect of hydrophobic mismatch seems to occur especially with α-helices that are too long to span the lipid bilayer, and therefore the resulting tilted conformation has been suggested to assist in helix packing [179].

Third, oligomerization of membrane proteins naturally depends on their concentration within the membrane, as exemplified by bacteriorhodopsin, which is monomeric at low molar ratios but trimerizes when the concentration is increased [180,

181]. In lipid bilayers containing separate domains of lipids in the fluid (liquid crystalline) and in the gel phase, transmembrane helices tend to be excluded from the gel-phase regions [41, 182]. This increases the local concentration of these proteins in the liquid-crystalline part of the bilayer and thus can lead to increased oligomer formation [41, 182, 183]. In lipid bilayers composed entirely of gel-state lipids, this can even lead to the formation of large, highly ordered peptide-containing domains [184]. The same local enrichment phenomenon might also apply to membranes containing liquid-ordered domains, also called rafts. The lipid acyl chains in such domains are tightly packed like gel-state lipids, but the lateral diffusion of lipids is nearly the same as in the fluid phase. Increasing concentrations of the raft lipids sphingomyelin and/or cholesterol lead to an increase in dimerization of transmembrane peptides [159, 172, 185]. This could be due to exclusion of the transmembrane peptides from liquid-ordered domains [186], although other membrane proteins are believed to be anchored within rafts via their TMSs [187–189]. Other effects of cholesterol, such as the ordering of the lipid acyl chains and the resulting increase in membrane thickness, could also play a role.

Not only the lipids but also the choice of detergents used to study transmembrane peptides can severely influence the extent of interaction that is observed. The negatively charged SDS is much more disruptive for transmembrane helix-helix interactions than are zwitterionic detergents [97, 101, 190]. In addition, the acyl chain length of the detergent is very important, since optimal packing of transmembrane peptides of cystic fibrosis transmembrane conductance regulator was observed with detergents with acyl chain of eight or nine carbon atoms. This may be related to the hydrophobic diameter of the detergent micelles [190].

Taken together, lipids may influence TMS-TMS interactions at various levels, although the extent to which this is relevant is currently not clear due to the paucity of experimental data. On the other hand, those few examples where TMS-TMS interactions have been compared in lipids and detergents [90, 91, 101] suggest that the environment may influence the fine structure and/or the association equilibrium [97, 100, 101], but not the identity of the interface [93, 95, 119, 134].

26.3 Conformational Flexibility of Transmembrane Segments

TM helices are often described as rather rigid structures, a notion that is reinforced by the stabilizing influence of the apolar environment of a lipid membrane (see Section 26.1.3.2). There is mounting evidence, however, that TMSs can exhibit considerable conformational flexibility and that this can be of significant functional relevance. Consistent with the fact that the network of intrachain hydrogen bonding is incomplete at helix termini, the internal flexibility of model helices was found to be greater at terminal positions than at the centers [191, 192]. Further, helix-destabilizing Pro and Gly residues appear to locally increase flexibilities of TM helices. Early indications for this were obtained by Deber and coworkers, who found that Pro residues are significantly over-represented in the TMSs of transporters and channels [193]. It was concluded that Pro residues facilitate the conformational changes during solute transport mediated by these proteins. As noted above (see Section 26.1.3.2), Pro residues in TM helices are frequently associated with local kinks, especially when a Gly residue is present in the vicinity, and it has been argued that these kinks may function as "molecular hinges," i.e., sites of increased local flexibility [63]. Indeed, a number of experimental and computational studies support these arguments. For example, Pro hinges have been reported to be important for gating of voltage-gated potassium channels [194] and for activation of G-protein-coupled receptors [195] or solute transporters [196]. Pro and Gly are frequently found in natural peptides, such as alamethicin and mellitin, that exhibit channel function and/or disrupt lipid bilayer structure. Replacing a Pro at a central location of the alamethicin sequence with alanine or shifting it to a different position influences conductance states and channel lifetimes [197, 198]. Alamethicin forms a helix that exhibits conformational flexibility around the Gly residue at the i-3 position of the central Pro residue in organic solvent as well in lipid bilayers [199, 200]. Mutational analysis of this GxxP pair suggested that the Pro residue by itself had little influence on bending of the helix, while concurrent mutation of Gly significantly reduced its flexibility [201].

Heck and coworkers studied the conformational flexibility of analogues of a non-natural model peptide reconstituted into liposomal membranes by hydrogen-deuterium exchange. They demonstrated that insertion of a Pro residue at the center of the TMS strongly enhanced the accessibility to deuterium of the amide groups of the entire central hydrophobic region that was largely inaccessible in the parental structure. This is likely to reflect an effect of Pro on helix flexibility that was, however, not seen upon insertion of Gly [202].

Conformational flexibility of TM helices has recently been proposed to play a role at the terminal stages of lipid membrane fusion [203–205]. Membrane fusion involves a restructuring of lipid bilayers brought into close proximity by single-span, membrane-anchored fusion proteins [110, 206, 207]. During the initial stages of the fusion process, these proteins mediate membrane apposition and undergo global conformational changes [208]. Late stages require the TMSs, as their replacement by lipid anchors or mutation abrogated the ability of various viral fusion proteins [110, 206, 207, 209–213] or SNAREs [214, 215] to mediate complete bilayer mixing. We and others showed that synthetic peptides corresponding to the TMSs of vesicular stomatitis virus (VSV) G-protein [204, 205] or of presynaptic SNAREs [203] induce unregulated sequence-specific membrane fusion upon reconstitution into liposomal membranes. In the case of the VSV G-protein, the peptide mimics partly reproduced the effects of point mutations affecting full-length VSV G-protein function in HeLa cells [204, 205]. Interestingly, these TMS mutations correspond to exchanges of Gly or of mostly β-branched residues for helix promoters. At the same time, these mutations increased the stability of the helical structures in isotropic solution [203, 205]. Gly and/or β-branched residues are generally over-represented in TMSs of viral fusion proteins [213] and SNAREs [203]. It was proposed, therefore, that the fusogenic activities of the TMSs at the final stage of lipid merger might require their conformational flexibility [203–205]. There are a few other cases known where TMS mutations suppressed the fusogenicity of full-length viral proteins. This is exemplified by influenza hemagglutinin, whose function was compromised by a Gly-to-Leu exchange [209], and by a murine leukemia virus glycoprotein where mutations of Pro and Phe residues had the strongest effects [211]. That fusogenic TMSs may be characterized by high internal flexibility is also consistent with hydrogen/deuterium exchange experiments conducted on the membrane-embedded TM helix of influenza hemagglutinin [216]. While the conformational effects of Gly and Pro residues are likely to result from local packing defects as discussed above, the situation is more complex with β-branched amino acids. The conformation of the latter has been extensively analyzed in the case of the TMS of lung surfactant protein C, which includes two segments of seven and four consecutive valyls. This TMS exists as a stable and very well-defined α-helix in phosphocholine micelles but refolds to β-sheet aggregates in organic solvent [217]. Its amide protons were virtually non-exchangeable in the helical state but accessible upon spontaneous unfolding. Thus, consecutive valyl side chains appear to protect a helical peptide backbone by tight interlocking [217]. Similarly, the amide protons of oligo-Leu helices were virtually non-exchangeable in bilayers and exchanged very slowly in organic solvent [218]. These and other data [59, 60] show that both Leu and Val residues form stable α-helices in membranes. One may wonder, therefore, whether the observed over-representation of β-branched residues in TMSs of fusogenic proteins [203, 213] is primarily related to their ability to self-interact or to fusogenicity. It will be interesting to determine the conformational and functional properties of β-branched residues in mixed sequences.

26.4 Experimental Protocols

In the following sections, we give an overview on the most widely used techniques for the characterization of TMS-TMS interactions. Whereas most of these techniques monitor protein-protein interaction upon solubilization with a suitable detergent that prevents unspecific aggregation [219], others are suited for analysis of membrane proteins reconstituted into lipid membranes. Experimental approaches not covered here include the use of electron paramagnetic resonance, Fourier transform infrared spectroscopy, small angle X-ray scattering, and nuclear magnetic resonance spectroscopy [220].

26.4.1 Biochemical and Biophysical Techniques

Most standard techniques developed for the characterization of interactions between soluble proteins can be adopted for the purpose of studying TMS-TMS interactions in the context of native or recombinant membrane proteins or of synthetic peptides. Recombinant membrane proteins are frequently produced by *E. coli*, which can be a difficult task due to toxic effects exerted by hydrophobic protein domains on the host cells. For high-level expression, therefore, it is often of considerable advantage to drive the recombinant protein into insoluble, and hence harmless, inclusion bodies that can later be detergent-solubilized upon cell lysis. Full-length membrane proteins or TMS sequences are frequently genetically fused to any of a variety of soluble proteins, such as glutathione-S-transferase or *Staphylococcus aureus* nuclease A, or to short epitope tags in order to facilitate expression and/or later detection as reviewed recently [221]. Expression in eukaryotic host systems, based on Sf9 insect cells, yeast, or mammalian cell lines, has the advantage of proper integration into eukaryotic membranes and proper posttranslational modifications but involves more cumbersome experimental procedures and frequently results in low protein yields [222, 223]. Alternatively, in vitro translation in the presence [153, 224–228] or absence [128] of microsomal membranes (derived from endoplasmatic reticulum) has been used to produce membrane protein subunits in sufficient quantity for investigation of their oligomeric structures. Translocation into microsomal membranes has the advantage that chaperoning factors and enzymes adding posttranslational modifications are present and active.

Short sequences representing TMSs from a variety of membrane proteins have also been synthesized by solid-phase chemical methods. Generally, synthesis of TMS peptides is a difficult task, as the sequences tend to aggregate nonspecifically on the resin, resulting in premature termination of the growing chains. Depending on the actual sequences, the standard 9-fluorenylmethoxycarbonyl (F-moc) methodology has been used successfully [147, 229, 230]. In certain instances, using the *tert*-butyloxycarbonyl (Boc) strategy appears to be better suited for production of pure peptides in satisfactory yields ([231] and our unpublished results).

26.4.1.1 Visualization of Oligomeric States by Electrophoretic Techniques

Denaturing Electrophoresis Sodium dodecyl sulfate polyacrylamide gel electrophoresis (SDS-PAGE) is generally used to determine the molecular weights of proteins based on comparison of their electrophoretic mobilities with a set of reference proteins [232]. Provided that the domains responsible for protein-protein interaction, e.g., TMSs, are not denatured by SDS, this technique is a simple and reliable tool for monitoring formation of oligomers. Stoichiometries and even subunit compositions of digomers can be investigated based on their molecular weights, and subsequent Western blotting reveals their identities. The preservation of noncovalent protein adducts in the presence of SDS is generally facilitated under mild conditions, i.e., using low SDS concentrations in sample buffer, omission of sample heating, and running of the gels in the cold room. Depending on the affinity of the TMS-TMS interaction under study and its competition by SDS binding, it is generally helpful to use protein concentrations in the micromolar range. Nevertheless, sample dilution during electrophoresis frequently results in partial dissociation of the protein complexes, thus producing smears on the gels. A major concern with this technique – as with any other method that monitors protein-protein interaction in detergent solution – is the issue of potential unspecific adduct formation.

Several criteria have been applied to exclude unspecific protein-protein interaction. First, inclusion of urea in sample buffer and gel significantly reduces unspecific interactions [233]. A critical aspect here is that the structure of the interaction domain must not be destroyed by urea, as seems to be the case with at least some TMSs [117, 119]. Second, the disruptive effect of certain point mutations within a TMS on adduct formation is a good indication of its specific nature and can reveal the identity of interfacial residues. By the same token, systematic point mutagenesis has been used to identify amino acid patterns of entire TMS-TMS interfaces (see Section 26.4.3). Third, competition of protein-protein interaction by added synthetic TMS peptides has been used to demonstrate the role of the TMSs in interaction and specificity [92, 147].

Natural TMSs whose interaction was previously studied by SDS-PAGE include those of glycophorin A [92, 93], Mas70p [234], phospholamban [127, 128], the major coat protein of M13 phage [235], and presynaptic SNAREs [117–119].

In cases where noncovalent protein-protein interaction cannot be detected in the presence of SDS, it has been proposed to substitute SDS with a novel detergent, perfluorooctanoic acid, which protects assembly of some membrane protein oligomers and allows their molecular weight determination [236]. Further, it is frequently helpful to stabilize weak interactions before SDS-PAGE analysis with homo- or heterobifunctional chemical cross-linking agents that are available in great variety with respect to chemical specificities, spacer lengths, and cleavability [237, 238].

Non-denaturing Electrophoresis The problem of denatured or masked TMSs upon SDS binding as encountered with SDS-PAGE is alleviated with non-denaturing or native PAGE. The membrane protein is solubilized with mild nonionic detergents and ε-aminocaproic acid. In the original version ("colorless native PAGE"), membrane protein migration in the electric field depended on the isoelectric point. Electrophoretic separation was greatly improved upon addition of the negatively charged dye Coomassie brilliant blue, which binds to membrane proteins, resulting in negatively charged dye-protein complexes at neutral pH. These complexes display reduced unspecific aggregation and travel towards the anode [239–242]. A problematic point with this technique, termed "blue native PAGE," is that precise determination of molecular weights is difficult since the ratio of negative charge to molecular weight is far less uniform than with SDS-PAGE. Further, its success depends on the nature of the respective protein and requires careful control of the ratio of detergent, dye, and protein concentrations.

26.4.1.2 **Hydrodynamic Methods**

Hydrodynamic methods such as gel-filtration chromatography and ultracentrifugation on density gradients have traditionally been used to study oligomeric complexes of soluble proteins and to calculate their molecular weights with reference to marker proteins. For membrane proteins, these techniques are carried out in the presence of non-denaturing detergents that may allow detection of TMS-TMS interactions that could be destroyed under the harsher conditions of SDS-PAGE

[134]. To make the molecular weight increment of the detergent molecules that are bound to the protein invisible, D_2O is added to the solution such as to adjust its density to the density of the detergent [243]. In analytical ultracentrifugation, the free-energy change of an interaction as well as the stoichiometry of a complex can be determined [10]. The method has been applied to the TMSs of glycophorin A [100] and of the influenza M2 protein [10] and to TMS peptide models engineered to self-interact [141, 147].

26.4.1.3 **Fluorescence Resonance Transfer**

Techniques based on fluorescence resonance energy transfer (FRET) between suitable chromophores enjoy widespread use in protein-protein interaction research. Their advantage for membrane proteins is that they are capable of monitoring interactions in detergent micelles, upon reconstitution into synthetic bilayers, or even in membranes of live cells depending on the experimental approach taken. The principle of FRET is that two interacting proteins must carry different fluorescent chromophores, termed "donor" and "acceptor," with complementary spectral characteristics. Fluorescent energy emitted by the donor can transfer to the acceptor in a radiation-less process, which is also called "Förster transfer," provided that both are sufficiently close to each other. Since the efficiency of radiation-less transfer, as determined by measuring the decreasing intensity of donor fluorescence, decreases with the sixth power of the distance, FRET is used to measure intermolecular distances and is also dubbed as "molecular ruler" [10, 244–246]. The molecular distances between interaction partners that are amenable to FRET analysis are typically in the tens of angstroms and depend on the type of FRET pair used. A variety of donor/acceptor pairs have been successfully employed for analyzing TMS-TMS interactions using recombinant proteins or synthetic peptides. Many chromophores that are commercially available are conjugated to *N*-hydroxysuccinimide esters or maleimides that react with free primary amine groups or sulfhydryl groups, respectively. Synthetic peptides are labeled either after or during synthesis. Some examples include the pairs carboxyfluorescein/tetramethylrhodamine [247], dansyl/dabcyl [230], pyrene acetic acid/7-dimethylcoumarin-4-acetic acid [97], and 7-nitro-2-1,3-benzoxadiazol-4-yl/5(6) carboxy tetramethylrhodamine [248]. In an alternative to synthetic chromophores, FRET was measured between synthetic peptides containing the pair tryptophan/pyrenylalanine [159]. Natural TMSs whose interaction was studied by the FRET technique include those of the pore-forming domain of *Bacillus thuringiensis* delta-endotoxin [249], phospholamban [250], the major coat protein of Ff bacteriophage [230], and glycophorin A [97, 251]. The latter case is particularly interesting since FRET measurements in detergent micelles have revealed a strong dependence of the TMS-TMS affinity, but not of helicity, on the type and concentration of detergent used [97].

Apart from small molecule chromophores, derivatives of the green fluorescent protein have been developed that are suitable as donor/acceptor pairs and can be genetically fused to the protein under study. While the first-generation derivatives suffered from low absorbances and spectral overlap [252], cyan and yellow fluorescent proteins [253–255] offer great hope for analysis of protein-protein interaction within the membranes of live cells [116, 256].

A technique based on bioluminescence resonance energy transfer (BRET) was recently described [257]. In that method, luciferase is genetically fused to the protein of interest; its luminescence is induced by addition of the membrane-permeable coelenterazine and excites yellow fluorescent protein fused to an interaction partner [258–260]. BRET has also been used to monitor membrane protein interactions in live cells and may have a number of advantages over FRET in that direct excitation of the acceptor and photobleaching of the donor are excluded; further, autofluorescence of endogenous cellular components does not interfere with the measurement [257].

26.4.2 Genetic Assays

A number of genetic assay systems have been developed for the analysis of isolated TMS sequences or full-length membrane proteins in natural membranes. Their unifying hallmark is that protein-protein interaction elicits expression of reporter genes whose products can be quantitated and qualitatively represent the efficiency of the underlying interaction.

26.4.2.1 The ToxR System

The single-span membrane protein ToxR transcription activator protein is the central regulator of virulence gene expression in *Vibrio cholerae*. Upon self-assembly in the inner membrane, the ToxR molecule activates the *ctx* promoter, thereby initiating transcription of downstream genes [261]. For a system that is suitable for studying TMS-TMS self-association, a tripartite protein consisting of the cytoplasmic ToxR domain, a TMS, and a periplasmatically located maltose-binding protein (MalE) has been engineered. Variants of this protein with TMSs of choice are expressed in *Escherichia coli* reporter strains, where reporter genes are under transcriptional control of the *ctx* promoter. While the original version of the ToxR system utilizes the *lacZ* gene as reporter [95, 262], the TOXCAT system is based on plasmid-borne chloramphenicol acetyltransferase [96]. Specific activities of the reporter enzymes in cell lysates reflect the degree of self-assembly of the ToxR proteins in the inner bacterial membrane. Since TMS-TMS interactions are concentration-dependent like any other protein-protein interaction, it is desirable to regulate the concentration of the ToxR proteins in the membrane. While ToxR expression in the first-generation vector is constitutive [95, 262], later versions make use of the regulatable *lac* [96] or arabinose [160] promoters, respectively. One advantage of regulatable promoters is that expression can be driven to levels that are high enough for protein solubilization, purification, and biochemical analysis (W. Ruan and D. L. Langosch, submitted).

It should be borne in mind that the orientation of the interacting face of a TMS relative to the DNA-binding ToxR domain is a critical factor since it influences the coupling of TMS-TMS interaction to transcription activation ([95, 274]). Therefore, the effect of inserting a TMS at different phases into ToxR chimeric proteins should initially be determined. Assuming α-helicity of the TMS, stepwise insertion of at least four additional residues at its N-terminus concurrent with stepwise deletion of four residues at its C-terminus rotates the potential TMS-TMS interface by up to $4 \times 100 = 400°$, i.e., more than a full helical turn, relative to the ToxR domain and thus reveals one construct whose orientation within the construct is suitable for further analysis.

Control experiments are advisable to ascertain similar levels of ToxR expression and correct membrane integration. Concentrations of ToxR proteins in inner bacterial membranes can be compared by determining their ability to complement the deficiency in maltose-binding protein (MalE) of an *E. coli* deletion strain (PD28) [103]. This strain cannot grow in minimal medium with maltose as the only carbon source unless the C-terminal MalE domain of expressed ToxR chimeric proteins is successfully translocated to the periplasmic space [263]. Further, correct vectoriality of membrane integration can be confirmed by proteolytic digestion of the MalE domain in spheroplast preparations [96, 107].

The ToxR system has been used to study self-assembly of TMSs derived from glycophorin A [95, 96, 103], E-cadherin [139], erythropoietin receptors [136], SNARE proteins [119], receptor epidermal growth factor receptors [107], etc. [134, 140]. A variation of the ToxR system has been described where synthetic TMS peptides added to the culture medium inhibit homodimerization of endogenously expressed ToxR proteins by heterophilic association [231].

Apart from investigating defined TMS-TMS interactions, ToxR-based systems, i.e., TOXCAT [96, 102, 161] and POSSYCCAT [160], have been most useful for the construction and selection of self-assembling TMSs from combinatorial libraries, since the interacting proteins and the corresponding genetic information are part of one particle, the bacterial cell. In other words, the physical coupling of phenotype to genotype allows for selection of functional properties (see Section 26.2.4).

Disadvantages of these systems in their present form are that heterophilic interactions cannot be determined and that no information on the stoichiometries of the ToxR complexes can be obtained.

26.4.2.2 **Other Genetic Assays**

Another in vivo genetic assay system for detecting homophilic TMS-TMS interactions is based on the bacteriophage lambda cI repressor C-terminal dimerization domain. There, the native C-terminal dimerization domain of the lambda repressor is replaced by candidate TMSs. The ability of the TMSs to drive dimerization of the lambda repressor headpiece is tested by measuring the effectiveness of the hybrid proteins in preventing infection by a lambda phage missing its repressor [264]. This system has been used to identify new self-interacting TMSs in a library of natural membrane proteins from *E. coli* [276, 277].

In an attempt to make heterophilic TMS-TMS interactions accessible to investigation, a system, termed GALLEX, was recently developed wherein the TMSs are fused to two LexA DNA-binding domains with different DNA-binding specificities.

Heterodimeric association of the TMSs in *E. coli* inner membranes induces repression of a β-galactosidase reporter gene [265]. As discussed by these authors, the potential disadvantage of this system is that homophilic TMS-TMS interactions that exist in parallel may indirectly influence reporter gene expression, since the equilibria of homo- and heterophilic interactions in the membrane are coupled.

To assess homo- and heterophilic membrane protein interactions in yeast membranes, a modification of the split-ubiquitin system [266] was presented. There, two potential interaction partners x and y are genetically fused to an altered N-terminal half ($N_{ub}G$) or the C-terminal half (C_{ub}) of ubiquitin, respectively. In addition, an epitope-tagged reporter protein R is linked to the C-terminus of C_{ub}. Upon xy interaction, $N_{ub}G$ and C_{ub}-R re-associate and thus allow for proteolytic release of R by ubiquitin-specific proteases in the cell [266]. For membrane protein interactions, the reporter consists of an artificial transcription factor consisting of LexA and the herpes simplex VP16. Assembly of x and y then releases the transcription factor, thus inducing activation of a reporter gene [267]. Alternatively, the reporter consists of the yeast Ura3 protein that converts added 5-fluoroorotic acid to a toxic substrate, thus killing the cells. Once $N_{ub}G$ and C_{ub}-*ura3* re-associate due to xy interaction, Ura3 is proteolytically released and quickly degraded in the cell. Interaction of x and y thus prevents substrate conversion and allows for the cell's survival [268].

Other methods for membrane interactions have been described that depend on genetic assays or complementation of enzyme function [269].

26.4.3 Identification of TMS-TMS Interfaces by Mutational Analysis

In principle, any of the above methods can be used to assess the effect of individual point mutations on the degree of interaction. Thereby, the critical residues constituting TMS-TMS interfaces can be mapped. Point mutations have been introduced by random mutagenesis using degenerate codons [93] or by successive replacement of the residues in different types of scanning mutagenesis. Alanine-scanning mutagenesis is frequently used for this purpose [95, 117], as exchange of most residue types to alanine creates voids and/or replaces side chains involved in hydrogen bonding, etc., thus resulting in incremental reductions of protein-protein affinity. Alternatively, asparagine-scanning mutagenesis has been developed, which is based on the observation that asparagine residues located within TMSs drive their self-interaction by hydrogen bond formation in apolar environments such as a lipid membrane [140, 141]. Systematic replacement of TMS residues by asparagine therefore results in different enhancements of TMS-TMS affinity depending on whether the mutated position is closely juxtaposed to its counterpart within the helix-helix interface or not [274]. Covalent linkages between TMSs have been introduced upon systematically replacing the residues by cysteine. If oxidation of the cysteine leads to formation of a disulfide cross-link, the respective residue is regarded as interfacial [124, 270, 271].

References

1 White, S. H. & Wimley, W. C. (1999). Membrane protein folding and stability: physical principles. *Annu. Rev. Biophys. Biomol. Struct.* 28, 319–365.

2 Lemmon, M. A. & Engelman, D. M. (1994). Specificity and promiscuity in membrane helix interactions. *Q. Rev. Biophys.* 27, 157–218.

3 Haltia, T. & Freire, E. (1995). Forces and factors that contribute to the structural stability of membrane proteins. *Biochim. Biophys. Acta* 1228, 1–27.

4 Fleming, K. G. (2000). Riding the wave: structural and energetic principles of helical membrane proteins. *Curr. Opin. Biotechnol.* 11, 67–71.

5 Ubarretxena-Belandia, I. & Engelman, D. M. (2001). Helical membrane proteins: diversity of functions in the context of simple architecture. *Curr. Opin. Struct. Biol.* 11, 370–376.

6 Arkin, I. T. (2002). Structural aspects of oligomerization taking place between the transmembrane alpha-helices of bitopic membrane proteins. *Biochim. Biophys. Acta* 1565, 347–363.

7 Popot, J.-L. & Engelman, D. M. (2000). Helical membrane protein folding, stability and evolution. *Annu. Rev. Biochem.* 69, 881–922.

8 Langosch, D., Lindner, E. & Gurezka, R. (2002). In vitro selection of self-interacting transmembrane segments – membrane proteins approached from a different perspective. *IUBMB Life* 54, 1–5.

9 Helms, V. (2002). Attraction within the membrane – Forces behind transmembrane protein folding and supramolecular complex assembly. *EMBO Rep.* 3(12), 1133–1138.

10 DeGrado, W. F., Gratkowski, H. & Lear, J. D. (2003). How do helix-helix interactions help determine the folds of membrane proteins? Perspectives from the study of homo-oligomeric helical bundles. *Prot. Sci.* 12(4), 647–65.

11 Chamberlain, A. K., Faham, S., Yohannan, S. & Bowie, J. U. (2003). Construction of helix-bundle membrane proteins. *Adv. Protein Chem.* 63, 19–46.

12 Shai, Y. (2001). Molecular recognition within the membrane milieu: Implications for the structure and function of membrane proteins. *J. Membr. Biol.* 182(2), 91–104.

13 Koebnik, R. (1999). Membrane assembly of the Escherichia coli outer membrane protein OmpA: exploring sequence constraints on transmembrane β-strands. *J. Mol. Biol.* 285, 1805–1810.

14 Koebnik, R. (1996). In vivo membrane assembly of split variants of the E. coli outer membrane protein OmpA. *EMBO J.* 15, 3529–3537.

15 Thompson, S. J., Kim, S. J. & Robinson, C. (1998). Sec-independent insertion of thylakoid membrane proteins – Analysis of insertion forces and identification of a loop intermediate involving the signal peptide. *J. Biol. Chem.* 273(30), 18979–18983.

16 Thompson, S. J., Robinson, C. & Mant, A. (1999). Dual signal peptides mediate the signal recognition particle Sec-independent insertion of a thylakoid membrane polyprotein, psbY. *J. Biol. Chem.* 274(7), 4059–4066.

17 Mant, A., Woolhead, C. A., Moore, M., Henry, R. & Robinson, C. (2001). Insertion of PsaK into the thylakoid membrane in a 'horse-shoe' conformation occurs in the absence of signal recognition particle, nucleoside triphosphates, or functional Albino3. *J. Biol. Chem.* 276(39), 36200–36206.

18 Ridder, A., Morein, S., Stam, J. G., Kuhn, A., de Kruijff, B. & Killian, J. A. (2000). Analysis of the role of interfacial tryptophan residues in controlling the topology of membrane proteins. *Biochemistry* 39(21), 6521–6528.

19 Muller, M., Koch, H. G., Beck, K. &

Schaefer, U. (2001). Protein traffic in bacteria: Multiple routes from the ribosome to and across the membrane. *Prog. Nucl. Acid Res. Mol. Biol.* 66(107), 107–157.

20 Chin, C. N., von Heijne, G. & de Gier, J. W. L. (2002). Membrane proteins: shaping up. *Trends Biochem. Sci.* 27(5), 231–234.

21 Ott, C. M. & Lingappa, V. R. (2002). Integral membrane protein biosynthesis: why topology is hard to predict. *J. Cell Sci.* 115(10), 2003–2009.

22 Drew, D., Froderberg, L., Baars, L. & de Gier, J. W. L. (2003). Assembly and overexpression of membrane proteins in Escherichia coli. *Biochim. Biophys. Acta* 1610(1), 3–10.

23 Popot, J.-L. & Engelman, D. M. (1990). Membrane Protein Folding and Oligomerization: the two-stage model. *Biochemistry* 29, 4031–4037.

24 Popot, J. L., Trewhella, J. & Engelman, D. M. (1986). Reformation of crystalline purple membrane from purified bacteriorhodopsin fragments. *Embo* 5, 3039–3044.

25 Ozawa, S., Hayashi, R., Masuda, A., Iio, T. & Takahashi, S. (1997). Reconstitution of bacteriorhodopsin from a mixture of a proteinase V_8 fragment and two synthetic peptides. *Biochim. Biophys. Acta* 1323, 145–153.

26 Marti, T. (1998). Refolding of bacteriorhodopsin from expressed polypeptide fragments. *J. Biol. Chem.* 273, 9312–9322.

27 Arkin, I. T. & Brünger, A. T. (1998). Statistical analysis of predicted transmembrane alpha-helices. *Biochim. Biophys. Acta* 1429, 113–128.

28 Senes, A., Gerstein, M. & Engelman, D. M. (2000). Statistical analysis of amino acid patterns in transmembrane helices: the GxxxG motif occurs frequently and in association with β-branched residues at neighboring positions. *J. Mol. Biol.* 296, 921–936.

29 Schneider, D., Liu, Y., Gerstein, M. & Engelman, D. M. (2002). Thermostability of membrane protein helix-helix interaction elucidated by statistical analysis. *FEBS Lett.* 532(1–2), 231–236.

30 von Heijne, G. (1986). The distribution of positively charged residues in bacterial inner membrane proteins correlates with the *trans*-membrane topology. *EMBO J.* 5, 3021–3027.

31 Landolt-Marticorena, C., Williams, K. A., Deber, C. M. & Reithmeier, R. A. F. (1993). Non-random distribution of amino acids in the transmembrane segments of human type I single span membrane proteins. *J. Mol. Biol.* 229, 602–608.

32 Jones, D. T., Taylor, W. R. & Thornton, J. M. (1994). A mutation data matrix for transmembrane proteins. *FEBS Lett.* 339(3), 269–275.

33 Jones, D. T., Taylor, W. R. & Thornton, J. M. (1994). A model recognition approach to the prediction of all-helical membrane protein structure and topology. *Biochemistry* 33(10), 3038–3049.

34 Stevens, T. J. & Arkin, I. T. (2000). The effect of nucleotide bias upon the composition and prediction of transmembrane helices. *Prot. Sci.* 9, 505–511.

35 von Heijne, G. (1989). Control of topology and mode of assembly of a polytopic membrane protein by positively charged residues. *Nature* 341(6241), 456–458.

36 Wallin, E., Tsukihara, T., Yoshikawa, S., von Heijne, G. & Elofsson, A. (1997). Architecture of helix bundle membrane proteins: an analysis of cytochrome c oxidase from bovine mitochondria. *Prot. Sci.* 6, 808–815.

37 Wimley, W. C. & White, S. H. (1996). Experimentally determined hydrophobicity scale for proteins at membrane interfaces. *Nat. Struct. Biol.* 3(10), 842–848.

38 Yau, W. M., Wimley, W. C., Gawrisch, K. & White, S. H. (1998). The preference of tryptophan for membrane interfaces. *Biochemistry* 37(42), 14713–14718.

39 Persson, S., Killian, J. A. & Lindblom, G. (1998). Molecular ordering of interfacially localized

tryptophan analogs in ester- and ether-lipid bilayers studied by 2H-NMR. *Biophys. J.* 75(3), 1365–1371.
40 Killian, J. A. & von Heijne, G. (2000). How proteins adapt to a membrane-water interface. *Trends Biochem. Sci.* 25, 429–434.
41 Mall, S., Broadbridge, R., Sharma, R. P., Lee, A. G. & East, J. M. (2000). Effects of aromatic residues at the ends of transmembrane alpha-helices on helix interactions with lipid bilayers. *Biochemistry* 39(8), 2071–2078.
42 de Planque, M. R., Bonev, B. B., Demmers, J. A., Greathouse, D. V., Koeppe, R. E., 2nd, Separovic, F., Watts, A. & Killian, J. A. (2003). Interfacial anchor properties of tryptophan residues in transmembrane peptides can dominate over hydrophobic matching effects in peptide-lipid interactions. *Biochemistry* 42(18), 5341–5348.
43 Kyte, J. & Doolittle, R. F. (1982). A simple method for displaying the hydropathic character of a protein. *J. Mol. Biol* 157, 105–132.
44 Engelman, D. M., Steitz, T. A. & Goldman, A. (1986). Identifying nonpolar transbilayer helices in amino acid sequences of membrane proteins. *Annu. Rev. Biophys. Biophys. Chem.* 15, 321–353.
45 Cserzo, M., Wallin, E., Simon, I., von Heijne, G. & Elofsson, A. (1997). Prediction of transmembrane alpha-helices in prokaryotic membrane proteins: the dense alignment surface method. *Protein Eng.* 10(6), 673–676.
46 Pasquier, C., Promponas, V. J., Palaios, G. A., Hamodrakas, J. S. & Hamodrakas, S. J. (1999). A novel method for predicting transmembrane segments in proteins based on a statistical analysis of the SwissProt database: the PRED-TMR algorithm. *Protein Eng.* 12(5), 381–385.
47 Rost, B., Casadio, R., Fariselli, P. & Sander, C. (1995). Prediction of helical transmembrane segments at 95% accuracy. *Prot. Sci.* 4, 521–533.
48 Chen, C. P., Kernytsky, A. & Rost, B. (2002). Transmembrane helix predictions revisited. *Protein Sci.* 11(12), 2774–2791.
49 von Heijne, G. (1992). Membrane protein structure prediction. Hydrophobicity analysis and the positive-inside rule. *J. Mol. Biol.* 225, 487–494.
50 Persson, B. & Argos, P. (1996). Topology prediction of membrane proteins. *Protein Sci.* 5(2), 363–371.
51 Persson, B. & Argos, P. (1997). Prediction Of Membrane Protein Topology Utilizing Multiple Sequence Alignments. *Journal Of Protein Chemistry* 16(5), 453–457.
52 Creighton, T. E. (1993). *Proteins: Structures and Molecular Properties*, Freeman, New York.
53 Padmanabhan, S., Marqusee, S., Ridgeway, T., Laue, T. M. & Baldwin, R. L. (1990). Relative helix-forming tendencies of nonpolar amino acids. *Nature* 344, 268–270.
54 O'Neil, K. T. & DeGrado, W. F. (1990). A thermodynamic scale for the helix-forming tendencies of the commonly occuring amino acids. *Science* 250, 646–651.
55 Blaber, M., Zhang, X. & Matthews, B. W. (1993). Structural basis for amino acid alpha helix propensity. *Science* 260, 1637–1640.
56 Smith, C. K., Withka, J. M. & Regan, L. (1994). A thermodynamic scale for the β-sheet forming tendencies of the amino acids. *Biochemistry* 33, 5510–5517.
57 Street, A. G. & Mayo, S. L. (1999). Intrinsic β-sheet propensities result from van der Waals interactions between side chains and the local backbone. *Proc. Natl. Acad. Sci. USA* 96, 9074–9076.
58 Li, S. C. & Deber, C. M. (1992). Glycine and beta-branched residues support and modulate peptide helicity in membrane environments. *FEBS Lett.* 311, 217–220.
59 Li, S. C. & Deber, C. M. (1994). A measure of helical propensity for amino acids in membrane environments. *Nature Struct. Biol.* 1, 368–373.

60 Liu, L.-P. & Deber, C. M. (1998). Uncoupling hydrophobicity and helicity in transmembrane segments. *J. Biol. Chem.* 273, 23645–23648.

61 Deber, C. M., Li, Z. M., Joensson, C., Glibowicka, M. & Xu, G. Y. (1992). Transmembrane Region of Wild-Type and Mutant M13 Coat Proteins – Conformational Role of Beta-Branched Residues. *J. Biol. Chem.* 267(8), 5296–5300.

62 Johansson, J., Szyperski, T., Curstedt, T. & Wüthrich, K. (1994). The NMR structure of pulmonary surfactant-associated polypeptide SP-C in a apolar solvent contains a valyl-rich α-helix. *Biochemistry* 33, 6015.

63 Cordes, F. S., Bright, J. N. & Sansom, M. S. P. (2002). Proline-induced Distortions of Transmembrane Helices. *J. Mol. Biol.* 323, 951–960.

64 Chakrabarti, P. & Chakrabarti, S. (1998). C–H–O hydrogen bond involving proline residues in alpha-helices. *J. Mol. Biol.* 284(4), 867–873.

65 von Heijne, G. (1991). Proline kinks in transmembrane alpha-helices. *J. Mol. Biol.* 218, 499–503.

66 Javadpour, M. M., Eilers, M., Groesbeek, M. & Smith, S. O. (1999). Helix packing in polytopic membrane proteins: Role of glycine in trans-membrane helix association. *Biophys. J.* 77(3), 1609–1618.

67 Daniel, J. M., Friess, S. D., Rajagopalan, S., Wendt, S. & Zenobi, R. (2002). Quantitative determination of noncovalent binding interactions using soft ionization mass spectrometry. *Int. J. Mass. Spec.* 216, 1–27.

68 Israelachvili, J. N. (1991). *Intermolecular and surface forces*, Academic Press, London.

69 Smith, S. O., Smith, C. S. & Bormann, B. J. (1996). Strong hydrogen bonding interactions involving a buried glutamic acid in the transmembrane sequence of the neu/erbB-2 receptor. *Nature Structural Biology* 3, 252–258.

70 Subczynski, W. K., Wisniewska, A., Yin, J.-J., Hyde, J. S. & Kusumi, A. (1994). Hydrophobic Barriers of Lipid Bilayer Membranes Formed by Reduction of Water Penetration by Alkyl Chain Unsaturation and Cholesterol. *Biochemistry* 33, 7670–7681.

71 White, S. H., King, G. I. & Cain, J. E. (1981). Location of hexane in lipid bilayers determined by neutron diffraction. *Nature* 290, 161–163.

72 Grasberger, B., Minton, A. P., DeLisi, C. & Metzger, H. (1986). Interaction between proteins localized in membranes. *Proc. Natl. Acad. Sci. USA* 83, 6258–6262.

73 Klibanov, A. M. (2001). Improving enzymes by using them in organic solvents. *Nature* 409, 241–246.

74 Bowie, J. U. (1997). Helix packing in membrane proteins. *J. Mol. Biol.* 272, 780–789.

75 Eilers, M., Shekar, S. C., Shieh, T., Smith, S. O. & Fleming, P. J. (2000). Internal packing of helical membrane proteins. *Proc. Natl. Acad. Sci.* 97, 5796–5801.

76 Senes, A., Ubarretxena-Belandia, I. & Engelman, D. M. (2001). The Ca–H···O hydrogen bond: A determinant of stability and specifity in transmembrane helix interactions. *Proc. Natl. Acad. Sci.* 98, 9056–9061.

77 Eilers, M., Patel, A. B., Liu, W. & Smith, S. O. (2002). Comparison of helix interactions in membrane and soluble alpha-bundle proteins. *Biophys. J.* 82(5), 2720–2736.

78 Ben-Tal, N. & Honig, B. (1996). Helix-helix interactions in lipid bilayers. *Biophys. J.* 71(6), 3046–3050.

79 Chothia, C., Levitt, M. & Richardson, D. (1981). Helix to helix packing in proteins. *J. Mol. Biol.* 145, 215–250.

80 Langosch, D. & Heringa, J. (1998). Interaction of transmembrane helices by a knobs-into-holes geometry characteristic of soluble coiled coils. *Proteins Struct. Funct. Genet.* 31, 150–160.

81 Stevens, T. J. & Arkin, I. T. (2001). Substitution rates in alpha-helical transmembrane proteins. *Protein Science* 10(12), 2507–2517.

82 Liu, Y., Engelman, D. M. &

Gerstein, M. (2002). Genomic analysis of membrane protein families: abundance and conserved motifs. *Genome Biol.* 3(10), RESEARCH0054.1–0054.12.

83 Pilpel, Y., Ben-Tal, N. & Lancet, D. (1999). kPROT: a knowledge-based scale for the propensity of residue orientation in transmembrane segments. Application to membrane protein structure prediction. *J. Mol. Biol.* 294, 921–935.

84 Adamian, L. & Liang, J. (2001). Helix-Helix Packing and Interfacial Pairwise Interactions of Residues in Membrane Proteins. *J. Mol. Biol.* 311, 891–907.

85 Wigley, W. C., Corboy, M. J., Cutler, T. D., Thibodeau, P. H., Oldan, J., Lee, M. G., Rizo, J., Hunt, J. F. & Thomas, P. J. (2002). A protein sequence that can encode native structure by disfavoring alternate conformations. *Nat. Struct. Biol.* 9(5), 381–388.

86 Lear, J. D., Gratkowski, H., Adamian, L., Liang, J. & DeGrado, W. F. (2003). Position-dependence of stabilizing polar interactions of asparagine in transmembrane helical bundles. *Biochemistry* 42(21), 6400–6407.

87 Samatey, F. A., Xu, C. & Popot, J.-L. (1995). On the distribution of amino acid residues in transmembrane α-helix bundles. *Proc. Natl. Acad. Sci.* 92, 4577–4581.

88 Adamian, L. & Liang, J. (2002). Interhelical hydrogen bonds and spatial motifs in membrane proteins: polar clamps and serine zippers. *Proteins-Structure Function and Genetics* 47, 209–218.

89 Adamian, L., Jackups, R., Jr., Binkowski, T. A. & Liang, J. (2003). Higher-order interhelical spatial interactions in membrane proteins. *J. Mol. Biol.* 327(1), 251–272.

90 MacKenzie, K. R., Prestegard, J. H. & Engelman, D. M. (1997). A transmembrane helix dimer: structure and implications. *Science* 276, 131–133.

91 Smith, S. O., Song, D., Shekar, S., Groesbeek, M., Ziliox, M. & Aimoto, S. (2001). Structure of the Transmembrane Dimer Interface of Glycophorin A in Membrane Bilayers. *Biochemistry* 40(22), 6553–6558.

92 Lemmon, M. A., Flanagan, J. M., Hunt, J. F., Adair, B. D., Bormann, B.-J., Dempsey, C. E. & Engelman, D. M. (1992). Glycophorin A dimerization is driven by specific interactions between transmembrane alpha-helices. *J. Biol. Chem.* 267, 7683–7689.

93 Lemmon, M. A., Flanagan, J. M., Treutlein, H. R., Zhang, J. & Engelman, D. M. (1992). Sequence specificity in the dimerization of transmembrane alpha-helices. *Biochemistry* 31, 12719–12725.

94 Lemmon, M. A., Treutlein, H. R., Adams, P. D., Brünger, A. T. & Engelman, D. (1994). A dimerization motif for transmembrane alpha-helices. *Nature Struct. Biol.* 1, 157–163.

95 Langosch, D. L., Brosig, B., Kolmar, H. & Fritz, H.-J. (1996). Dimerisation of the glycophorin A transmembrane segment in membranes probed with the ToxR transcription activator. *J. Mol. Biol.* 263, 525–530.

96 Russ, W. P. & Engelman, D. M. (1999). TOXCAT: A measure of transmembrane helix association in a biological membrane. *Proc. Natl. Acad. Sci. USA* 96, 863–868.

97 Fisher, L. E., Engelman, D. M. & Sturgis, J. N. (1999). Detergents modulate dimerization but not helicity, of the glycophorin A transmembrane domain. *J. Mol. Biol.* 293(3), 639–651.

98 Treutlein, H. R., Lemmon, M. A., Engelman, D. M. & Brünger, A. T. (1992). The glycophorin A transmembrane domain dimer: sequence-specific propensity for a right-handed supercoil of helices. *Biochemistry* 31, 12726–12733.

99 Adams, P. D., Engelman, D. M. & Brünger, A. T. (1996). Improved prediction for the structure of the dimeric transmembrane domain of glycophorin A obtained through global searching. *Proteins* 26, 257–261.

100 Fleming, K. G., Ackerman, A. L. & Engelman, D. M. (1997). The Effect Of Point Mutations On the Free Energy Of Transmembrane Alpha Helix Dimerization. *J. Mol. Biol.* 272, 266–275.

101 Fleming, K. G. & Engelman, D. M. (2001). Specificity in transmembrane helix-helix interactions can define a hierarchy of stability for sequence variants. *Proceedings of the National Academy of Sciences of the United States of America* 98(25), 14340–14344.

102 Russ, W. P. & Engelman, D. M. (2000). The GxxxG motif: a framework for transmembrane helix-helix association. *J. Mol. Biol.* 296, 911–919.

103 Brosig, B. & Langosch, D. (1998). The dimerization motif of the glycophorin A transmembrane segment in membranes: importance of glycine residues. *Protein Sci.* 7, 1052–1056.

104 Mingarro, I., Whitley, P., Lemmon, M. A. & von Heijne, G. (1996). Ala-insertion scanning mutagenesis of the glycophorin A transmembrane helix: a rapid way to map helix-helix interactions in integral membrane proteins. *Protein Sci.* 5, 1339–1341.

105 Williams, K. A., Glibowicka, M., Li, Z., Li, H., Khan, A. R., Chen, Y. M. Y., Wang, J., Marvin, D. A. & Deber, C. M. (1995). Packing of coat protein amphipathic and transmembrane helices in filamentous bacteriophage M13: role of small residues in protein oligomerization. *J. Mol. Biol.* 252, 6–14.

106 Asundi, V. K. & Carey, D. J. (1995). Self-association of N-syndecan (syndecan-3) core protein is mediated by a novel structural motif in the transmembrane domain and ecto-domain flanking region. *J. Biol. Chem.* 270, 26404–26410.

107 Mendrola, J. M., Berger, M. B., King, M. C. & Lemmon, M. A. (2002). The single transmembrane domains of ErbB receptors self-associate in cell membranes. *J. Biol. Chem.* 277(7), 4704–4712.

108 Kleiger, G., Grothe, R., Mallick, P. & Eisenberg, D. (2002). GXXXG and AXXXA: Common alpha-helical interaction motifs in proteins, particularly in extremophiles. *Biochemistry* 41(19), 5990–5997.

109 Orzaez, M., Perez-Paya, E. & Mingarro, I. (2000). Influence of the C-terminus of the glycophorin A transmembrane fragment on the dimerization process. *Prot. Sci.* 9, 1246–1253.

110 Jahn, R., Lang, T. & Südhof, T. C. (2003). Membrane Fusion. *Cell* 112, 519–533.

111 Söllner, T., Bennett, M. K., Whiteheard, S. W., Scheller, R. H. & Rothman, J. E. (1993). A protein assembly-disassembly pathway in vitro that may correspond to sequential steps of synaptic vesicle docking, activation, and fusion. *Cell* 75, 409–418.

112 Sutton, R. B., Fasshauer, D., Jahn, R. & Brünger, A. T. (1998). Crystal structure of a SNARE complex involved in synaptic exocytosis at 2.4 A resolution. *Nature* 395, 347–353.

113 Calakos, N. & Scheller, R. H. (1994). Vesicle-associated membrane protein and synaptophysin are associated on the synaptic vesicle. *J. Biol. Chem.* 269, 24534–24537.

114 Washbourne, P., Schiavo, G. & Montecucco, C. (1995). Vesicle-associated membrane protein-2 (synaptobrevin-2) forms a complex with synaptophysin. *Biochem. J.* 305, 721–724.

115 Edelman, L., Hanson, P. I., Chapman, E. R. & Jahn, R. (1995). Synaptobrevin binding to synaptophysin: a potential mechanism for controlling the exocytotic fusion machine. *EMBO J.* 14, 224–231.

116 Pennuto, M., Dunlap, D., Contestabile, A., F., B. & Valtorta, F. (2002). Fluorescence Resonance Energy Transfer Detection of Synaptophysin I and Vesicle-associated Membrane Protein 2 Interactions during Exocytosis from Single Live Synapses. *Mol. Biol. Cell* 13, 2706–2717.

117 Laage, R. & Langosch, D. (1997). Dimerization of the synaptic vesicle

protein synaptobrevin/VAMP II depends on specific residues within the transmembrane segment. *Eur. J. Biochem.* 249, 540–546.

118 Margittai, M., Otto, H. & Jahn, R. (1999). A stable interaction between syntaxin 1a and synaptobrevin 2 mediated by their transmembrane domains. *FEBS Lett.* 446, 40–44.

119 Laage, R., Rohde, J., Brosig, B. & Langosch, D. (2000). A conserved membrane-spanning amino acid motif drives homomeric and supports heteromeric assembly of presynaptic SNARE proteins. *J. Biol. Chem.* 275, 17481–17487.

120 Fleming, K. G. & Engelman, D. M. (2001). Computation and mutagenesis suggest a right-handed structure for the synaptobrevin transmembrane dimer. *Proteins* 45(4), 313–317.

121 Poirier, M. A., Hao, J. C., Malkus, P. N., Chan, C., Moore, M. F., King, D. S. & Bennett, M. K. (1998). Protease resistance of syntaxin – SNAP-25 – VAMP complexes – Implications for assembly and structure. *J. Biol. Chem.* 273, 11370–11377.

122 Holsinger, L. J., Nichani, D., Pinto, L. H. & Lamb, R. A. (1994). Influenza – a Virus M(2) Ion-Channel Protein – a Structure-Function Analysis. *J. Virol.* 68(3), 1551–1563.

123 Pinto, L. H., Dieckmann, G. R., Gandhi, C. S., Papworth, C. G., Braman, J., Shaugnessy, M. A., Lear, J. D., Lamb, R. A. & DeGrado, W. F. (1997). A functionally defined model for the M2 proton channel of influenza A virus suggests a mechanism for its ion selectivity. *Proc. Natl. Acad. Sci. USA* 94, 11301–11306.

124 Bauer, C. M., Pinto, L. H., Cross, T. A. & Lamb, R. A. (1999). The Influenza Virus M2 Ion Channel Protein: Probing the Structure of the Transmembrane Domain in Intact Cells by Using Engineered Disulfide Cross-Linking. *Virology* 254, 196–209.

125 Dieckmann, G. R. & DeGrado, W. F. (1997). Modeling transmembrane helical oligomers. *Curr. Opin. Struct. Biol.* 7, 486–494.

126 Arkin, I. T., Adams, P. D., Brünger, A. T., Smith, S. & Engelman, D. M. (1997). Structural perspectives of phospolamban, a helical transmembrane pentamer. *Annu. Rev. Biophys. Biomol. Struct.* 26, 157–179.

127 Arkin, I. T., Adams, P. D., MacKenzie, K. R., Lemmon, M. A., Brünger, A. T. & Engelman, D. M. (1994). Structural organization of the pentameric transmembrane alpha-helices of phospholamban, a cardiac ion channel. *EMBO J.* 13, 4757–4764.

128 Simmerman, H. K. B., Kobayashi, Y. M., Autry, J. M. & Jones, L. R. (1996). A leucine zipper stabilizes the pentameric membrane domain of phospholamban and forms a coiled-coil pore structure. *J. Biol. Chem.* 271, 5941–5946.

129 Arkin, I. T., Rothman, M., Ludlam, C. F. C., Aimoto, S., Engelman, D., Rothschild, K. J. & Smith, S. O. (1995). Structural Model of the phospholamban ion channel complex in phospholipid membranes. *J. Mol. Biol.* 248, 824–834.

130 Karim, C. B., Stamm, J. D., Karim, J., Jones, L. R. & Thomas, D. D. (1998). Cysteine reactivity and oligomeric structures of phospholamban and its mutants. *Biochemistry* 37(35), 12074–12081.

131 Torres, J., Kukol, A. & Arkin, I. T. (2001). Mapping the energy surface of transmembrane helix-helix interactions. *Biophysical Journal* 81(5), 2681–2692.

132 Livnah, O., Stura, E. A., Middleton, S. A., Johnson, D. L., Jolliffe, L. K. & Wilson, I. A. (1999). Crystallographic evidence for preformed dimers of eryhtropoietin receptor before ligand activation. *Science* 283, 987–990.

133 Remy, I., Wilson, I. A. & Michnick, S. W. (1999). Erythropoietin receptor activation by a ligand-induced conformation change. *Science* 283, 990–993.

134 Gurezka, R., Laage, R., Brosig, B. & Langosch, D. (1999). A Heptad Motif of Leucine Residues Found in Membrane Proteins Can Drive Self-

Assembly of Artificial Transmembrane Segments. *J. Biol. Chem.* 274, 9265–9270.

135 Constantinescu, S. N., Keren, T., Socolovsky, M., Nam, H. S., Henis, Y. I. & Lodish, H. F. (2001). Ligand-independent oligomerization of cell-surface erythropoietin receptor is mediated by the transmembrane domain. *Proc. Natl. Acad. Sci. U.S.A.* 98, 4379–4384.

136 Kubatzky, K. F., Ruan, W., Gurezka, R., Cohen, J., Ketteler, R., Watowich, S. S., Neumann, D., Langosch, D. & Klingmüller, U. (2001). Self-assembly of the transmembrane domain is a crucial mediator for signalling through the erythropoietin receptor. *Current Biology* 11, 110–115.

137 Constantinescu, S. N., Liu, X., Beyer, W., Fallon, A., Shekar, S., Henis, Y. I., Smith, S. O. & Lodish, J. F. (1999). Activation of the erythropoietin receptor by the gp55-P viral envelope protein is determined by a single amino acid in its transmembrane domain. *EMBO J.* 18, 3334–3347.

138 Shapiro, L. & Colman, D. R. (1998). Structural biology of cadherins in the nervous system. *Curr. Opin. Neurobiol.* 8, 593–599.

139 Huber, O., Kemmler, R. & Langosch, D. (1999). Mutations affecting transmembrane segment interaction impair adhesiveness of E-cadherin. *J. Cell Sci.* 112, 4415–4423.

140 Zhou, F. X., Cocco, M. J., Russ, W. P., Brunger, A. T. & Engelman, D. M. (2000). Interhelical hydrogen bonding drives strong interactions in membrane proteins. *Nature Struct. Biol.* 7, 154–160.

141 Choma, C., Gratkowski, H., Lear, J. D. & DeGrado, W. F. (2000). Asparagine-mediated self-association of a model transmembrane helix. *Nature Struct. Biol.* 7, 161–166.

142 O'Shea, E. K., Klemm, J. D., Kim, P. S. & Alber, T. (1991). X-ray structure of the GCN4 leucine zipper, a two-stranded, parallel coiled coil. *Science* 243, 539–544.

143 Alber, T. (1992). Structure of the leucine zipper. *Corr. Op. Genet. and Developm.* 2, 205–210.

144 Harbury, P. B., Zhang, T., Kim, P. S. & Alber, T. (1993). A switch between two-, three-, and four-stranded coiled coils in GCN4 leucine zipper mutants. *Science* 262, 1401–1406.

145 Gonzalez, L., Plecs, J. J. & Alber, T. (1996). An engineered allosteric switch in leucine-zipper oligomerization. *Nature Struct. Biol.* 3, 510–515.

146 Harbury, P. B., Kim, P. S. & Alber, T. (1994). Crystal structure of an isoleucine-zipper trimer. *Nature* 371, 80–83.

147 Gratkowski, H., Lear, J. D. & DeGrado, W. F. (2001). Polar side chains drive the association of model transmembrane peptides. *Proc. Natl. Acad. Sci. USA* 98, 880–885.

148 Zhou, F. X., Merianos, H. J., Brünger, A. T. & Engelman, D. M. (2001). Polar residues drive association of polyleucine transmembrane helices. *Proc. Natl. Acad. Sci. USA.*

149 Weiner, D. B., Liu, J., Cohen, J. A., Williams, W. V. & Greene, M. I. (1989). A point mutation in the neu oncogene mimics ligand induction of receptor aggregation. *Nature* 339, 230–231.

150 Webster, M. & Donoghue, J. (1996). Constitutive activation of fibroblast growth factor receptor 3 by the transmembrane domain point mutation found in achondroplasia. *EMBO J.* 15, 520–527.

151 Onishi, M., Mui, A. L. F., Morikawa, Y., Cho, L., Kinoshita, S., Nolan, G. P., Gorman, D. M., Miyajima, A. & Kitamura, T. (1996). Identification of an oncogenic form of the thrombopoietin receptor MPL using retrovirus-mediated gene transfer. *Blood* 88(4), 1399–1406.

152 Forbes, L. V., Gale, R. E., Pizzey, A., Pouwels, K., Nathwani, A. & Linch, D. C. (2002). An activating mutation in the transmembrane domain of the granulocyte colony-stimulating factor receptor in patients with acute myeloid leukemia. *Oncogene* 21(39), 5981–5989.

153 Call, M. E., Pyrdol, J., Wiedmann, M. & Wucherpfennig, K. W. (2002). The Organizing Principle in the Formation of the T Cell Receptor-CD3 Complex. *Cell* 111, 967–979.

154 Alcover, A., Mariuzza, R. A., Ermonval, M. & Acuto, O. (1990). Lysine 271 in the Transmembrane Domain of the T-Cell Antigen Receptor Beta-Chain Is Necessary for Its Assembly with the Cd3 Complex but Not for Alpha Beta-Dimerization. *J. Biol. Chem.* 265(7), 4131–4135.

155 Blumberg, R. S., Alarcon, B., Sancho, J., McDermott, F. V., Lopez, P., Breitmeyer, J. & Terhorst, C. (1990). Assembly and Function of the T-Cell Antigen Receptor – Requirement of Either the Lysine or Arginine Residues in the Transmembrane Region of the Alpha-Chain. *J. Biol. Chem.* 265(23), 14036–14043.

156 Cosson, P., Lankford, S., Bonifacino, J. S. & Klausner, R. D. (1991). Membrane protein association by potential intramembrane charge pairs. *Nature* 351, 414–416.

157 Bonifacino, J. S., Suzuki, C. K. & Klausner, R. D. (1990). A peptide sequence confers retention and rapid degradation in the endoplasmic reticulum. *Science* 247, 79–82.

158 Heinrich, S. U., Mothes, W., Brunner, J. & Rapoport, T. A. (2000). The Sec61p complex mediates the integration of a membrane protein by allowing lipid partitioning of the transmembrane domain. *Cell* 102(2), 233–244.

159 Shigematsu, D., Matsutani, M., Furuya, T., Kiyota, T., Lee, S., Sugihara, G. & Yamashita, S. (2002). Roles of peptide-peptide charge interaction and lipid phase separation in helix-helix association in lipid bilayer. *Biochim. Biophys. Acta* 1564, 271–280.

160 Gurezka, R. & Langosch, D. (2001). In vitro selection of membrane-spanning leucine zipper protein-protein interaction motifs using POSSYCCAT. *J. Biol. Chem.* 276, 45580–45587.

161 Dawson, J. P., Weinger, J. S. & Engelman, D. M. (2002). Motifs of serine and threonine can drive association of transmembrane helices. *J. Mol. Biol.* 316(3), 799–805.

162 Pebay-Peyroula, E. & Rosenbusch, J. P. (2001). High-resolution structures and dynamics of membrane protein–lipid complexes: a critique. *Curr. Opin. Struct. Biol.* 11(4), 427–432.

163 Fyfe, P. K., McAuley, K. E., Roszak, A. W., Isaacs, N. W., Cogdell, R. J. & Jones, M. R. (2001). Probing the interface between membrane proteins and membrane lipids by X-ray crystallography. *Trends Biochem. Sci.* 26(2), 106–112.

164 Lee, A. G. (2003). Lipid-protein interactions in biological membranes: a structural perspective. *Biochim. Biophys. Acta* 1612(1), 1–40.

165 Essen, L., Siegert, R., Lehmann, W. D. & Oesterhelt, D. (1998). Lipid patches in membrane protein oligomers: crystal structure of the bacteriorhodopsin-lipid complex. *Proc. Natl. Acad. Sci. USA* 95(20), 11673–11678.

166 Cartailler, J. P. & Luecke, H. (2003). X-ray crystallographic analysis of lipid-protein interactions in the bacteriorhodopsin purple membrane. *Annu. Rev. Biophys. Biomol. Struct.* 32, 285–310.

167 Sedlak, E. & Robinson, N. C. (1999). Phospholipase A(2) digestion of cardiolipin bound to bovine cytochrome c oxidase alters both activity and quaternary structure. *Biochemistry* 38(45), 14966–14972.

168 Lange, C., Nett, J. H., Trumpower, B. L. & Hunte, C. (2001). Specific roles of protein-phospholipid interactions in the yeast cytochrome bc(1) complex structure. *EMBO J.* 20(23), 6591–6600.

169 Camara-Artigas, A., Brune, D. & Allen, J. P. (2002). Interactions between lipids and bacterial reaction centers determined by protein crystallography. *Proc. Natl. Acad. Sci. USA* 99(17), 11055–11060.

170 Jormakka, M., Tornroth, S., Byrne, B. & Iwata, S. (2002). Molecular basis

of proton motive force generation: structure of formate dehydrogenase-N. *Science* 295(5561), 1863–1868.

171 Petrache, H. I., Grossfield, A., MacKenzie, K. R., Engelman, D. M. & Woolf, T. B. (2000). Modulation of glycophorin A transmembrane helix interactions by lipid bilayers: molecular dynamics calculations. *J. Mol. Biol.* 302(3), 727–746.

172 Mall, S., Broadbridge, R., Sharma, R. P., East, J. M. & Lee, A. G. (2001). Self-association of model transmembrane alpha-helices is modulated by lipid structure. *Biochemistry* 40(41), 12379–12386.

173 Gil, T., Ipsen, J. H., Mouritsen, O. G., Sabra, M. C., Sperotto, M. M. & Zuckermann, M. J. (1998). Theoretical analysis of protein organization in lipid membranes. *Biochim. Biophys. Acta* 1376(3), 245–266.

174 Cantor, R. S. (1999). Lipid composition and the lateral pressure profile in bilayers. *Biophys. J.* 76(5), 2625–2639.

175 Cantor, R. S. (2002). Size distribution of barrel-stave aggregates of membrane peptides: influence of the bilayer lateral pressure profile. *Biophys. J.* 82(5), 2520–2525.

176 Killian, J. A. (1998). Hydrophobic mismatch between proteins and lipids in membranes. *Biochim. Biophys. Acta* 1376(3), 401–415.

177 Dumas, F., Lebrun, M. C. & Tocanne, J. F. (1999). Is the protein/lipid hydrophobic matching principle relevant to membrane organization and functions? *FEBS Lett.* 458(3), 271–277.

178 Lewis, B. A. & Engelman, D. M. (1983). Bacteriorhodopsin remains dispersed in fluid phospholipid bilayers over a wide range of bilayer thicknesses. *J. Mol. Biol.* 166(2), 203–210.

179 Ren, J., Lew, S., Wang, J. & London, E. (1999). Control of the transmembrane orientation and interhelical interactions within membranes by hydrophobic helix length. *Biochemistry* 38, 5905–5912.

180 Dencher, N. A., Kohl, K. D. & Heyn, M. P. (1983). Photochemical cycle and light-dark adaptation of monomeric and aggregated bacteriorhodopsin in various lipid environments. *Biochemistry* 22(6), 1323–1334.

181 Piknova, B., Perochon, E. & Tocanne, J. F. (1993). Hydrophobic mismatch and long-range protein/lipid interactions in bacteriorhodopsin/phosphatidylcholine vesicles. *Eur. J. Biochem.* 218(2), 385–396.

182 Horowitz, A. D. (1995). Exclusion of SP-C, but not SP-B, by gel phase palmitoyl lipids. *Chem. Phys. Lipids* 76(1), 27–39.

183 Zhang, Y. P., Lewis, R. N., Hodges, R. S. & McElhaney, R. N. (2001). Peptide models of the helical hydrophobic transmembrane segments of membrane proteins: interactions of acetyl-K2-(LA)12-K2-amide with phosphatidylethanolamine bilayer membranes. *Biochemistry* 40(2), 474–482.

184 Rinia, H. A., Kik, R. A., Demel, R. A., Snel, M. M., Killian, J. A., van Der Eerden, J. P. & de Kruijff, B. (2000). Visualization of highly ordered striated domains induced by transmembrane peptides in supported phosphatidylcholine bilayers. *Biochemistry* 39(19), 5852–5858.

185 Jones, D. H., Rigby, A. C., Barber, K. R. & Grant, C. W. M. (1997). Oligomerization Of the EGF Receptor Transmembrane Domain: a H^2 NMR Study In Lipid Bilayers. *Biochemistry* 36, 12616–12624.

186 Van Duyl, B. Y., Rijkers, D. T. S., Kruijff, B. d. & Killian, J. A. (2002). Influence of hydrophobic mismatch and palmitoylation on the association of transmembrane a-helical peptides with detergent-resistant membranes. *FEBS Lett.*, 79–84.

187 Scheiffele, P., Roth, M. G. & Simons, K. (1997). Interaction Of Influenza Virus Haemagglutinin With Sphingolipid Cholesterol Membrane Domains Via Its Transmembrane Domain. *EMBO J.* 16, 5501–5508.

188 Perschl, A., Lesley, J., English, N., Hyman, R. & Trowbridge, I. S.

(1995). Transmembrane domain of CD44 is required for its detergent insolubility in fibroblasts. *J. Cell Sci.* 108(Pt 3), 1033–1041.

189 Field, K. A., Holowka, D. & Baird, B. (1999). Structural aspects of the association of FcepsilonRI with detergent-resistant membranes. *J. Biol. Chem.* 274(3), 1753–1758.

190 Therien, A. G. & Deber, C. M. (2002). Interhelical packing in detergent micelles – Folding of a cystic fibrosis transmembrane conductance regulator construct. *J. Biol. Chem.* 277, 6067–6072.

191 Vogel, H. (1992). Structure and dynamics of polypeptides and proteins in lipid membranes. *Q. Rev. Biophys.* 25, 433–457.

192 Belohorcova, K., Davis, J. H., Woolf, T. B. & Roux, B. (1997). Structure and Dynamics of an Amphiphilic Peptide in a Lipid Bilayer: A Molecular Dynamics Study. *Biophys. J.* 73, 3039–3055.

193 Brandl, C. J. & Deber, C. M. (1986). Hypothesis About the Function of Membrane-Buried Proline Residues in Transport Proteins. *Proc. Natl. Acad. Sci. U.S.A.* 83(4), 917–921.

194 Labro, A. J., Raes, A. L., Ottschytsch, N. & Snyders, D. J. (2001). Role of the S6 tandem proline motif in gating of Kv channels. *Biophys. J.* 80(1), 441A–441A.

195 Sansom, M. S. P. & Weinstein, H. (2000). Hinges, swivels and switches: the role of prolines in signalling via transmembrane a-helices. *Trends. Pharmacol. Sci.* 21, 445–451.

196 Shelden, M. C., Loughlin, P., Tierney, M. L. & Howitt, S. M. (2001). Proline residues in two tightly coupled helices of the sulphate transporter, SHST1, are important for sulphate transport. *Biochem. J.* 356, 589–594.

197 Duclohier, H., Molle, G., Dugast, J. Y. & Spach, G. (1992). Prolines Are Not Essential Residues in the Barrel-Stave Model for Ion Channels Induced by Alamethicin Analogs. *Biophys. J.* 63(3), 868–873.

198 Kaduk, C., Duclohier, H., Dathe, M., Wenschuh, H., Beyermann, M., Molle, G. & Bienert, M. (1997). Influence of proline position upon the ion channel activity of alamethicin. *Biophys. J.* 72(5), 2151–2159.

199 Gibbs, N., Sessions, R. B., Williams, P. B. & Dempsey, C. E. (1997). Helix bending in alamethicin: molecular dynamics simulations and amide hydrogen exchange in methanol. *Biophys. J.* 72, 2490–2495.

200 Tieleman, D. P., Sansom, M. S. P. & Berendsen, H. J. C. (1999). Alamethicin Helices in a Bilayer and in Solution: Molecular Dynamics Simulations. *Biophys. J.* 76, 40–49.

201 Jacob, J., Duclohier, H. & Cafiso, D. S. (1999). The Role of Proline and Glycine in Determining the Backbone Flexibility of a Channel-Forming Peptide. *Biophys. J.* 76, 1367–1376.

202 Demmers, J. A. A., Duijn, E. v., Haverkamp, J., Greathouse, D. V., II. R. E., K., Heck, A. J. R. & Kilian, J. A. (2001). Interfacial Positioning and Stability of Transmembrane Peptides in Lipid Bilayers Studied by Combining Hydrogen/Deuterium Exchange and Mass Spectrometry. *J. Biol. Chem.* 276, 34501–34508.

203 Langosch, D., Crane, J. M., Brosig, B., Hellwig, A., Tamm, L. K. & Reed, J. (2001). Peptide mimics of SNARE transmembrane segments drive membrane fusion depending on their conformational plasticity. *J. Mol. Biol.* 311, 709–721.

204 Langosch, D., Brosig, B. & Pipkorn, R. (2001). Peptide mimics of the vesicular stomatitis virus G-protein transmembrane segment drive membrane fusion in vitro. *J. Biol. Chem.* 276, 32016–32021.

205 Dennison, S. M., Greenfield, N., Lenard, J. & Lentz, B. R. (2002). VSV transmembrane domain (TMD) peptide promotes PEG-mediated fusion of liposomes in a conformationally sensitive fashion. *Biochemistry* 41(50), 14925–14934.

206 Li, L. & Chin, L. S. (2003). The molecular machinery of synaptic vesicle exocytosis. *Cell. Mol. Life Sci.* 60(5), 942–960.

207 Scales, S. J., Bock, J. B. & Scheller, R. H. (2000). Cell biology – The specifics of membrane fusion. *Nature* 407, 144–146.

208 Blumenthal, R., Clague, M. J., Durell, S. T. & Epand, R. M. (2003). Membrane Fusion. *Chem. Rev.* 103, 53–69.

209 Melikyan, G. B., Lin, S. S., Roth, M. G. & Cohen, F. S. (1999). Amino acid sequence requirements of the transmembrane and cytoplasmic domains of influenza virus hemagglutinin for viable membrane fusion. *Mol. Biol. Cell* 6, 1821–1836.

210 Owens, R. J., Burke, C. & Rose, J. K. (1994). Mutations in the membrane-spanning domain of the human immunodeficiency virus envelope glycoprotein that affect fusion activity. *J. Virol.* 68, 570–574.

211 Taylor, G. M. & Sanders, D. A. (1999). The role of the membrane-spanning domain sequence in glycoprotein-mediated membrane fusion. *Mol. Biol. Cell* 10, 2803–2815.

212 Odell, D., Wanas, E., Yan, J. & Ghosh, H. P. (1997). Influence of membrane anchoring and cytoplasmic domains on the fusogenic activity of vesicular stomatitis virus glycoprotein G. *J. Virol.* 71, 7996–8000.

213 Cleverley, D. Z. & Lenard, J. (1998). The transmembrane domain in viral fusion: essential role for a conserved glycine residue in vesicular stomatitis virus G protein. *Proc. Natl. Acad. Sci. USA* 95, 3425–3430.

214 Grote, E., Baba, M., Ohsumi, Y. & Novick, P. J. (2000). Geranylgeranylated SNAREs are dominant inhibitors of membrane fusion. *J. Cell Biol.* 151, 453–465.

215 Rohde, J., Dietrich, L., Langosch, D. & Ungermann, C. (2003). The transmembrane domain of Vam3 affects the composition of *cis-* and *trans-*SNARE complexes to promote homotypic vacuole fusion. *J. Biol. Chem.* 278(3), 1656–1662.

216 Tatulian, S. A. & Tamm, L. K. (2000). Secondary structure, orientation, oligomerization, and lipid interactions of the transmembrane domain of influenza hemagglutinin. *Biochemistry* 39, 496–507.

217 Szyperski, T., Vandenbussche, G., Curstedt, T., Ruysschaert, J. M., Wüthrich, K. & Johansson, J. (1998). Pulmonary surfactant-associated polypeptide C in a mixed organic solvent transforms from a monomeric alpha-helical state into insoluble beta-sheet aggregates. *Protein Sci.* (12), 2533–40.

218 Zhang, Y.-P., Lewis, R. N. A. H., Hodges, R. S. & McElhaney, R. N. (1992). FTIR Spectroscopic studies of the conformation and amide hydrogen exchange of a peptide model of the hydrophobic transmembrane α-helices of membrane proteins. *Biochemistry* 31, 11572–11578.

219 Helenius, A., McCaslin, D. R., Fries, E. & Tanford, C. (1979). Properties of Detergents. *Meth. Enzymol.* 56, 734–749.

220 Torres, J., Stevens, T. J. & Samso, M. (2003). Membrane proteins: the 'Wild West' of structural biology. *Trends Biochem. Sci.* 28(3), 137–144.

221 Laage, R. & Langosch, D. (2001). Strategies for prokaryotic expression of eukaryotic membrane proteins. *Traffic* 2, 99–104.

222 Grisshammer, R. & Tate, C. G. (1995). Overexpression of integral membrane proteins for structural studies. *Q. Rev. Biophys.* 28, 315–422.

223 Tate, C. G. & Grisshammer, R. (1996). Heterologous expression of G-protein-coupled receptors. *TIBTECH* 14, 426–430.

224 Hackam, A. S., Wang, T.-L., Guggino, W. B. & Cutting, G. R. (1997). The N-terminal domain of human GABA receptor ρ1 subunits contains signals for homooligomeric and heterooligomeric interaction. *J. Biol. Chem.* 272, 13750–13757.

225 Szczesna Skorupa, E. & Kemper, B. (1991). Cell-free analysis of targeting of cytochrome P450 to microsomal membranes. *Methods Enzymol.* 206, 64–75.

226 Rosenberg, R. L. & East, J. E. (1992). Cell-free expression of functional

Shaker potassium channels. *Nature* 360, 166–169.

227 Rusinol, A. E., Jamil, H. & Vance, J. E. (1997). In vitro reconstitution of assembly of apolipoprotein B48-containing lipoproteins. *J. Biol. Chem.* 272, 8019–8025.

228 Huppa, J. B. & Ploegh, H. L. (1997). In vitro translation and assembly of a complete T cell receptor-CD3 complex. *J. Exp. Med.* 186, 393–403.

229 Fisher, L. E. & Engelman, D. M. (2001). High-yield synthesis and purification of an alpha-helical transmembrane domain. *Analytical Biochemistry* 293(1), 102–108.

230 Melnyk, R. A., Partridge, A. W. & Deber, C. M. (2002). Transmembrane domain mediated self-assembly of major coat protein subunits from Ff bacteriophage. *J. Mol. Biol.* 315(1), 63–72.

231 Gerber, D. & Shai, Y. (2001). In vivo detection of hetero-association of glycophorin-A and its mutants within the membrane. *J. Biol. Chem.* 276(33), 31229–31232.

232 Laemmli, U. K. (1970). Cleavage of structural proteins during the assembly of the head of bacteriophage T4. *Nature* 227, 680–685.

233 Soulié, S., Moller, J. V., Falson, P. & Maire, l. M. (1996). Urea Reduces the Aggregation of Membrane Proteins on Sodium Dodecyl Sulfate-Polyacrylamide Gel Electrophoresis. *Anal. Biochem.* 236, 363–364.

234 Millar, D. G. & Shore, G. C. (1993). The signal anchor sequence of mitochondrial Mas70p contains an oligomerization domain. *J. Biol. Chem.* 268, 18403–18406.

235 Deber, C. M., Khan, A. R., Zuomei, L., Joensson, C., Glibowicka, M. & Wang, J. (1993). Val-Ala mutations selectively alter helix-helix packing in the transmembrane segment of phage M13 coat protein. *Proc. Natl. Acad. Sci. USA* 90, 11648–11652.

236 Ramjeesingh, M., Huan, L., Garami, E. & Bear, C. E. (1999). Novel method for evaluation of the oligomeric structure of membrane proteins. *Biochem. J.* 342, 119–123.

237 Peters, K. & Richards, M. (1977). Chemical Cross-linking: Reagents and Problems in Studies of Membrane Structure. *Annu. Rev. Biochem.* 46, 523–551.

238 Gaffney, B. J. (1985). Chemical and biochemical crosslinking of membrane components. *Biochim. Biophys. Acta* 822, 289–317.

239 Schägger, H., Cramer, W. A. & von Jagow, G. (1994). Analysis of molecular masses and oligomeric states of protein complexes by blue native electrophoresis and isolation of membrane protein complexes by two-dimensional native electrophoresis. *Anal. Biochem.* 217, 220–230.

240 Schägger, H. & von Jagow, G. (1991). Blue native electrophoresis for isolation of membrane protein complexes in enzymatically active form. *Anal. Biochem.* 199, 223–231.

241 Poetsch, A., Neff, D., Seelert, H., Schägger, H. & Dencher, N. A. (2000). Dye removal, catalytic activity and 2D crystallization of chloroplast H^+-ATP synthase purified by blue native electrophoresis. *Biochim. Biophys. Acta* 1466, 339–349.

242 Rexroth, S., Meyer zu Tittingdorf, J. M. W., Krause, F., Dencher, N. A. & Seelert, H. (2003). Thylakoid membrane at altered metabolic state: Challenging the forgotten realms of the proteome. *Electrophoresis* 24, 2814–2823.

243 Clarke, S. (1975). Size and Detergent Binding of Membrane Proteins. *J. Biol. Chem.* 250(14), 5459–5469.

244 Selvin, P. R. (1995). Fluorescence resonance energy transfer. *Meth. Enzymol.* 246, 300–334.

245 Clegg, R. M. (1995). Fluorescence resonance energy transfer. *Curr. Op. Biotech.* 6, 103–110.

246 Wu, P. a. B., L. (1994). Resonance energy transfer: methods and applications. *Anal. Biochem.* 218, 1–13.

247 Peled, H. & Shai, Y. (1993). Membrane interaction and self-assembly within phospholipid membranes of synthetic segments corresponding to the H-5 region of the

shaker K+ channel. *Biochemistry* 32, 7879–7885.

248 Yano, Y., Takemoto, T., Kobayashi, S., Yasui, H., Sakurai, H., Ohashi, W., Niwa, M., Futaki, S., Sugiura, Y. & Matsuzaki, K. (2001). Topological Stability and Self-Association of a Completely Hydrophobic Model Transmembrane Helix in Lipid Bilayers. *Biochemistry* 41, 3073–3080.

249 Gazit, E. & Shai, Y. (1995). The assembly and organization of the alpha5 and alpha7 helices from the pore-forming domain of Bacillus thuringiensis delta-endotoxin. *J. Biol. Chem.* 270, 2571–2578.

250 Li, M., Reddy, L. G., Bennett, R., Silva, N. D., Jones, L. R. & Thomas, D. D. (1999). A fluorescence energy transfer method for analyzing protein oligomeric structure: application to phospholamban. *Biophys. J.* 76, 2587–2599.

251 Adair, B. D. & Engelman, D. M. (1994). Glycophorin A helical transmembrane domains dimerize in phospholipid bilayers: a resonance energy transfer study. *Biochemistry* 33, 5539–5544.

252 Mitra, R. D., Silva, C. M. & Youvan, D. C. (1996). Fluorescence resonance energy transfer between blue-emitting and red-shifted excitation derivatives of the green fluorescent protein. *Gene* 173, 13–17.

253 Heim, R. & Tsien, R. Y. (1996). Engineering green fluorescent protein for improved brightness, longer wavelenghts and fluorescence resonance energy transfer. *Curr. Biol.* 6, 178–182.

254 Ellenberg, J., Lippincott-Schwartz, J. & Presley, J. F. (1999). Dual-color imaging with GFP variants. *Trends. Cell Biol.* 9, 52–56.

255 Lippincott-Schwartz, J. & Patterson, G. H. (2003). Development and Use of Fluorescent Protein Markers in Living Cells. *Science* 300, 87–91.

256 Majoul, I., Straub, M., Duden, R., Hell, S. W. & Söling, H.-D. (2002). Fluorescence resonance energy transfer analysis of protein-protein interactions in single living cells by multifocal multiphoton microscopy. *Rev. Mol. Biotech.* 82, 267–277.

257 Couturier, C., Ayoub, M. A. & Jockers, R. (2002). BRET ermöglicht die Messung von Protein-Interaktionen in lebenden Zellen. *Biospektrum* 5, 612–615.

258 Ayoub, M. A., Couturier, C., Lucas-Meunier, E., Angers, S., Fossier, P., Bouvier, M. & Jockers, R. (2002). Monitoring of ligand-independent dimerization and ligand-induced conformational changes of melatonin receptors in living cells by bioluminescence resonance energy transfer. *J. Biol. Chem.* 277(24), 21522–21528.

259 Kroeger, K. M., Hanyaloglu, A. C. & Eidne, K. A. (2001). Applications of BRET to study dynamic G-protein coupled receptor interactions in living cells. *Letters Pep. Sci.* 8(3–5), 155–162.

260 Eidne, K. A., Kroeger, K. M. & Hanyaloglu, A. C. (2002). Applications of novel resonance energy transfer techniques to study dynamic hormone receptor interactions in living cells. *Trends Endocrin. Metabol.* 13(10), 415–421.

261 DiRita, V. (1992). Co-ordinate expression of virulence genes by ToxR in *Vibrio cholerae*. *Mol. Microbiol.* 6, 451–458.

262 Kolmar, H., Fritsch, C., Kleeman, G., Götze, K., Stevens, F. J. & Fritz, H. J. (1994). Dimerization of bence jones proteins: linking the rate of transcription from an Escherichia coli promoter to the association constant of REIv. *Biol. Chem. Hoppe-Seyler* 375, 61–69.

263 Bedouelle, H. & Duplay, P. (1988). Production in Escherichia coli and one-step purification of bifunctional hybrid proteins which bind maltose – export of the Klenow polymerase into the periplasmic space. *Eur. J. Biochem.* 171, 541–549.

264 Leeds, J. A. & Beckwith, J. (1998). Lambda repressor n-terminal DNA-binding domain as an assay for protein transmembrane segment interactions in vivo. *J. Mol. Biol.* 280(5), 799–810.

265 Schneider, D. & Engelman, D. M. (2003). GALLEX: a measurement of heterologous association of transmembrane helices in a biological membrane. *J. Biol. Chem.* 278, 3105–3111.

266 Johnsson, N. & Varshavsky, A. (1994). Split ubiquitin as a sensor of protein interactions in vivo. *Proc. Natl. Acad. Sci. USA* 91, 10340–10344.

267 Stagljar, I., Korostensky, C., Johnsson, N. & TeHeesen, S. (1998). A genetic system based on split-ubiquitin for the analysis of interactions between membrane proteins in vivo. *Proc. Natl. Acad. Sci. USA* 95, 5187–5192.

268 Wittke, S., Lewke, N., Muller, S. & Johnsson, N. (1999). Probing the molecular environment of membrane proteins in vivo. *Mol. Biol. Cell* 10(8), 2519–2530.

269 Stagljar, I. & Fields, S. (2002). Analysis of membrane protein interactions using yeast-based technologies. *Trends. Biochem. Sci.* 27, 559–563.

270 Lee, G. F. & et al. (1994). Deducing the organization of a transmembrane domain by disulfide crosslinking. *J. Biol. Chem.* 269, 29920–29927.

271 Pakula, A. A. & Simon, M. I. (1992). Determination of transmembrane protein structure by disulfide crosslinking: the *Escherichia coli* Tar receptor. *Proc. Natl. Acad. Sci. USA* 89, 4144–4148.

272 Doyle, D. A., Cabral, J. M., Pfuetzner, R. A., Kuo, A., Gulbis, J. M., Cohen, S. L., Chait, B. T. & MacKinnon, R. (1998). The structure of the potassium channel: molecular basis of K^+ conduction and selectivity. *Science* 280, 69–77.

273 Cowan, S. W., Schirmer, T., Rummel, G., Steiert, M., Ghosh, R., Pauptit, R. A., Jansonius, J. N. & Rosenbusch, J. P. (1992). Crystal structures explain functional properties of two E. coli porins. *Nature* 358, 727–733.

274 Ruan, W., Becker, V., Klingmuller, U., Langosch, D. The interface between self-assembling erythropoietin receptor transmembrane segments corresponds to a membrane-spanning leucine zipper. *J. Biol. Chem.* 279(5), 3273–3279.

275 Ruan, W., Lindner, E., and Langosch, D. (2004). The interface of a membrane-spanning leucine zipper mapped by asparagine-scanning mutagenesis. *Prot. Sci.* 13, 555–559.

276 Toutain, C. M., Clarke, D. J., Leeds, J. A., Kuhn, J., Beckwith, J., Holland, I. B., Jacq, A. (2003). The transmembrane domain of the DnaJ-like protein DjlA is a dimerisation domain. *Mol. Genet. Genom.* 268(6), 761–770.

277 Leeds, J. A., Boyd, D., Huber, D. R., Sonoda, G. K., Luu, H. T., Engelman, D. M., Beckwith, J. (2001). Genetic selection for and molecular dynamic modeling of a protein transmembrane domain multimerization motif from a random Escherichia coli genomic library. *J. Mol. Biol.* 313(1), 181–195.